Techniques and Standards for Image, Video, and Audio Coding

Techniques and Standards for Image, Video, and Audio Coding

K. R. RAO
University of Texas at Arlington

J. J. HWANG
Kunsan National University
Republic of Korea

Prentice Hall PTR, Upper Saddle River, New Jersey 07458

Library of Congress Cataloging-in-Publication Data

Rao, K. Ramamohan (Kamisetty Ramamohan)
Techniques and standards for image, video, and audio coding/K. R. Rao, J. J. Hwang.
p. cm.
Includes bibliographical references and index.
ISBN 0-13-309907-5
1. Coding theory—Standards. 2. Image processing—Digital techniques—Standards. 3. Signal processing—Digital techniques—Standards. I. Hwang, J. J. II. Title.
TK5102.92.R365 1996 96-15550
621.389'7—dc 20 CIP

Acquisitions editor: *Karen Gettman*
Production supervision: *Kathleen M. Lafferty/Roaring Mountain Editorial Services*
Cover designer: *Design Source*
Cover design director: *Jerry Votta*
Manufacturing manager: *Alexis R. Heydt*

The publisher offers discounts on this book when ordered in bulk quantities. For more information, contact:
Corporate Sales Department
Prentice Hall PTR
One Lake Street
Upper Saddle River, NJ 07458

Phone: 800-382-3419
Fax: 201-236-7141
email: corpsales@prenhall.com

Printed in the United States of America

10 9 8 7 6 5

ISBN 0-13-309907-5

Prentice-Hall International (UK) Limited, *London*
Prentice-Hall of Australia Pty. Limited, *Sydney*
Prentice-Hall Canada Inc., *Toronto*
Prentice-Hall Hispanoamericana, S.A., *Mexico*
Prentice-Hall of India Private Limited, *New Delhi*
Prentice-Hall of Japan, Inc., *Tokyo*
Prentice-Hall Asia Pte. Ltd., *Singapore*
Editora Prentice-Hall do Brasil, Ltda., *Rio de Janerio*

Contents

Chapter 2
Color Formats 9

Chapter 3
Quantization 17

Chapter 4
Predictive Coding 31

Chapter 5
Transform Coding 43

Chapter 6
Hybrid Coding and Motion Compensation 85

Chapter 7
Vector Quantization and Subband Coding 97

PART II
International Standards for Image, Video, and Audio Coding

Chapter 8
JPEG Still-Picture Compression Algorithm 127

Chapter 9
ITU-T H.261 Video Coder 163

Chapter 10
MPEG-1 Audiovisual Coder for Digital Storage Media 201

Chapter 11
MPEG-2 Generic Coding Algorithms 273

Chapter 12
MPEG-4 and H.263 Very Low Bit-Rate Coding 323

Chapter 13
High-Definition Television Services 357

Bibliography 489

Index 557

Preface

Techniques and Standards for Image, Video, and Audio Coding provides a comprehensive coverage of various international standards (in different stages: approved, recommended, finalized, emerging, developing) related to digital video, image, and audio coding. These standards are aimed at interactive video/audio communication services with applications in medicine, consumer electronics, education, training, defense, desktop publishing, entertainment, sports, multimedia, videophone, videoconferencing, and so forth. The emphasis is on redundancy reduction so that single images or a sequence (black and white or color at any resolution) can be stored with reduced memory at a lower bit rate. Generically, this approach is called compression, and it is applicable also to audio, graphics, binary data, and so forth. The spectrum of consumer electronics ranges from electronic digital still-image cameras, video jukeboxes, and mobile videophones to computer video games. The marriage of communications and computers has thus opened a new era in interactive audiovisual applications.

The objective of this book is not only to familiarize the reader with the international standards but also to provide the underlying theory, concepts, and principles related to bit-rate reduction. The first part of the book thus focuses on the compression concepts such as predictive coding, transform coding, motion compensation, vector quantization, subband coding, and their combinations. These techniques are supplemented by human visual sensitivity and variable-length coding. Strengthened by the familiarity with the fundamental principles underlying the video/audio coding, the reader is now ready to grasp a wide spectrum of standards that are under various stages of development. This is the major thrust of this book, and Part II provides an exposition of these standards. The various standard groups such as International Telecommunications Union (Telecommunication standardization sector, ITU-T), Radiocommunication sector (ITU-R), Committee for Mixed Telephone and Television (CMTT), Joint Photographic Experts Group (JPEG), Moving Picture Experts Group (MPEG), International Organization of Standards (ISO), Multimedia Hypermedia Experts Group (MHEG), Joint Binary Image Group (JBIG), and so forth have either finalized or are in the process of recommending standards for specific application areas, such as Videophone and Videoconferencing (H.261), still-frame image compression (JPEG), Broadcast TV: Distribution and Contribution (ITU-T), Terrestrial HDTV Transmission (FCC), Digital Storage Media: Video and Audio (MPEG), Multimedia and Hypermedia (MHEG), Binary Image Coding (JBIG), and Video Coding for ATM Networks (ITU-T). Algorithmic description of these standards constitutes the major emphasis of Part II. This is supplemented by suggestions for further improvements and enhancements.

The goal of these standards is that systems developed by different industries worldwide can communicate with one another; that is, a decoder manufactured by one company can interpret the bit stream transmitted by a coder manufactured by another company. Connectivity and compatibility among different services (i.e., videophone/videoconferencing, MPEG-1, and MPEG-2) are other factors that influence these standards. So as not to stifle the innovation and entrepreneurship, the standards deal with only the basic services (actual hardware implementation is left to the designer), leaving room for various additional features (i.e., freeze, split screen, zoom, and pan) that can be incorporated by the manufacturers as marketing tools. The reader now (based on Parts I and II) has not only a thorough understanding of the algorithmic description of the standards but also has an appreciation for the particular techniques that have converged into the encoding and decoding processes. We include more than 260 figures, 80 tables, and 1030 references in this book. The reader now has the capability to simulate these algorithms for analysis and evaluation and also to investigate the effects of alternative algorithms.

Intentionally, software is not provided, as it is widely available in the open market (email, computer clubs, ftp, WWW, networks, and so forth). The scope of this book, as stated earlier, is not only to familiarize the reader with the algorithmic description of the standards cited earlier, but also to provide the fundamentals and principles that can provide the necessary insight and intuition for understanding and appreciating of these algorithms. For additional reading, an extensive list of references is provided at the end of the book. Appendices that describe briefly the chips, chip sets, PC boards, codecs, systems, and software packages (related to these standards) available from various vendors conclude the book. This list is made as comprehensive and as current as possible. Although some books have been published that cater to one or the other standard, such as JPEG and MPEG, to our knowledge no book has dealt with the family of standards being developed in the audio/video field. Thus this book is a rich resource in the discipline of digital image, video, and audio coding.

K. R. Rao
J. J. Hwang

Acknowledgments

Several people have helped in countless ways to make *Techniques and Standards for Image, Video, and Audio Coding* possible. Some reviewed portions of the manuscript; others provided valuable information related to development of the standards, their implementation (hardware/software), manuscript preparation in electronic form, and so forth. Special thanks are due to the following persons:

D. Anastassiou	Columbia University
S. J. M. Ann	Seoul National University
R. L. Baker	Picturetel
M. Barbero	RAI—Centro Ricerche
M. Boliek	Ricoh
D. Bye	SGS Thomson Microelectronics
B. Chitprasert	Shinawatra Satellite Co.
C. S. Choi	Myongji University
F. Colaitis	CCETT
K. Dejhan	King Mongkut's Institute of Technology
H. Fujiwara	Graphics Communication Labs
D. Galby	C-Cube Microsystems
H. Gharavi	University of Technology
E. Hamilton	C-Cube Microsystems
H. M. Hang	National Chiao Tung University
M. A. Haque	Hyundai Electronics
D. Hein	VTEL
T. R. Hsing	Bellcore
Y. Huh	University of Texas at Arlington
M. Isnardi	David Sarnoff Research Labs

A. Jalali	Wiltel
B. W. Jeon	Samsung Electronics
J. H. Jeon	Korea Telecom
J. Jeong	Hanyang University
J. K. Kim	KAIST
S. D. Kim	KAIST
T. Koga	NEC
R. Lancini	CEFRIEL
B. G. Lee	Seoul National University
B. R. Lee	ETRI
M. H. Lee	Chonbuk National University
W. Lee	University of Washington
M. L. Liou	Hong Kong University of Science and Technology
T. Maldonado	University of Texas at Arlington
H. Malvar	Picturetel
K. Matsuda	Fujitsu
J. Y. Nam	Geimyung University
M. A. Narasimhan	General Instruments
K. N. Ngan	University of Western Australia
Y. Ninomiya	NHK
K. O'Connell	Motorola
S. Oh	Texas Instruments
N. Ohta	NTT
S. Okubo	Graphics Communication Labs
K. K. Pang	Monash University
K. Panusopone	University of Texas at Arlington
J. Park	Samsung Electronics
E. D. Petajan	AT & T Bell Labs
P. Priebjrivat	TAC
P. Pirsch	University of Hanover
R. De Queiroz	Xerox Corp.
S. Rajala	North Carolina State University
A. A. Rodriguez	Scientific Atlanta
J. Sefcik	ITT
M. T. Sun	Bellcore
Y. Tadokoro	Toyohashi University of Technology
R. Talluri	Texas Instruments
V. Thomas	CCETT
C. Todd	Dolby Research Labs
T. Watanabe	Toshiba
H. R. Wu	Monash University
K. H. Yang	Hanryo Sanup University

H. Yasuda	NTT
S. Yun	Hanyoung Technical College

This book was prepared with LaTeX. Typeset chapters were translated into PostScript using dvips, resulting in 75 Mbytes in size. Files were exchanged back and forth between the two sites, the University of Texas at Arlington and Kunsan National University, via Internet email. We wish to acknowledge K. E. Baek at the Kunsan National University's computer center, who contributed in processing the huge files on the network.

The secretarial help in all facets of desktop publishing has been exceptional. Special thanks go to Doan Phuong Kieu Tuan, HoangVan Thi, Nguyen Nancy Tiet Thi, Teresa Phuong Nguyen, Hong Chu Hoang, Hyun Ju Lee, and So Young Park.

Without the patience and forbearance of our families, preparation of this book would have been impossible. We appreciate their constant and continuous support and understanding.

Acronyms

AC	Audio Coding
ACATS	Advisory Committee on Advanced Television Services
ACM	Association for Computing Machinery
ADPCM	Adaptive Differential Pulse Code Modulation
AES	Audio Engineering Society
ASIC	Application-Specific Integrated Circuits
ASPEC	Audio Spectral Perceptual Entropy Coding
ATAC	Adaptive Transform Aliasing Cancellation
ATC	Adaptive Transform Coding
ATRC	Advanced Television Research Consortium
ATM	Asynchronous Transfer Mode
ATTC	Advanced Television Test Center
ATV	Advanced Television
AVC	ATM Video Coding experts group
BCH	Bose-Chaudhuri-Hocquenghem
BD	Block Difference
BDPCM	Signal-blurred DPCM
BMA	Block-Matching Algorithm
BOPS	Billion Operations Per Second
BSVQ	Binary Search Vector Quantization
BTC	Block Truncation Coding
CBP	Coded Block Pattern
CBPC	Coded Block Pattern for Chrominance
CBPY	Coded Block Pattern for Luminance
CCDC	Channel Compatible DigiCipher
CCITT	International Telephone and Telegraph Consultative Committee

CCIR	International Radio Consultative Committee
CD	Committee Draft, or Compact Disc
CDS	Conjugate Direction Search
CELP	Code Excited Linear Prediction
CICC	Custom Integrated Circuits Conference
CIF	Common Intermediate Format
CMCI	Conditional Motion-Compensated Interpolation
CMOS	Complimentary Metal Oxide Semiconductor
CMTT	Committee for Mixed Telephone and Television
COD	Coded Macroblock Indication
CRC	Cyclic Redundancy Check
CRT	Cathode-Ray Tube
CSVT	Circuits and Systems for Video Technology
DAT	Digital Audio Tape
DBD	Displaced Block Difference
DBS	Direct Broadcast Satellite
DCT	Discrete Cosine Transform
DFD	Displaced Frame Difference
DIS	Draft International Standard
DMA	Direct Memory Access
DMV	Differential MV
DPCM	Differential Pulse Code Modulation
dpi	dots per inch
DQ	Dequantization
DRAM	Dynamic Random-Access Memory
DSC	Digital Spectrum Compatible
DSM	Digital Storage Media
DSP	Digital Signal Process
DVB	Digital Video Broadcasting
EBU	European Broadcasting Union
ECSA	Exchange Carrier Standards Association
EDTV	Extended-Definition Television
ELT	Extended Lapped Transform
EOB	End of Block
EUSIPCO	European Signal Processing Conference
FDCT	Forward DCT
FCC	Federal Communications Commission
FDM	Frequency Division Multiplexing
FEC	Forward Error Correction
FFT	Fast Fourier Transform
FLC	Fixed-Length Coding
FTP	File Transfer Protocol

GA	Grand Alliance
GBSC	GOB Start Code
GLOBECOM	IEEE Global Telecommunications Conference
GOB	Group of Blocks
GOP	Group of Pictures
GSM	General Special Mobile
H.261(H.263)	ITU-T Recommendations
HDP	High-Definition Progressive
HDTV	High-Definition Television
HPC	Hierarchical Predictive Coding
HVS	Human Visual System
ICASSP	IEEE International Conference on Acoustics, Speech, and Signal Processing
ICC	IEEE International Conference on Communications
ICCE	IEEE International Conference on Consumer Electronics
ICIP	IEEE International Conference on Image Processing
ICSPAT	International Conference on Signal Processing Applications & Technology
IDCT	Inverse DCT
IEC	International Electrotechnical Commission
IEEE	Institute of Electrical and Electronics Engineers
IFS	Iterated Function System
IMBE	Improved Multiband Excitation Coder
IQ	Inverse Quantization
IRE	Institute of Radio Engineers
IS	International Standard
ISCAS	International Symposium on Circuits and Systems
ISDN	Integrated Services Digital Network
ISO	International Organization for Standardization
IS&T	Imaging Science & Technology
ITU	International Telecommunications Union
ITU-R	ITU Radiocommunication Sector
ITU-T	ITU Telecommunication Standardization Sector
JBIG	Joint Binary Image Group
JDC	Japanese Digital Cellular
JPEG	Joint Photographic Experts Group
JTC	Joint Technical Committee
Kbps	Kilobits per second
KLT	Karhunen-Loève Transform
LD-CELP	Low-Delay CELP
LF	Loop Filter
LFE	Low-Frequency Enhancement

LPF	Low Pass Filter
LMS	Least Mean Square
LOT	Lapped Orthogonal Transform
LTP	Long-Term Predictor
MAC	Multiplexed Analog Components
MAE(D)	Mean Absolute Error (Difference)
MAQE	Mean Absolute Quantization Error
MB	Macroblock
Mbps	Megabits per second
MBA	Macroblock Address
MC	Motion Compensation, or Model Compliance
MCBPC	Macroblock Type & Coded Block Pattern for Chrominance
MCU	Minimum Coded Unit
MDCT	Modified Discrete Cosine Transform
MDST	Modified Discrete Sine Transform
ME	Motion Estimation
MF	Model Failure
MHEG	Multimedia Hypermedia Experts Group
MIPS	Million Instructions per Second
MLT	Modulated La,pped Transform
MPEG	Moving Picture Experts Group
ms	millisecond
MSDL	MPEG-4 Syntactic Description La,nguage
MSE	Mean Square Error
Msps	Megasamples (or Symbols) per second
MSQE	Mean Square Quantization Error
MUSICAM	Masking-pattern Universal Subband Integrated Coding and Multiplexing
MV	Motion Vector
MVD	Motion Vector Data
MVZS	Maximum Variance Zonal Sampling
NMR	Noise-to-Mask Ratio
NTC	National Telecommunications Union
NTSC	National Television System Committee
OCF	Optimum Coding in the Frequency domain
OTS	One-at-a-Time Search
PACS	Picture Archiving and Communica,tion Systems
PAL	Phase Alternating Line
PCM	Pulse Code Modulation
PE	Psychoa,coustic Entropy
PES	Packetized Elementary Stream
PIT	Progressive Image Transmission

POCS	Projection Onto Convex Sets
PPD	Proposal Package Description
PRA	Pel-Recursive Algorithm
PSD	Power Spectral Density
PSTN	Public Switched Telephone Network
PTS	Piecewise Transformation System, or Presentation Time Stamps
PVQ	Predictive Vector Quantization
Q	Quantization
QAM	Quadrature Amplitude Modulation
QCIF	Quarter CIF
QMF	Quadrature Mirror Filter
RAM	Random-Access Memory
RISC	Reduced Instruetion Set Command
RLC	Run-Length Coding
RM	Reference Model (H.261)
ROM	Read-Only Memory
RPE	Regular Pulse Excitation
RS	Reed-Solomon
SAC	Syntax-based Arithmetic Coding
SAR	Synthetic Aperture Radar
SBC	Subband Coding
SC	Study Committee
SCR	System Clock Reference
SECAM	Sequential Couleur Avec Memoire
SG	Study Group
SIF	Source Input Format
SM	Simulation Model (MPEG-1)
SMPTE	Society of Motion Picture and Television Engineers
SMR	Signal-to-Mask Ratio
SNR	Signal-to-Noise Ratio
SONET	Synchronous Optical Network
SPA	Significant Pel Area
SPIE	Society of Photo-optical and Instrumentation Engineers
SPIFF	Still-Picture Interchange File Format
SPL	Sound Pressure Level
SQ	Scalar Quantization
SRG	Special Rapporteur Group
STC	System Time Clock
STS	Self-Transformation System
TCQ	Trellis Coded Quantization
TDAC	Time Domain Aliasing Cancellation
TDM	Time Division Multiplexing

TI	Texas Instruments
TM	Test Model (MPEG-2)
TMN	Test Model for Near-term Sollltion (H.263)
TSVQ	Tree Search Vector Quantization
UHF	Ultra High Frequency
VBR	Variable Bit Rate
VBV	Video Buffering Verifier
VCIP	Visual Communication and Image Processing
VHF	Very High Frequency
VLC(D)	Variable Length Coding (Decoding)
VLSI	Very Large Scale Integration
VM	Verification Model (MPEG-4)
VQ	Vector Quantization
VRP	Video Risc Processor
VSB	Vestigial Side Band
VSELP	Vector Sum Excited Linear Predietion
VSPC	Visual Signal Processing and Communications
VWL	Variable Word Length
VXC	Vector Excited Coding
WD	Working Draft
WFTA	Winograd Fourier Transform Algorithm
WG	Working Group
WHT	Walsh-Hadamard Transform
WT	Wavelet Transform
WWW	World Wide Web

Techniques and Standards for Image, Video, and Audio Coding

Part I
Digital Coding Techniques

Chapter 1
Introduction

Summary

The philosophy behind the international standards related to digital image, video, and audio coding is outlined. This is followed by the description of the chapters. An outline of the appendices and references concludes this chapter.

1.1 Overview

The development of standards (emerging and established) by the International Organization for Standardization (ISO), the International Telecommunications Union (ITU), and the International Electrotechnical Commission (IEC), for audio, image, and video, for both transmission and storage, has led to worldwide activity in developing the products/systems (both hardware and software) applicable to a number of diverse disciplines. Although the standards implicitly address the basic encoding operations, there is freedom and flexibility in the actual design and development of devices. In fact, only the syntax and semantics of the bit stream for decoding are specified by standards. There is, thus, much room for innovation and ingenuity. Some of the operations are optional and some others, such as the human visual system (HVS) and weighting matrices (Tables 8.1 and 8.2), are given as examples. The entire preprocessing that precedes the encoder and the postprocessing that follows the decoder are outside the domain of the standards (Fig. 5.21). As such, the standards do not stifle the research and development, the main objective being the compatibility and interoperability among the systems (hardware/software) manufactured by different companies.

By ensuring these characteristics, the market potential for these products is assured. The industry can incorporate additional options in their products, providing a competitive atmosphere. The list of manufacturers and vendors (although not complete) described in Appendix A is a testimony to the phenomenal growth in this field. In addition, the industry is assured of a multifunction/multiservice capability of the VLSI chips by selecting some of the basic operations such as discrete cosine transform (DCT), block motion estimation, and interframe prediction enhanced by motion compensation common to at least some of the standards. Another equally important feature is the backward/forward compatibility of these systems, so that the consumer is assured of the new services. Also, the modularity, expandability, and interoperability of the systems can be maintained.

With the proliferation of personal computers (PCs) and the evolution of digital networks, it is only natural that video be integrated with audio and data over the information superhighway. Whereas the telephone and, more recently, digital cellular mobile radio have revolutionized instant communication (voice), the incorporation of video in the computer/communication fields is another milestone in human history. Examples include the videophone (at present a novelty, but destined to replace the ordinary phone), videoconferencing, multimedia, video on CD (compact disc), movies on high-density digital video disc, video mail, and HDTV. The market, however, is not limited to consumer electronics. Other fields impacted by the standards are medicine, education, training, defense, sports, stock market, desktop video, entertainment, restaurants, travel, computers, and related industries. Some services such as direcTV™ and PrimeStar, using a small satellite dish, can provide a plethora of TV channels. Video on demand, pay-per-view, jukebox video, karaoke video, video disks, and other similar services are the natural end products of the standards, which have already set their imprints on personal communication services.

By integrating audio (both speech and sound), images (still frame), video (motion/sequence), text, graphics, and data into a multimedia format, the communication arena has become instant and interactive. Multimedia computers (in which CD-ROM drives are an integral part) have now become the rule rather than the exception. Laptops with multimedia capability are already on the market. The desktop PC with multitask/multimedia with ubiquitous videoconferencing, the next step in the electronics explosion, is already on the horizon. Already, set-top boxes that can decode (and thereby display on a monitor) from storage media, such as video CD, are easily available. MPEG-1 decoders on multimedia PCs are offered by a number of vendors.

Video communication by mobile cellular radio is another exciting possibility. HDTV broadcasting over the air is expected by early 1998. Although this is based on FCC recommendations/rulings, the impact is profound as other media (satellite, fiber optics, coaxial cable, and laser disk) may follow the same format. Also, by default, this may lead to a worldwide standard. Thus, the standards have set the stage for a concerted and coordinated approach. The field of computers-communications-consumer electronics has advantageously utilized these standards in offering a multitude of services to the public. The growth of this field has no limit.

So far we have liberally used the term *standards*. The ones of interest to us and, hence, addressed in this book are in the domain of the interactive multimedia communication services. These can be listed as follows:

JPEG:	Joint Photographic Experts Group
ITU-T H.261:	Video codec for audiovisual services at $p \times 64$ Kbps
MPEG-1:	Moving Picture Experts Group: Digital storage media, up to 1.5 Mbps
MPEG-2:	Generic coding of moving pictures and associated audio
MPEG-4:	Multiple bit-rate audiovisual coding (up to 1024 Kbps for video and up to 64 Kbps for audio/speech)
ITU-T H.263:	Experts group on very low bit-rate video telephony
Digital HDTV:	Grand Alliance, FCC
ITU-R:	CMTT.723 and CMTT.721, Committee for Mixed Telephone and Television

Some other related standards such as those developed by MHEG (Multimedia Hypermedia Experts Group) and JBIG (Joint Binary Image Group) although equally important, are reluctantly left out due to space constraints. The latter addresses encoding and, indirectly, decoding of bi-level images. The algorithm can also be used effectively for coding of greyscale and color images by decomposing these into bit planes. The MHEG standard provides an encoding format for multimedia/hypermedia information that can be used and interchanged by applications in a wide range of domains.

1.2 Organization of the Book

This book is organized into two parts. Part I deals with the various redundancy reduction techniques that are utilized in Part II. Chapter 2 describes the various color formats and the relationships (conversion) among them. Also some sampling formats (Fig. 2.3) and ITU-R 601 parameters are listed. Chapter 3 addresses quantization with emphasis on the Lloyd-Max quantizers [27, 33]. The quantizer design is quite general and can be extended to minimization of any distortion. Compression techniques are discussed in Chapter 4, which describes predictive coding. The standard feedback predictor and its adaptive features are described, followed by techniques to minimize the effects of channel error. A recently developed double-predictor DPCM (differential pulse code modulation) and signal-blurred DPCM conclude this chapter.

The concept of transform coding is illustrated via the discrete cosine transform (DCT) in Chapter 5. Definitions, properties, fast algorithms, and applications of DCT to video coding are addressed. Incorporation of HVS in the transform domain, layered coding, and progressive image transmission are also outlined. After describing predictive coding and transform coding, the next logical step is to describe hybrid (mixed) coding. Whereas hybrid coding is much more general, Chapter 6 deals with the transform-DPCM combination. Enhancement of the encoding process by motion compensation has been widely adopted by both the standards groups and the industry. Hence, motion estimation is a vital part of the encoder and, therefore, is described in detail. Other compression schemes such as vector quantization (VQ) and subband coding are the topics of Chapter 7. VQ and many of its variations are outlined. This chapter (and Part I as well) concludes with subband coding. Some of the possible hybrid schemes involving VQ-subband-transform-DPCM are also addressed.

Having described the different compression techniques, we have set the stage for the main emphasis of this book, i.e., Part II, which begins with Chapter 8. The focus is on the various international standards related to coding of digital image/video/audio. The development of these standards is governed not only by bit-rate reduction, but also by other important factors such as compatibility, extensibility, interoperability, and cost complexity. The coding schemes applicable to continuous-tone still images are addressed in Chapter 8. Extensions to JPEG and implementation aspects conclude this chapter. Although the ITU-T H-series recommendations relate to frame structure, control, terminals, systems, and coding (see Section 9.1), our emphasis is entirely on the video coding related to H.261, which is useful for videotelephony and videoconferencing. Whereas H.261 deals only with video coding, both audio and video coding are the domain of MPEG. MPEG-1 (Chapter 10) discusses coding algorithms designed for storage media such as CD, DAT, and magnetic hard disk. Because of the need for interactive features beyond the traditional transform-DPCM-MC (Fig. 9.8), a group-of-pictures concept (Fig. 10.9) is introduced in Chapter 10. A deviation from transform coding is the subband coding selected for MPEG audio. Implementations of MPEG-1 conclude this chapter.

MPEG-2 (ITU-T H.262) addresses generic coding of moving video and associated audio and is described in Chapter 11. Whereas MPEG-1 is exclusively designed for digital storage media, the application arena for MPEG-2 extends also to television broadcasting and communication. Novel features such as profiles, levels, scalability, field/frame coding, interframe HVS weighting, and dual prime motion estimation are introduced. MPEG-2 audio, which deals with surround sound, multichannel/multilingual services, additional bit rates, and sampling rates besides MPEG-1 audio, concludes Chapter 11.

Chapter 12 deals with MPEG-4 and H.263. The latter is aimed at a very high compression ratio leading to a generic video/audio coding system optimized for very low bit rates. MPEG-4 is part of ISO/IEC with a schedule to finalize an international standard by November 1998. MPEG-4 addresses a wide range of applications, services, bit rates, and functionalities. H.263 is a draft recommendation by ITU-T SG 15 for video coding for narrow telecommunication channels at less than 64 Kbps that was finalized in December 1995. The video encoder is similar to H.261. Most of the variations from H.261 have been formulated by the ITU-T experts group for very low bit-rate visual telephony whose long-term recommendations are expected in 1998, coinciding with the scheduled MPEG-4 completion.

Chapter 13 focuses almost exclusively on the terrestrial broadcasting standard of HDTV being approved by the FCC. The proposals by the four digital proponents led the foundation to the Grand Alliance (GA) digital HDTV system. GA video is fully MPEG-2 compliant and uses main profile at a high level. Dolby AC-3 audio compression selected by the GA concludes this chapter. The popular interframe MC hybrid (DCT/DPCM) is adopted as ITU-R CMTT.723 for coding of component television signals in the 34 to 45 Mbps range (Chapter 14). Variations from the H.261 algorithm include HVS matrices, VWL codes, MV range, and scanning the DCT coefficients (Fig. 14.5). The chapter ends with a description of ITU-R CMTT.721-1 for coding the component TV signals at contribution quality at nearly 140 Mbps. This high quality is met by fixed DPCM.

1.3 Appendices

The appendices start with Appendix A, which lists the manufacturers and vendors related to video/audio codecs. Although this list is extensive, it is by no means complete. Every effort is made to provide the information as accurately as possible. The international nature of this list demonstrates the worldwide interest in this rapidly growing market. Separate categories are provided as follows: A.1, JPEG, H.261, MPEG Video; A.2, MPEG Audio; A.3, Motion Estimation; A.4, Image Processing Software; A.5, Image Processing Systems; and A.6, Image Capture/Digitizer Boards. FTP sites on the Internet are given in Appendix B. Standards documentation, source codes, test images/sequences, news groups, frequently asked questions, bibliographies, and any other related resources can be accessed over the Internet. Some topics such as fractals and wavelets, although not covered in the text, are also included. VLC tables for coding the DCT coefficients in JPEG, H.261, and so forth, are provided in Appendix C.

1.4 References

The references are grouped under the various standards: JPEG, H.261, MPEG-1, MPEG-2, and MPEG-4. Additional classifications include ghost cancellation, subjective assessment of video, subjective assessment of audio, BMA ME hardware, VQ (hardware/software), HDTV, and CMTT. Special efforts have been taken to make this list as up-to-date and exhaustive as possible. This has made it difficult to cite in the text some of the latest references.

Chapter 2
Color Formats

Summary

Various color formats are used in the standards described in Part II of this book. The basic color format is *RGB* color space, and all the other formats in use can be derived from the *RGB* information provided by input devices such as cameras or scanners.

YIQ, *YUV*, YC_bC_r, and SMPTE 240M have been developed for broadcast and television systems, and CMY_eK is used in color printing. The primary goal here is to understand the main color formats that are being adopted in the worldwide standards.

2.1 *RGB* Color Coordinate

The *RGB* (red, green, and blue) color space is the basic choice for computer graphics and image frame buffers because color CRTs use red, green, and blue phosphors to create the desired color. Red, green, and blue are the three primary additive colors. Individual components are added together to form a color and an equivalent addition of all components produces white, as shown in Fig. 2.1.

However, *RGB* is not very efficient for representing real-world images, since equal bandwidths are required to describe all three color components. The equal bandwidths result in the same pixel depth and display resolution for each color component. Moreover, the sensitivity of the color component of the human eye is

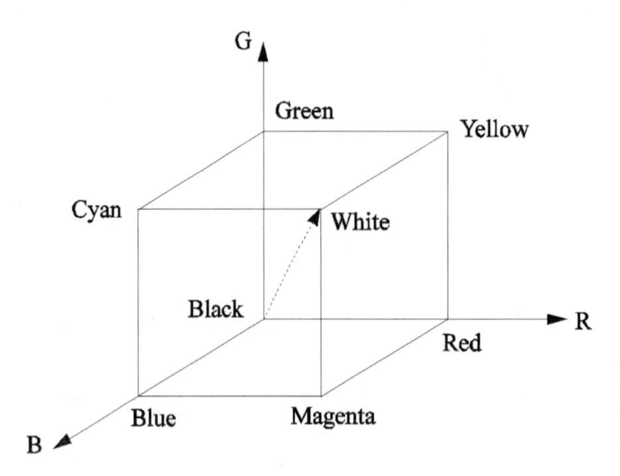

Figure 2.1 Tristimulus *RGB* color cube. The dotted line represents gray levels.

less than that of the luminance component. For these reasons, many image-coding standards and broadcast systems use luminance and color-difference signals. These are the *YUV*, *YIQ*, YC_bC_r, and SMPTE 240M color formats.

Gamma correction: Color components appear as voltage waveforms whose instantaneous values are directly proportional to the illumination falling on the corresponding spots of a camera tube target. Such a video signal is not directly suitable for CRT display on a television receiver. The main reason for this is that cathode-ray tubes are not linear devices; the transfer function of a CRT produces light intensity, I, that is proportional to some power of the signal voltage, V by an equation of the form

$$I_{display} \propto V_{received}^{\gamma} \tag{2.1}$$

where γ, the display gamma, is in the range 2.0 to 3.0 depending on display tube type. The typical gamma factors are 2.2 for NTSC (Fig. 2.2) and 2.8 for PAL/SECAM [13]. Note that input voltage is normalized to have a range of 0 to 1.

To compensate for the nonlinear processing at the display, linear *RGB* data are *gamma-corrected* prior to transmission (in the camera) rather than in the receiver, by an inverse process of Eq. (2.1), e.g., $I_{transmit} = {I_{input}}^{0.45}$ when the gamma factor is 2.2, as in the upper curve shown in Fig. 2.2. As a result, the darker areas of the picture have a higher gain applied to them than do the lighter areas. The benefits of gamma correction in the transmitter are that the cost of television receivers is minimized and the transmission process is reduced in dark areas of the picture, resulting in the elimination of visual artifacts.

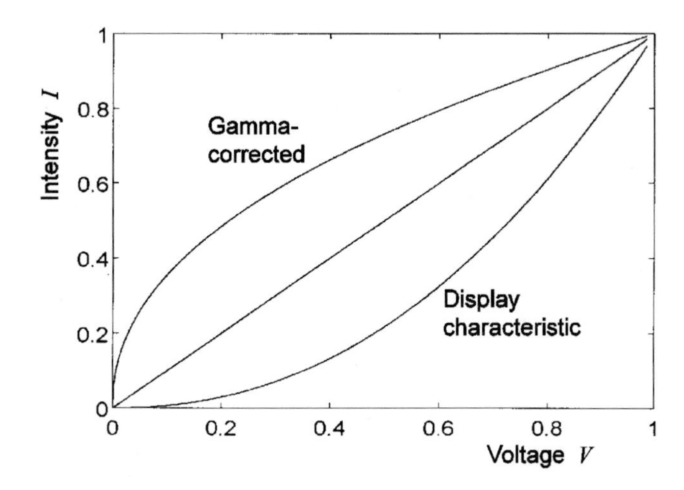

Figure 2.2 The effect of gamma correction with a factor of 2.2.

2.2 *YUV* Color Coordinate

The *YUV* color coordinate is the basic color format used by the NTSC, PAL and SECAM composite color TV standards. Y represents the black-and-white component and color information (U and V) is added to display a color picture. A black-and-white receiver would still work using the color composite signal. *YUV* signals are derived by using the gamma-corrected RGB signal as

$$Y = 0.299R' + 0.587G' + 0.114B'$$

$$U = -0.147R' - 0.289G' + 0.436B' = 0.492(B' - Y)$$

$$V = 0.615R' - 0.515G' - 0.100B' = 0.877(R' - Y) \tag{2.2}$$

where the prime indicates gamma-corrected *RGB* signal.

In a digital television system such as the one proposed for HDTV (see Section 13.2), the *YUV* signal is obtained by an *RGB*-to-*YUV* matrix. Usually, color-difference signals (U and V) are subsampled by a factor of two to four in the spatial dimensions, because these are much less sensitive to the human visual system than the luminance Y is. This is helpful in reducing the bit rate in video compression techniques.

2.3 *YIQ* Color Coordinate

The *YIQ* color coordinate is derived from the *YUV* format and is optionally used by the NTSC composite TV standard. The I stands for *in-phase* and the Q for *quadrature-phase*, which are the modulation methods separated by a phase angle of 90 degrees [14].

The basic equations to derive from gamma-corrected *RGB* are

$$\begin{aligned} Y &= 0.299R' + 0.587G' + 0.114B' \\ I &= 0.596R' - 0.275G' - 0.321B' \\ &= 0.736(R' - Y) - 0.268(B' - Y) \\ Q &= 0.212R' - 0.523G' + 0.311B' \\ &= 0.478(R' - Y) + 0.413(B' - Y) \end{aligned} \tag{2.3}$$

There are obtained by advancing the phase of the $(B' - Y)$ and $(R' - Y)$ signals by 33 degrees as a simple rotation from the $(B' - Y)$ and $(R' - Y)$ axes, which is given by the notation

$$\begin{bmatrix} I \\ Q \end{bmatrix} = \begin{bmatrix} 0 & 1 \\ 1 & 0 \end{bmatrix} \begin{bmatrix} \cos(33^\circ) & \sin(33^\circ) \\ -\sin(33^\circ) & \cos(33^\circ) \end{bmatrix} \begin{bmatrix} U \\ V \end{bmatrix} \tag{2.4}$$

2.4 YC_bC_r Color Coordinate

The YC_bC_r color coordinate was developed as part of ITU-R BT.601 [898] during the establishment of a worldwide digital video component standard. The standard yields a compatible digital approach between the two different systems (525-line and 625-line systems). The YC_bC_r signals are scaled and offset versions of the *YUV* format. Y is defined to have a nominal range of 16 to 235; C_b and C_r are defined to have a range of 16 to 240, with zero signal corresponding to level 128.

Most image compression standards as discussed in Chapters 8 through 14 adopt this color format as an input image signal. We prefer the notation YC_bC_r rather than YC_rC_b, since the block multiplexing in the application standards follows the order YC_bC_r.

There are several YC_bC_r sampling formats, such as 4:4:4, 4:2:2, and 4:1:1 (4:2:0). The basic features of the worldwide standard for digital television studios, ITU-R BT.601, are shown in Table 2.1 for the 4:2:2 format. The sampling format 4:2:2 implies that the sampling rates of C_b and C_r are one-half that of Y. Also, there are alternate Y samples only sandwiched between cosited Y, C_b, and C_r samples. In a consecutive run of four samples there are 4 Y samples and 2 samples each of C_b and C_r. This sampling pattern is signified by 4:2:2 [18].

Table 2.1 Basic parameters for the 4:2:2 format of ITU-R 601

	525-line, 60 field/s systems	625-line, 50 field/s systems
Coded signals	Y, C_b, C_r	
Number of samples per line:		
• luminance (Y)	858	864
• color-difference (C_b, C_r)	429	432
Sampling structure	Orthogonal, line, field and frame repetitive. C_b and C_r samples cosited with odd (1st, 3rd, 5th, etc.) Y samples in each line.	
Sampling frequency:		
• luminance	13.5 MHz	
• color-difference	6.75 MHz	
Form of coding	Uniformly quantized PCM, 8 bits/sample, for Y, C_b, C_r.	
Number of samples per digital active line:		
• luminance	720	
• color-difference	360	
Quantization levels:		
• scale	0 to 255	
• luminance	220 quantization levels (16 (black level) to 235 (white level))	
• color-difference	224 quantization levels (16 to 240, 128 corresponding to zero on a scale 0 to 255)	

A luminance sampling frequency of 13.5 MHz (about four times the color subcarrier frequency 3.58 MHz for the NTSC signal) was adopted. For the 4:2:2 format, the color-difference signals are sampled at about two times the color subcarrier of the NTSC system. This is a compromise between picture quality and practical realization [14]. Also, the frequency 13.5 MHz is the only one in the range of 12 to 14.3 MHz (studied during the standardization) that permits an integer number of samples per line for both systems. The 525-line system has a line period of 63.56 μseconds and the 625-line system has a line period of 64 μseconds. Hence, 858 (13.5×63.56) and 864 (13.5×64) samples are obtained, respectively. Samples per active line, however, are 720 for both systems, which have the same active line period of 53 μseconds. The active lines per frame are 486 (21 to 263 and 283 to 525) for the 525-line system and 576 (23 to 310 and 336 to 623) for the 625-line system.

The positioning of YC_bC_r samples or pixels for the 4:4:4, 4:2:2, 4:1:1, and 4:2:0 formats is illustrated in Fig. 2.3. Each component is typically quantized

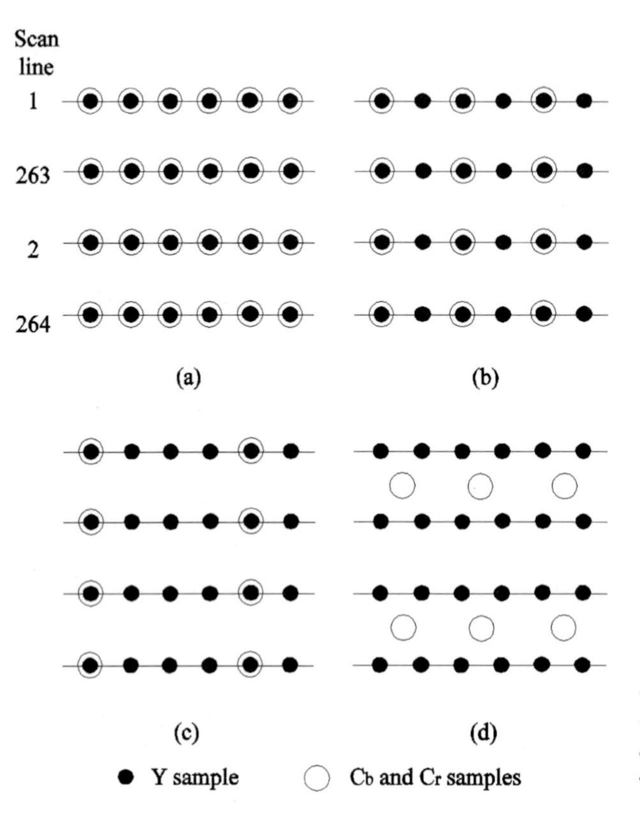

Figure 2.3 Orthogonal sampling on the scan lines of an interlaced system: (a) 4:4:4; (b) 4:2:2; (c) 4:1:1; and (d) 4:2:0 format.

with 8-bit resolution. Each sample therefore requires 24 bits for the 4:4:4 format and 16 bits for the 4:2:2 format. Four samples require 6 times 8 bits for the 4:1:1 format.

In the 4:2:0 format, the C_b and C_r samples are offset, although the Y samples conform to the 4:1:1 format. The video compression standards (such as H.261 and MPEG) adopt slightly different sampling formats, as shown in Figs. 9.4 and 11.10. The chrominance samples are centered among the four luminance samples or between the first vertical pair of luminance samples.

YC_bC_r signals are obtained from digital gamma-corrected RGB signals as follows:

$$\begin{aligned} Y &= 0.299R' + 0.587G' + 0.114B' \\ C_b &= -0.169R' - 0.331G' + 0.500B' \\ C_r &= 0.500R' - 0.419G' - 0.081B' \end{aligned} \tag{2.5}$$

The color-difference signals are given by

$$(B - Y) = -0.299R' - 0.587G' + 0.886B'$$
$$(R - Y) = 0.701R' - 0.587G' - 0.114B' \tag{2.6}$$

where the values for $(B - Y)$ have a range of ± 0.886 and for $(R - Y)$ a range of ± 0.701; the values for Y have a range of 0 to 1. To restore the signal excursion of the color-difference signals to unity (-0.5 to $+0.5$), $(B - Y)$ is multiplied by a factor of 0.564 (0.5 divided by 0.886) and $(R - Y)$ is multiplied by a factor of 0.713 (0.5 divided by 0.701). Thus, the C_b and C_r are the renormalized blue and red color-difference signals, respectively.

Given that the luminance signal is to occupy 220 levels (16 to 235), the luminance signal has to be scaled to obtain the decimal value, $\overline{Y}$. Similarly, the color difference signals are to occupy 224 levels and the zero level is to be level 128. The decimal values for the three components are expressed as

$$\overline{Y} = 219Y + 16$$
$$\overline{C}_b = 224[0.564(B - Y)] + 128 = 126(B - Y) + 128$$
$$\overline{C}_r = 224[0.713(R - Y)] + 128 = 160(R - Y) + 128 \tag{2.7}$$

where the corresponding level number after quantization is the nearest integer. This is useful for most of the applications where 8-bit binary encoding is adopted.

2.5 SMPTE 240M Color Coordinate

The SMPTE 240M (1988) color coordinate standard [16] was developed to standardize the production of 1125/60 high-definition source material in the US. The main systems proposed for the all-digital HDTV system in the US and the Grand Alliance system adopt a similar format (see Tables 13.1 and 13.6). The signal parameters for the HDTV system are shown in Table 2.2 (the Eureka 95 HDP format proposed in Europe is also shown).

The color components (Y, P_b, P_r) are derived from gamma-corrected (factor of 2.2) RGB signals as follows:

$$Y = 0.212R' + 0.701G' + 0.087B'$$
$$P_b = -0.116R' - 0.384G' + 0.500B'$$
$$= (B' - Y)/1.826$$
$$P_r = 0.500R' - 0.445G' - 0.055B'$$
$$= (R' - Y)/1.576 \tag{2.8}$$

Table 2.2 Signal parameters for HDTV systems

	U.S.	Europe
Total scan lines per frame	1125	1250
Active lines per frame	1035	1152
Field rate (Hz)	60	50
Aspect ratio	16:9	16:9
Scanning format	2:1 interlaced	2:1 interlaced
Bandwidth (MHz)	30	30
Active pixels per line:		
• luminance	1920	1920
• chrominance	960	960
Sampling frequency (MHz):		
• luminance	74.25	72
• chrominance	37.125	36

2.6 CMY_eK Color Coordinate

The CMY_eK (cyan, magenta, yellow, black) color format is widely used for color printing. It is based on the subtractive properties of inks, as opposed to the additive properties of light. Cyan, magenta, and yellow are the subtractive primaries and are the complements of red, green, and blue. Color is specified by what is subtracted from white light (the sum of red, green, and blue); cyan subtracts red from white, and so on. In other words, cyan is green plus blue, magenta is red plus blue, and yellow is red plus green, as shown in Fig. 2.1. If all the *RGB* colors are subtracted, there is black. Therefore, white can only be generated on a white paper.

Since ideally CMY_e is the complement of *RGB*, the following linear equations are initially used to convert between *RGB* and CMY_e:

$$\begin{bmatrix} C \\ M \\ Y_e \end{bmatrix} = \begin{bmatrix} 1 \\ 1 \\ 1 \end{bmatrix} - \begin{bmatrix} R \\ G \\ B \end{bmatrix} \tag{2.9}$$

where the element 1 means the sum of *RGB*.

However, more sophisticated transformations account for the dependency of the inks and paper used. A separate black ink is used, rather than printing cyan, magenta, and yellow to generate black, to maintain black color purity.

Chapter 3
Quantization

Summary

A brief description of quantization technique is presented, followed by the design of both uniform and nonuniform quantizers. The reader can design these quantizers for any probability distribution based on minimizing any specified distortion criteria. This design can be modified such that the quantization error is below the visibility thresholds (21, 48). Quantizer design can also be made adaptive (46, 47).

3.1 Introduction

Digital signal processing involves sampling of an analog signal and quantization of the sampled data into a finite number of levels. It is assumed that the sampling is uniform and that the sampling rate is above the Nyquist rate so that there is no aliasing in the frequency domain. Quantization (Fig. 3.1) [19, 50] involves representing the sampled data by a finite number of levels based on some criteria such as minimization of the quantizer distortion. The distortion is to be meaningful. Quantizer design includes input (decision) levels and output (representation) levels as well as the number of levels. The design can be enhanced by psychovisual or psychoacoustic perception.

Quantizers can be classified as memoryless or with memory. The former assumes that each sample is quantized independently; the latter takes into account

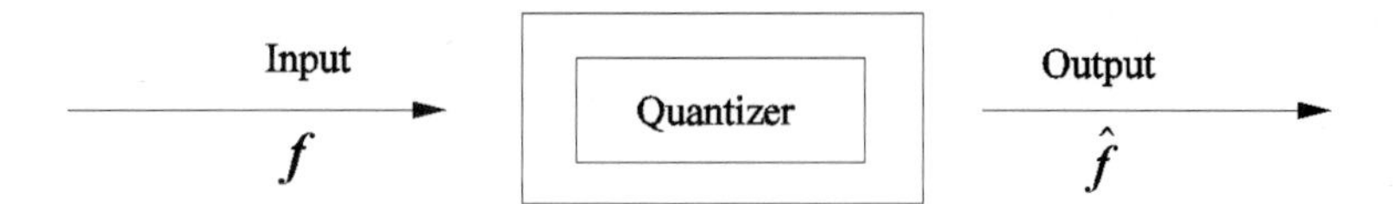

Figure 3.1 The quantized output of f is $\hat{f}$. The quantization error is $\hat{f} - f$.

previous sample(s). We will limit our discussion to the memoryless quantizers. Another classification is uniform or nonuniform. A uniform quantizer is completely defined by the number of levels it has, its step size, and whether it is a midriser or a midtreader. Our discussion also is limited to symmetric quantizers, i.e., the input and output levels in the third quadrant are negatives of the corresponding levels in the first quadrant. A nonuniform quantizer implies that the step sizes are not constant. Hence, the nonuniform quantizer has to be specified by the input and output levels (first quadrant or third quadrant).

Uniform quantizers are shown in Fig. 3.2a (midtreader) and Fig. 3.2b (midriser), while nonuniform quantizers are shown in Fig. 3.3. If the input f lies in the interval $[d_2, d_3)$, then it is quantized as r_2. Quantization inherently is a lossy process as r_2 represents the range $[d_2, d_3)$. A nearly uniform quantizer is shown in Fig. 3.4. Except for the deadzone (input range for which the output is zero), the stepsize is constant. Such a nearly uniform quantizer has been specified in H.261 (Chapter 9) , MPEG-1 video (Chapter 10), MPEG-2 video (Chapter 11), and H.263 (Chapter 12).

3.2 Quantizer Design

Given the number of input or output levels, quantizer design involves minimizing any meaningful quantizer distortion, such as:

1. Mean square quantization error (MSQE) [27]:

$$E\left[(f - \hat{f})^2\right] = \int_{a_L}^{a_U} (f - \hat{f})^2 \, p(f) \, df \tag{3.1}$$

   where input f ranges from a_L to a_U and $p(f)$ is the probability density function of f.

2. Mean absolute quantization error (MAQE) [27]:

$$E\left[|f - \hat{f}|\right] = \int_{a_L}^{a_U} |f - \hat{f}| \, p(f) \, df \tag{3.2}$$

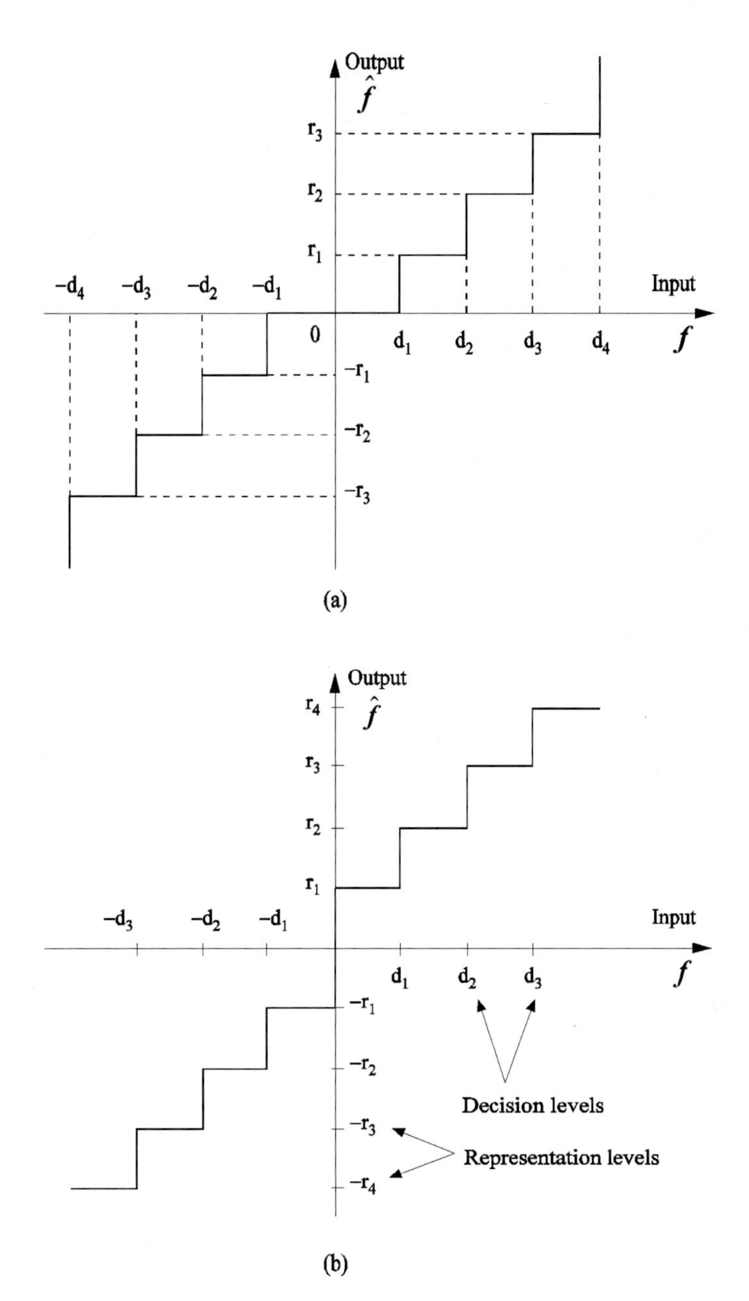

Figure 3.2 Uniform symmetric quantizers: (a) midtreader; (b) midriser. Step size is constant. The r_i are output levels; the d_i are input levels.

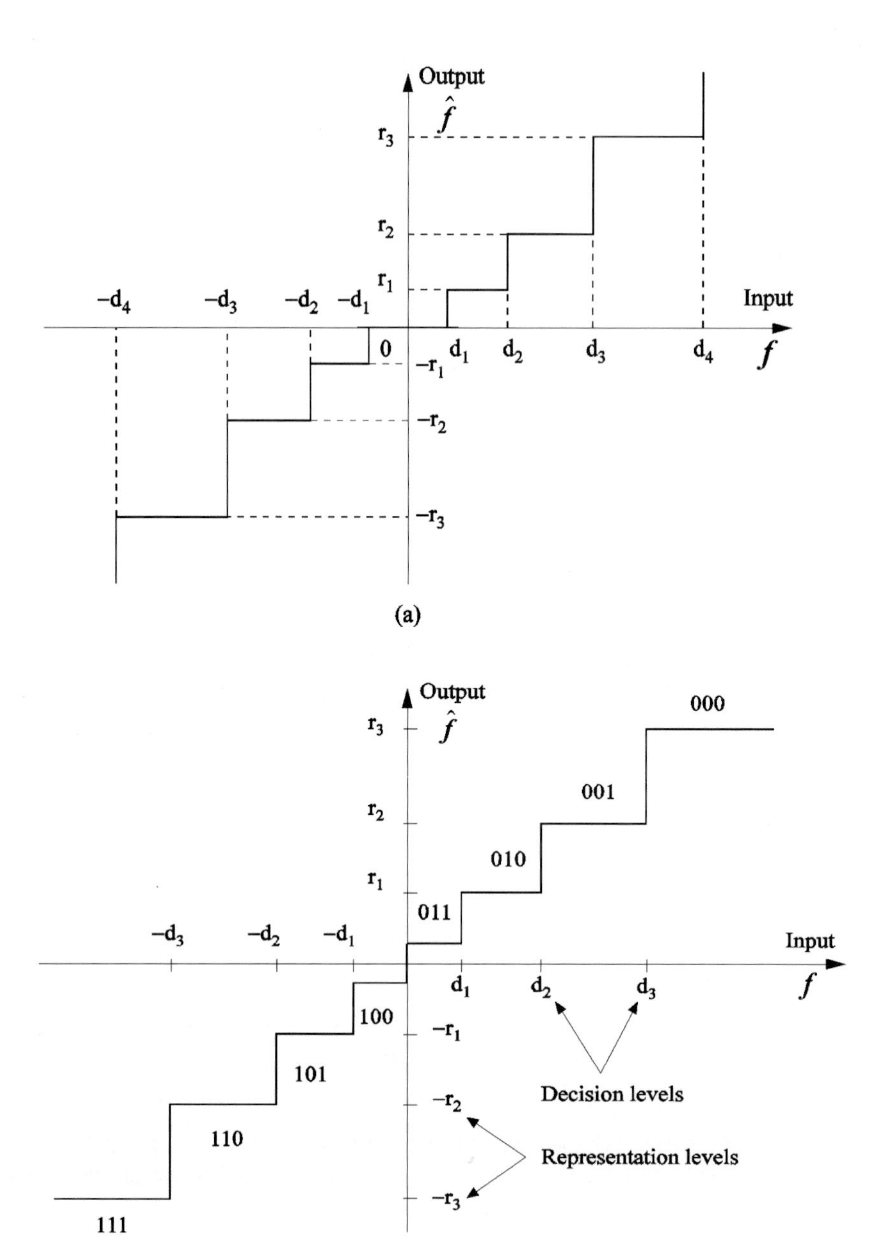

Figure 3.3 Nonuniform symmetric quantizers: (a) midtreader (this has a zero output level); (b) midriser (this has no zero output level). The r_i are output levels; the d_i are input levels.

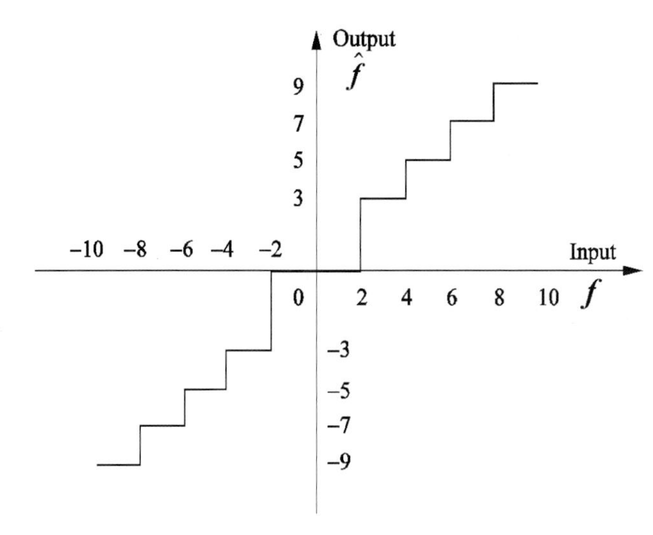

Figure 3.4 A nearly uniform symmetric quantizer. Except for the deadzone ($-2 \leq f < 2$), the stepsize is constant. When ($-2 \leq f < 2$), the output is zero.

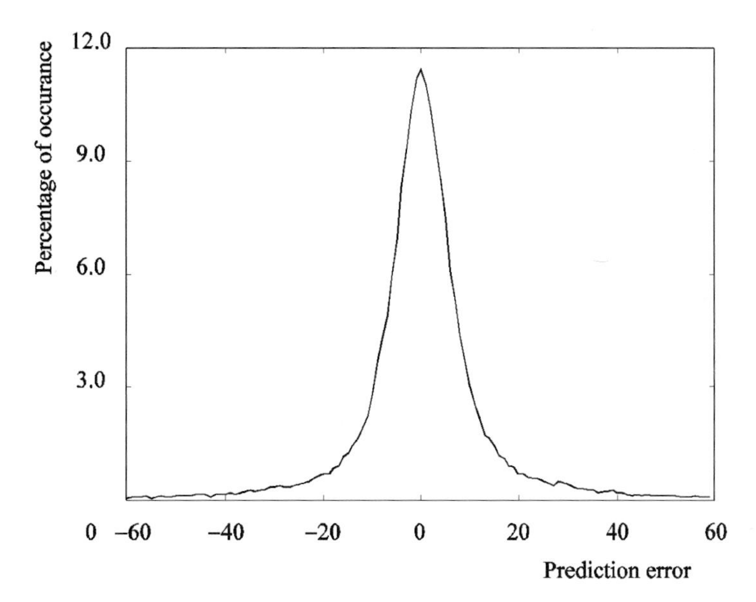

Figure 3.5 A histogram of the prediction error based on some test images. Prediction errors tend to follow Laplacian distribution.

3. Mean L_N norm quantization error:

$$E\left[|f - \hat{f}|^N\right] = \int_{a_L}^{a_U} |f - \hat{f}|^N \, p(f) \, df \tag{3.3}$$

where N is any positive integer.

4. Weighted quantization error:

$$\int_{a_L}^{a_U} w(f)\,|f - \hat{f}|\,p(f)\,df \tag{3.4}$$

where $w(f)$ is the weighting function.

Max-Lloyd quantizer

Max [27] and Lloyd [33] independently designed the quantizers minimizing the MSQE and developed tables for input governed by standard distribution functions such as gamma, Laplacian [36], Gaussian, Rayleigh, and uniform [35, 36, 49]. Quantizers can also be designed tailored to histograms [51]. For a well-designed predictor, the histogram of the prediction error tends to follow the Laplacian distribution (Fig. 3.5). Reininger and Gibson [51] have also shown that the ac coefficients (2D-DCT) of images tend to follow this distribution (see also [118]).

Given the range of input f as from a_L to a_U and the number of output levels as J, design the quantizer so that the MSQE is minimized.

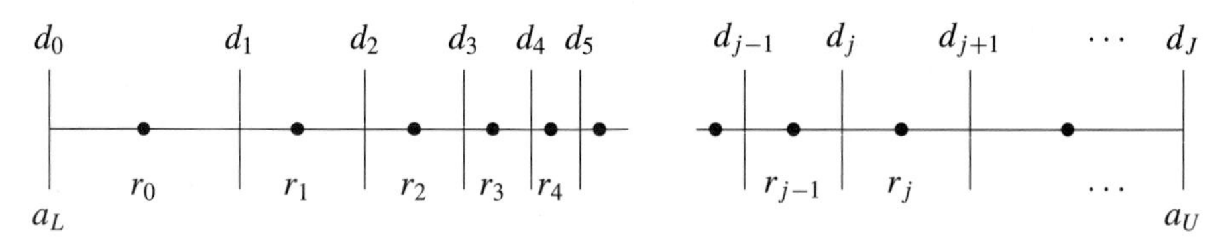

$$d_i = \text{decision level (input level)}$$

$$r_i = \text{reconstruction level (output level)}$$

If $d_j \le f < d_{j+1}$, then $Q(f) = r_j$ and

$$\text{MSQE} = \varepsilon = E\left[(f - \hat{f})^2\right] = \int_{a_L}^{a_U} (f - \hat{f})^2\,p(f)\,df \tag{3.5}$$

where $p(f)$ = the probability density function of the random variable f. It can be rewritten as

$$\varepsilon = \sum_{l=0}^{J-1} \int_{d_l}^{d_{l+1}} (f - r_l)^2\,p(f)\,df \tag{3.6}$$

Because r_k and d_k are variables, to minimize ε set

$$\frac{\partial \varepsilon}{\partial r_k} = \frac{\partial \varepsilon}{\partial d_k} = 0 \tag{3.7}$$

$$\frac{\partial \varepsilon}{\partial d_k} = \frac{\partial}{\partial d_k}\left[\int_{d_{k-1}}^{d_k} (f - r_{k-1})^2\, p(f)\, df + \int_{d_k}^{d_{k+1}} (f - r_k)^2\, p(f)\, df\right]$$

$$= (d_k - r_{k-1})^2\, p(d_k) - (d_k - r_k)^2\, p(d_k) = 0 \tag{3.8}$$

which yields $(d_k - r_{k-1}) = \pm(d_k - r_k)$.

Since $(d_k - r_{k-1}) > 0$ and $(d_k - r_k) < 0$, of the two possible solutions only the following is valid:

$$(d_k - r_{k-1}) = -(d_k - r_k) \tag{3.9}$$

$$d_k = \frac{r_k + r_{k-1}}{2} \tag{3.10}$$

This means that the input level is the average of the two adjacent output levels.

Also,

$$\frac{\partial \varepsilon}{\partial r_k} = \frac{\partial}{\partial r_k}\left[\int_{d_k}^{d_{k+1}} (f - r_k)^2\, p(f)\, df\right]$$

$$= -2\int_{d_k}^{d_{k+1}} (f - r_k)\, p(f)\, df = 0 \tag{3.11}$$

Hence,

$$r_k = \frac{\int_{d_k}^{d_{k+1}} f\, p(f)\, df}{\int_{d_k}^{d_{k+1}} p(f)\, df} \tag{3.12}$$

The output level is the centroid of the adjacent input levels.

This solution is not in closed form. To find the input level d_k one has to find the output level r_k and vice versa. However, by iterative techniques (Newton's method) both r_k and d_k can be evaluated. Techniques to speed up the iterative process and other variations have also been reported [37, 38, 41].

When the number of output levels J is very large, the quantizer design can be approximated as follows. Assuming that $p(f)$ is constant over each quantization level, $p(f) \approx p(r_j)$,

$$\varepsilon = \sum_{j=0}^{J-1} p(r_j) \int_{d_j}^{d_{j+1}} (f - r_j)^2\, df$$

$$= \frac{1}{3}\sum_{j=0}^{J-1} p(r_j)\left[(d_{j+1} - r_j)^3 - (d_j - r_j)^3\right] \tag{3.13}$$

$$\frac{\partial \varepsilon}{\partial r_l} = 0 = p(r_l)\left[(d_{l+1} - r_l)^2 - (d_l - r_l)^2\right] \tag{3.14}$$

results in $(d_{j+1} - r_j)^2 = (d_j - r_j)^2$.

d_j r_j d_{j+1}

Hence,

$$\begin{aligned}(d_{j+1} - r_j) &= -(d_j - r_j), \qquad d_{j+1} > r_j > d_j \\ d_{j+1} + d_j &= 2\, r_j \\ r_j &= \frac{d_{j+1} + d_j}{2}\end{aligned} \tag{3.15}$$

Each reconstruction level r_j is midway between its two adjacent decision levels d_j and d_{j+1}.

Substitute Eq. (3.15) in Eq. (3.13) to get

$$\begin{aligned}\varepsilon &= \frac{1}{3}\sum_{j=0}^{J-1} p(r_j)\left[\left(\frac{d_{j+1} - d_j}{2}\right)^3 - \left(\frac{d_j - d_{j+1}}{2}\right)^3\right] \\ &= \frac{1}{12}\sum_{j=0}^{J-1} p(r_j)\,(d_{j+1} - d_j)^3\end{aligned} \tag{3.16}$$

Minimize ε further with respect to d_j.

Set $d_{j+1} - d_j = \Delta d_j$. Then

$$\begin{aligned}\varepsilon &= \frac{1}{12}\sum_{j=0}^{J-1} p(r_j)\,(\Delta d_j)^3 \\ &= \sum_{j=0}^{J-1} \varepsilon_j, \quad \text{where } \varepsilon_j = \frac{1}{12}\, p(r_j)\,(\Delta d_j)^3\end{aligned} \tag{3.17}$$

This is minimum when ε_j is constant independent of the jth level.

Set

$$\frac{1}{12}\sum_{j=0}^{J-1} \left[p(r_j)\right]^{1/3}\,\Delta d_j \approx \frac{1}{12}\int_{a_L}^{a_U} p^{1/3}(f)\,df = k \tag{3.18}$$

where k = constant.

Let $\mu_j = [p(r_j)]^{1/3}\ \Delta d_j$; then

$$\sum_{j=0}^{J-1} \frac{1}{12}\ p(r_j)\ (\Delta d_j)^3 = \varepsilon = \frac{1}{12} \sum_{j=0}^{J-1} \mu_j^3$$

Thus, $\frac{1}{12} \sum_{j=0}^{J-1} \mu_j = k$. Constraint $\left[\frac{1}{12} \sum_{j=0}^{J-1} \mu_j - k\right] = 0$. Minimize $\varepsilon + \lambda \left(\frac{1}{12} \sum_{j=0}^{J-1} \mu_j - k\right)$, where λ is the Lagrange variable.

Set

$$\frac{\partial}{\partial \mu_l}\left[\frac{1}{12} \sum_{j=0}^{J-1} \mu_j^3 + \lambda\left(\frac{1}{12} \sum_{j=0}^{J-1} \mu_j - k\right)\right] = 0$$

$$\frac{3}{12}\mu_l^2 + \frac{\lambda}{12} = 0, \quad \lambda = -3\mu_l^2$$

that is,

$$\mu_0 = \mu_1 = \cdots = \mu_{J-1} = [p(r_j)]^{1/3}\ \Delta d_j$$

$$\therefore \quad \frac{1}{12}\ \mu_l\ J = k, \quad \mu_l = 12\,\frac{k}{J}$$

$$[p(r_j)]^{1/3}\ \Delta d_j = \mu_j = 12\frac{k}{J}$$

Hence, from Eq. (3.17), we obtain

$$\varepsilon = \frac{1}{12} \sum_{j=0}^{J-1} \mu_j^3 = (12)^2 \frac{k^3}{J^2}$$

$$= \frac{(12)^2}{J^2} \left(\frac{1}{12}\right)^3 \left[\int_{a_L}^{a_U} p^{1/3}(f)\, df\right]^3 \tag{3.19}$$

From Eq. (3.18),

$$\Delta d_j = 12 \frac{k}{J\,[p(r_j)]^{1/3}}$$

$$d_0 = a_L$$
$$d_1 = a_L + (d_1 - d_0)$$
$$d_2 = a_L + (d_1 - d_0) + (d_2 - d_1)$$

Generalizing,

$$d_j = a_L + \sum_{m=1}^{j} (d_m - d_{m-1}) = a_L + \sum_{m=0}^{j-1} \Delta d_m$$

Substituting for Δd_m yields

$$\begin{aligned} d_j &= a_L + \frac{12k}{J} \sum_{m=0}^{j-1} \left[\frac{1}{[p(r_j)]^{1/3}} \right] \\ &= a_L + \frac{12k}{J} \int_{a_L}^{d_j} [p(r_j)]^{-1/3} \, df \\ a_U &= d_J = a_L + \frac{12k}{J} \int_{a_L}^{a_U} [p(r_j)]^{-1/3} \, df \end{aligned}$$

Hence,

$$\frac{12k}{J} = (a_U - a_L) \Big/ \int_{a_L}^{a_U} [p(r_j)]^{-1/3} \, df. \tag{3.20}$$

The decision level is

$$d_j = a_L + \frac{(a_U - a_L) \int_{a_L}^{d_j} [p(r_j)]^{-1/3} \, df}{\int_{a_L}^{a_U} [p(r_j)]^{-1/3} \, df}, \quad j = 0, 1, \ldots, J \tag{3.21}$$

Quantizers using Eq. (3.20) can be designed for any probability density function.

3.3 Uniform Quantizer

When $p(f)$ is a uniform density function (Fig. 3.6), i.e.,

$$p(f) = \frac{1}{a_U - a_L} = \frac{1}{A} \tag{3.22}$$

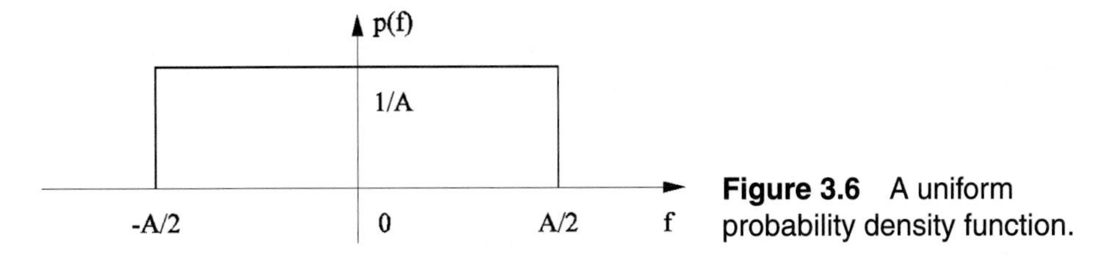

Figure 3.6 A uniform probability density function.

then

$$r_k = \frac{\int_{d_k}^{d_{k+1}} f\, p(f)df}{\int_{d_k}^{d_{k+1}} p(f)df} = \frac{1}{2}\left(d_{k+1} + d_k\right) \tag{3.23}$$

Since

$$d_{(k)} = \frac{r_k + r_{k-1}}{2} = \frac{d_{k+1} + d_{k-1}}{2}, \tag{3.24}$$

then

$$\begin{aligned} d_k - d_{k-1} &= d_{k+1} - d_k \\ &= \text{constant step size} \\ &= \frac{a_U - a_L}{J} \end{aligned} \tag{3.25}$$

Because the quantization error is distributed uniformly over each step size (SS), the MSQE is

$$\varepsilon = \frac{1}{SS}\int_{-SS/2}^{SS/2} f^2\, df = \frac{(SS)^2}{12} \tag{3.26}$$

Let the range of f be A. Its variance σ_f^2 is

$$\sigma_f^2 = \frac{1}{A}\int_{-A/2}^{A/2} f^2\, df = \frac{A^2}{12} \tag{3.27}$$

A b bit quantizer can represent J output levels, i.e., $2^b = J$. The stepsize is $SS = \frac{A}{2^b}$.

The signal-to-noise ratio (SNR) for a uniform quantizer is

$$\text{SNR} = \frac{\text{Variance}}{\text{MSQE}} = \frac{A^2/12}{\left(\frac{A^2}{2^{2b}}/12\right)} = 2^{2b} \tag{3.28}$$

$$(\text{SNR})\ \text{dB} = 10\log_{10}(\text{SNR}) = 20\,b\,\log_{10}2 \cong 6\,b\ \text{dB} \tag{3.29}$$

SNR for a uniform mean square quantizer increases by 6 dB per bit.

3.4 Quantization of Transform Coefficients in a Block

In transform coding (Chapter 5), assuming the transform is orthogonal, the transform coefficients can be quantized based on their variances. It can be shown easily that the average reconstruction error variance is equal to the quantization error of the transform coefficients. A minimization of this error leads to the optimum bit allocation among the transform coefficients [26, 29, 30, 31, 40, 51], i.e.,

$$b_{ij} = b_{av} + \frac{1}{2}\log_2 \tilde{\sigma}_{ij}^2 - \frac{1}{2N^2}\sum_{k=0}^{N-1}\sum_{l=0}^{N-1}\log_2 \tilde{\sigma}_{kl}^2 \tag{3.30}$$

$$i,\ j = 0,\ 1, \ldots, N-1$$

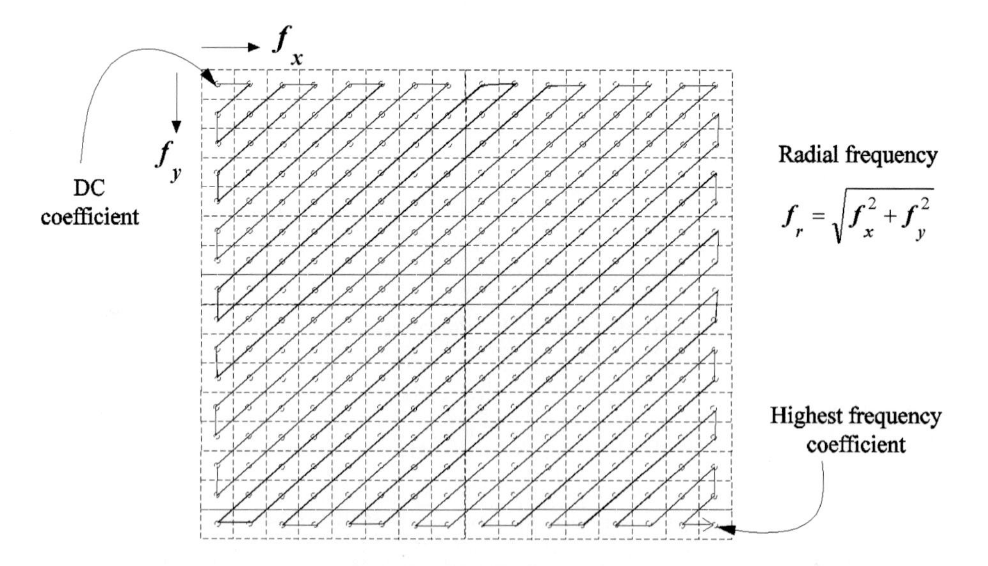

Figure 3.7 A zigzag scan of the 2D (16 x 16) DCT coefficients.

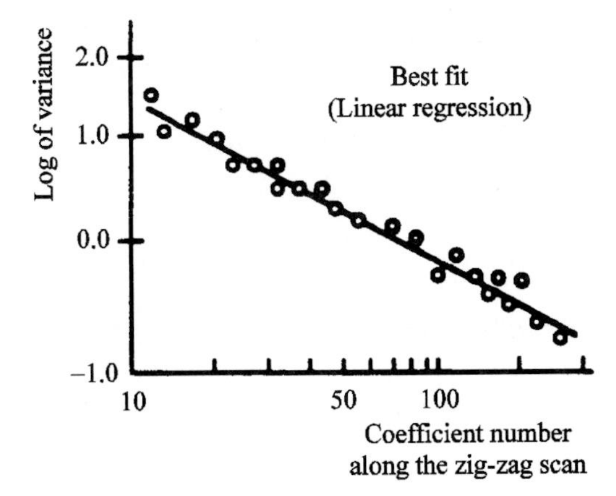

Figure 3.8 Variances of 2D-DCT coefficients along the zigzag scan (Fig. 3.7). Transmit only the slope and intercept. Estimate the variances at the receiver [52]. © 1992 IEE.

where $\tilde{\sigma}_{ij}^2$ is the variance of the transform coefficient $X^{c2}(i, j)$, b_{ij} is the number of bits assigned to the transform coefficient $X^{c2}(i, j)$, and b_{av} is the average bit rate such that $\sum_{i=0}^{N-1} \sum_{j=0}^{N-1} b_{ij} = N^2 b_{av}$ is satisfied by rounding b_{ij} to the nearest integer. Here an $(N \times N)$ image block is mapped to the 2D $(N \times N)$ DCT domain [Eq. (5.36) and Figs. 5.11 and 5.14].

Simulation of test image sequences has shown that the variances of the 2D-DCT coefficients along the zigzag scan (Fig. 3.7) tend to decrease linearly on a log-log scale [52] (Fig. 3.8). The DCT coefficients are numbered along the zigzag scan. By transmitting only the slope and intercept of this variance distribution and the average bit rate, the receiver can reconstruct the bit allocation matrix for the DCT coefficients. This approach avoids a large amount of overhead.

Chapter 4
Predictive Coding

Summary

The concept of DPCM is presented. Design of linear predictors is outlined. Techniques to minimize prediction error accumulation such as leak factor and periodic refresh are presented. Adaptive predictors, although more complex and sensitive to channel noise, reduce prediction errors. Techniques to minimize the overhead needed for adaptive predictor selection are presented.

4.1 Introduction

The underlying philosophy behind predictive coding is if the present sampled signal, say a current pixel (pixel or pel means picture element), can be predicted reasonably well based on the previous neighborhood samples (causal predictor), then the prediction error has a smaller entropy than the original sampled signal [59, 60]. Hence the prediction error can be quantized with fewer quantization levels than can the sampled signal. The prediction can be based on the statistical distribution of the signal (the signal can be audio, speech, image, video, text, graphics, etc.). The most common predictive coding is called differential pulse code modulation (DPCM) shown in Fig. 4.1.

Inverse DPCM at the decoder side is shown in Fig. 4.2. In the absence of channel noise, the reconstruction error is equal to the quantization error resulting in a lossy coder. The feedback loop (closed-loop) avoids the accumulation of

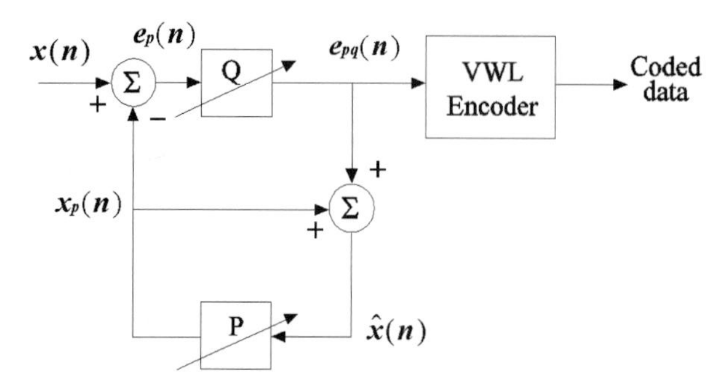

Figure 4.1 A closed-loop DPCM encoder. *P* (predictor) and/or *Q* (quantizer) can be adaptive.

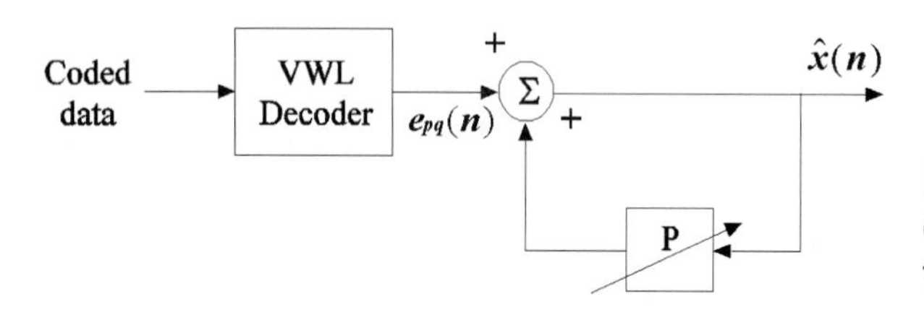

Figure 4.2 The DPCM decoder corresponding to Fig. 4.1.

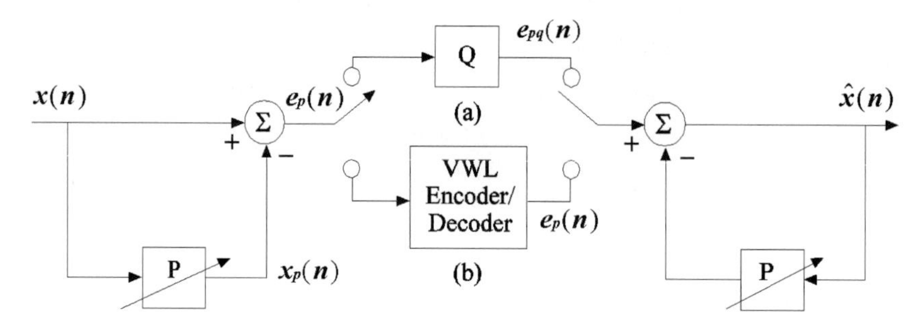

Figure 4.3 A feedforward predictor used for (a) lossy and (b) lossless predictive coding. The lossless mode is part of the Joint Photographic Experts Group (JPEG) specifications (Section 8.6) applicable to continuous-tone gray-level images.

the quantization error during successive prediction cycles. The predictor at the decoder is an exact replica of the predictor at the encoder. In practice, because only reconstructed pels (not the original pels) are available at the decoder the prediction is based only on the previously reconstructed pels. Also, the prediction process can be improved by adaptive prediction and/or quantization [54, 56, 62, 64, 66]. In a feedforward predictor, the quantization error can accumulate during successive cycles (Fig. 4.3a). Such a scheme, however, is useful in lossless coding (Fig. 4.3b).

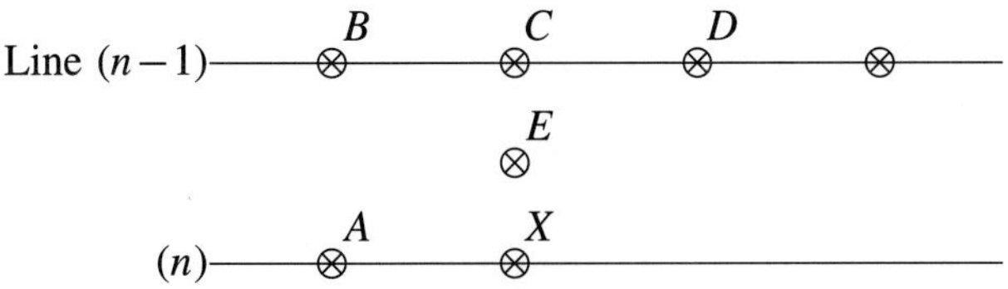

Figure 4.4 Predictor dimension and order. Pel X is under prediction. Pel E comes from previous field.

The order of the predictor is the number of previously reconstructed pels used in the prediction. $x_p(n) = f[\hat{x}(n-1), \hat{x}(n-2), \ldots, \hat{x}(n-k)]$ is a kth order predictor. The dimension of the predictor (Fig. 4.4) is dependent on the locations of the previously reconstructed pels (causal): pels in the same line imply a 1D predictor; pels in the horizontal and vertical directions imply a 2D predictor; pels in the horizontal, vertical, and temporal domains imply a 3D predictor [64]. Interfield or interframe prediction requires storage of field(s) or frame(s) respectively. From Fig. 4.1 the following relations are valid:

$$e_p(n) = \text{Predicton error} = x(n) - x_p(n)$$

$$q(n) = \text{Quantization error} = e_p(n) - e_{pq}(n)$$

$$x(n) = \text{Input signal}$$

$$\hat{x}(n) = \text{Reconstructed signal} = e_{pq}(n) + x_p(n)$$

$$x_p(n) = \text{Predicted value of } x(n)$$

$$e_{pq}(n) = \text{Quantized prediction error}$$

$$\begin{aligned} \text{Reconstruction error} &= x(n) - \hat{x}(n) \\ &= x_p(n) + e_p(n) - [e_{pq}(n) + x_p(n)] \\ &= e_p(n) - e_{pq}(n) = q(n) \\ &= \text{quantization error} \end{aligned}$$

Quantized prediction error is defined in the Z-transform domain as follows:

$$e_{pq}(z) = (1 - p(z))(x(z) + q(z)) \tag{4.1}$$

Predictor	Predictor dimension	Predictor order
$f(A, C, E)$	3D	3
A (previous pel)	1D	1
$(A + C)/2$	2D	2
C (previous line)	1D	1
E (previous field)	1D	1
$f(A, B, C, D)$	2D	4

4.2 Linear Predictor

In general, linear predictors are used where the prediction is based on the linear combination of the previously reconstructed pels:

$$x_p(n) = \sum_{l=1}^{k} a_l \, \hat{x}(n-l) \tag{4.2}$$

where a_l = predictor weights (coefficients). To ensure that the predicted pel is within the allowable range, set

$$\sum_{l=1}^{k} a_l = 1 \tag{4.3}$$

Predictor weights a_l generally reflect the correlation of $x(n)$ with the neighborhood pels. DPCM design involves both predictor and quantizer. Designing both the predictor and quantizer together is complicated as each is dependent on the other. To simplify the design, the predictor is designed initially independent of (ignoring) the quantizer. This is followed by the quantizer design based on the prediction error statistics. Any meaningful distortion (metric) [57] can be used for optimizing the predictor. Similar operation (not necessarily the same metric) can be implemented for the quantizer. We will illustrate this with an example.

Predictor design Minimize the mean square prediction error, i.e.,

$$E[e_p^2(n)] = E\left[\{x(n) - x_p(n)\}^2\right] \tag{4.4}$$

Assume $x(n) \cong \hat{x}(n)$ (negligible quantization error). Choose a kth-order linear predictor

$$x_p(n) = \sum_{l=1}^{k} a_l \, x(n-l) \tag{4.5}$$

where $x(n-l)$, $l = 1, 2, \ldots, k$ are previous pels.

Hence,

$$E[e_p^2(n)] = E\left[\left(x(n) - \sum_{l=1}^{k} a_l \, x(n-l)\right)^2\right] \tag{4.6}$$

Because the predictor weights a_l are variable, set

$$\frac{\partial}{\partial a_i}\left(E\left[e_p^2(n)\right]\right) = 0, \quad i = 1, 2, \ldots, k$$

and assume $E[x(n)] = 0$, variance $= \sigma_x^2$.

The minimization procedure yields

$$\underset{(k\times 1)}{\underline{a}} = \underset{(k\times k)}{\underline{R}^{-1}} \underset{(k\times 1)}{\underline{r}} \tag{4.7}$$

where

$$\underset{(k\times 1)}{\underline{a}} = [a_1, a_2, \ldots, a_k]^T = \text{predictor weights vector}$$

$$\underset{(k\times 1)}{\underline{r}} = \left(E[x(n)x(n-1), E[x(n)x(n-2), \ldots, E[x(n)x(n-k)]]\right)^T$$

$$\underset{(k\times k)}{\underline{R}} = E\begin{bmatrix} x(n-1)x(n-1) & x(n-1)x(n-2) & \cdots & x(n-1)x(n-k) \\ \vdots & \vdots & \vdots & \vdots \\ x(n-k)x(n-1) & x(n-k)x(n-2) & \cdots & x(n-k)x(n-k) \end{bmatrix}$$

$$= \text{autocorrelation matrix} \tag{4.8}$$

The prediction error variance can be derived as

$$\sigma_{pe}^2 = \sigma_x^2 - \underset{(1\times k)}{\underline{a}^T} \underset{(k\times 1)}{\underline{r}}$$

$$= \sigma_x^2 - \sum_{l=1}^{k} a_l \; E[x(n)x(n-l)] \tag{4.9}$$

An alert reader can observe that the predictor can be based on minimizing other distortion criteria such as mean absolute prediction error or even mean-weighted prediction error. Psychovisual [65] or psychoacoustic (audio coding) [61] effects can influence the choice of weights. DPCM is sensitive to image statistics and to channel errors. Performance becomes poor at high compression ratios (i.e., low bit rates). To reduce or eliminate the effects of channels errors, a *leak* factor [54] is introduced after the predictor output (Fig. 4.5). From an implementation viewpoint, a leak factor (close to 1) of 15/16, or 31/32, or 7/8, etc., is preferred. The function of *clip* is to clip the reconstructed pel to the allowable range [53]. Also, the prediction process can be periodically refreshed, i.e., the DPCM operation is bypassed. For a previous horizontal pel predictor (Fig. 4.6), the pels along the first horizontal line and the leftmost pels along each horizontal scan are transmitted (quantized and coded) without going through the prediction loop [63]. In this case channel noise, if any, does not propagate

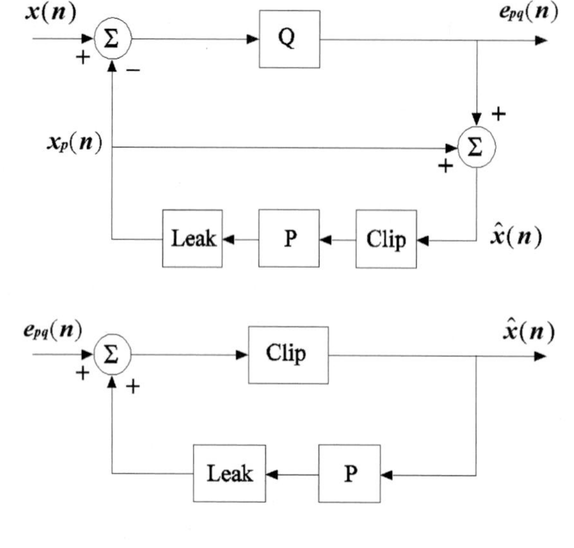

Figure 4.5 A DPCM encoder with leak and clip functions and corresponding decoder.

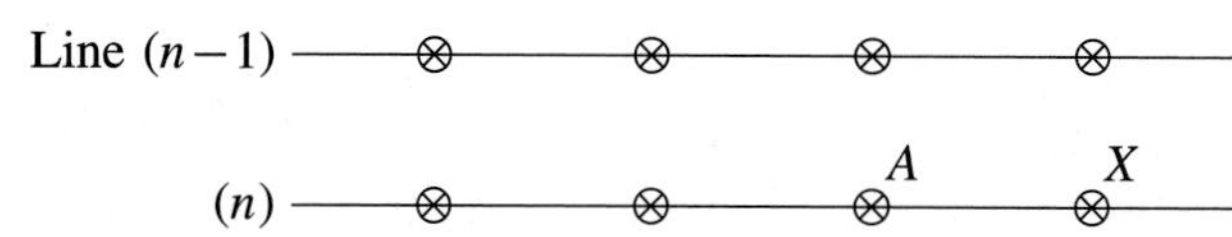

Figure 4.6 Previous horizontal pel predictor, $x_p = A$.

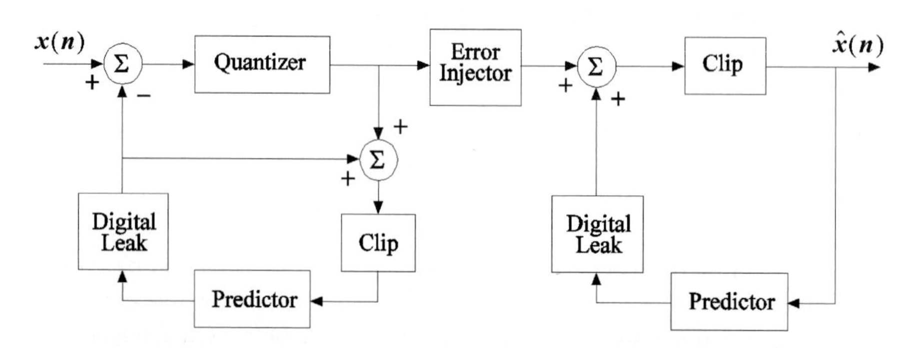

Figure 4.7 A system for studying the effects of channel noise on a feedback predictor.

beyond a horizontal line. The effects of channel noise for a given predictor can be studied by injecting random or burst noise and observing the signal at the decoder (Fig. 4.7).

Adaptive prediction As stated earlier, adaptive prediction can improve the overall prediction process. This adaptivity can be based on local image activity, motion, scene change, or any other meaningful criterion. For example, the

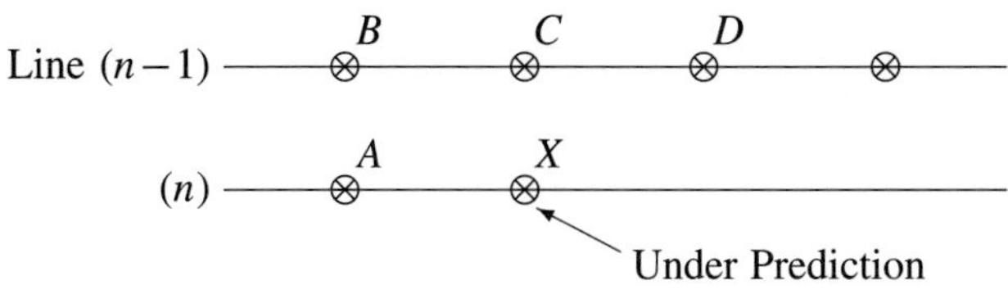

Figure 4.8 Pel positioning for adaptive prediction.

prediction can be based on a number of different predictors. A particular predictor that yields the minimum absolute prediction error can then be chosen. This, of course, increases the complexity and necessitates an overhead to indicate the predictor selection to the decoder. Some specific examples follow:

1. Compute $(|x-A|,\ |x-B|,\ |x-C|,\ |x-D|)$. See Fig. 4.8. Choose $\min(|x-A|,\ |x-B|,\ |x-C|,\ |x-D|)$ for the prediction of x. To avoid overhead choose a prediction of the present pel based on the previous pel prediction.
2. For a block of pels, say (4×4), (4×8), (8×8), (8×16), or (16×16), apply various predictors and choose the predictor that yields the minimum of the absolute sum of prediction errors of the block. This needs an overhead to indicate the predictor chosen. This overhead in terms of bits/pel is considerably reduced, however, because the predictor selection is based on a group of pels. In a practical coder [62] the following scheme was adopted.

Three adaptive predictors (i) intrafield, (ii) interfield, and (iii) motion-compensated interframe are used. Predictor selection is based on 12 luminance samples and 6 samples of each color-difference signal.

← 12 → ← 6 → ← 6 →

Y + C_b + C_r

This coder was proposed by KDD (a Japanese company) for digital transmission of the NTSC color TV signal at DS-3 rate (44.736 Mbps) to T1.Y1.1. T1 is the standards committee within the Exchange Carrier Standards Association (ECSA).T1.Y1 is related to video coding.

In another study, of the seven predictors (Table 4.1) applied to test images [54], predictors 4 and 7 were found to yield the minimum MSE.

The DS-3 coder proposed to T1.Y1.1 by NEC (a Japanese company) chooses one of the four predictors in Table 4.2. DPCM is applied to composite color TV signal (NTSC) sampled at 3 f_{sc} where f_{sc} is the color subcarrier frequency, 3.58 MHz.

Table 4.1 Seven predictors applied to test images (pixels A, B, C, and D are shown in Fig. 4.8) [52]

Predictor number	Predictor weights			
	A	B	C	D
1	1	–	–	–
2	1	−1/2	1/2	–
3	3/4	−3/8	5/8	–
4	13/16	−5/16	1/2	–
5	7/8	−1/2	1/2	1/8
6	1/2	1/8	1/4	1/8
7	3/4	−1/4	3/8	1/8

Source: © 1990 IEEE.

Table 4.2 Four predictors applied to the NEC DS-3 coder (the predictor weights are shown in Fig. 4.9)

Predictor number	Prediction
1	$x = a$ (previous pel prediction)
2	$x = 0.5a + c - 0.5d$ (higher-order prediction)
3	$x = e$ (two-line delay)
4	$x = f$ (one-field delay)

Predictors 1–4 are appropriate for black-and-white signals, composite NTSC signal, high correlation in the vertical direction, and relatively slow motion, respectively. To avoid overhead in indicating the predictor selection, the predictor choice of a current pel is based on the optimum prediction of the previous pel (past information). The decoder implements the same logic and selects the corresponding predictor.

The prediction process is not limited to original samples such as pels or audio samples. Predictive coding can be extended to transform coefficients (Chapter 5), vectors (PVQ: predictive vector quantization, Fig. 7.13), subband signals, and so forth. Also, DPCM efficiency can be enhanced by motion compensation wherein interframe/interfield prediction is improved by estimating motion within a frame/field interval (Section 6.3).

These techniques have been extensively investigated and implemented in a number of codecs (coders/decoders). Also, motion-compensated interframe prediction has been adopted in a number of standards such as H.261 (Chapter 9),

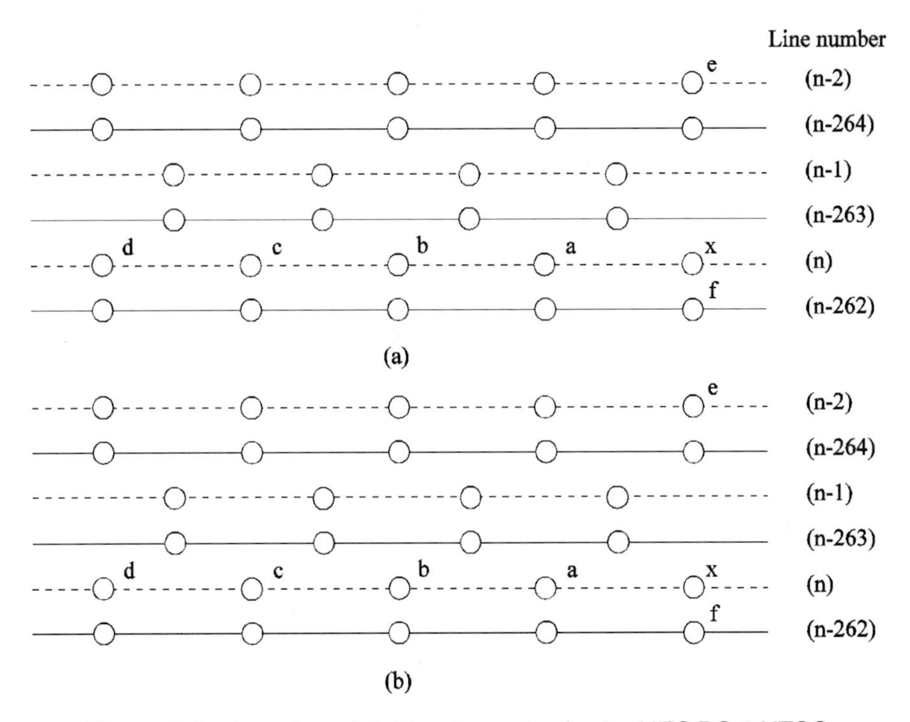

Figure 4.9 Location of digitized samples for the NEC DS-3 NTSC color TV coder (—— odd field, - - - - even field). (a) Mth frame; (b) $(M+1)$th frame.

MPEG-1 (Chapter 10), MPEG-2 (Chapter 11), H.263 (Chapter 12), FCC/HDTV (Chapter 13), and CMTT (Chapter 14).

4.3 Variants of DPCM

Double-predictor DPCM Recently, double-predictor DPCM [58] for image coding has been developed (Fig. 4.10). This scheme involves an inner loop operating on the prediction error of the outer loop. Based on the optimal design of the two predictors and the quantizer, it is shown that the new DPCM system is robust to changes in the input image statistics and is less sensitive to channel noise compared with single-loop DPCM (Fig. 4.1). For a noise-free channel,

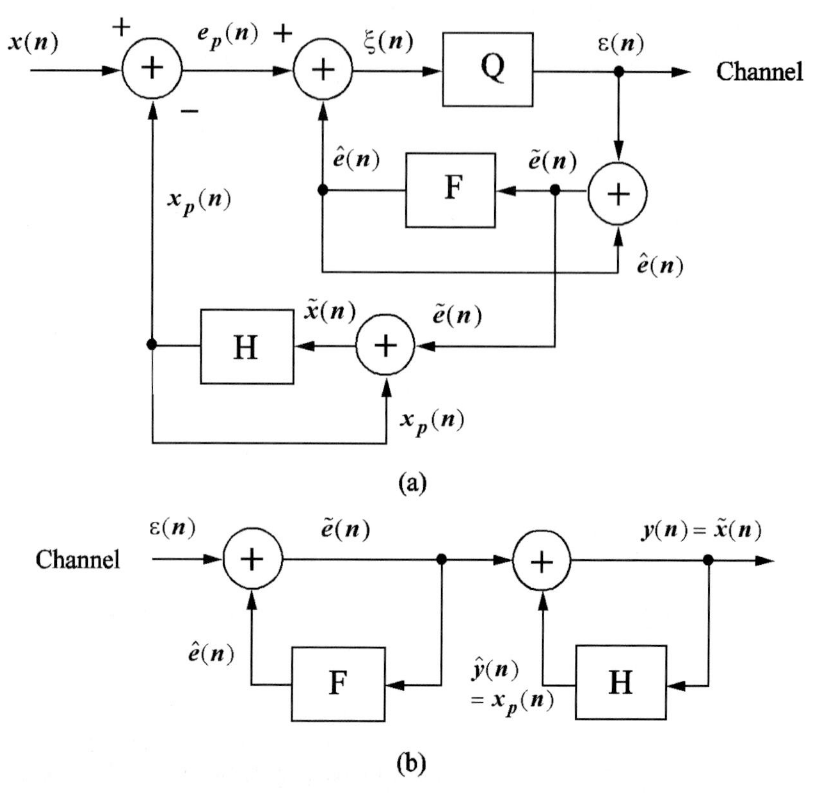

Figure 4.10 A block diagram of a double-predictor DPCM system: (a) system encoder; (b) system decoder. F and H are inner and outer predictors [58].

$$\begin{aligned} \xi(n) - \varepsilon(n) &= Q(n) \\ &= x(n) - y(n) \\ &= \text{reconstruction error} \end{aligned}$$

In Fig. 4.10, H is a first prediction filter and F is a secondary prediction filter. Q is a PDF-optimal quantizer of either a uniform or nonuniform structure. The output of the first subtractor, $e(n)$, is input to a second subtractor whose other input is a local estimate of the differential signal itself. The corresponding decoder consists of two cascaded prediction filters. Filter coefficients are chosen under constraints that guarantee system stability.

Signal-blurred DPCM In the BDPCM coder (Fig. 4.11) [71, 72], a recursive prefilter is added before quantizing and coding. The prefilter acts as a blurring

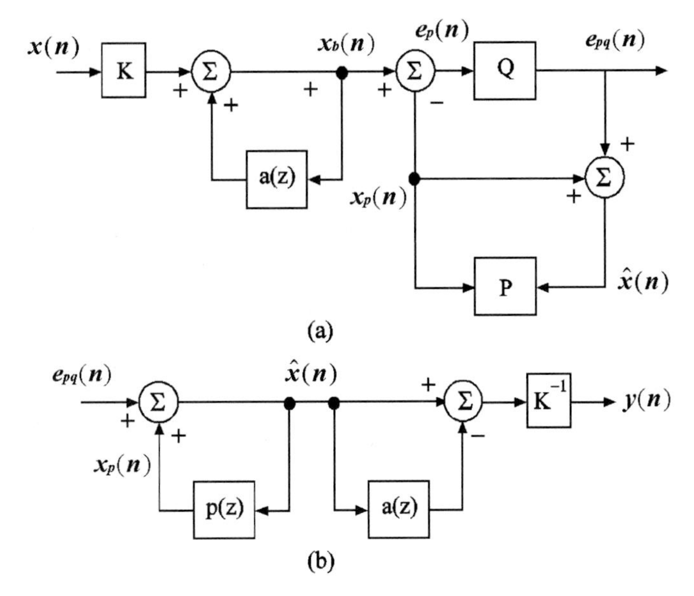

Figure 4.11 BDPCM (a) encoder and (b) decoder [72] © 1993 IEEE.

aperture in order to reduce the dynamic range, variance, and entropy of the input signal. The compressed signal is coded by the conventional DPCM, which reduces irrelevant information and produces near white noise. Correspondingly, the nonrecursive postfilter of the decoder is the deblurring filter.

The signal blurred by the filter $a(z)$ (recursive and causal) and amplified by constant gain K is input to the DPCM loop. Its transfer function is given by $K(1 - a(z))^{-1}$. The spectrum $a(z)$ of the recursive loop can be realized by a transversal filter that has the same weight at all taps of the surrounding pixels. This equi-weighting operation yields a dc-invariance characteristic; that is, any constant input sequence is not changed by the filtering.

The encoder and decoder operations are defined in Z-transform domain as follows:

$$e_{pq}(z) = (1 - P(z))\left(\frac{K}{1 - a(z)}x(z) + q(z)\right) \tag{4.10}$$

$$y(z) = x(z) + K^{-1}(1 - a(z))q(z) \tag{4.11}$$

Assume that quantization noise is additive to the signal. In comparison to the conventional DPCM in Eq. (4.1), the dynamic range of the signal $x(n)$ is reduced by the factor K, but quantizer noise is unchanged. Prediction errors of BDPCM decrease as shown in Fig. 4.12. At the decoder, the signal $x(n)$ is completely

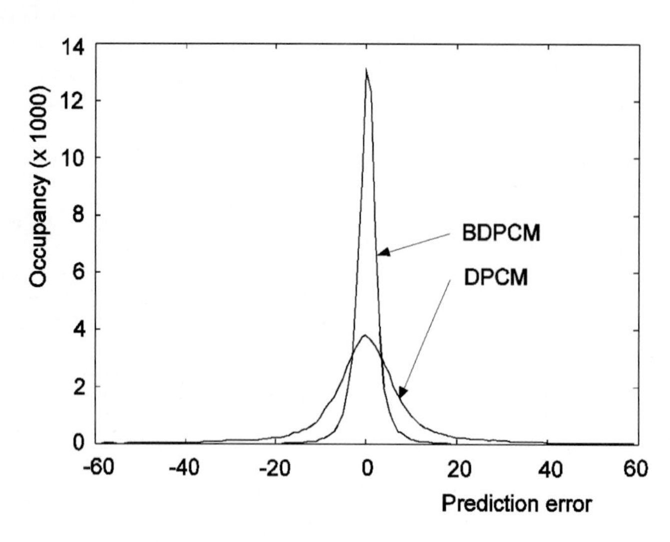

Figure 4.12 Prediction error of DPCM and BDPCM [72]. © 1993 IEEE.

reconstructed with increased quantization noise. Thus, the filter is designed to produce invisible high-frequency noise (blue noise).

Noise-shaping DPCM [74] is another variant of DPCM coding. Quantizer output is fed back for coloring the quantization noise.

Chapter 5
Transform Coding

Summary

Although transform coding is a much more general subject, we have focussed mainly on the discrete cosine transform (DCT). The definitions, properties, fast algorithms, and performance of the DCT as related to image coding are described. Because of space constraints, many other efficient algorithms (tailored mainly from an architectural viewpoint) for the DCT, including other interesting applications, have been reluctantly left out. A prominent application of DCT, is in speech encryption ((132) through (136)). A basic transform image coding scheme is outlined. This is extended to hybrid coding in the following chapter.

5.1 Introduction

In transform coding, a signal is mapped from one domain (usually spatial or temporal) into the transform domain [78, 185, 186]. The signal can be one-dimensional or multidimensional. Our discussion is limited to orthogonal transforms; the mapping is therefore unique and reversible. In this case, the energy is preserved in the transform domain, and the signal can be recovered completely by the inverse transform (inverse mapping). We will also consider only separable transforms that have some beneficial properties such as implementation of a multidimensional (MD) transform, which involves a series of one-dimensional (1D) transforms. Also, fast algorithms for efficient implementation of 1D transforms, recursive relations, and so forth can be directly extended to MD transforms.

Because the transforms are orthogonal (unitary, if the transform has complex basis functions), implementing the inverse transform is essentially the same as implementing the forward transform. Hence, all the properties (fast algorithms, recursive structure, etc.) are preserved in the inverse transformation. In fact, the same software or hardware (VLSI chip) designed for the forward transform can be used with minor modification for implementing the inverse transform. Taking advantage of the recursive relations, the same chip can implement different-size transforms (forward or inverse) such as (4×4), (4×8), (8×4), (8×8), (8×16), (16×8), and (16×16) at video rates. The integration is highly advanced, enabling a single chip to implement the entire decoding operation, as the MPEG-2 audio/video decoder does. This also implies our discussion is limited to discrete transforms.

Whereas the application of the discrete transforms (hereafter transforms imply discrete transforms) is quite diverse, our focus will be mainly on coding (audio or video) from a compression viewpoint. Additional applications relevent to coding, such as filtering (decimation in the transform domain), will also be covered. Mapping the signal (speech, audio, image, video, etc.) into the transform domain by itself does not lead to bit-rate reduction. In general, mapping results in energy compaction or concentration in the low-frequency zone of the transform domain. By cleverly quantizing those coefficients that carry significant information (energy), and at the same time coarsely quantizing or dropping (setting them to zero) the remaining coefficients, the average bit rate can be reduced. The compression ratio can be increased further by using classification, run-length coding (RLC), variable length coding (VLC), and so forth. Other features, such as quantization reflecting the human visual system (HVS), motion compensated prediction, and so forth, can further contribute to the overall compression ratio. All this, of course, is accomplished with a price. The penalty is that the encoder complexity increases (as does the decoder complexity), and the encoding scheme becomes much more vulnerable to channel noise, requiring sophisticated error detection and correction techniques. In general, the decoder is much less complex (hence lower cost) than the encoder because most of the decision-making processes (adaptive modes) are carried out at the encoder. The decoder has in turn only to track the selected modes for reconstructing the signal. Also, in motion-compensated hybrid (transform-DPCM) coding (Chapter 6), motion estimation and quantization control need to be implemented at the encoder only. This complex encoder–simple decoder scenario is also appropriate because the decoder can be designed as a mass consumer electronics item, such as video CD (MPEG-1) or HDTV-VCR.

The number of discrete transforms is innumerable. We will, however, concentrate on the discrete cosine transform (DCT) for one simple reason [77, 78].

All the standard groups (ISO, ITU, IEC, FCC/HDTV, etc.) have so far selected in image and video coding DCT as the primary factor in achieving compression. As stated earlier, the DCT, or for that matter any other transform by itself, does not lead to the redundancy reduction. Many other complementary and contributing factors such as quantization, classification, HVS weighting, motion-compensated prediction, VLC, RLC, loop filtering, packetization, color-format conversion, and preprocessing play equally important roles in the compression cuisine.

The object is to make this cuisine palatable (imperceptible impairments) and widely acceptable. DCT is by no means an optimal transform. Karhunen-Loeve transform (KLT), although statistically optimal, has limited applications. Its main drawbacks are that it neither is a fixed transform nor has a fast algorithm. DCT, on the other hand, comes very close to the KLT (especially for first-order Markov signals) in performance. Added to this are the fast algorithms and recursive structures that weigh heavily in its favor. One major disadvantage of the DCT is the block structure that dominates at very low bit rates. This feature, of course, is a characteristic of all block transforms. Although there are techniques to minimize this impairment, other transforms, such as lapped orthogonal transform (LOT) along with its variations (ELT: extended lapped transforms; MLT: modulated lapped transforms), and wavelets are devoid of these disparities. It is entirely conceivable that in the future these transforms may replace the DCT in the international standards. At present, however, the industry worldwide has completely immersed itself in adopting these standards because of the interoperability and compatibility. Some companies have developed codecs based on proprietary algorithms (these can stand alone or can complement the established standards).

Apart from the JBIG (Joint Binary Image Group) aimed at binary image coding, the only areas conspicuous by the absence of DCT are lossless image coding in JPEG (Chapter 8), MPEG audio (Chapter 10) and FCC/HDTV (Chapter 13, Dolby AC-3 audio-coding algorithm). While DCT was selected for JPEG (baseline and extended systems) and H.261 (Chapter 9), video coding algorithms for MPEG-1 and 2 were apparently influenced by H.261 (motion-compensated hybrid DCT-DPCM coding). Similarly, H.263 (Chapter 12) (very low bit-rate video coding), being developed by ITU-T, is a modified version of H.261. Some of the issues were backward-forward compatibilities, multifunctional-multipurpose codecs, and incentives for the industry to develop the chips for various applications (increasing the marketability) that have similar functionalities, i.e., DCT, motion estimation, motion-compensated prediction, HVS weighting, zigzag scan, VLC, macroblock addressing (MBA), coded block pattern (CBP), and so forth.

The properties of the DCT can be summarized as follows:

- The energy of the signal/picture is packed in a few coefficients.

- It has fast implementation, forward and inverse.
- It has a recursive structure (easier to implement multiple size transforms).
- It uses real arithmetic, orthogonal and separable: extension to multiple dimensions is simple.
- It is a close to statistically optimal transform (KLT).
- There is minimal residual correlation.
- There is minimal MSE in scalar Wiener filtering.
- ASIC DCT chips are available. A number of functions such as on-chip addition/subtraction, zigzag scan, and so forth, besides DCT are integrated into a single chip.
- It has been adopted for different applications by ITU-T, ISO, IEC, JPEG, MPEG, and HDTV.
- There is minimal rate distortion and maximum $\left(\frac{\text{Transform}}{\text{DPCM}}\right)$ coding gain.
- It has decimation and filtering properties in the DCT domain. Adopted in frequency scalable layered video coding.
- MDCT (modified DCT) is utilized to decompose a signal into equal subbands.
- DCT is used for VQ codebook design.

The applications of the DCT are listed below:

Speech Coding

LMS Filters

Digital Filter Banks (Transmultiplexers) (TDM/FDM)

Ceptstral Analysis

Image Coding (Bandwidth Compression)

Low Bit Rate Codecs, Videoconferencing

Progressive Image Transmission

Hierarchical Image Retrieval

Videotext

Image Enhancement

Pattern Recognition

Image Filtering

DCT/Mandala Transform [79]

Speech Recognition

Packet Video

Texture Analysis

Image-Compression Boards

Videophone

HDTV, Compatible Coding, Layered Coding

Broadcast TV

VQ Codebook Design

Infrared Image Coding

SAR Image Coding

Pattern Classification

Printed Color Image Coding

Digital Storage Media (CD-ROM, DAT, Optical Disk, Magnetic Hard Disk, Laser Disk, Magneto-optical Disk)

Decimation (Subsampling)

Subband Decomposition (Filter Banks)

Speech Encryption

5.2 Discrete Cosine Transform

A family of DCTs (types I–IV) have been developed. They are very similar, and type II, called DCT-II, is the most widely used. The basis functions and the forward/inverse transform are defined as

$$\left[C_{N+1}^{\mathrm{I}}\right] = \sqrt{\frac{2}{N}}\left[K_m\, K_n \cos\left(\frac{mn\pi}{N}\right)\right], \quad m, n = 0, 1, \ldots, N \tag{5.1}$$

$$\left[C_{N}^{\mathrm{II}}\right] = \sqrt{\frac{2}{N}}\left[K_m \cos\left(m\left(n+\frac{1}{2}\right)\frac{\pi}{N}\right)\right], \quad m, n = 0, 1, \ldots, N-1$$

$$\left[C_{N}^{\mathrm{III}}\right] = \sqrt{\frac{2}{N}}\left[K_n \cos\left(\left(m+\frac{1}{2}\right)\frac{n\pi}{N}\right)\right], \quad m, n = 0, 1, \ldots, N-1$$

$$\left[C_{N}^{\mathrm{IV}}\right] = \sqrt{\frac{2}{N}}\left[\cos\left(\left(m+\frac{1}{2}\right)\left(n+\frac{1}{2}\right)\frac{\pi}{N}\right)\right], \quad m, n = 0, 1, \ldots, N-1$$

where

$$K_j = \begin{cases} 1, & j \neq 0 \text{ or } N \\ 1/\sqrt{2}, & j = 0 \text{ or } N \end{cases}$$

The unitarity of DCTs results in

$$\left[C_{N+1}^{\mathrm{I}}\right]^{-1} = \left[C_{N+1}^{\mathrm{I}}\right], \quad \text{Forward} = \text{Inverse}$$
$$\left[C_{N}^{\mathrm{II}}\right]^{-1} = \left[C_{N}^{\mathrm{III}}\right] = \left[C_{N}^{\mathrm{II}}\right]^{T}$$
$$\left[C_{N}^{\mathrm{III}}\right]^{-1} = \left[C_{N}^{\mathrm{II}}\right] = \left[C_{N}^{\mathrm{III}}\right]^{T}$$
$$\left[C_{N}^{\mathrm{IV}}\right]^{-1} = \left[C_{N}^{\mathrm{IV}}\right], \quad \text{Forward} = \text{Inverse} \tag{5.2}$$

The family of forward and inverse DCTs can be defined as follows:

DCT-I

$$X^{c1}(m) = \sqrt{\frac{2}{N}}\, K_m \sum_{n=0}^{N} K_n\, x(n) \left[\cos\left(\frac{mn\pi}{N}\right)\right], \tag{5.3}$$
$$m = 0, 1, \ldots, N$$

IDCT-I

$$x(n) = \sqrt{\frac{2}{N}}\, K_n \sum_{m=0}^{N} K_m\, X^{c1}(m) \cos\left(\frac{mn\pi}{N}\right),$$
$$n = 0, 1, \ldots, N$$

DCT-II

$$X^{c2}(m) = \sqrt{\frac{2}{N}}\, K_m \sum_{n=0}^{N-1} x(n) \cos\left[\frac{m(2n+1)\pi}{2N}\right], \tag{5.4}$$
$$m = 0, 1, \ldots, N-1$$

IDCT-II

$$x(n) = \sqrt{\frac{2}{N}} \sum_{m=0}^{N-1} K_m\, X^{c2}(m) \cos\left[\frac{m(2n+1)\pi}{2N}\right],$$
$$n = 0, 1, \ldots, N-1$$

DCT-III

$$X^{c3}(m) = \sqrt{\frac{2}{N}} \sum_{n=0}^{N-1} K_n \, x(n) \, \cos\left[\frac{n(2m+1)\pi}{2N}\right], \qquad (5.5)$$

$$m = 0, 1, \ldots, N-1$$

IDCT-III

$$x(n) = \sqrt{\frac{2}{N}} \sum_{m=0}^{N-1} K_n \, X^{c3}(m) \cos\left[\frac{n(2m+1)\pi}{2N}\right],$$

$$n = 0, 1, \ldots, N-1$$

DCT-IV

$$X^{c4}(m) = \sqrt{\frac{2}{N}} \sum_{n=0}^{N-1} x(n) \, \cos\left[\frac{(2n+1)(2m+1)\pi}{4N}\right], \qquad (5.6)$$

$$m = 0, 1, \ldots, N-1$$

IDCT-IV

$$x(n) = \sqrt{\frac{2}{N}} \sum_{m=0}^{N-1} X^{c4}(m) \cos\left[\frac{(2n+1)(2m+1)\pi}{4N}\right],$$

$$n = 0, 1, \ldots, N-1$$

DCTs (types I–IV) can be sparse-matrix factorized based on decomposition of $\left[C_N^{\mathrm{IV}}\right]$. These fast algorithms are recursive and are based on real arithmetic. (All these DCTs are orthogonal, separable, and real, and they all have fast algorithms.)

Example 5.1

$$\underset{(N\times N)}{\left[C_N^{\mathrm{II}}\right]} = [P_N] \left[\begin{array}{c|c} C_{N/2}^{\mathrm{II}} & 0 \\ \hline 0 & \overline{\overline{C}}{}_{N/2}^{\mathrm{IV}} \end{array}\right] \left[A_N^{\mathrm{II}}\right] \qquad (5.7)$$

where

$$\underset{(N\times N)}{\left[A_N^{\mathrm{II}}\right]} = \frac{1}{\sqrt{2}} \left[\begin{array}{c|c} I_{N/2} & \bar{I}_{N/2} \\ \hline \bar{I}_{N/2} & -I_{N/2} \end{array}\right] \qquad (5.8)$$

$$[P_N] = (N \times N) \quad \textit{unit matrix with columns (rows) permuted}$$

$$I_N = (N \times N), \quad \textit{unit matrix}$$

$$\bar{I}_N = \begin{bmatrix} 0 & & 1 \\ & \cdot^{\cdot^{\cdot}} & \\ 1 & & 0 \end{bmatrix}, \quad \textit{opposite diagonal unit matrix} \tag{5.9}$$

$$\underset{(N\times N)}{\bar{\bar{C}}^{\text{IV}}_{N/2}} = \bar{I}_{N/2}\left[C^{\text{IV}}_{N/2}\right]\bar{I}_{N/2}$$

DCT-II was originally developed in 1974 [77]. Subsequently, several fast algorithms via other transforms such as discrete Fourier transform (DFT), discrete Hartley transform (DHT), sparse-matrix factorization, and so forth, have been developed. Most of these involve real arithmetic only and have recursive structure. The basis functions of DCT-II for $N = 16$ are shown in Fig. 5.1. From now on, DCT implies DCT-II. The N-point forward and inverse DCTs can be defined as follows.

Forward DCT (*N*-point DCT)

$$X^{c2}(k) = \frac{2}{N} c_k \sum_{n=0}^{N-1} x(n) \cos\left[\frac{(2n+1)k\pi}{2N}\right], \tag{5.10}$$

$$k = 0, 1, \ldots, N-1$$

Inverse DCT (*N*-point IDCT)

$$x(n) = \sum_{k=0}^{N-1} c_k X^{c2}(k) \cos\left[\frac{(2n+1)k\pi}{2N}\right],$$

$$n = 0, 1, \ldots, N-1$$

where

$$c_k = \begin{cases} 1/\sqrt{2}, & k = 0 \\ 1, & k \neq 0 \end{cases}$$

Normalized DCT

$$X^{c2}(k) = \sqrt{\frac{2}{N}} c_k \sum_{n=0}^{N-1} x(n) \cos\left[\frac{(2n+1)k\pi}{2N}\right], \tag{5.11}$$

$$k = 0, 1, \ldots, N-1$$

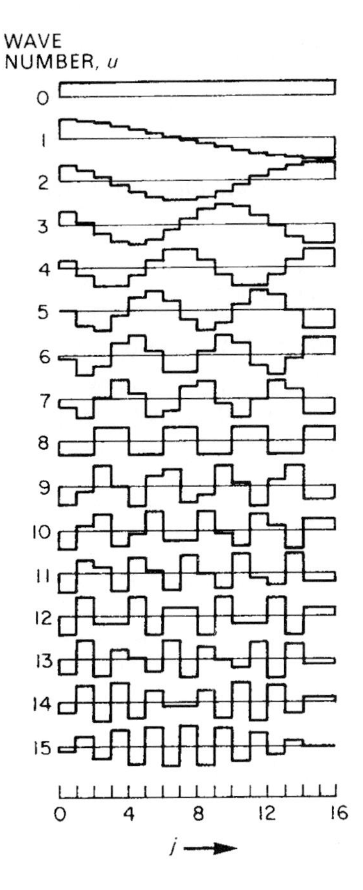

Figure 5.1 Basis functions of DCT-II, $N = 16$.

Normalized IDCT

$$x(n) = \sqrt{\frac{2}{N}} \sum_{k=0}^{N-1} c_k \, X^{c2}(k) \cos\left[\frac{(2n+1)k\pi}{2N}\right],$$

$$n = 0, 1, \ldots, N-1$$

From Eq. (5.10), the $k = 0$ coefficient is

$$X^{c2}(0) = \frac{2}{N}\left(\frac{1}{\sqrt{2}}\right) \sum_{n=0}^{N-1} x(n) \tag{5.12}$$

$$= \frac{\sqrt{2}}{N} \sum_{n=0}^{N-1} x(n) \quad \text{dc coefficient, mean or average of the sequence}$$

$$X^{c2}(k) = \text{ ac coefficients}, \; k = 0, 1, \ldots, N-1$$

As k increases, $X^{c2}(k)$ represents increasing frequencies.

$$x(n), \; k = 0, 1, \ldots, N-1, \quad \text{Data sequence (sampled uniformly in time or spatial domain)}$$

$$X^{c2}(k), \; k = 0, 1, \ldots, N-1, \quad \text{DCT sequence}$$

The DCT pair defined in Eq. (5.10) can be symbolically denoted as $x(n) \Leftrightarrow X^{c2}(k), \; n = 0, 1, \ldots, N-1, \;\; k = 0, 1, \ldots, N-1$.

The DCT/IDCT can be expressed in vector-matrix form as

$$\underset{\substack{(N \times 1) \\ \text{(Transform Vector)}}}{\overset{X^{c2}(k)}{\begin{bmatrix} 0 \\ 1 \\ 2 \\ 3 \\ \vdots \\ N-1 \end{bmatrix}}} = \left(\frac{2}{N}\right) \begin{matrix} n \rightarrow \\ k \;\; 0 \\ \downarrow \;\; 1 \\ 2 \\ 3 \\ \vdots \\ N-1 \end{matrix} \underset{\substack{(N \times N) \\ \text{(DCT Matrix)}}}{\overset{0\;1\;2\;3\;\ldots\;N-1}{\begin{bmatrix} & & \\ & c_k \cos\left[\dfrac{(2n+1)k\pi}{2N}\right] & \\ & & \end{bmatrix}}} \underset{\substack{(N \times 1) \\ \text{(Data Vector)}}}{\overset{x(n)}{\begin{bmatrix} 0 \\ 1 \\ 2 \\ 3 \\ \vdots \\ N-1 \end{bmatrix}}} \tag{5.13}$$

$$c_k = \begin{cases} 1/\sqrt{2}, & k = 0 \\ 1, & k \neq 0 \end{cases}$$

$$\underset{\substack{(N \times 1) \\ \text{(Data Vector)}}}{\overset{x(n)}{\begin{bmatrix} 0 \\ 1 \\ 2 \\ 3 \\ \vdots \\ N-1 \end{bmatrix}}} = \begin{matrix} k \rightarrow \\ n \;\; 0 \\ \downarrow \;\; 1 \\ 2 \\ 3 \\ \vdots \\ N-1 \end{matrix} \underset{\substack{(N \times N) \\ \text{(IDCT Matrix)}}}{\overset{0\;1\;2\;3\;\ldots\;N-1}{\begin{bmatrix} & & \\ & c_k \cos\left[\dfrac{(2n+1)k\pi}{2N}\right] & \\ & & \end{bmatrix}}} \underset{\substack{(N \times 1) \\ \text{(Transform Vector)}}}{\overset{X^{c2}(k)}{\begin{bmatrix} 0 \\ 1 \\ 2 \\ 3 \\ \vdots \\ N-1 \end{bmatrix}}} \tag{5.14}$$

DCT is an orthogonal matrix, i.e.,

$$\frac{2}{N}\underset{(N\times N)}{[\text{DCTmatrix}][\text{IDCTmatrix}]} = I_N$$

$$[\text{DCT}][\text{DCT}]^{-1} = I_N$$

$$\therefore\ [\text{DCT}]^{-1} = \frac{2}{N}[\text{IDCT}]$$

$$= \frac{2}{N}[\text{DCT}]^T$$

$$[\text{IDCT}] = [\text{DCT}]^T$$

$$\frac{2}{N}[\text{DCT}][\text{DCT}]^T = I_N \tag{5.15}$$

The basis vectors of the normalized DCT are

$$c(k,n) = 1/\sqrt{N}, \quad \text{for } k = 0,\ 0 \le n \le N-1$$

$$c(k,n) = \sqrt{\frac{2}{N}}\cos\left[\frac{(2n+1)k\pi}{2N}\right], \quad \text{for } \begin{array}{l} 0 \le n \le N-1 \\ 1 \le k \le N-1 \end{array} \tag{5.16}$$

$$= \sqrt{\frac{2}{N}}\, C_{2N}^{k(2n+1)}, \quad \text{where } C_a^b = \cos\left(\frac{b\pi}{a}\right)$$

Normalized DCT matrix:

$$\begin{array}{ll} & \text{Column} \rightarrow n \\ & \quad 0\ \ 1\ \ 2\ \ 3\ \ \ldots\ldots\ \ N-1 \\ \begin{array}{c}\text{Row} \\ k\downarrow\end{array} \begin{array}{c} 0 \\ 1 \\ 2 \\ 3 \\ \vdots \\ N-1 \end{array} & \left[\begin{array}{c} (1/\sqrt{N})\ (1/\sqrt{N})\ \ldots\ldots\ (1/\sqrt{N}) \\ \sqrt{\frac{2}{N}}\left[\begin{array}{c} \cos\left[\frac{(2n+1)k\pi}{2N}\right] \\ 0 \le n \le N-1 \\ 1 \le k \le N-1 \end{array}\right] \end{array}\right] \end{array} \tag{5.17}$$

Normalized IDCT matrix:

$$
\begin{array}{ll}
 & \text{Column} \rightarrow k \\
 & \quad 0 \quad 1 \quad 2 \quad 3 \quad \ldots\ldots \quad N-1 \\
\begin{array}{c} \text{Row} \\ n \downarrow \end{array}
\begin{array}{c} 0 \\ 1 \\ 2 \\ 3 \\ \vdots \\ N-1 \end{array}
&
\left[\begin{array}{c|c}
\begin{array}{c} \frac{1}{\sqrt{N}} \\ \frac{1}{\sqrt{N}} \\ \vdots \\ \frac{1}{\sqrt{N}} \end{array}
&
\left[\sqrt{\frac{2}{N}} \cos\left[\frac{(2n+1)k\pi}{2N}\right] \quad \begin{array}{c} 0 \le n \le N-1 \\ 1 \le k \le N-1 \end{array} \right]
\end{array} \right]
\end{array}
\tag{5.18}
$$

DCT diagonalizes a tridiagonal matrix $[Q_c]$, i.e.,

$$
\underset{(N \times N)}{\left[\hat{C}_N^{\mathrm{II}}\right]^{\mathrm{T}}} \quad \underset{(N \times N)}{[Q_c]} \quad \underset{(N \times N)}{\left[\hat{C}_N^{\mathrm{II}}\right]} = \Lambda \text{ (diagonal matrix)}
\tag{5.19}
$$

where $\hat{C}_N^{\mathrm{II}}$ is a normalized version of DCT-II.

$$
\underset{(N \times N)}{Q_c} = \begin{bmatrix}
1-\alpha & -\alpha & 0 & 0 & 0 \\
-\alpha & 1 & -\alpha & 0 & 0 \\
0 & -\alpha & 1 & \ddots & 0 \\
0 & 0 & \ddots & \ddots & -\alpha \\
0 & 0 & 0 & -\alpha & 1-\alpha
\end{bmatrix}
\tag{5.20}
$$

First-order Markov process, $R = \left[\rho^{|l-k|}\right]$,

where ρ = Adjacent correlation coefficient

$= E[x_j \, x_{j+1}]$

$$
\underset{(N \times N)}{R} = \begin{bmatrix}
1 & \rho & \rho^2 & \cdots & \rho^{N-1} \\
\rho & 1 & \rho & \cdots & \rho^{N-2} \\
\vdots & \vdots & \vdots & \vdots & \vdots \\
\rho^{N-1} & \rho^{N-2} & \rho^{N-3} & \cdots & 1
\end{bmatrix}
\tag{5.21}
$$

= Correlation Matrix

$$\underset{(N \times N)}{\beta^2 R^{-1}} = \begin{bmatrix} 1-\rho\alpha & -\alpha & 0 & 0 & 0 \\ -\alpha & 1 & -\alpha & 0 & 0 \\ 0 & -\alpha & 1 & \ddots & 0 \\ 0 & 0 & \ddots & \ddots & -\alpha \\ 0 & 0 & 0 & -\alpha & 1-\rho\alpha \end{bmatrix} \tag{5.22}$$

where $\beta^2 = (1-\rho^2)/(1+\rho^2)$, $\alpha = \rho/(1+\rho^2)$.

When $\rho \cong 1$, $(1-\rho\alpha) \cong (1-\alpha)$, $\underset{(N \times N)}{\beta^2 R^{-1}} \cong \underset{(N \times N)}{Q_c}$.

Eigenvectors of R and eigenvectors of Q_c, i.e., the DCT, are very close. Hence, the near optimal performance of the DCT is obtained when ρ, the adjacent correlation coefficient, is very high (close to 1). Eigenvectors of R are the basis vectors of KLT.

The DCT and its inverse for $N = 4$ can be expressed as

$$\underset{(4 \times 1)}{\overset{X^{c2}(\ )}{\begin{bmatrix} 0 \\ 1 \\ 2 \\ 3 \end{bmatrix}}} = \frac{2}{4} \underset{(4 \times 4)}{\begin{bmatrix} \frac{1}{\sqrt{2}}(1 & 1 & 1 & 1) \\ \cos\frac{\pi}{8} & \cos\frac{3\pi}{8} & \cos\frac{5\pi}{8} & \cos\frac{7\pi}{8} \\ \cos\frac{\pi}{4} & \cos\frac{3\pi}{4} & \cos\frac{5\pi}{4} & \cos\frac{7\pi}{4} \\ \cos\frac{3\pi}{8} & \cos\frac{9\pi}{8} & \cos\frac{15\pi}{8} & \cos\frac{21\pi}{8} \end{bmatrix}} \underset{(4 \times 1)}{\overset{x(\ )}{\begin{bmatrix} 0 \\ 1 \\ 2 \\ 3 \end{bmatrix}}} \tag{5.23a}$$

$$\overset{x(\ )}{\begin{bmatrix} 0 \\ 1 \\ 2 \\ 3 \end{bmatrix}} = \begin{bmatrix} \frac{1}{\sqrt{2}} & \cos\frac{\pi}{8} & \cos\frac{\pi}{4} & \cos\frac{3\pi}{8} \\ \frac{1}{\sqrt{2}} & \cos\frac{3\pi}{8} & \cos\frac{3\pi}{4} & \cos\frac{9\pi}{8} \\ \frac{1}{\sqrt{2}} & \cos\frac{5\pi}{8} & \cos\frac{5\pi}{4} & \cos\frac{15\pi}{8} \\ \frac{1}{\sqrt{2}} & \cos\frac{7\pi}{8} & \cos\frac{7\pi}{4} & \cos\frac{21\pi}{8} \end{bmatrix} \overset{X^{c2}(\ )}{\begin{bmatrix} 0 \\ 1 \\ 2 \\ 3 \end{bmatrix}} \tag{5.23b}$$

The column vectors in Eq. (5.23a) represent the elements of $x(n)$ and $X^{c2}(k)$.

It is easy to verify that the matrix multiplication of forward and inverse DCT matrices shown in Eq. (5.23) yields I_4. The DCT matrix for $N = 8$ is shown below. The multiplier $(2/N)$ is not included.

$$
\begin{array}{ll}
 & \text{Column} \rightarrow n \\
 & \begin{array}{cccccccc} 0 & 1 & 2 & 3 & 4 & 5 & 6 & 7 \end{array} \\
\begin{array}{c} \text{Row } 0 \\ k \downarrow 1 \\ 2 \\ 3 \\ 4 \\ 5 \\ 6 \\ 7 \end{array} &
\begin{bmatrix}
\frac{1}{\sqrt{2}}(1 & 1 & 1 & 1 & 1 & 1 & 1 & 1) \\
c^1 & c^3 & c^5 & c^7 & c^9 & c^{11} & c^{13} & c^{15} \\
c^2 & c^6 & c^{10} & c^{14} & c^{18} & c^{22} & c^{26} & c^{30} \\
c^3 & c^9 & c^{15} & c^{21} & c^{27} & c^{33} & c^{39} & c^{45} \\
c^4 & c^{12} & c^{20} & c^{28} & c^{36} & c^{44} & c^{52} & c^{60} \\
c^5 & c^{15} & c^{25} & c^{35} & c^{45} & c^{55} & c^{65} & c^{75} \\
c^6 & c^{18} & c^{30} & c^{42} & c^{54} & c^{66} & c^{78} & c^{90} \\
c^7 & c^{21} & c^{35} & c^{49} & c^{63} & c^{77} & c^{91} & c^{105}
\end{bmatrix} \\
 & \text{Here } c^m = \cos(m\pi/16)
\end{array}
\tag{5.24}
$$

To show that DCT is orthogonal, substitute the forward DCT in the inverse DCT [see Eq. (5.10)], which yields

$$
\begin{aligned}
x_n &= \frac{2}{N} \sum_{k=0}^{N-1} c_k^2 \sum_{m=0}^{N-1} x_m \cos\left[\frac{(2m+1)k\pi}{2N}\right] \cos\left[\frac{(2n+1)k\pi}{2N}\right] \\
&= \frac{2}{N} \sum_{m=0}^{N-1} x_m \sum_{k=0}^{N-1} c_k^2 \cos\theta_m \cos\theta_n
\end{aligned}
\tag{5.25}
$$

where

$$
\begin{aligned}
\theta_m &= \frac{(2m+1)k\pi}{2N}, \quad \theta_n = \frac{(2n+1)k\pi}{2N} \\
\cos\theta_m \cos\theta_n &= \frac{1}{2}\left[\cos(\theta_m + \theta_n) + \cos(\theta_m - \theta_n)\right] \\
&= \frac{1}{2}\left[\frac{e^{j(\theta_m+\theta_n)} + e^{-j(\theta_m+\theta_n)}}{2}\right] + \frac{1}{2}\left[\frac{e^{j(\theta_m-\theta_n)} + e^{-j(\theta_m-\theta_n)}}{2}\right]
\end{aligned}
$$

Here $j = \sqrt{-1}$. Equation (5.25) becomes

$$
\begin{aligned}
x_n = \frac{1}{2N} \sum_{m=0}^{N-1} x_m \sum_{k=0}^{N-1} c_k^2 \Bigg(&\exp\left[j\frac{2k\pi}{2N}(m+n+1)\right] + \exp\left[-j\frac{2k\pi}{2N}(m+n+1)\right] \\
&+ \exp\left[j\frac{2k\pi}{2N}(m-n)\right] + \exp\left[-j\frac{2k\pi}{2N}(m-n)\right]\Bigg)
\end{aligned}
$$

$$
c_k = \begin{cases} 1, & k \neq 0 \\ 1/\sqrt{2}, & k = 0 \end{cases}
\tag{5.26}
$$

$$= \frac{1}{2N} \sum_{m=0}^{N-1} x_m \sum_{k=0}^{N-1} c_k^2 \left(W_N^{-(m+n+1)\frac{k}{2}} + W_N^{(m+n+1)\frac{k}{2}} + W_N^{-(m-n)\frac{k}{2}} + W_N^{(m-n)\frac{k}{2}} \right) \tag{5.27}$$

where $W_N = \exp\left(\frac{-j2\pi}{N}\right) = N$th root of unity.

Equation (5.27) reduces to the following:

When $m = n$ and $k \neq 0$, $\left(\frac{1}{2N} x_n \, 2N\right) = x_n$

When $m = n$ and $k = 0$, $\left(\frac{1}{2N} x_n \left(\frac{1}{\sqrt{2}}\right)^2 4N\right) = x_n$

Note that $\sum_{k=0}^{N-1} W_N^{kl} = N\delta(l)$, where $\delta(l) = \begin{cases} 0, & l \neq 0 \\ 1, & l = 0 \end{cases}$

As can be observed from Eq. (5.10), an N-point DCT or IDCT requires approximately N^2 multiplies and N^2 adds. This is a *brute force* or *direct* approach. Various fast algorithms that require about $N \log_2 N$ arithmetic operations (multiplies and adds) have been developed for implementing N-point DCT or IDCT [75, 76, 77, 78, 80, 82, 83, 86, 87, 89, 91, 92, 97, 100, 101, 106, 115, 119, 120, 124, 127, 128, 129, 158, 159]. These include hardware-oriented algorithms leading to VLSI chips. Added advantages are reduced memory, reduced error due to finite wordlength (bit size) arithmetic, recursive structure, and straightforward extension for implementing MD-DCT/IDCT. These algorithms are similar to those for the DFT and are listed in the following section.

5.3 Fast Algorithms for DCT

N-point DCT via $2N$-point FFT

N-point DCT via N-point FFT

Recursive (fast) algorithms

Sparse-matrix factorization

PFA (prime factor algorithm) for DCT

DIT and DIF algorithms for DCT

DCT via arcsine transform

DCT via WHT

DCT via Hartley transform

DCT via polynomial transforms

2D-DCT algorithms

WFTA: Winograd Fourier transform algorithm

Planar rotations index mapping

Mixed radix algorithms

Split-radix algorithms

Vector-radix algorithms

Radix-4 DIT/DIF algorithms

Polynomial transforms

LMS algorithm

Vector-split-radix algorithms

Fast recursive algorithm of Hou [106]

Algorithm of Duhamel and Guillemot [108]

Algorithm of Toshiba [189]

Scaled DCT algorithms

This list of fast algorithms for DCT, although extensive, is by no means exhaustive. Several new algorithms specially from a hardware viewpoint are being developed.

Sparse-matrix factorization of DCT (or any transform) leads directly to its efficient implementation.

$$(N \times N) \text{ DCT matrix } \Rightarrow \left(c_k \cos\left[\frac{(2n+1)k\pi}{2N}\right]\right), \quad k, n = 0, 1, \ldots, N-1$$

$$\underset{(N \times N)}{[\text{DCT}]} = A_1 A_2 \ldots A_n \quad \begin{pmatrix} A_i = \text{Sparse matrix factor} \\ i = 0, 1, \ldots, n \end{pmatrix}$$

$$[\text{IDCT}] = [\text{DCT}]^T = A_n^T A_{n-1}^T \ldots A_1^T$$

DCT is a real matrix. (All elements are real.)

$\therefore$ $A_1, A_2, \ldots, A_n$ are real matrices.

DCT implementation requires real arithmetic only.

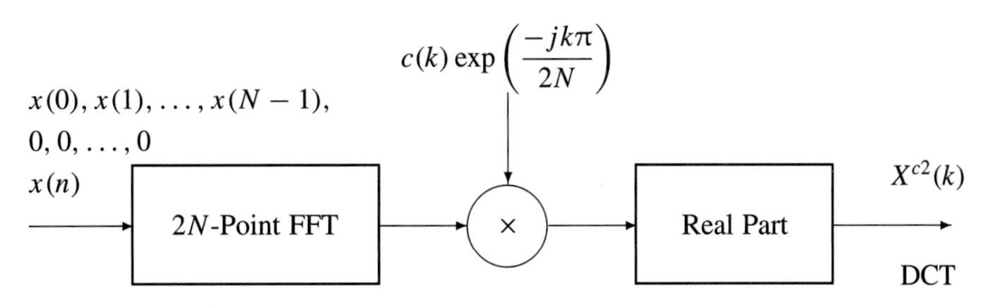

Figure 5.2 N-point DCT via $2N$-point FFT.

***N*-point DCT via *2N*-point FFT** [77] Because FFT (fast Fourier transform) was developed earlier, initially the fast algorithmic approach for DCT was based on FFT [77] (Fig. 5.2). Equation (5.10) can be expressed as

$$X^{c2}(k) = \frac{2c(k)}{N} \operatorname{Re}\left[\exp\left(\frac{-jk\pi}{2N}\right) \sum_{m=0}^{2N-1} x(m)\, W_{2N}^{mk}\right] \tag{5.28}$$

$$(j = \sqrt{-1}), \quad k = 0, 1, \ldots, N-1$$

where $W_{2N} = \exp\left[\frac{-j2\pi}{2N}\right]$, $x(m) = 0$, for $m = N, N+1, \ldots, 2N-1$.

$$\begin{aligned} X^{c2}(k) &= \frac{2c(k)}{N} \operatorname{Re}\left[\sum_{m=0}^{2N-1} x(m) \exp\left(\frac{-j(2m+1)k\pi}{2N}\right)\right] \\ &= \frac{2c(k)}{N} \sum_{m=0}^{2N-1} x(m) \cos\left(\frac{(2m+1)k\pi}{2N}\right) \end{aligned} \tag{5.29}$$

Because $x(m) = 0$, for $m = N, N+1, \ldots, 2N-1$,

$$X^{c2}(k) = \frac{2c(k)}{N} \sum_{m=0}^{N-1} x(m) \cos\left(\frac{(2m+1)k\pi}{2N}\right), \tag{5.30}$$

$$k = 0, 1, \ldots, N-1$$

Another algorithm is based on N-point DCT via N-point FFT [103]. Rewrite Eq. (5.10) as

$$\hat{X}^{c2}(k) = \frac{2}{N} \sum_{n=0}^{N-1} x(n) \cos\left(\frac{(2n+1)k\pi}{2N}\right), \tag{5.31}$$

$$k = 0, 1, \ldots, N-1$$

where $c_k = \begin{cases} 1, & k \neq 0 \\ 1/\sqrt{2}, & k = 0 \end{cases}$ is absorbed in the DCT coefficient.

Let

$$\begin{pmatrix} y(n) = x(2n) \\ y(N-1-n) = x(2n+1) \end{pmatrix}, \quad n = 0, 1, \ldots, N/2 - 1$$

Equation (5.31) can be written as

$$\hat{X}^{c2}(k) = \frac{2}{N} \sum_{m=0}^{\frac{N}{2}-1} x(2m) \cos\left(\frac{(4m+1)k\pi}{2N}\right)$$
$$+\frac{2}{N} \sum_{m=0}^{\frac{N}{2}-1} x(2m+1) \cos\left(\frac{(4m+3)k\pi}{2N}\right), \qquad (5.32a)$$
$$k = 0, 1, \ldots, N-1$$

$$\hat{X}^{c2}(k) = \frac{2}{N} \sum_{m=0}^{\frac{N}{2}-1} y(m) \cos\left(\frac{(4m+1)k\pi}{2N}\right)$$
$$+\frac{2}{N} \sum_{m=0}^{\frac{N}{2}-1} y(N-1-m) \cos\left(\frac{(4m+3)k\pi}{2N}\right), \qquad (5.32b)$$
$$k = 0, 1, \ldots, N-1$$

In the second term of Eq. (5.32b), let $N-1-m = n$, $N-1-n = m$. This becomes

$$\frac{2}{N} \sum_{n=N-1}^{\frac{N}{2}} y(n) \cos \frac{[4(N-1-n)+3]k\pi}{2N}$$
$$= \frac{2}{N} \sum_{n=N-1}^{\frac{N}{2}} y(n) \cos\left(\frac{(4n+1)k\pi}{2N}\right) \qquad (5.33)$$

Using Eq. (5.33), Eq. (5.32b) can be simplified to

$$\hat{X}^{c2}(k) = \frac{2}{N} \sum_{n=0}^{N-1} y(n) \cos\left(\frac{(4n+1)k\pi}{2N}\right) \qquad (5.34)$$

Let $\hat{X}^{c2}(k) = \text{Re}[H(k)]$. Then

$$H(k) = \left(e^{j\pi k/2N}\right)\left[\frac{2}{N}\sum_{n=0}^{N-1} y(n)\, e^{j2\pi nk/N}\right]$$

$$= \frac{2}{N}\left(e^{j\pi k/2N}\right) Y^F(k) \tag{5.35}$$

where $Y^F(k) = \sum_{n=0}^{N-1} y(n)\, e^{j2\pi nk/N}$ is the IDFT of N-point $y(n)$.

Also, $H(N-k) = j\, H^*(k)$.

Proof:

$$j H^*(k) = j\,\frac{2}{N}\sum_{n=0}^{N-1} y(n)\, e^{-j(4n+1)\pi k/2N}$$

$$H(k) = \frac{2}{N}\sum_{n=0}^{N-1} y(n)\, e^{j(4n+1)\pi k/2N}$$

Hence,

$$H(N-k) = \frac{2}{N}\sum_{n=0}^{N-1} y(n) e^{j(4n+1)(N-k)\pi/2N}$$

$$= \frac{2}{N}\sum_{n=0}^{N-1} y(n) e^{-j(4n+1)k\pi/2N} e^{j(4n+1)\pi/2}$$

$$= j\left\{\frac{2}{N}\sum_{n=0}^{N-1} y(n)\left[e^{-j(4n+1)k\pi/2N}\right]\right\}$$

Therefore, $H(N-k) = j\, H^*(k)$.

Hence,

$$\hat{X}^{c2}(k) = \text{Re}\left[H(k)\right], \qquad k = 0, 1, \ldots, N/2$$

$$\hat{X}^{c2}(N-k) = \text{Im}\left[H(k)\right]$$

$$\hat{X}^{c2}(N-k) = \text{Re}\left[H(N-k)\right] = \text{Re}\left[j\, H^*(k)\right]$$

Let $H(N-k) = a(N-k) + j\, b(N-k) = j\, H^*(k) = j\,[a(k) - j\, b(k)] = b(k) + j\, a(k)$. Then

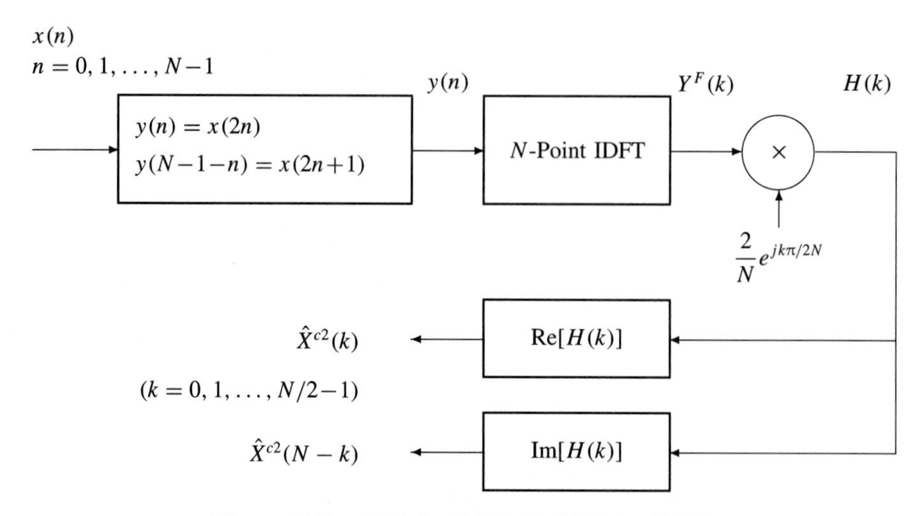

Figure 5.3 N-Point DCT via N-Point DFT.

$$\text{Im}[H(k)] = b(k) = \text{Re}[j\, H^*(k)]$$

where $H(k) = a(k) + j\, b(k)$, and $\text{Im}[H(k)] = b(k) = \text{Re}[H(N-k)]$. The above development is shown in a block diagram format in Fig. 5.3. □

Chen et al. [188] developed a fast recursive (real arithmetic) algorithm for DCT. The flowgraph shown in Fig. 5.4 implements the DCT recursively for $N =$ 4, 8, and 16. It is obvious that this can be extended to $N = 32$, 64, and so forth. This same flowgraph can be used for implementing the IDCT by reversing the direction of the arrows (keeping the multipliers as they are) and by considering the input as the transform sequence (on the right side) and output as the data sequence (on the left side). This property is a characteristic of all the orthogonal transforms. If the transform is unitary such as DFT then, besides the above process, the multipliers have to be replaced by their complex conjugates. The recursivity and the fast algorithmic property are retained in the IDCT.

Lee [104] has developed an efficient algorithm (recursive and real arithmetic) similar to the DIF-FFT (decimation in frequency–fast Fourier transform). Both the forward and inverse DCTs are shown in Fig. 5.5a and b respectively. A DCT chip based on this algorithm has been developed by Thomson Microelectronics.

Several other fast algorithms for the DCT (hence for the IDCT) have been developed. The algorithm of Suehiro and Hatori [189] for 8-point IDCT is illustrated in Figs. 5.6 and 5.7. Loeffler, Ligtenberg, and Moschytz [190], using rotators, have developed a practical 8-point algorithm that requires only 11 multiplies and 29 adds. Needless to say, all these algorithms can be extended

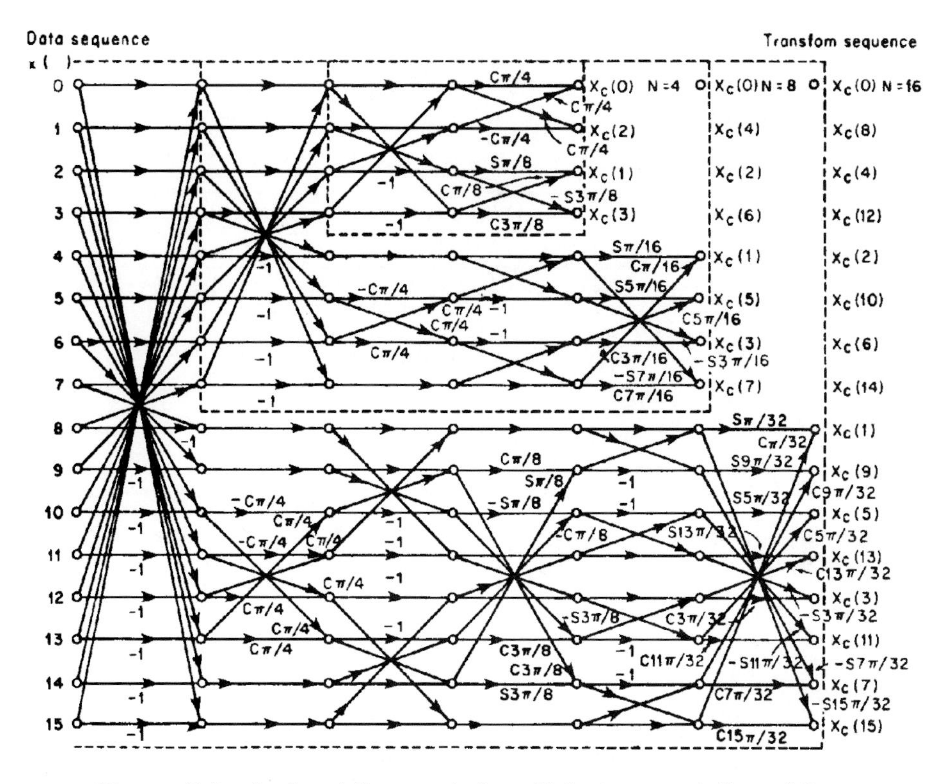

Figure 5.4 A signal flowgraph for efficient computation of the DCT for $N = 4, 8, 16$. For notational simplicity, the multipliers $c\,\theta$ and $s\,\theta$ stand for $\cos\theta$ and $\sin\theta$ [188]. © 1977 IEEE.

to DCTs of length 16, 32, and so forth. The domain of fast algorithms is not limited to integer powers of two (lengths of DCT). Prime factor algorithms (PFA) for $N = 3, 5, 7, 11, 13, 17, 19$, and so forth, have also been developed [110, 120]. N_1N_2-point DCT can be decomposed into the cascade of N_1-point DCTs and N_2-point DCTs, when N_1 and N_2 are relatively prime to each other [105]. Compared to brute-force techniques for computation, the number of real multiplications reduces from $(N_1N_2)^2$ to $N_1^2N_2 + N_1N_2^2$, which amounts to an $N_1N_2/(N_1+N_2)$-to-1 reduction. These can easily be extended to DCTs of lengths resulting in multiplication of these prime numbers (or any integers). From a hardware viewpoint, the computational complexity (number of multiplies and adds) is not the only criterion. Other factors, such as memory requirements, load, transfer, shift, and recursivity, influence the overall algorithmic structure.

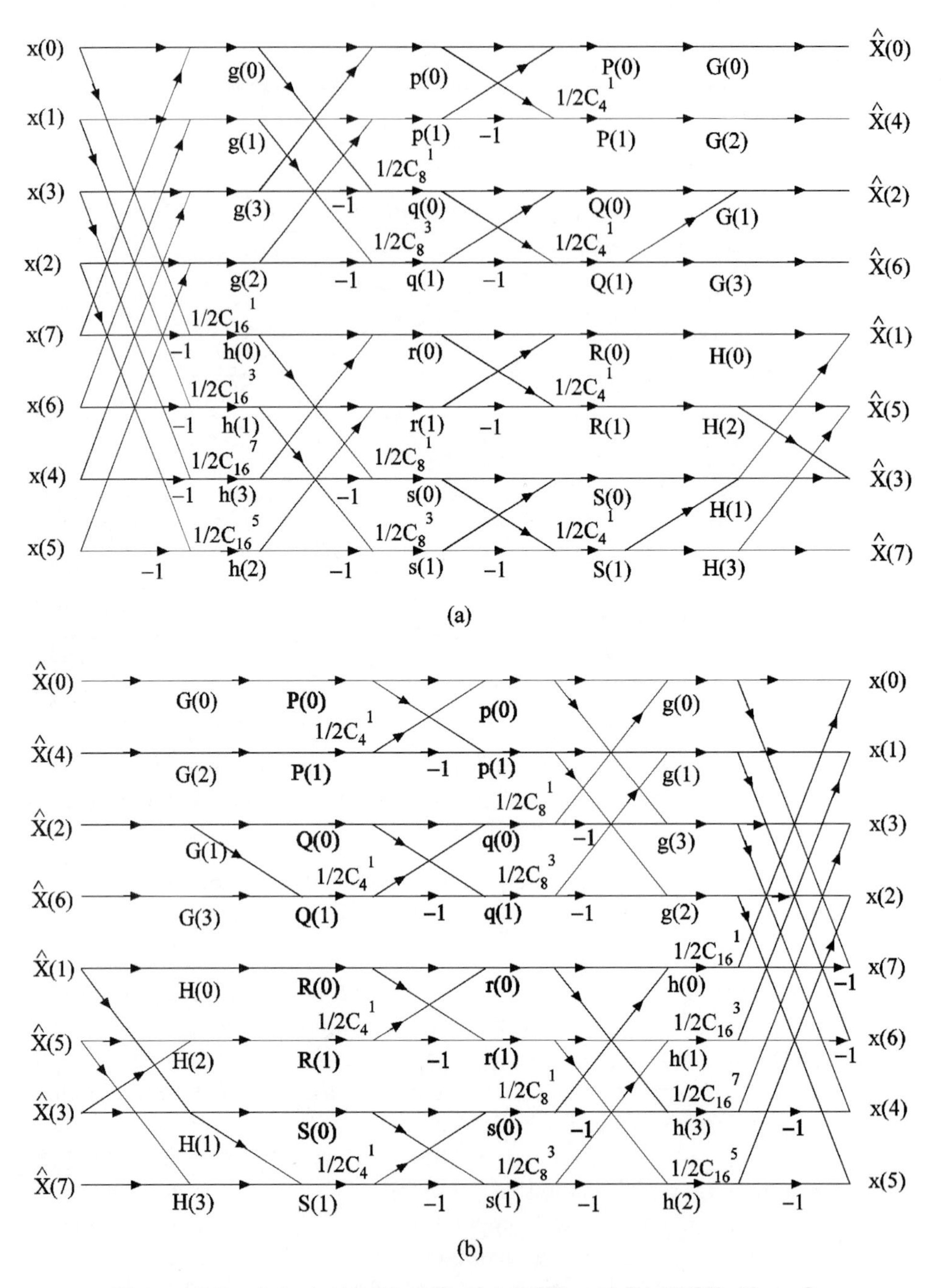

Figure 5.5 A fast algorithm for (a) DCT and (b) IDCT, $N = 8$. $\hat{X}(n) = c(n)X(n)$, where $c(n) = 1/\sqrt{2}$ for $n = 0$, $c(n) = 1$ for $n \neq 0$, and $C_b^a = \cos\left(\frac{a\pi}{b}\right)$ [104]. © 1984 IEEE.

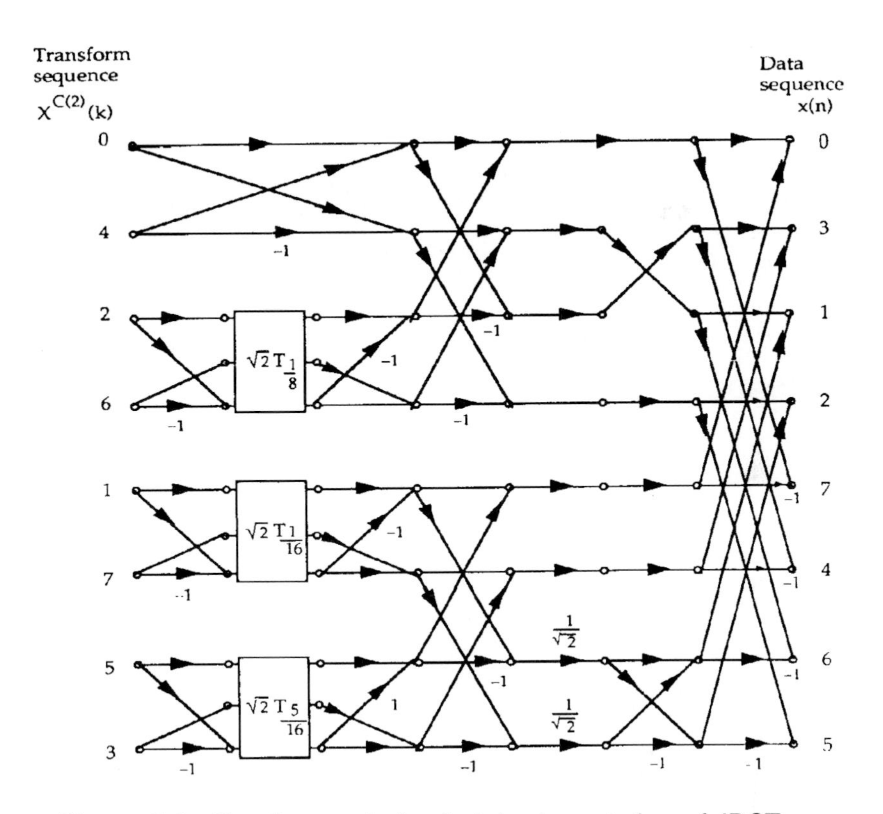

Figure 5.6 The flowgraph for fast implementation of IDCT, $N = 8$, based on [189]. © 1986 IEEE.

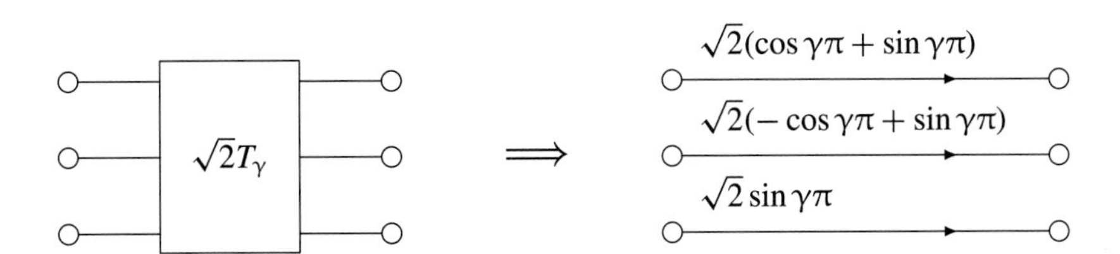

Figure 5.7 Description of notation shown in Fig. 5.6.

5.4 Multidimensional DCT

As DCT is a separable transform, 1D-DCT, defined in Eq. (5.10), can be extended to 2D-DCT as follows:

2D-DCT (separable)

$$X^{c2}_{u,v} = \frac{4}{NM} c_u\, c_v \sum_{n=0}^{N-1} \sum_{m=0}^{M-1} x_{n,m} \cos\left[\frac{(2n+1)u\pi}{2N}\right] \cos\left[\frac{(2m+1)v\pi}{2M}\right], \quad (5.36)$$

$$u = 0,\ 1,\ \ldots,\ N-1, \qquad c_l = \begin{cases} 1/\sqrt{2}, & l = 0 \\ 1, & l \neq 0 \end{cases}$$
$$v = 0,\ 1,\ \ldots,\ M-1,$$

2D-IDCT

$$x_{n,m} = \sum_{u=0}^{N-1} \sum_{v=0}^{M-1} c_u\, c_v\, X^{c2}_{u,v} \cos\left[\frac{(2n+1)u\pi}{2N}\right] \cos\left[\frac{(2m+1)v\pi}{2M}\right], \quad (5.37)$$

$$n = 0,\ 1,\ \ldots,\ N-1, \qquad m = 0,\ 1,\ \ldots,\ M-1$$

Similar to Eq. (5.11), the normalized version of 2D-DCT is:

2D-DCT (normalized)

$$X^{c2}_{u,v} = c_u\, c_v \frac{2}{\sqrt{NM}} \sum_{n=0}^{N-1} \sum_{m=0}^{M-1} x_{n,m} \cos\left[\frac{(2n+1)u\pi}{2N}\right] \cos\left[\frac{(2m+1)v\pi}{2M}\right] \quad (5.38)$$

$$= \sqrt{\frac{2}{N}} \sum_{n=0}^{N-1} c_u \left[\sqrt{\frac{2}{M}} c_v \sum_{m=0}^{M-1} x_{n,m} \cos \frac{(2m+1)v\pi}{2M}\right] \cos \frac{(2n+1)u\pi}{2N},$$

$$u = 0,\ 1,\ \ldots,\ N-1, \qquad c_l = \begin{cases} 1/\sqrt{2}, & l = 0 \\ 1, & l \neq 0 \end{cases}$$
$$v = 0,\ 1,\ \ldots,\ M-1,$$

2D-IDCT (normalized)

$$x_{n,m} = \frac{2}{\sqrt{NM}} \sum_{u=0}^{N-1} \sum_{v=0}^{M-1} c_u\, c_v X^{c2}_{u,v} \cos\left[\frac{(2n+1)u\pi}{2N}\right] \cos\left[\frac{(2m+1)v\pi}{2M}\right], \quad (5.39)$$

$$n = 0,\ 1,\ \ldots,\ N-1, \qquad m = 0,\ 1,\ \ldots,\ M-1$$

DCT is a separable transform, as is IDCT. An implication of this is that 2D-DCT can be implemented by a series of 1D-DCTs, i.e., 1D-DCTs along rows (columns) of a 2D array followed by 1D-DCTs along columns (rows) of the semi-transformed array (see Fig. 5.8). Theoretically, both are equivalent. All the properties of the 1D-DCT (fast algorithms, recursivity, etc.) extend automatically to the MD-DCT. The separability property can be observed by rewriting Eq. (5.36) as follows:

$$X^{c2}_{u,v} = \frac{2}{N}\sum_{n=0}^{N-1} c_u \left[\frac{2}{M}\sum_{m=0}^{M-1} c_v x_{n,m} \cos\frac{(2m+1)v\pi}{2M}\right] \cos\frac{(2n+1)u\pi}{2N} \tag{5.40}$$

$$= \frac{2}{M}\sum_{m=0}^{M-1} c_v \left[\frac{2}{N}\sum_{n=0}^{N-1} c_u x_{n,m} \cos\frac{(2n+1)u\pi}{2N}\right] \cos\frac{(2m+1)v\pi}{2M},$$

$$u = 0, 1, \ldots, N-1, \qquad v = 0, 1, \ldots, M-1$$

A similar manipulation on Eq. (5.37) yields the separability property for the 2D-IDCT. This property is illustrated in Fig. 5.8. Fast algorithms, software, architecture, and VLSI implementation for 2D-DCT and its inverse have been extensively developed [81, 84, 85, 88, 93, 96, 99, 107, 108, 109, 111, 112, 114, 117, 158, 161, 162, 169, 172, 173, 175, 177].

Since DCT is a separable transform, it can be expressed in a matrix form as follows.

2D-DCT

$$\underset{(N\times N)}{\left[\underline{X}^{c2}\right]} = \frac{2}{N}\underset{(N\times N)}{\left[\underline{C}^{II}_N\right]}\;\underset{(N\times N)}{\left[\underline{x}\right]}\;\frac{2}{N}\underset{(N\times N)}{\left[\underline{C}^{II}_N\right]^T} \tag{5.41}$$

2D-IDCT

$$\underset{(N\times N)}{\left[\underline{x}\right]} = \underset{(N\times N)}{\left[\underline{C}^{II}_N\right]^T}\;\underset{(N\times N)}{\left[\underline{X}^{c2}\right]}\;\underset{(N\times N)}{\left[\underline{C}^{II}_N\right]} \tag{5.42}$$

$$\frac{2}{N}\underset{(N\times N)}{\left[\underline{C}^{II}_N\right]}\;\underset{(N\times N)}{\left[\underline{C}^{II}_N\right]^T} = \frac{2}{N}\underset{(N\times N)}{\left[\underline{C}^{II}_N\right]^T}\;\underset{(N\times N)}{\left[\underline{C}^{II}_N\right]} = \underset{(N\times N)}{\underline{I}_N}$$

For the 2D-DCT, the sizes (dimensions) along each coordinate need not be the same.

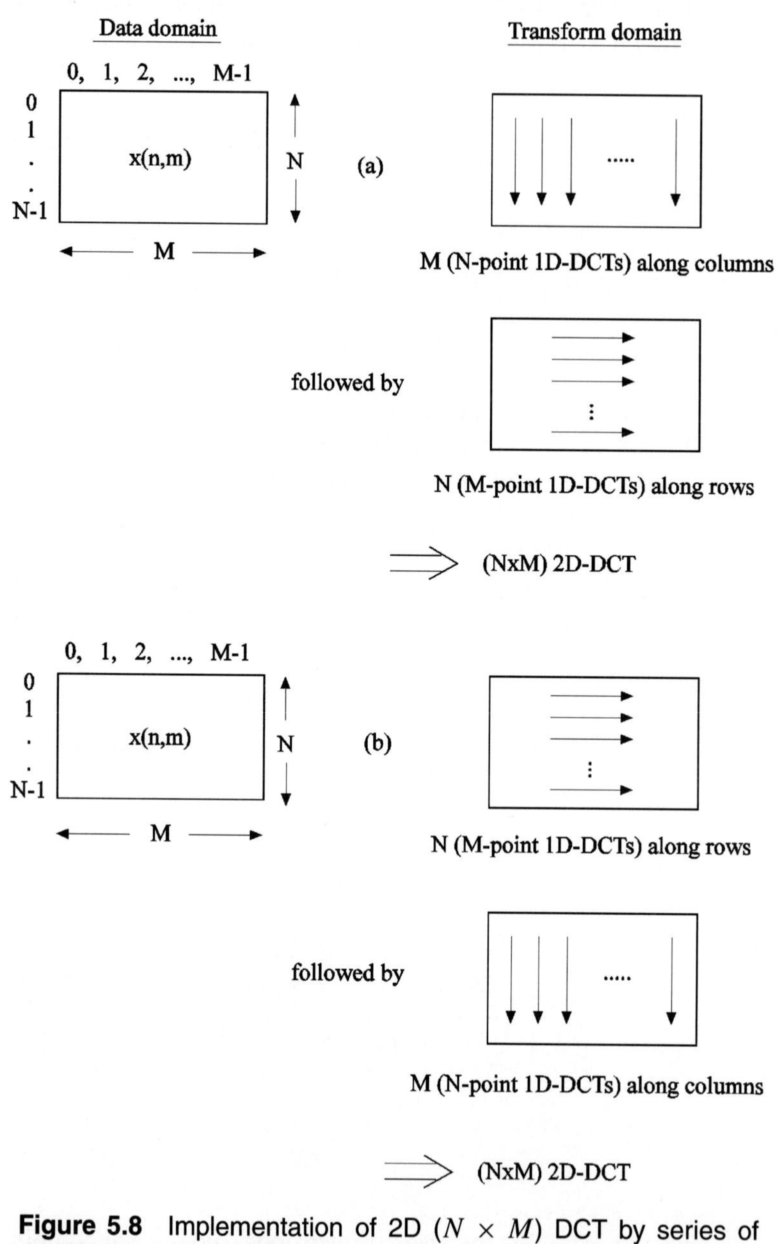

Figure 5.8 Implementation of 2D ($N \times M$) DCT by series of 1D-DCTs: (a) 1D-DCTs along columns followed by 1D-DCTs along rows, (b) 1D-DCTs along rows followed by 1D-DCTs along columns.

2D-DCT

$$\underset{(N \times M)}{\left[\underline{X}^{c2}\right]} = \frac{2}{N} \underset{(N \times N)}{\left[\underline{C}_N^{II}\right]} \underset{(N \times M)}{\left[\underline{x}\right]} \frac{2}{M} \underset{(M \times M)}{\left[\underline{C}_M^{II}\right]^T}$$

2D-IDCT

$$\underset{(N \times M)}{\left[\underline{x}\right]} = \underset{(N \times N)}{\left[\underline{C}_N^{II}\right]^T} \underset{(N \times M)}{\left[\underline{X}^{c2}\right]} \underset{(M \times M)}{\left[\underline{C}_M^{II}\right]}$$

$$\frac{2}{N}\left[\underline{C}_N^{II}\right]\left[\underline{C}_N^{II}\right]^T = \frac{2}{N}\left[\underline{C}_N^{II}\right]^T\left[\underline{C}_N^{II}\right] = \underline{I}_N$$

$$\frac{2}{M}\left[\underline{C}_M^{II}\right]\left[\underline{C}_M^{II}\right]^T = \underline{I}_M$$

The extension of DCT/IDCT to 3D is straightforward.

3D-DCT

$$X_{u,v,p}^{c2} = \frac{8}{NML} c_u c_v c_p \sum_{n=0}^{N-1} \sum_{m=0}^{M-1} \sum_{l=0}^{L-1} x_{n,m,l} \cos\left[\frac{(2n+1)u\pi}{2N}\right] \cos\left[\frac{(2m+1)v\pi}{2M}\right] \cos\left[\frac{(2l+1)p\pi}{2L}\right],$$

$$u = 0, 1, \ldots, N-1, \quad v = 0, 1, \ldots, M-1, \quad p = 0, 1, \ldots, L-1, \qquad c_k = \begin{cases} 1/\sqrt{2}, & k = 0 \\ 1, & k \neq 0 \end{cases}$$

3D-IDCT

$$x_{n,m,l} = \sum_{u=0}^{N-1} \sum_{v=0}^{M-1} \sum_{p=0}^{L-1} X_{u,v,p}^{c2} c_u c_v c_p \cos\left[\frac{(2n+1)u\pi}{2N}\right] \cos\left[\frac{(2m+1)v\pi}{2M}\right] \cos\left[\frac{(2l+1)p\pi}{2L}\right]$$

Normalized versions of 3D-DCT/IDCT can also be expressed. Three-dimensional-DCT and 3D-IDCT can be implemented by series of 1D-DCTs, since DCT is a separable transform (Fig. 5.9).

Eigen or basis images Like the 1D-DCT, which has basis functions, the 2D-DCT has basis images. Mapping of a 2D data array into the 2D-DCT domain

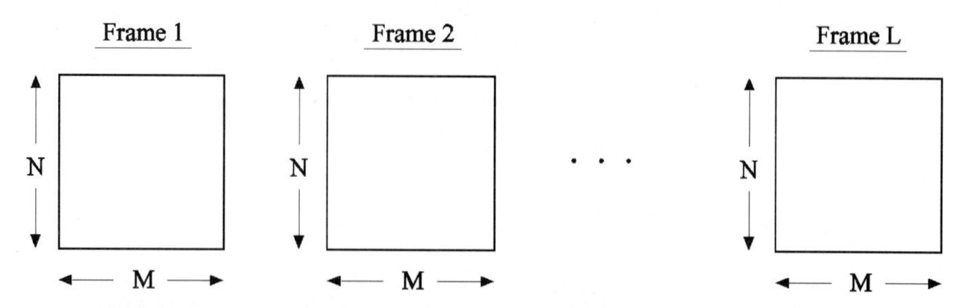

Figure 5.9 The 3D-DCT of L frames each of size $(N \times M)$. Take the 2D-DCT of each frame using 1D-DCT technique as before. Follow this with L-point 1D-DCTs of corresponding 2D-DCT coefficients of the L frames (Series of 1D-DCTs).

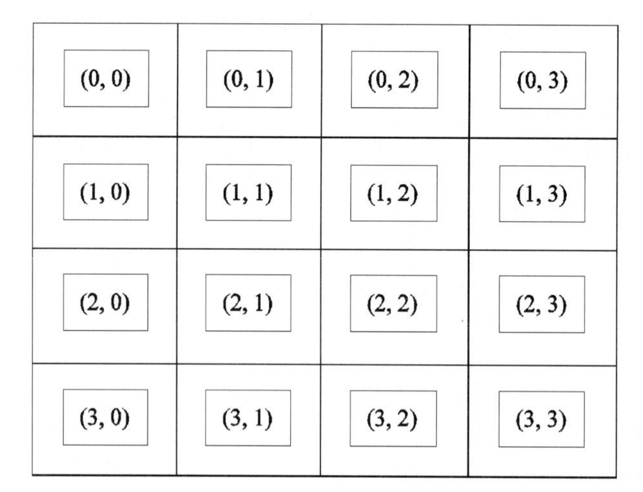

Figure 5.10 Labelling of the 2D (4×4) DCT basis images.

implies decomposing the 2D data array into the basis images of the DCT. This can be illustrated with 2D (4×4) DCT. The basis images are numbered in Fig. 5.10.

(4×4) 2D-DCT

$$X_{u,v}^{c2} = \frac{1}{4} c_u \, c_v \sum_{n=0}^{3} \sum_{m=0}^{3} x_{n,m} \cos\left[\frac{(2n+1)u\pi}{8}\right] \cos\left[\frac{(2m+1)v\pi}{8}\right],$$

$$u, v = 0, 1, 2, 3, \qquad c_k = \begin{cases} 1/\sqrt{2}, & k = 0 \\ 1, & k \neq 0 \end{cases}$$

The 2D (4 × 4) DCT of the 2D (4 × 4) data array results in 2D-DCT coefficients. These coefficients represent the weights of the corresponding basis images inherent in the 2D (4 × 4) data array. If the 2D (4 × 4) data array is uniform, then only $X^{c2}_{0,0}$ is present (nonzero). All other coefficients are zero.

The (4 × 4) DCT matrix is

$$\begin{bmatrix} \frac{1}{\sqrt{2}}\,(1 & 1 & 1 & 1) \\ C^1 & C^3 & C^5 & C^7 \\ C^2 & C^6 & C^{10} & C^{14} \\ C^3 & C^9 & C^{15} & C^{21} \end{bmatrix}, \qquad C^k = \cos\left(\frac{k\pi}{8}\right)$$

The basis images are obtained by the outer (vector) product of each basis vector with all the basis vectors. Two-dimensional (4 × 4) DCT implies decomposing a 2D (4 × 4) data array into 16 basis images, (4 × 4) image array. The lowest frequency (top left) basis image is

$$\left[\frac{1}{\sqrt{2}}\underset{(4\times 1)}{\begin{pmatrix}1\\1\\1\\1\end{pmatrix}}\underset{(1\times 4)}{(1\ 1\ 1\ 1)}\frac{1}{\sqrt{2}}\right] = \frac{1}{2}\begin{bmatrix}1&1&1&1\\1&1&1&1\\1&1&1&1\\1&1&1&1\end{bmatrix}$$

$$= \text{basis image } (0, 0)$$

The highest frequency (bottom right) basis image is

$$\left[\begin{pmatrix}C^3\\C^9\\C^{15}\\C^{21}\end{pmatrix}(C^3\ C^9\ C^{15}\ C^{21})\right] = \begin{bmatrix} C^3\ (C^3\ C^9\ C^{15}\ C^{21}) \\ C^9\ (C^3\ C^9\ C^{15}\ C^{21}) \\ C^{15}\ (C^3\ C^9\ C^{15}\ C^{21}) \\ C^{21}\ (C^3\ C^9\ C^{15}\ C^{21}) \end{bmatrix}$$

$$= \text{basis image } (3, 3),$$

$$C^k = \cos(k\pi/8)$$

The basis images for 2D (8 × 8) DCT are shown in Fig. 5.11. The top left image represents the mean intensity of the (8 × 8) spatial block. The remaining basis images on the first row (column) represent the vertical (horizontal) edges. A 2D-DCT of a 2D data array is basically a structural decomposition in terms of its basis images.

This is similar to the decomposition of 1D data in terms of the DCT basis functions (Fig. 5.1). This decomposition property, of course, is valid for any orthogonal transform. The structural representation and the frequency distribution in the DCT domain are shown in Fig. 5.12. One way of determining the features in

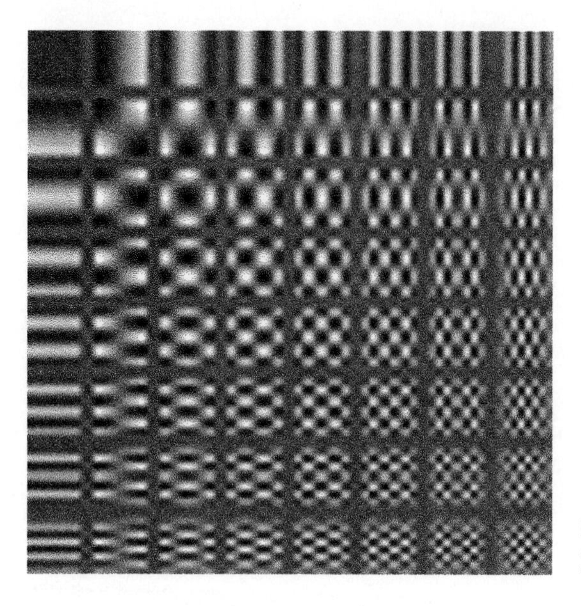

Figure 5.11 Basis images of 2D (8 × 8) DCT.

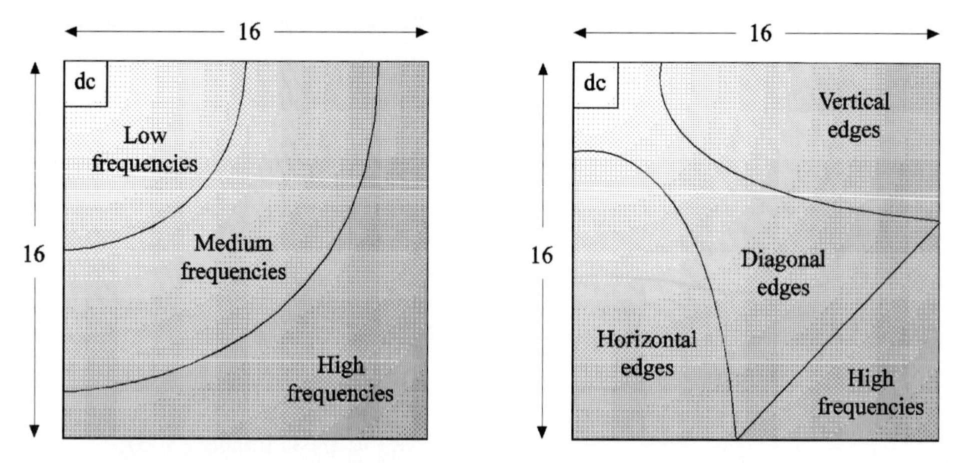

Figure 5.12 Two-dimensional (16 × 16) DCT: (a) frequency distribution, (b) structural decomposition.

the spatial block is by looking into the energy distribution in the 2D-DCT domain (Fig. 5.13) [141]. Recall that the energy is invariant to orthogonal transformation. The sum of the squares of the DCT coefficients (or the spatial data values) is the energy of the block, i.e.,

$$\sum_{u=0}^{N-1}\sum_{v=0}^{N-1}\left(X_{u,v}^{c2}\right)^2 = \sum_{n=0}^{N-1}\sum_{m=0}^{N-1} x_{n,m}^2$$

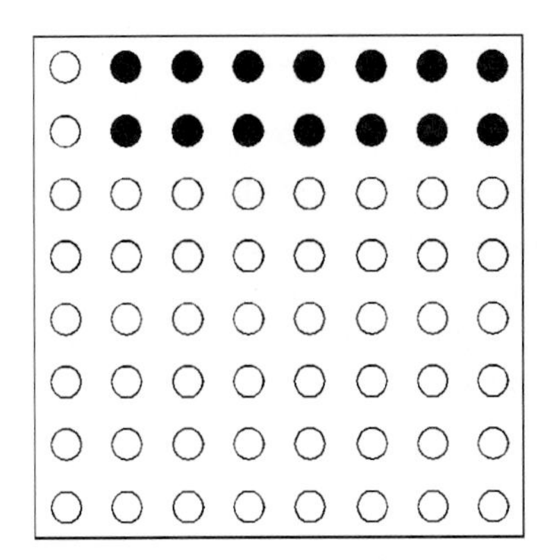

H = Energy in horizontal region, Vertical edges

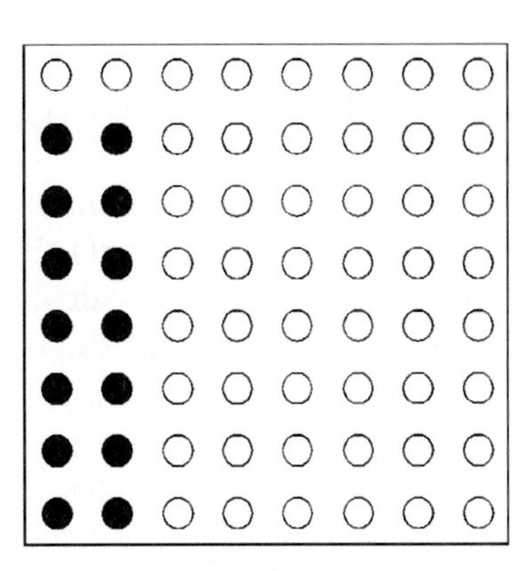

V = Energy in vertical region, Horizontal edges

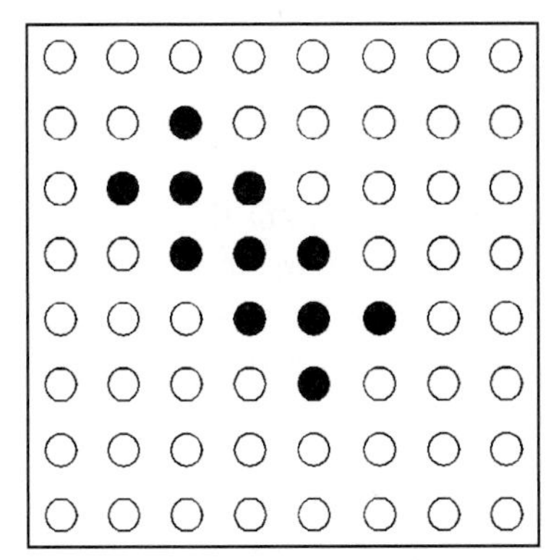

D = Energy in diagonal region, Diagonal edges

M = MAX (H, V, D)
IF M < Threshold, CLASS = L
ELSE CLASS = dir (M)

Figure 5.13 DCT block classification. Energy in each region is calculated by the sum of the squares of the ac coefficients represented by the solid circles. *L* represents low activity class. © 1989 IEEE.

Here $\left(X_{0,0}^{c2}\right)^2$ is the dc energy and the remaining energy is called the ac energy.

5.5 Transform Image Coding

An $(L \times L)$ image is divided into integer number of blocks (non-overlapping), each of size $(N \times N)$ (Fig. 5.14), which are mapped into the 2D $(N \times N)$ DCT domain. A simple version of the transform image coding is shown in Fig. 5.15. The original image can be $(P \times L)$ and the block can be $(M \times N)$. Of the $(N \times N)$ DCT coefficients, only a few are significant. These are selected for quantization. The rest are set to zero. The classification process can be based on coefficient variances, magnitudes, the block structure (Figs. 5.12 and 5.13) or any other valid criterion. In general, the coefficients go through uniform quantizers and then are variable-length coded. In this case, for transmission over a constant bit rate channel, a buffer is required. Other ingredients are buffer control of the quantizer, multiplexer (to multiplex video, audio, and data), FEC, channel coder, and so on.

Overhead may be required to indicate the coefficient selection because the decoder needs to track the entire coding process. Block transform coding, therefore, is conducive to classification (adaptive process) and is also simpler from an implementation viewpoint. In general, most of the international standards favor (8 × 8) blocks for the 2D-DCT, as can be seen in the later chapters. There is no significant gain in choosing block sizes larger than (8 × 8), as can be inferred from Fig. 5.16. Also, the human visual system (HVS) can be incorporated into the coding scheme, reflecting the visual perception of the 2D-DCT coefficients.

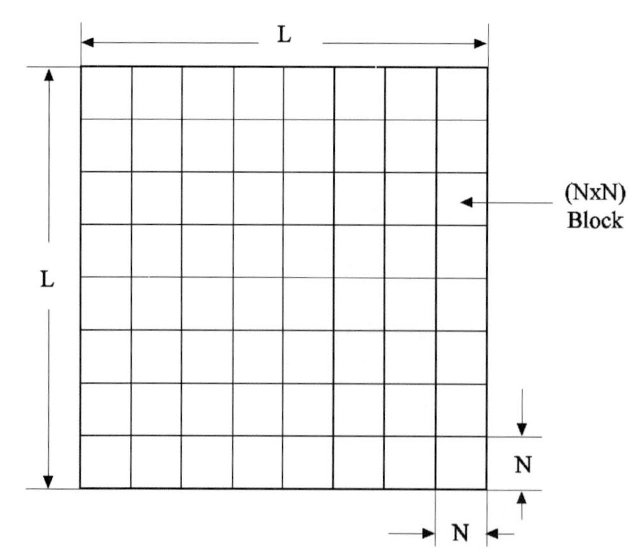

Figure 5.14 Subblock division in transform image coding.

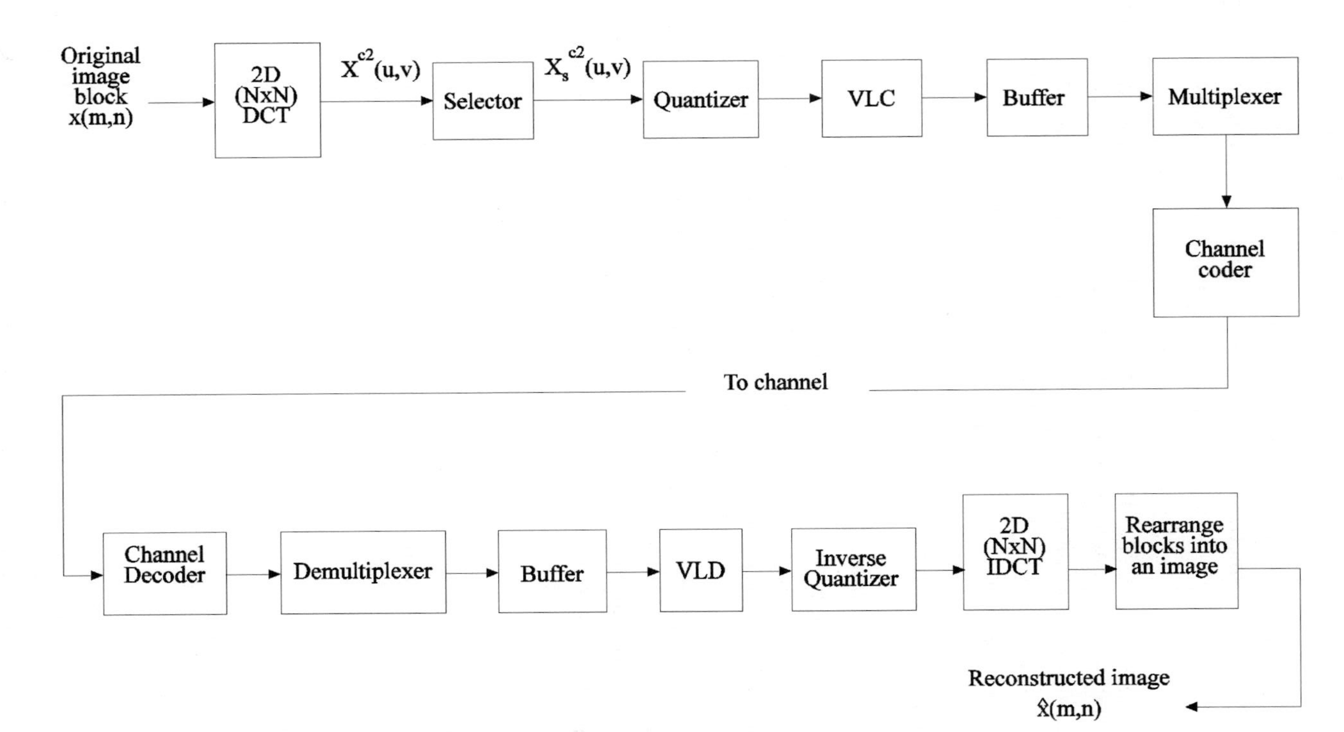

Figure 5.15 Transform image encoder/decoder.

A detailed description of Fig. 5.16 is in order. Assuming that the image statistics are governed by a separable first-order Markov process, the variances of the transform coefficients (for any orthogonal transform) can be easily derived. For an $(N \times N)$ block size, a fraction of the N^2 coefficients representing large variances can be retained with the rest set to zero, followed by an inverse 2D transform, yielding the reconstructed $(N \times N)$ block. This is called the maximum variance zonal sampling (MVZS) (Fig. 5.17).

The MVZS can be replaced by threshold sampling, i.e., retain 1/4 of the DCT coefficients having large magnitudes (the rest are set to zero). This requires a large overhead indicating the coefficient selections. While KLT is optimal for any block size (Fig. 5.16), there is no significant improvement in increasing the block size beyond (16×16). Of all the transforms, DCT comes very close to the KLT. In view of this, apart from other considerations (fixed transform, fast algorithms, separability, recursive structure, filter banks, filtering property, etc.), it is easy to see why DCT has been chosen in most of the image-coding standards. The cumulative contribution of various operations that follow the 2D transform, such as scanning the transform coefficients along a particular pattern such as the zigzag scan (Fig. 5.18), HVS weighted quantization, 2D or 3D VLC (run-length coding of zero coefficients followed by the level of the nonzero coefficient), leads to a substantial bit-rate reduction with insignificant visually perceptible impairments. By classifying the blocks into different categories and coding them accordingly, a further increase in compression ratio can be achieved. So far, we have discussed transform coding applied to still (intraframe) images, the objective being to take advantage of spatial correlation.

Additional bandwidth reduction can be achieved by extending the transform coding to a sequence of images, i.e., video. In general, from a complexity viewpoint, temporal correlation can be reduced by predictive coding (prediction along the time axis between adjacent frames). The prediction process can be enhanced by motion estimation. These techniques, which have been adopted by the standards groups for video coding, are discussed in Chapter 6.

Zigzag scan The 2D-DCT coefficients can be scanned in a predetermined fashion. One popular pattern (heavily used) is the *zigzag scan* (Fig. 5.18). This results in transform coefficients in increasing order of frequency starting with the dc coefficient at the top left and ending with $X^{c2}_{7,7}$ representing the highest frequency. This scanning mechanism is advantageous to 2D VLC. Large numbers of coefficients along this scan, after quantization, reduce to zero, resulting in efficient use of RLC. Also, an end of block (EOB) code signifying that all the coefficients that follow this code along the zigzag scan are zero, can effectively reduce the number of bits needed to code the block. This concept has been adopted in most of the standards as can be observed from Part II of this book.

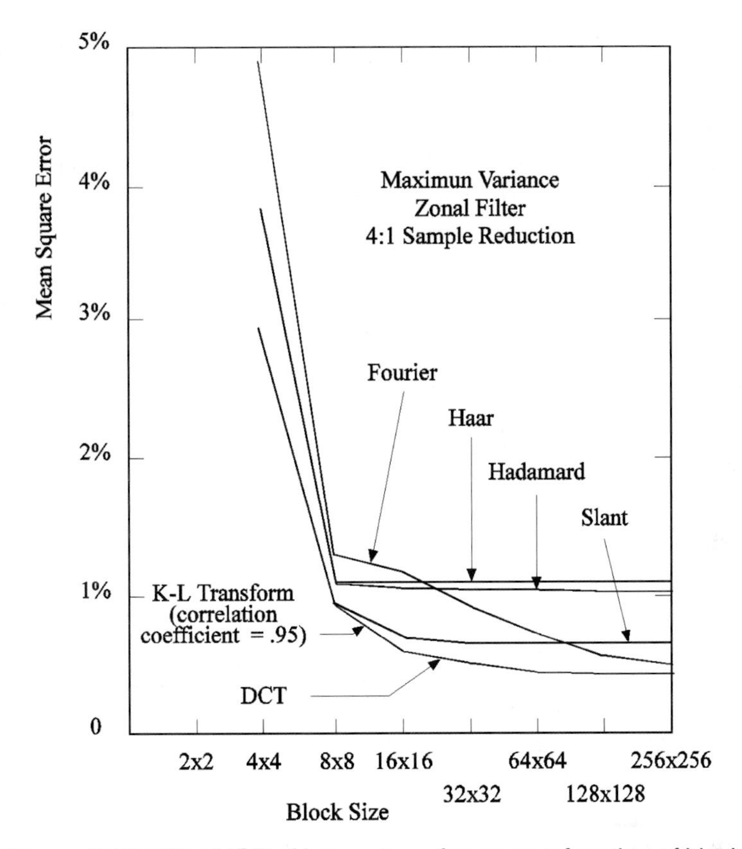

Figure 5.16 The MSE of image transforms as a function of block size. The image statistics along both rows and columns are assumed to be first-order Markov process with correlation coefficient $\rho = 0.95$. No quantization and coding. © 1974 IEEE.

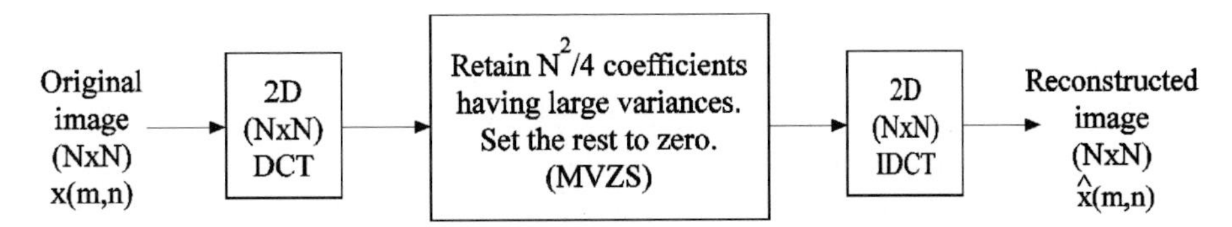

Figure 5.17 Maximum variance zonal sampling (4:1 sample reduction). The MSE between the original and reconstructed blocks based on MVZS can be expressed as

$$\text{MSE} = \frac{1}{N^2} \sum_{n=1}^{N} \sum_{m=1}^{N} (x_{n,m} - \hat{x}_{n,m})^2$$

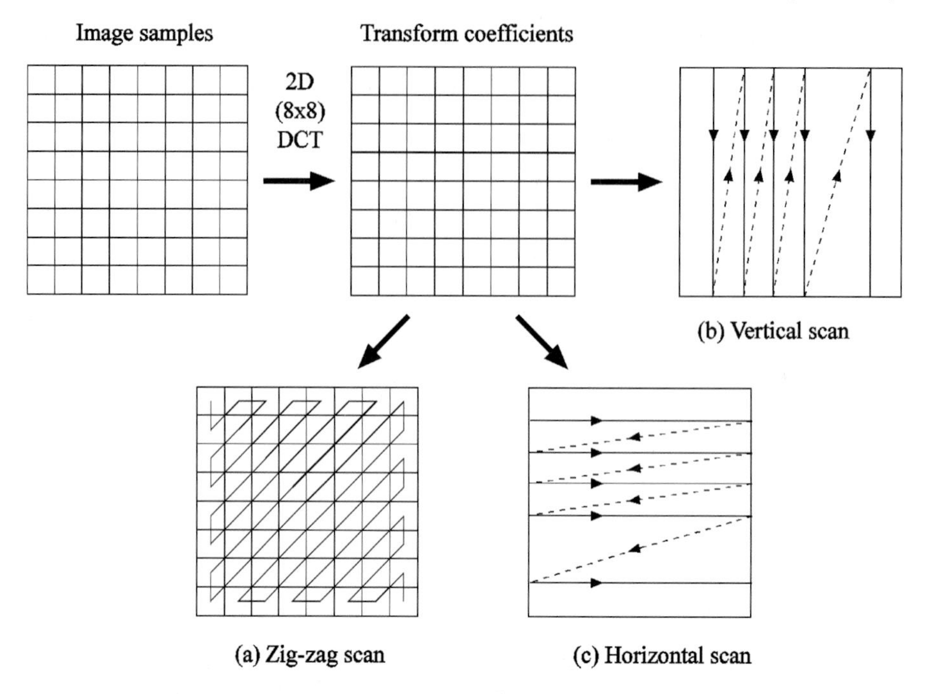

Figure 5.18 Scanning of 2D (8 × 8) DCT coefficients: (a) zigzag scan; (b) vertical scan; (c) horizontal scan. (See Fig. 11.20b for an alternative scan.)

An alternative zigzag scan for interlaced video (Fig. 11.20b) has been adopted by both MPEG-2 (Chapter 11) and the Grand Alliance (Chapter 13).

HVS weighting The 2D transform coefficients, in general, have different visual sensitivities, i.e., from a perception viewpoint, they have unequal significance [142, 145, 146, 165, 187, 191]. This can be analyzed easily for the 2D discrete Fourier transform (DFT) because the 2D-DFT coefficients of an image represent spatial frequencies (cycles/meter). By defining a human visual model in terms of the modulation transfer function (MTF), the HVS as a function of the frequency in cycles/degree of visual angle subtended can be developed (Fig. 5.19). Because the spatial frequency can be related to "visual angle" frequency, the HVS weighting for the 2D-DCT (or for any other discrete transform) can be derived (Table 5.1). Because the human visual model is ad hoc (not precise), several HVS weighting matrices for the DCT are reported in the literature. In this case, the 2D (8 × 8) DCT coefficients are multiplied by the corresponding elements before uniform quantization.

Spatial frequencies in an (8 × 8) block are closely related to distance from human eye to screen and image resolution (sampling density). For example, the

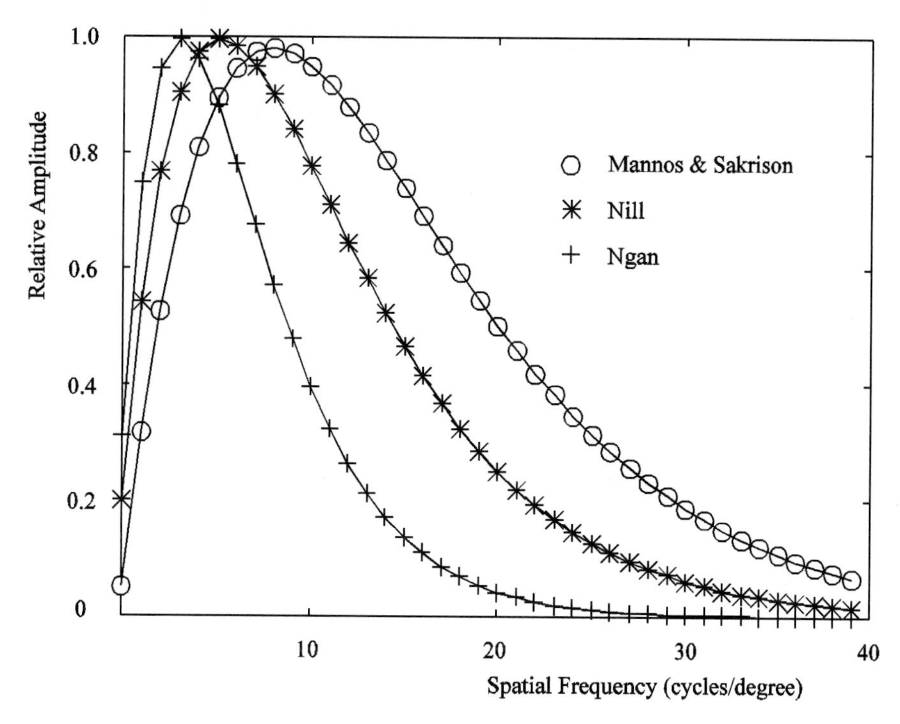

Figure 5.19 Comparison of the MTFs [142]. © 1990 IEEE.

Table 5.1 An HVS weighting matrix based on a sampling density of 64 pels/degree [142]

0.4942	1.0000	0.7023	0.3814	0.1856	0.0849	0.0374	0.0160
1.0000	0.4549	0.3085	0.1706	0.0845	0.0392	0.0174	0.0075
0.7023	0.3085	0.2139	0.1244	0.0645	0.0311	0.0142	0.0063
0.3814	0.1706	0.1244	0.0771	0.0425	0.0215	0.0103	0.0047
0.1856	0.0845	0.0645	0.0425	0.0246	0.0133	0.0067	0.0032
0.0849	0.0392	0.0311	0.0215	0.0133	0.0075	0.0040	0.0020
0.0374	0.0174	0.0142	0.0103	0.0067	0.0040	0.0022	0.0011
0.0160	0.0075	0.0063	0.0047	0.0032	0.0020	0.0011	0.0006

Source: © 1990 IEEE.

lowest spatial frequency in an (8×8) block is about 4 cycles/degree, assuming that the sampling density is 64 pels/degree and 512×512 pels in an image [192]. That the blocking effect of an image coder using (8×8) block transform is so visible (sensitive) can thus be explained by this spatial frequency concept. Similarly, the spatial frequency weighting can be adjusted by the luminance intensity of pixels, since the sensitivity of the human eye is not a linear function of the luminance intensity.

In general, from a visual perception viewpoint, the low-frequency coefficients are much more sensitive than the high-frequency coefficients. By incorporating HVS weighting in the DCT coefficients, a number of high-frequency coefficients can be coarsely quantized.

5.6 DCT Filtering and Layered Coding

Bacause the DCT involves mapping into the frequency domain, another application is filtering. By reconstructing the signal/image based on only the low-frequency coefficients, a subsampled (decimated) signal/image can be obtained. Such a technique has been extensively investigated and also has been proposed to MPEG-2 (Chapter 11) as part of scalability. This was not, however, adopted because of the compatibility problems. Based on the frequency scalability ([121, 123, 126, 131, 137, 138, 147, 148, 150, 152, 153, 155, 157], [179] thru [185]) video services at different spatial resolutions (layered coding) can be offered (Fig. 5.20). Bits representing the DCT coefficients and motion vectors at both layers need to be transmitted.

The layered coding is part of the hybrid coding addressed in Chapter 6. A third layer can be added (at the bottom of Fig. 5.20b) in which all the operations will be on (2×2) blocks. If the topmost layer involves (16×16) blocks, then a four-layered coding scheme with successively decreasing spatial resolutions can be developed. The decoders corresponding to this frequency scalability are inherently simpler because they are devoid of some functions such as motion estimation, forward 2D-DCT, and so forth (Chapter 9).

5.7 Pre- and Postprocessing

Figure 5.15 represents a highly simplified version of transform image coding. Neither preprocessing nor postprocessing is shown. The image/video may be acquired from video camera, scanner, VCR, optical disk, or any other image source. The video may be in *RGB* format, in which case it can be mapped into *YIQ* or YC_bC_r (See Chapter 2). As the codecs process digital signals, A/D conversion is a necessary step. If the video source is composite color, then it can be decomposed into the desired components. After the sync information is removed, the active portion can be transformed into the format needed for a specific application. For example, CIF (common intermediate format) and quarter-CIF (QCIF) belong to H.261 (Fig. 9.3). Similarly, SIF (source input format) is the domain of MPEG-1 (Fig. 10.15). Thus, spatial subsampling may be needed. A number of other formats are described in Tables 11.3 and 11.4. Also, both MPEG-1 and 2 require rearranging the frames/pictures into a coding order that is different from display order (Fig. 10.8 and Table 10.2). Another preprocessing operation is arranging

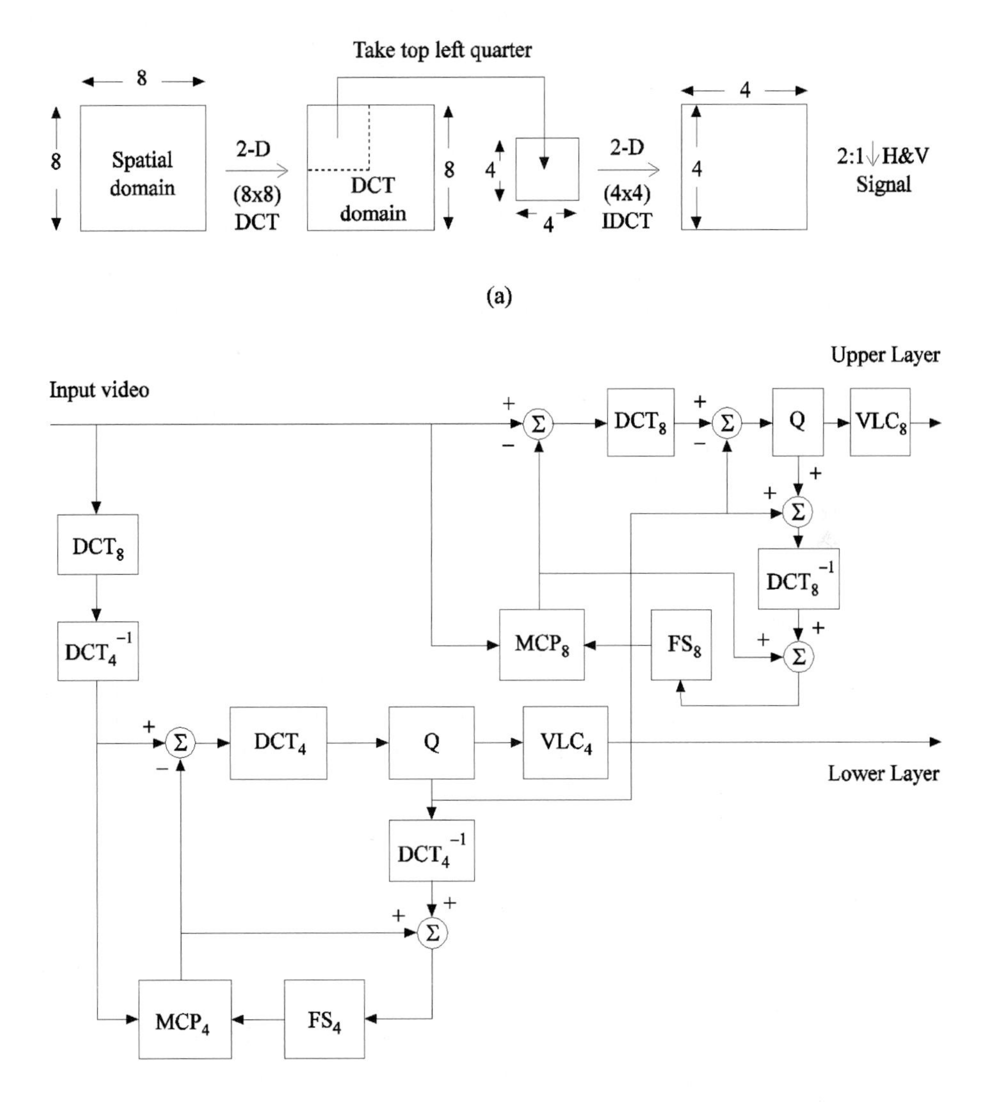

Figure 5.20 (a) Subsampling using filtering in the DCT domain and (b) a two-layer frequency scalable coder. Spatial resolution of the lower layer is 1/2 (both horizontally and vertically) that of the upper layer.

the frame/picture into nonoverlapping blocks, e.g., (8 × 8) or (16 × 8). The major preprocessing operations can be summarized as follows:

1. Format conversion and A-to-D (this may involve conversion among NTSC, PAL, and SECAM TV systems)
2. Spatial and/or temporal down-sampling/up-sampling (decimation and/or interpolation filters)
3. Rearrangement of a group of pictures into a coding order
4. Rearranging the picture/frame into nonoverlapping blocks
5. Any other operation(s) required for a particular coder

It is evident that postprocessing in the reverse order is required at the ouput of the decoder, so as to display the reconstructed image/video on the monitor. Also, proper sychronization between video and audio is necessary. The exact implementation of both pre- and postprocessing (Fig. 5.21) is not part of any of

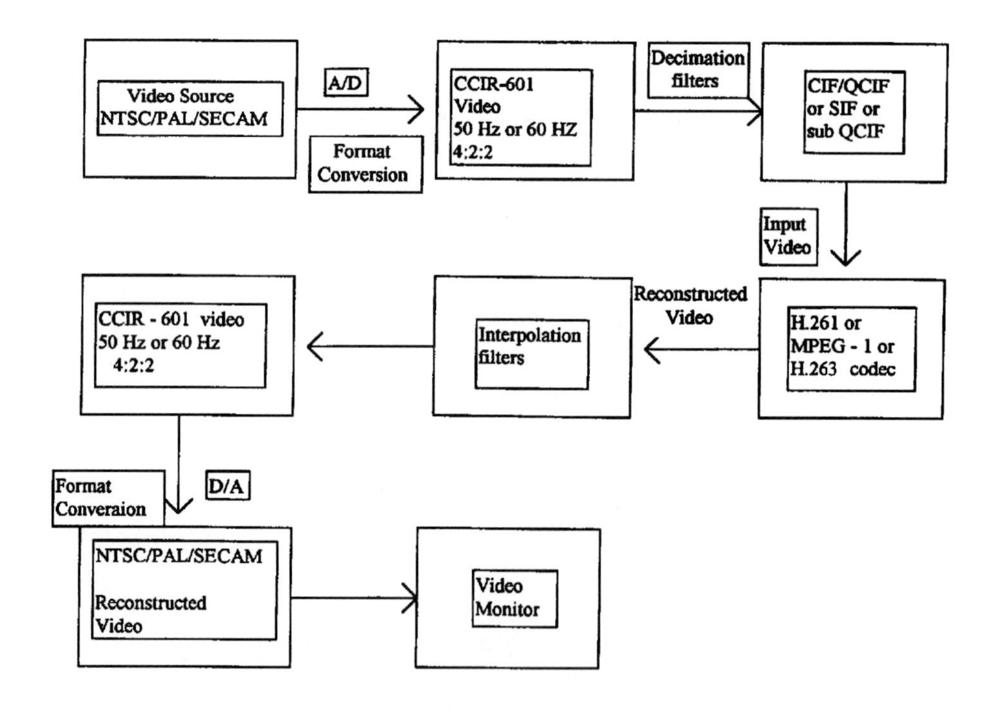

Figure 5.21 An example of pre- and postprocessing for a video codec. Standards address only the codec. Pre- and postprocessing may accept (output) multiple formats.

the standards specifications. The designer can rely on his/her DSP expertise to come up with ingenious schemes that execute these operations most efficiently.

5.8 Progressive Image Transmission

By transmitting and decoding either the 2D-DCT coefficients in stages (generally from low frequency to high frequency) or the bit planes of all these coefficients, the image quality can be built up progressively at the decoder [78, 122, 143]. Such a scheme is part of the JPEG extended system (Section 8.5). The former is called spectral selection (Fig. 8.11) as the DCT coefficients represent spatial frequencies (in case of intraframe coding). Conceivably, in the case of 2D (8×8) DCT, the image quality can be enhanced in 64 stages by transmitting one coefficient per stage. Bit-plane coding in the DCT domain is called the successive approximation (Fig. 8.11). Clever combination of these two techniques can result in reaching reasonable quality during the early stages of progression. Several adaptive and interactive techniques in the progressive image transmission (PIT) have been developed [140, 142].

Chapter 6
Hybrid Coding and Motion Compensation

Summary

This chapter concentrates on motion-compensated predictive-transform coding, one of several possible techniques in hybrid coding. The emphasis in motion compensation (MC) is on block motion estimation (ME) because this has been the favorite approach of the standards groups. Some of the block matching algorithms (BMA) described in this chapter have been implemented in software/hardware. The following chapter addresses vector quantization (VQ) and subband coding (SBC).

6.1 Introduction

As stated in Chapter 5, coding efficiency can be improved by incorporating a number of different redundancy reduction techniques. The primary objective is to increase the compression ratio (or reduce the bit rate) while maintaining the desired visual quality. A number of other factors, such as implementation complexity, robustness to channel noise, compatibility, scalability, and interoperability, also influence the overall coding process.

The previous chapters have described transform coding and predictive coding. Each of these techniques has its own advantages and disadvantages. In general, transform coding (Chapter 5) is less susceptible to channel errors, can achieve higher compression ratios, and is amenable to a number of adaptive features, but is more complex. On the other hand, predictive coding (Chapter 4) is quite sensitive to channel noise and is useful only for small compression ratios, but is much less complex.

A combination of the two (transform-DPCM) can lead to an efficient encoder. Such a technique is called hybrid coding. As hybrid implies mixed, the domain of hybrid coding is not limited to the transform-DPCM combination. Various other tools, such as vector quantization (VQ), subband coding (Chapter 7 describes both these techniques), and wavelets, play significant roles in video/audio coding. In fact, different possible combinations of these techniques, i.e., transform-VQ, transform-predictive VQ, subband-VQ, subband-transform-VQ, subband-transform, wavelet-VQ, predictive-VQ, and so forth, have been extensively investigated, and some have been implemented in hardware.

We will, however, limit our discussion mainly to transform-DPCM [225, 250, 252, 254, 256, 257] as this has been the favorite candidate among all the standards groups. This process has been enhanced by motion compensation, as can be observed from Chapter 9 (H.261), Chapter 10 (MPEG-1 video), Chapter 11 (MPEG-2 video), Chapter 12 (H.263), Chapter 13 (Digital HDTV Grand Alliance), and Chapter 14 (CMTT). In view of the overwhelming adoption of the transform(DCT)-DPCM-motion compensation method for interframe (video) coding, both the hardware (chips, chipsets, boards, systems) and software development has accelerated at a fast pace. Set-top boxes, add-in cards, decoders and other consumer-oriented electronics (karaoke, video jukebox, video CD, etc.) have proliferated (Appendix A).

6.2 Hybrid (Transform-DPCM) Coding

The most practical application of the transform-DPCM encoder is shown in Fig. 6.1. This is part of H.261 (some functions are optional) described in Chapter 9. The decoder corresponding to this is inherently simpler (Fig. 9.10). In this case the transform is in the spatial domain and the prediction is in the temporal domain (prediction between adjacent frames or pictures). Here, frame implies interlaced image and picture denotes progressive or noninterlaced. In MPEG-1 and -2 (Chapters 10 and 11) and GA HDTV systems (Chapter 13) prediction based on two- or three-frame/picture interval is also utilized (Fig. 10.8). Conceptually, this hybrid scheme can also be extended to intraframe images, i.e., transform along rows (columns) followed by prediction along columns (rows) (Fig. 6.2). This process can also be reversed. Another example is transform of nonoverlapping blocks followed by prediction between adjacent blocks (Fig. 6.3).

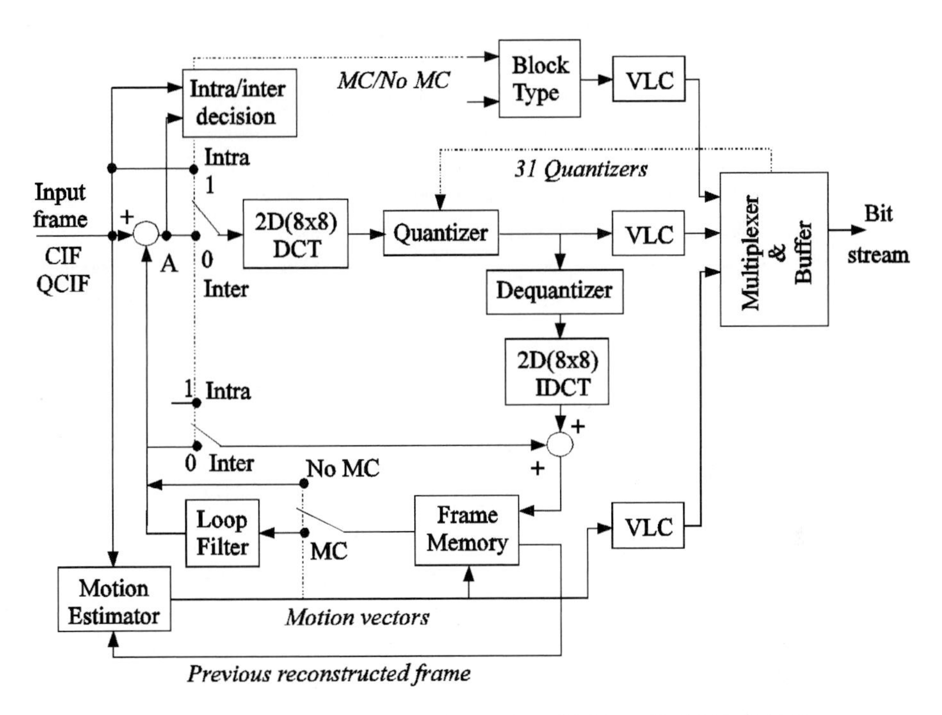

Figure 6.1 Video encoder based on H.261. © 1990 ITU-T.

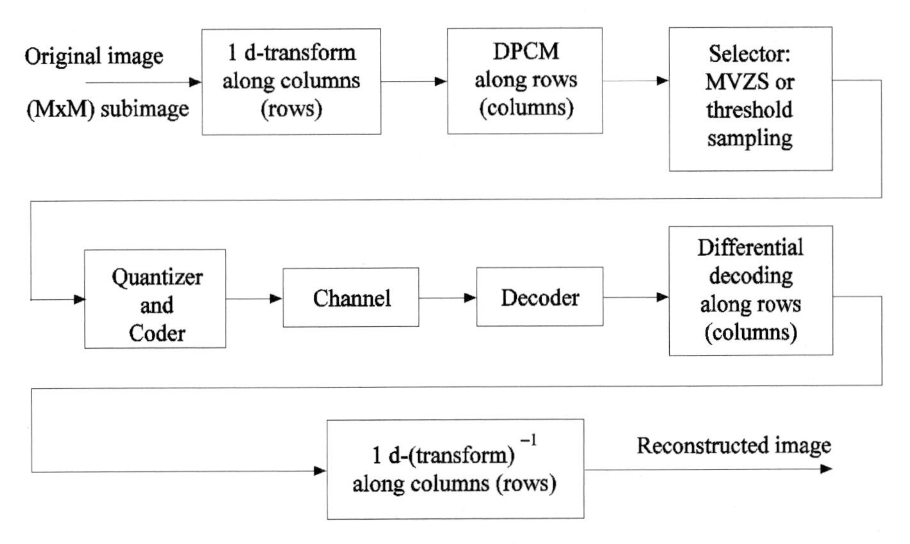

Figure 6.2 1D-transform/DPCM image coding.

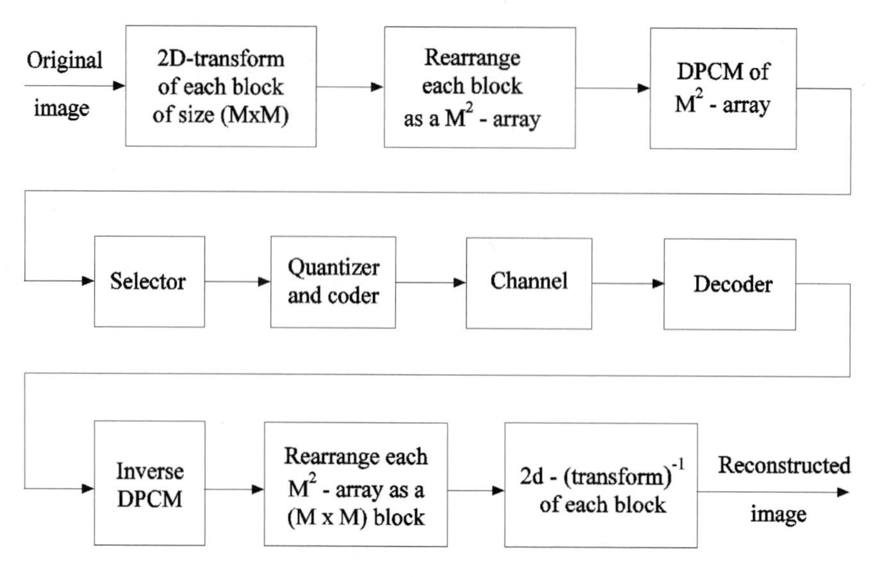

Figure 6.3 2D-transform/DPCM image coding.

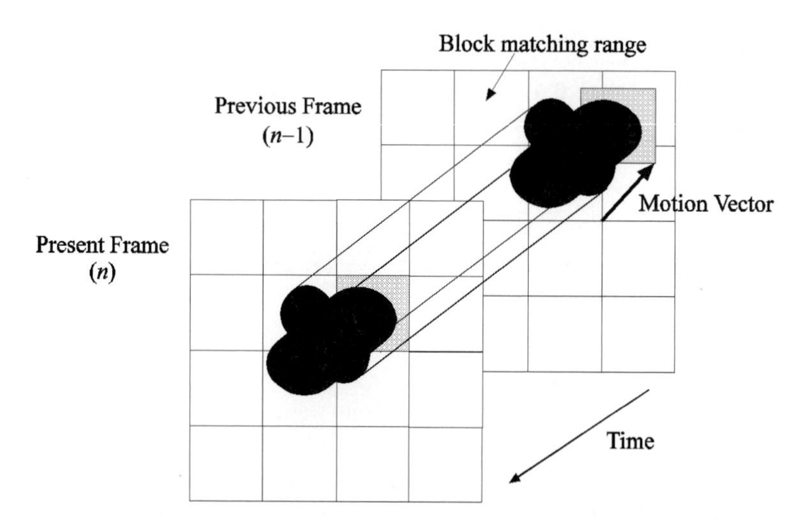

Figure 6.4 Object-based motion estimation.

We will, however, limit our discussion to an H.261-like coder (Fig. 6.1). The philosophy behind this coder is quite simple. If a pel or group of pels in the present frame/picture can be reasonably well predicted based on pel(s) in the previous frame/picture, then the prediction error entropy is quite small. Besides reducing the temporal correlation, the hybrid technique applied to the interframe/interpicture prediction errors can further reduce/remove the spatial correlation. Coefficient selection, quantization, VLC, and so forth, then follow the prediction/transform operation. The former are similar to those described in Chapter 5. Of course, if prediction is exact, assuming no changes due to illumination, reflection, and so forth, then transform operation is redundant. On the other hand, when prediction fails, such as during scene change, fast motion, and special effects, then the only redeemable alternative is intraframe transform coding. This is an adaptive feature (intra or inter), inherently built into the system as a fail-safe feature. (See Fig. 9.9 for intra/inter mode and Fig. 9.12 for MC/no MC.) Detailed operations of the hybrid coder (prediction in the temporal domain followed by transform in the spatial domain) including the adaptive modes are described in Chapter 9 (H.261), Chapters 10 and 11 (MPEG-1 and -2), Chapter 12 (H.263), Chapter 13 (HDTV), and Chapter 14 (CMTT).

As stated earlier, the prediction process can be significantly enhanced by motion estimation, i.e., predicting where the pels or group of pels or even objects in the present frame/picture were in the previous frame/picture based on motion estimation. This leads us to the next section where the concepts of motion estimation are presented.

6.3 Motion Estimation

Motion estimation (ME), in general, can improve the prediction accuracy between adjacent frames/pictures [245, 247]. This technique falls into two categories:

1. Pel-by-pel ME, called pel-recursive algorithms (PRA) [251, 259]
2. Block-by-block ME, called block-matching algorithms (BMA) [225, 250]

PRAs have been rarely used because they are inherently complex, and the ME algorithms sometimes run into convergence problems. Object-based ME (Fig. 6.4) is tailored for image coding. This, however, at present, is not practical because the ME is dependent on identifying (or locating) the object, or at least its boundaries, in both frames/pictures. Also, the overhead representing the object can be significant, thus nullifying any reduction in the bit rate resulting from improved prediction. (Model-based motion estimation in image sequences has also been investigated [262].) As a compromise, BMA, even though not optimal, has been universally used [237, 248, 252, 256, 257] in interframe motion-compensated (MC) hybrid coding. Instead of the 2D transform, subband, VQ (Chapter 7), wavelets, or any other valid technique can be applied to the MC prediction errors. We will, from a standards viewpoint, concentrate on the transform approach.

In BMA, motion of a block of pels, say $(M \times N)$, within a frame/field interval is estimated. The range of the motion vector is constrained by the search window. BMA also ignores rotational motion and assumes all pels within the $(M \times N)$ block have the same uniform motion. Under these limitations, the goal is to find the *best match* (or the least distortion) between the $(M \times N)$ block in the present frame/field and a corresponding block in the previous frame/field within a search window say of size $[(M + 2m_2) \times (N + 2n_1)]$ (Fig. 6.5). Hence, the MV range is $\pm n_1$ pels and $\pm m_2$ lines. In case of fast motion or scene change, ME may not be effective. Also, the pels in the $(M \times N)$ block can conceivably be moving in different directions. A control mechanism such as the absolute sum of the MC prediction errors of the $(M \times N)$ block can be introduced to investigate the effectiveness of BMA. When the BMA is no longer useful in the ME process, one can switch to intraframe coding, i.e., no prediction, just a 2D transform of the $(M \times N)$ block as described in Section 5.5. This adaptive mode requires overhead as the decoder has to track the exact mode of operation. ME of a small block size is much more meaningful than large block size ME. The penalty, of course, is the increased number of bits needed to represent the large number of motion vectors. In H.261 (Chapter 9), MPEG-1 (Chapter 10), and H.263 (Chapter 12), ME is based on a (16×16) luminance block. In MPEG-2 (Chapter 11), both (16×16) and (16×8) block sizes are allowed for ME. In CMTT (Chapter 14), a (16×8) luminance block is used for ME. Also, ME for luminance and color components can be independently evaluated. Techniques that can handle the complex motion of a block of pels have been developed [261]. Block-matching technique is basically a pattern matching process. The words *best match* or *least distortion* are very generic, but the BMA implementation requires a meaningful definition. As in the case of VQ (Chapter 7), the distortion between the block in the present frame and the displaced block in the previous frame (within the search window) can be defined as follows:

Mean squared error (MSE)

$$M_1(i, j) = \frac{1}{MN} \sum_{m=1}^{M} \sum_{n=1}^{N} \left(X_{m,n} - X^R_{m+i,n+j} \right)^2, \qquad |i| \le m_2, |j| \le n_1 \tag{6.1}$$

This is an L_2 norm. $X_{m,n}$ is the pel intensity at row m and column n in the present frame/field. $X^R_{m+i,n+j}$ refers to row $m + i$ and column $n + j$ in the reference (previous) frame/field.

Mean absolute error (MAE)

$$M_2(i, j) = \frac{1}{MN} \sum_{m=1}^{M} \sum_{n=1}^{N} \left| X_{m,n} - X^R_{m+i,n+j} \right|, \qquad |i| \le m_2, |j| \le n_1 \tag{6.2}$$

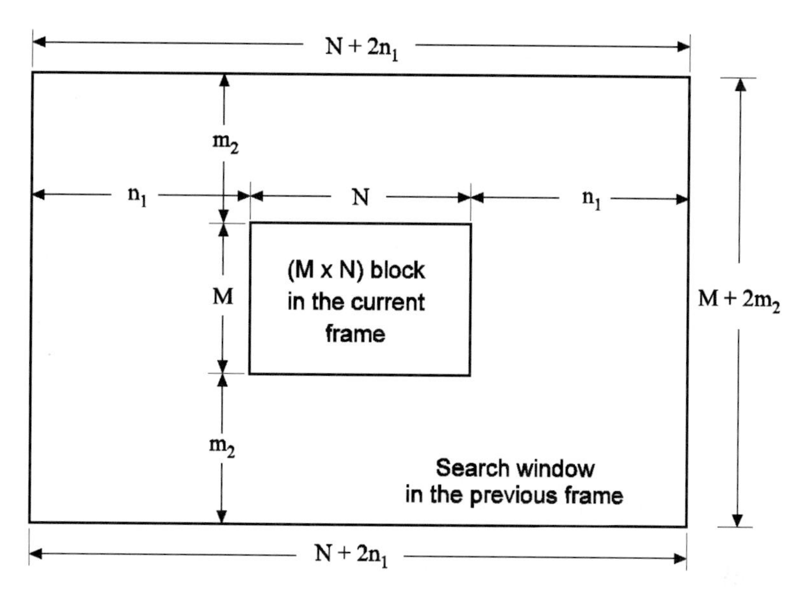

Figure 6.5 Motion estimation of an $(M \times N)$ block in the previous frame within an $[(M + 2m_2) \times (N + 2n_1)]$ search window (full search). MV range is $\pm n_1$ pels and $\pm m_2$ lines per frame interval.

This is an L_1 norm.

Some other distortion functions have also been tried. The best match can also be defined in terms of correlation functions (maximum correlation) as follows:

Cross-correlation function

$$M_3(i, j) = \frac{\sum_{m=1}^{M} \sum_{n=1}^{N} X_{m,n} X^R_{m+i,n+j}}{\left[\sum_{m=1}^{M} \sum_{n=1}^{N} X^2_{m,n}\right]^{1/2} \left[\sum_{m=1}^{M} \sum_{n=1}^{N} \left(X^R_{m+i,n+j}\right)^2\right]^{1/2}}, \tag{6.3}$$

$$|i| \leq m_2, \; |j| \leq n_1$$

Brute (full) search implies that for every pel and line displacement within the search window, the cost function (distortion, correlation, etc.) has to be computed and compared so as to find the location corresponding to the optimal cost function. For example, choosing the MSE as the cost function, it has to be computed $(2m_2 + 1) \times (2n_1 + 1)$ times (for every possible displacement constrained by the MV range) before locating the position of the minimum MSE. The MV range in

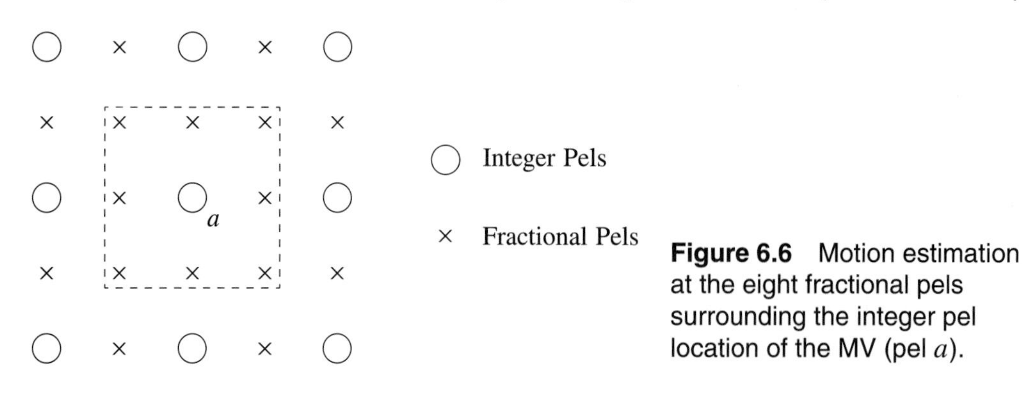

Figure 6.6 Motion estimation at the eight fractional pels surrounding the integer pel location of the MV (pel a).

H.261 (Chapter 9) is ±15 pels or lines/picture (ME is optional in the encoder). Brute search therefore for integer pel/line MV resolution requires computation and comparison of (31 × 31) cost functions. Extensive investigations have shown that the MAE performs as well as the MSE in MC prediction [253]. The industry therefore has designed and developed the hardware based on the MAE as the cost function.

Once the ME is carried out based on the integer pel/line MV resolution, surrounding this best match, the MV resolution can be extended to 1/2 pel/line, by interpolating the fractional pels (Fig. 6.6), or to even finer resolutions.

The reader can easily observe that ME is computationally intensive. Hence, researchers have developed several simplified algorithms using adaptive search processes, leading to reasonably good ME. In some cases, the ME may be locally optimal rather than globally optimal, as in the brute search.

6.3.1 Simplified ME algorithms

Most of the hardware/software offered by the industry ([221] through [228], [231, 232, 233, 234, 236, 237, 238, 239, 240, 242]) addresses BMA–ME based on brute search. Also enough flexibility is provided to vary the block size and search window. Increased block size and increased MV range can be accomplished by cascading ME chips. A few of them, however, implement simplified algorithims. Hence, a description of some of these is in order.

Logarithmic search [250]

To the authors' knowledge, this is the first simplified search published in the literature. The logarithmic search tracks block motion along the direction of minimum distortion (Fig. 6.7). The software for BMA–ME offered by both Optibase and North Valley Research follows this technique.

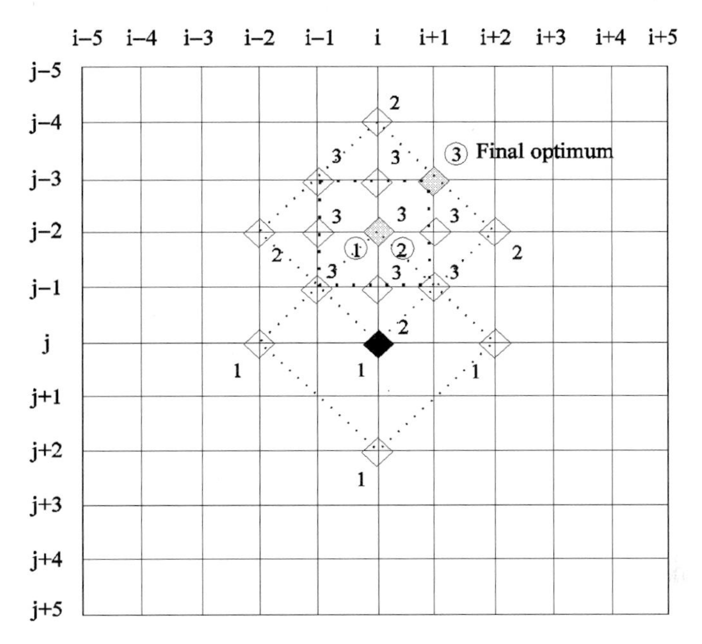

Figure 6.7 The logarithmic search algorithm [250] implemented by the Optibase MPEG encoder. © 1981 IEEE.

Conjugate direction search and one-at-a-time search [244, 253]

One-at-a-time search (OTS) is a simplified version of the conjugate direction search (CDS). OTS tracks the motion alternately horizontally and vertically (Fig. 6.8). A modified and improved version of this has recently been reported (Fig. 6.9) [246].

Three-step search [225]

Originally proposed by Koga et al. [225], this is a fine-coarse search mechanism (Fig. 6.10). The first step (circle 1) involves ME based on 4-pel/4-line resolution at the nine locations, with the center point corresponding to zero MV. The second step (circle 2) involves ME based on 2-pel/2-line resolution around the location determined by the first step. This is repeated in the third step (circle 3) with 1-pel/1-line resolution. The last step yields the MV.

The MV search range, therefore, is ± 7 pels/± 7 lines. MV resolution can be improved by adding a fourth step, i.e., $\frac{1}{2}$-pel/$\frac{1}{2}$-line resolution, corresponding to fractional pels around the third step location of the MV.

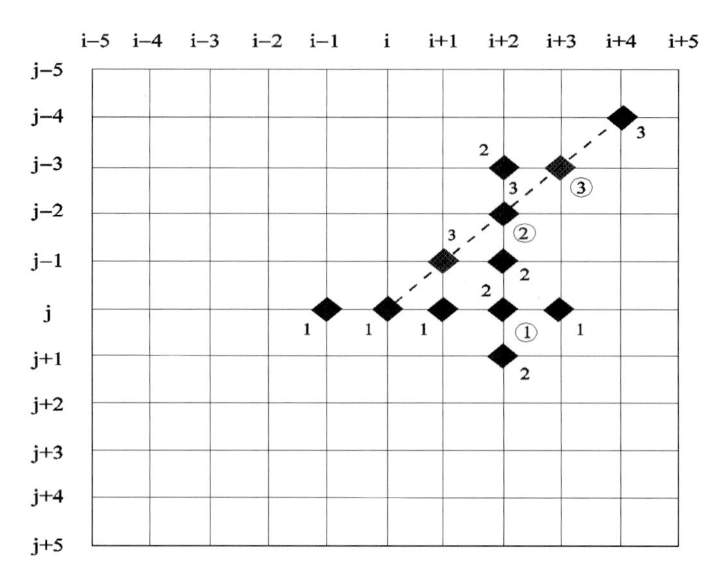

Figure 6.8 A one-at-a-time search and conjugate direction search [253]. © 1985 IEEE.

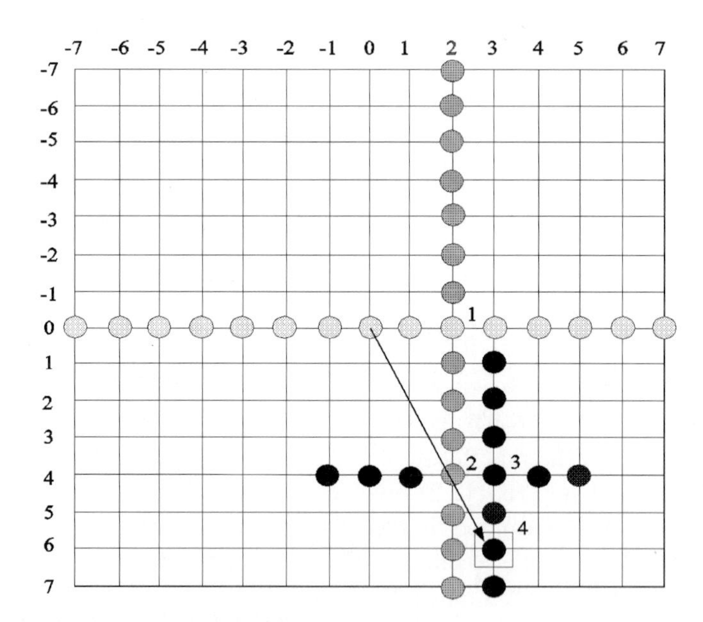

Figure 6.9 A one-dimensional full search motion-estimation algorithm [246]. © 1994 IEEE.

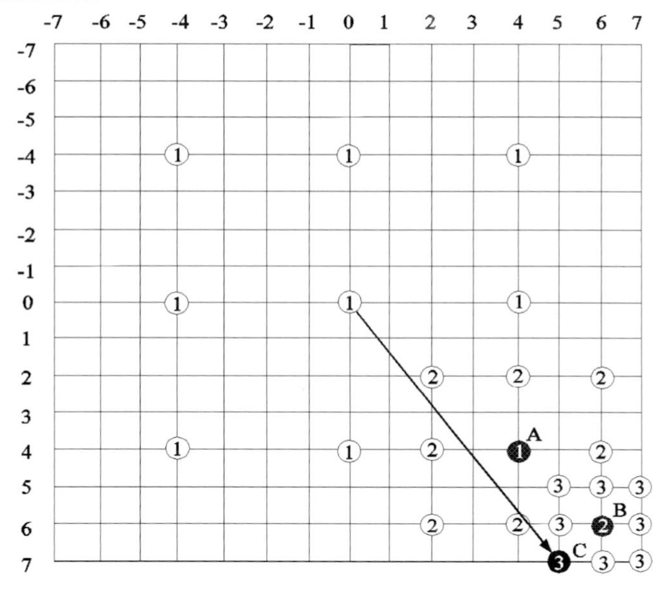

Figure 6.10 A three-step search motion-estimation algorithm. [225]. © 1981 IEEE.

Hierarchical search [220, 229, 235, 1035]

This process (Fig. 6.11) involves decimating (subsampling) the present image and the reference image successively by 2:1 both horizontally and vertically. The search process starts with the lowest resolution images, using a small block size. The MV estimated at this first stage is used as the starting point for ME at the next stage. Note that block size is now doubled along both directions. This process is repeated until the original resolution images are reached. The HDTV codec proposed by Zenith/AT & T uses a hierarchical search for ME (Fig. 13.11).

The field of simplified search is an ongoing process, as can be observed from the literature. The major objective in all these techniques is to simplify the implementation by sidestepping the brute search while obtaining ME closest to that of brute search. Another observation at this stage is in order. The ME (Fig. 6.1) is based on the previous reconstructed frame rather than the previous original frame. The reason for this is that the receiver (decoder) has only the reconstructed frames and not the original frames. The prediction here is in the temporal domain instead of in the transform domain (Fig. 6.12) that has been explored in [250]. The latter is much more complex because additional 2D-DCTs are needed both at the encoder and decoder. The function of the loop filter (Fig. 6.1) is outlined in Section 9.3.4. The standards, in general, do not specify how to implement any of the functions, e.g., DCT, quantization, ME, and so forth. There is, however, one

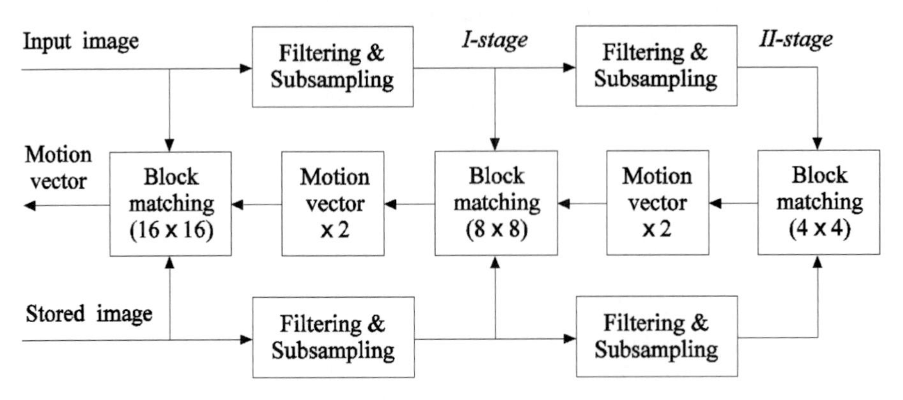

Figure 6.11 Hierarchical block-matching algorithm for motion estimation [1035].

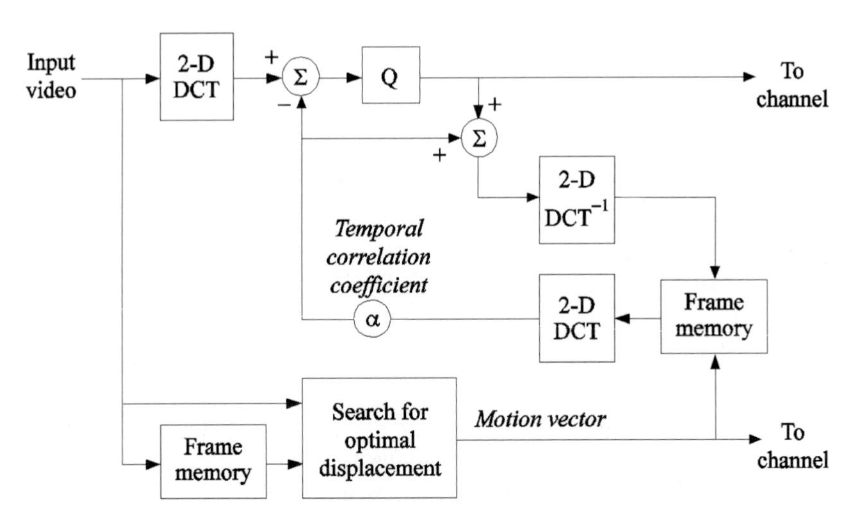

Figure 6.12 MC interframe DCT/DPCM coding with prediction in the transform domain [250]. © 1981 IEEE.

exception: the 2D-IDCT in the feedback loop of Fig. 6.1. The minimum accuracy specifications for the 2D-IDCT are specified in H.261 (it is also an IEEE standard, P1180–1190 [448]) to reduce the accumulation of the DCT/IDCT mismatch error [265] (Section 9.3.1 describes this in detail).

The BMA described so far are based on nonoverlapping blocks. The transform (DCT) is also applied to (8×8) non overlapping blocks based on the ME of a (16×16) luminance block. At low bit rates, this can result in blocking artifacts that are visually unpleasant. An overlapped MC technique that involves processing of overlapping windowed data blocks (ME is based on complex lapped transform) eliminates the blocking structure [248, 249].

Chapter 7
Vector Quantization and Subband Coding

Summary

The concepts of vector quantization (VQ) and subband coding (SBC) are introduced. The applications of both VQ and SBC in data compression (video/-audio) are illustrated. Their integration with other redundancy reduction techniques such as transform or DPCM with or without MC are outlined. As with other methods, adaptability, HVS, and so forth can be incorporated in any of these algorithms to further improve the coding efficiency, at the cost of increased complexity.

7.1 Introduction to Vector Quantization

Scalar quantization (SQ), i.e., quantization of individual samples, has been addressed in Chapter 3. Based on the rate distortion, Shannon has shown that vector quantization (VQ), i.e., quantization of groups of samples or vectors, is much more efficient. The limiting process, i.e., vector dimension reaching infinity, is, of course, impractical. From a practical viewpoint, vector dimensions of 4, 5, 9, 12, 16, 20, 25, and so forth have been used. The vector can be composed of video or audio samples, prediction errors, transform coefficients, transform coefficients of prediction errors, subband samples, and so forth. Even though VQ had been developed much earlier, an empirical approach outlined by Linde, Buzo, and Gray [283] popularly called the *LBG algorithm* has activated researchers into applying VQ in video/audio coding. The LBG algorithm based on a large training sequence representative of the test sequences culminates in the codebook design. The codebook consisting of codevectors or members of the alphabet is a significant component of the VQ process. The objective is to represent the input vector with a member of the codebook based on some matching or distortion criteria. Several enhancements, modifications, and additions to the original LBG algorithm [315, 316] have been suggested. Subsequently, efficient methods for codebook design independent of the LBG algorithm have also been developed [284, 285, 286, 290, 291, 292, 293, 299, 311]. Several codebook design techniques have been compared in [305]. Review papers [316, 317], special issues [307], and books [288, 323] dealing solely with VQ have been published. VQ by itself and with other redundancy reduction techniques (transform, DPCM, subband, etc.) has been extensively explored in video/audio coding, resulting in hardware and codecs ([1010] through [1020]). VQ is destined to play an increasing and important role in compression as processing speeds increase and memory costs go down.

7.2 Vector Quantization

The concept of VQ is quite simple, i.e., identifying or representing an input vector with a member of the codebook (assuming this has already been designed and stored) based on some valid criterion (best match, least distortion, etc). Basically, this is a pattern-matching procedure (Fig. 7.1). There are, of course, several ways of choosing the representative codevector. An index (not the codevector) is sent to the receiver where an exact replica of the codebook (lookup table) is stored. The decoder, using this index, retrieves the codevector from the lookup table and outputs it as the reconstructed vector. As such the decoder is inherently much simpler. VQ, hence, is ideal in a single-encoder/multiple-decoder scenario.

Matching or distortion criteria

At the encoder the objective is to find the best match among the code-vectors (also called representative vectors, patterns, alphabet, etc.) for the input vector. The term *best match* or *least distortion* has to be meaningful. Several distortion metrics have been formulated and applied in VQ. These are as follows (assume the vector dimension is K and codebook size is M).

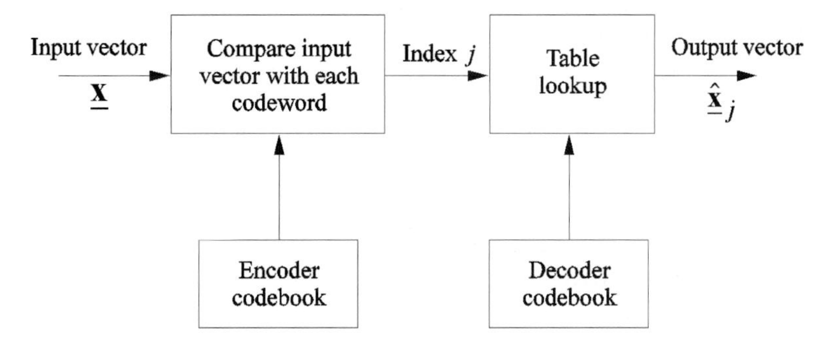

Figure 7.1 A block diagram of a simple vector quantizer.

Distortion metric

MSE (mean square error)

$$d_1(\underline{x}, \hat{\underline{x}}_i) = \frac{1}{K} \sum_{m=1}^{K} \left[x(m) - \hat{x}_i(m)\right]^2 \tag{7.1}$$

MAE (mean absolute error) [295]

$$d_2(\underline{x}, \hat{\underline{x}}_i) = \frac{1}{K} \sum_{m=1}^{K} \left|x(m) - \hat{x}_i(m)\right| \tag{7.2}$$

ℓ_n holder norm

$$d_3(\underline{x}, \hat{\underline{x}}_i) = \left[\sum_{m=1}^{K} |x(m) - \hat{x}_i(m)|^n\right]^{\frac{1}{n}} \tag{7.3}$$

$$d_4(\underline{x}, \hat{\underline{x}}_i) = \left[d_3(\underline{x}, \hat{\underline{x}}_i)\right]^n \tag{7.4}$$

Weighted distortion [308]

$$d_5(\underline{x}, \hat{\underline{x}}_i) = \sum_{m=1}^{K} w_m\left[x(m) - \hat{x}_i(m)\right]^2 \text{ or } \sum_{m=1}^{K} w_m\left|x(m) - \hat{x}_i(m)\right| \tag{7.5}$$

where

K = dimension of the input vector

$x(m)$ = mth component of input vector

$m = 1, 2, \ldots, K$

w_m = weight on the mth component of $\underline{x}$

$\hat{x}_i(m)$ = mth component of ith codevector

A weighted square error distortion has been applied to image coding in [308].

General quadratic distortion

$$d_6(\underline{x}, \hat{\underline{x}}_i) = (\underline{x} - \hat{\underline{x}}_i)\Big[W\Big](\underline{x} - \hat{\underline{x}}_i)^T \quad (7.6)$$

where

$\underline{x} = \Big[x(1), x(2), \ldots, x(K)\Big]$ is the input vector

$\Big[W\Big]$ = $(K \times K)$ positive definite symmetric weighting matrix

$\hat{\underline{x}}_i = [\hat{x}_i(1), \hat{x}_i(2), \ldots, \hat{x}_i(K)]$

$i = 1, 2, 3, \ldots, M$ is the ith codevector of the codebook

Maximum distortion

$$d_7(\underline{x}, \hat{\underline{x}}_i) = \max_m \Big|x(m) - \hat{x}_i(m)\Big| \quad (7.7)$$

= maximum absolute error among the corresponding K components of $\underline{x}$ and $\hat{\underline{x}}_i$

A subjective distortion measure has been developed in [309]. An example of a VQ encoder is shown in Fig. 7.2.

***K*-dimensional vector** Let

$\underline{x} = \Big[x(1), x(2), \ldots, x(K)\Big]$ be the input vector

$d(\underline{x}, \hat{\underline{x}}_i)$ = distortion measure

number of bits for the codeword index = $\log_2 M$

K = vector dimension

Then the *coding rate* (number of bits/sample) is $r = (\log_2 M)/K$.

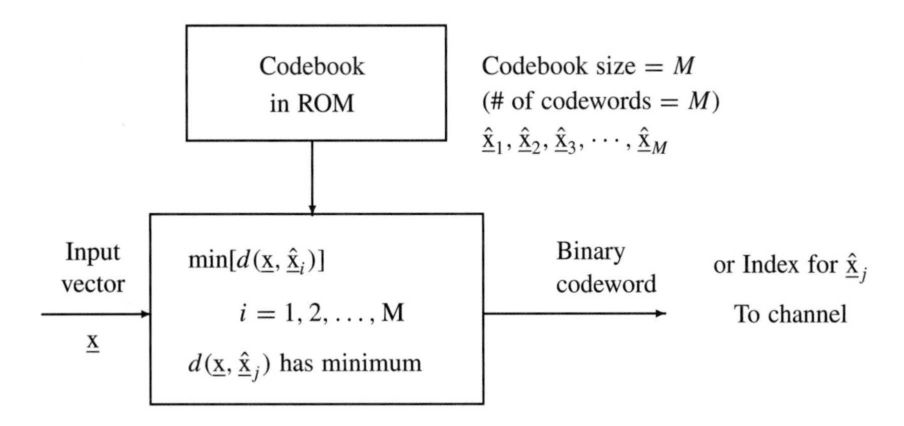

Figure 7.2 Illustration of a VQ encoder.

Example 7.1 *Let* $K = 16$, *(*$\underline{x} = [x(1), x(2), \ldots, x(16)]$*), and the number of codewords* $M = 256$ *(*$\log_2 M = 8$ *bits, the index for each codeword). Then the number of bits/sample* $= \frac{8}{16} = 0.5$ *bits/sample.*

To increase the compression ratio, either the vector dimension K has to be increased or the codebook size M has to be decreased. However, K needs to be small from adaptivity and correlation viewpoints, whereas for a better match M has to be as large as possible. Increasing M implies large storage and added search complexity. A compromise, hence, is needed among these conflicting requirements. By using a VWL scheme instead of FLC for coding the index for $\hat{\underline{x}}_j$ (Fig. 7.2), additional bit-rate reduction can be achieved. Codebook design, however, needs to be altered, reflecting the VLC of the VQ output [288].

7.3 Codebook Design

Once the codebook size is chosen, the objective is to design the codebook in such a way that it is optimal in some fashion. One criterion, for the optimality, is that the selected codebook result in the lowest possible distortion among all the possible codebooks of the same size. Using a probabilistic model for the input vectors, an analytical technique for codebook design can be developed. This by itself is quite complex. A more reasonable and empirical approach is that formulated by Linde, Buzo, and Gray [283]. This process starts with a large training sequence that is representative of the (test) input vectors and is illustrated with the following example. (The parameters $K = 2$, $M = 4$, and $N = 12$ are chosen to describe the design. In general, these parameters are much larger. In particular, N needs to be significantly larger to arrive at a desirable codebook.)

Example 7.2 *Codebook design starting with a uniform quantizer (© 1980 IEEE [283]).*

$$\begin{pmatrix} K = 2, \\ \textit{Vector dimension} \end{pmatrix}, \begin{pmatrix} M = 4, \\ \textit{Codebook size} \end{pmatrix}, \begin{pmatrix} N = 12, \textit{Number of vectors} \\ \textit{in the training sequence} \end{pmatrix}$$

$$\begin{array}{ll}
\underline{x}_1(-0.37, 0.99) & \underline{x}_7(-0.6, 0.18) \\
\underline{x}_2(-0.64, -0.12) & \underline{x}_8(0.14, 1.76) \\
\underline{x}_3(-0.83, 0.61) & \underline{x}_9(0.71, -0.35) \\
\underline{x}_4(-0.71, -1.22) & \underline{x}_{10}(0.3, 0.8) \\
\underline{x}_5(-0.29, -0.95) & \underline{x}_{11}(0.3, 1.07) \\
\underline{x}_6(1.1, 0.52) & \underline{x}_{12}(0.38, -0.33)
\end{array}$$

Start with a uniform quantizer codebook $\hat{A}_0$.

$$(\hat{\underline{x}}_1, \hat{\underline{x}}_2, \hat{\underline{x}}_3, \hat{\underline{x}}_4) = \begin{pmatrix} S_1 \\ 2, 2 \end{pmatrix}, \begin{pmatrix} S_2 \\ 2, -2 \end{pmatrix}, \begin{pmatrix} S_3 \\ -2, 2 \end{pmatrix}, \begin{pmatrix} S_4 \\ -2, -2 \end{pmatrix}$$

Find $P(\hat{A}_0) = [S_1, S_2, S_3, S_4]$, *where* $P(\hat{A}_0)$ *is the minimum distortion partition of the training sequence corresponding to* $\hat{A}_0$. *The distortion metric defined by Eq. (7.1) is used.*

$$\underline{x}_j \in S_i, \quad \textit{if } d(\underline{x}_j, \hat{\underline{x}}_i) \le d(\underline{x}_j, \hat{\underline{x}}_m)$$

$$\textit{for all } m \ne i$$

$$S_1 = [\underline{x}_6, \underline{x}_8, \underline{x}_{10}, \underline{x}_{11}]$$

$$S_2 = [\underline{x}_2, \underline{x}_9]$$

$$S_3 = [\underline{x}_1, \underline{x}_3, \underline{x}_7]$$

$$S_4 = [\underline{x}_4, \underline{x}_5, \underline{x}_{12}]$$

$$\textit{Choose } \epsilon = 0.001.$$

When $[(D_{m-1} - D_m)/D_m] \le \epsilon$, *halt the iterative process.*
D_m *is the average distortion at stage* m.
Set $D_{-1} \longrightarrow \infty$, $m = 0$.
$P(\hat{A}_0) = [S_1, S_2, S_3, S_4]$

A $m = 0$

The distortion $D_0 = \frac{1}{12} \sum_{j=1}^{12} \min_{\hat{\underline{x}} \in \hat{A}_0} d(\underline{x}_j, \hat{\underline{x}})$

$= 2.0172$

$[(D_{-1} - D_0)/D_0] > \epsilon$; *continue the process.*

Find the optimal reproduction alphabet.

$$\hat{\underline{x}}(S_1) = [(\underline{x}_6 + \underline{x}_8 + \underline{x}_{10} + \underline{x}_{11})/4] = (.46, 1.04)$$

$$\hat{\underline{x}}(S_2) = [(\underline{x}_2 + \underline{x}_9)/2] = (.67, -.23)$$

$$\hat{\underline{x}}(S_3) = [(\underline{x}_1 + \underline{x}_3 + \underline{x}_7)/3] = (-.6, .59)$$

$$\hat{\underline{x}}(S_4) = [(\underline{x}_4 + \underline{x}_5 + \underline{x}_{12})/3] = (-.46, -.83).$$

This is the new codebook $\hat{A}_1$.

If any $S_i = 0$, *just use the old codeword.*

$m = 1$

Find $P(\hat{A}_1)$, *this leads to* $P(\hat{A}_1) = P(\hat{A}_0)$, *go to* **A**

No change in partition.

Compute D_1, *the total distortion.*

$$D_1 = \frac{1}{12} \sum_{j=1}^{12} \min_{\hat{\underline{x}} \in \hat{A}_1} d(\underline{x}_j, \hat{\underline{x}}) = 0.1 \tag{7.8}$$

$$[(D_0 - D_1)/D_1] = \left(\frac{2.02 - 0.1}{0.1}\right) = 19 > \epsilon \tag{7.9}$$

Set $m = 2$. *Repeat the process from* **A**

Find $(\hat{A}_2)$, $\hat{A}_2 = \hat{\underline{x}}\, P(\hat{A}_1) = \hat{A}_1$.

$D_2 = D_1$, $[(D_1 - D_2)/D_2] = 0 < \epsilon$

Halt. Final quantizer is described by $[\hat{A}_1,\ P(\hat{A}_1)]$.

Codebook: (0.46, 1.04), (0.67, −0.23), (−0.6, 0.59), (−0.46, −0.83)

Another codebook design is based on the *splitting algorithm*.

Choose (select) a large training sequence (data base). Let $\underline{x}_i$ be a member of the training sequence.

$$\underline{x}_i = [x_i(1), x_i(2), \ldots, x_i(K)], \quad i = 1, 2, \ldots, N$$

Suppose $N = 10,000$ vectors.

$$\underline{x}_i = K\text{-dimensional vector.}$$

Step 1. Take the average (centroid) of the training sequence, i.e.,

$\frac{1}{N}\sum_{i=1}^{N}\underline{x}_i = \underline{y}_1$.Perturb (split) $\underline{y}_1$ into two vectors $\underline{y}_1 + \underline{\varepsilon}_1$, $\underline{y}_1 - \underline{\varepsilon}_1$,

where $\underline{\varepsilon}_1$ is the perturbation vector.

Based on the distortion criterion (MSE, MAE, WMSE, WMAE), divide the training sequence into two groups each belonging to $\underline{y}_1 + \underline{\varepsilon}_1$, or $\underline{y}_1 - \underline{\varepsilon}_1$:

$[\underline{x}_{iA}]$ belonging to $\underline{y}_1 + \underline{\varepsilon}_1$, $i = 1, 2, \cdots, N_1$, and
$[\underline{x}_{iB}]$ belonging to $\underline{y}_1 - \underline{\varepsilon}_1$, $i = 1, 2, \cdots, N_2$, where $N_1 + N_2 = N$

Step 2. Take the centroid (mean) of $[\underline{x}_{iA}]$, i.e.,

$$\frac{1}{N_1}\sum_{i=1}^{N_1}\underline{x}_{iA} = \underline{y}_2$$

Perturb (split) $\underline{y}_2$ into two vectors $\underline{y}_2 + \underline{\varepsilon}_2$, $\underline{y}_2 - \underline{\varepsilon}_2$.

Based on the distortion criterion, divide $[\underline{x}_{iA}]$, $i = 1, 2, \ldots, N_1$, into two groups each belonging to $\underline{y}_2 + \underline{\varepsilon}_2$, or $\underline{y}_2 - \underline{\varepsilon}_2$:

$[\underline{x}_{iC}]$ belonging to $\underline{y}_2 + \underline{\varepsilon}_2$, $i = 1, 2, \ldots, N_3$, and
$[\underline{x}_{iD}]$ belonging to $\underline{y}_2 - \underline{\varepsilon}_2$, $i = 1, 2, \ldots, N_4$, where $N_3 + N_4 = N_1$

Repeat this process for $[\underline{x}_{iB}]$, $i = 1, 2, \ldots, N_2$ to create two more groups:

$[\underline{x}_{iE}]$ belonging to $\underline{y}_3 + \underline{\varepsilon}_3$, $i = 1, 2, \ldots, N_5$, and
$[\underline{x}_{iF}]$ belonging to $\underline{y}_3 - \underline{\varepsilon}_3$, $i = 1, 2, \ldots, N_6$, such that $N_5 + N_6 = N_2$,

where $\frac{1}{N_2}\sum_{i=1}^{N_2}\underline{x}_{iB} = \underline{y}_3$

Also, $N_3 + N_4 + N_5 + N_6 = N_1 + N_2 = N$.

Now there are four codevectors (members of the alphabet), i.e., $\underline{y}_2 \pm \underline{\varepsilon}_2$, $\underline{y}_3 \pm \underline{\varepsilon}_3$.

Step 3. Repeat this process until the desired number of codevectors (M) in the codebook is obtained. It is easy to see that the codebook size is an integer power of two.

In the splitting process it may happen that very few members of the training sequence belong to a codevector (or some codevectors), i.e., there may be an empty cell. Delete this codevector from the codebook. Find the codevector that is most heavily populated, i.e., that has a large number of the vectors from the training sequence. Now split this codevector into two so that each of these two code vectors is equally (nearly) populated.

This splitting algorithm is useful in generating the codebook for binary-search VQ (BSVQ), which is a member of the tree-search VQ (TSVQ) (Fig. 7.3). Similar to the binary-search VQ, a quad-search VQ codebook can be designed. As in BSVQ, find the centroid of the training sequence, i.e.,

$$\frac{1}{N}\sum_{i=1}^{N}\underline{x}_i = \underline{y}_1,\ \text{perturb (split)}\ \underline{y}_1\ \text{into four vectors}\ \underline{y}_1 \pm \underline{\varepsilon}_1,\ \underline{y}_1 \pm \underline{\varepsilon}_2$$

Based on the distortion criterion, divide the training sequence into four groups each belonging to one of the four vectors $\underline{y}_1 \pm \underline{\varepsilon}_1$, $\underline{y}_1 \pm \underline{\varepsilon}_2$. Repeat this process as in BSVQ.

7.4 Types of VQ

Tree-search VQ A modification of the full-search VQ is the tree-search VQ (TSVQ) (Fig. 7.3), in which the codebook search starts from the roots of a tree ending at the top branches. The arrow at each stage in Fig. 7.3 indicates the best match at that stage. TSVQ requires fewer searches than the full-search VQ. The former, however, requires more storage. It is also locally optimal rather than globally optimal. A variation of the TSVQ is the unbalanced TSVQ [326]. In this scheme, at each stage, only the node that corresponds to the largest distortion reduction is split (Fig. 7.4).

An observant reader may notice that the limiting process of TSVQ (binary, quad, octal, ...) is a full-search VQ.

There are many variations to the basic VQ scheme [288]. These can be illustrated as follows.

Multistage VQ A two-stage cascaded VQ is shown in Fig. 7.5. Multistage VQ reduces memory size and encoding complexity. Codebooks at each stage have to be carefully designed, and their sizes must be carefully chosen. The error vector from the first stage is the input to the second stage. The selected codevectors $\hat{\underline{x}}_j$ and $\hat{\underline{e}}_j$ are to be added at the decoder to reconstruct the input vector $\underline{x}$.

Separating mean VQ The quantized mean of the input vector $\underline{x}$ is subtracted before going through the VQ (Fig. 7.6). The shape codebook has zero mean codevectors. The mean is scalar quantized (N output levels). The quantized mean is added to the selected codeword at the decoder.

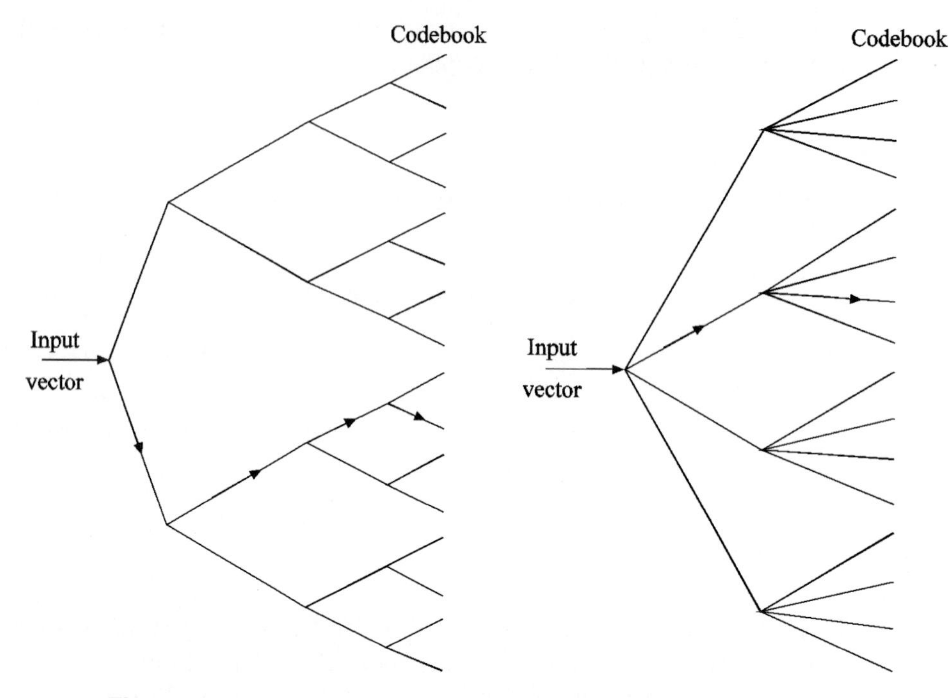

Figure 7.3 Tree-search VQ (TSVQ): Binary search (left) and quad search (right).

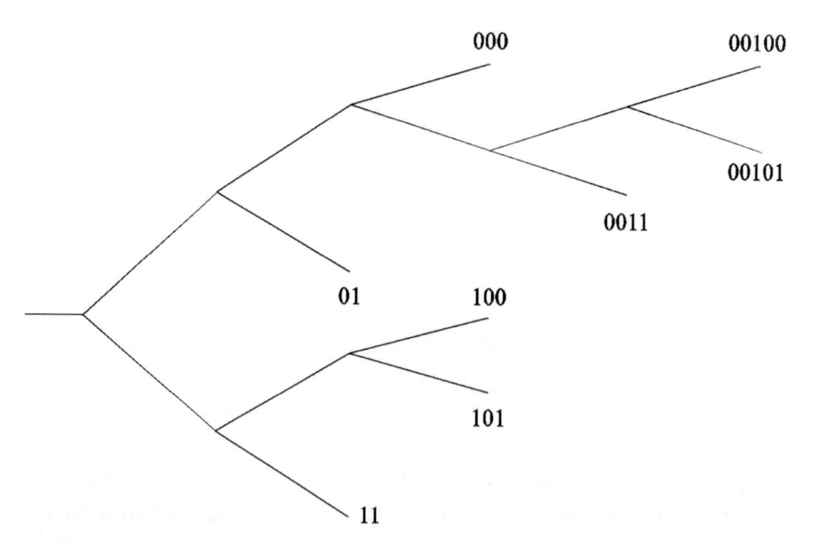

Figure 7.4 An example of unbalanced tree-structured VQ [326]. © 1991 IEEE.

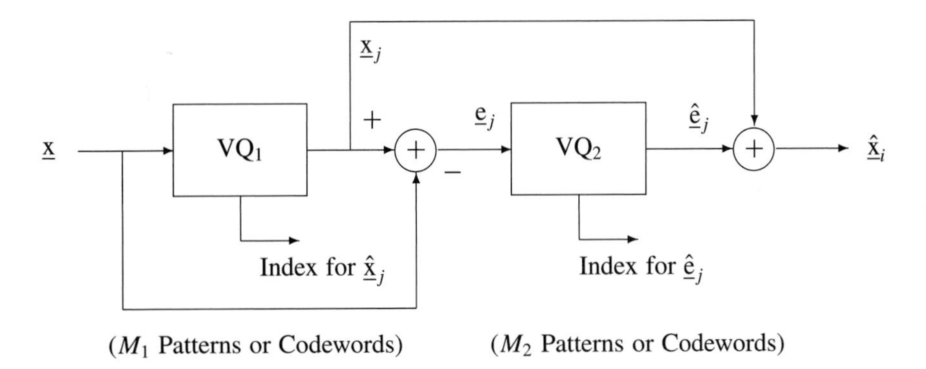

This is equivalent to

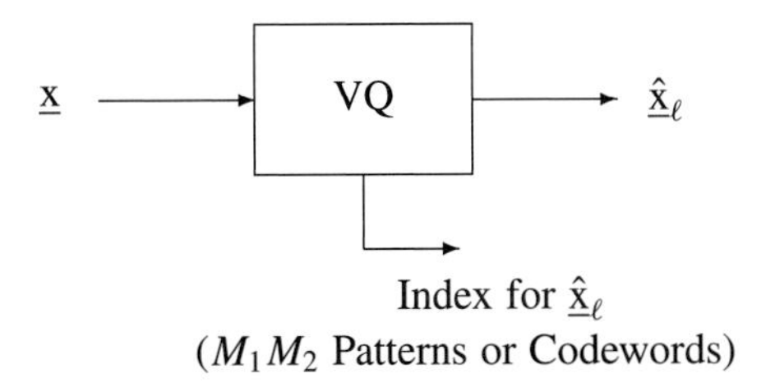

Figure 7.5 Multistage VQ.

Gain-shape VQ (A) From the separating-mean VQ, another VQ technique is based on extracting the mean value (gain) and normalizing by the signal variance (shape). The concept of gain-shape VQ is that the same pattern of variation in a vector may occur again with different gain values. Mean and variance are defined as follows:

$$\text{Mean:} \qquad \mu = \frac{1}{K}\sum_{i=1}^{K} x(i) \tag{7.10}$$

$$\text{Variance:} \qquad \sigma^2 = \frac{1}{K}\sum_{i=1}^{K} [x(i) - \mu]^2 \tag{7.11}$$

The mean separated-gain normalized input vector $(\underline{x} - \mu)/\sigma^2$ goes through VQ.

$$(\underline{x} - \mu) = \left[x(1) - \mu,\ x(2) - \mu,\ \ldots,\ x(K) - \mu\right]$$

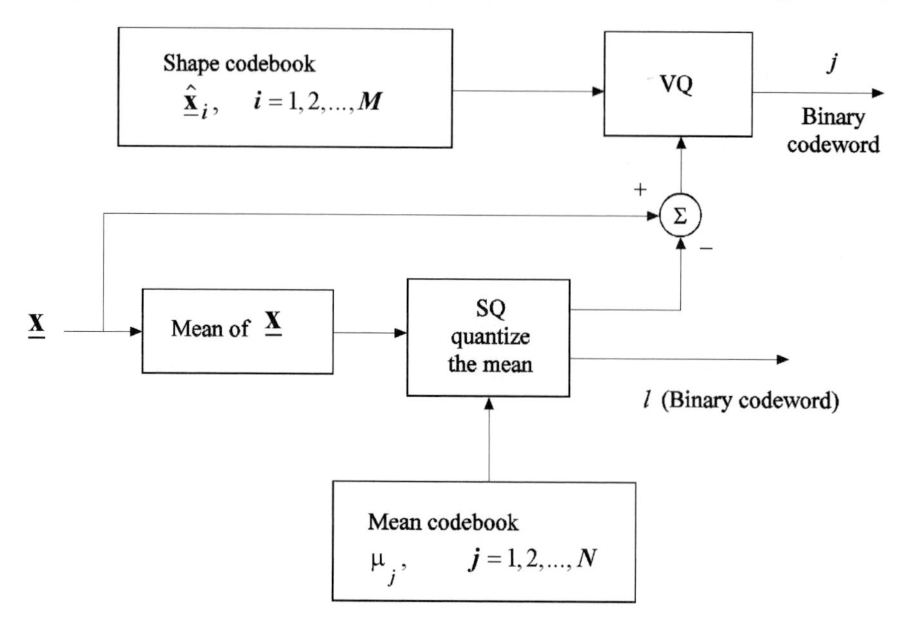

Figure 7.6 Separating-mean VQ.

Apply SQ to both μ and σ^2 and send their codes indicating their quantized values through the channel. At the decoder the selected codeword is multiplied by the quantized value of σ^2 followed by the addition of the quantized mean.

Gain-shape VQ (B) Another gain-shape VQ (Fig. 7.7) involves minimizing the MSE

$$d_1(\underline{x},\ g_q,\ \underline{s}_i)$$

where g_q is the quantized value of the gain g defined as

$$g = \left[\sum_{m=1}^{K} x^2(m)\right]^{1/2} \tag{7.12}$$

and $\underline{s}_i$ is the shape codevector with unit gain, i.e.,

$$\left[\sum_{m=1}^{K} s_i^2(m)\right]^{1/2} = 1 \tag{7.13}$$

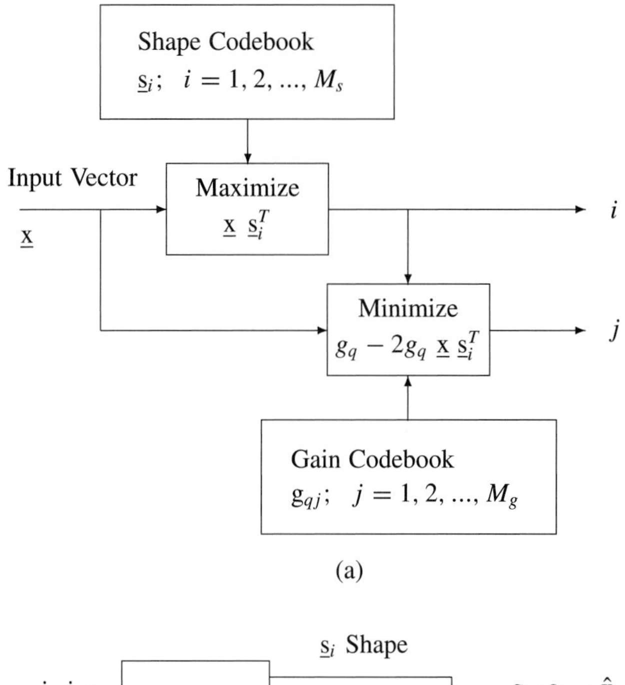

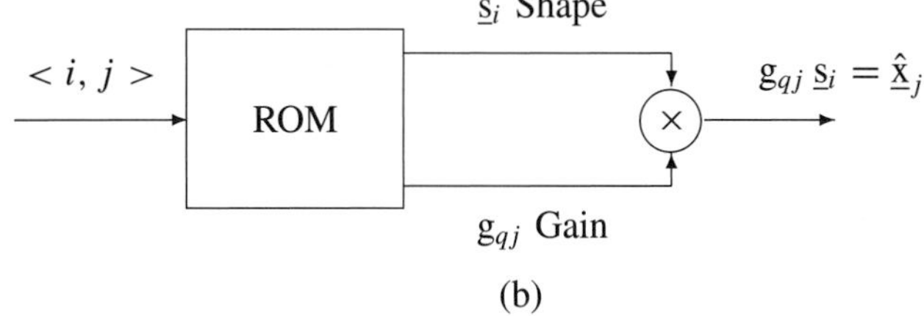

Figure 7.7 Gain/shape VQ [317]. (a) Encoder; (b) decoder. © 1984 IEEE.

Hierarchical VQ (Fig. 7.8) Compute the variance of an $(N \times N)$ block. If it is above a given threshold, divide it into four $(N/2 \times N/2)$ blocks. Compute the variances of the four $(N/2 \times N/2)$ blocks. If all or some of the variances of the $(N/2 \times N/2)$ blocks are above another threshold, then divide each $(N/2 \times N/2)$ block (that is, those above the threshold) into four $(N/4 \times N/4)$ blocks. This is a variable block size and, hence, a variable vector dimension VQ reflecting the image activity. Separate codebooks for each block size are to be designed and, of course, stored at the encoder and decoder. Overhead bits indicating the actual coding mode need to be transmitted. This adaptivity improves the coding efficiency at the cost of increased complexity. A corresponding quad-tree representation is shown in Fig. 7.9.

Interpolative VQ Interpolative VQ (Fig. 7.10) is a combination of PCM and VQ. A pel in each block of small size, say (2×2) or (3×3), is transmitted as it is, and the remaining pels in the block are interpolated using the neighboring pels,

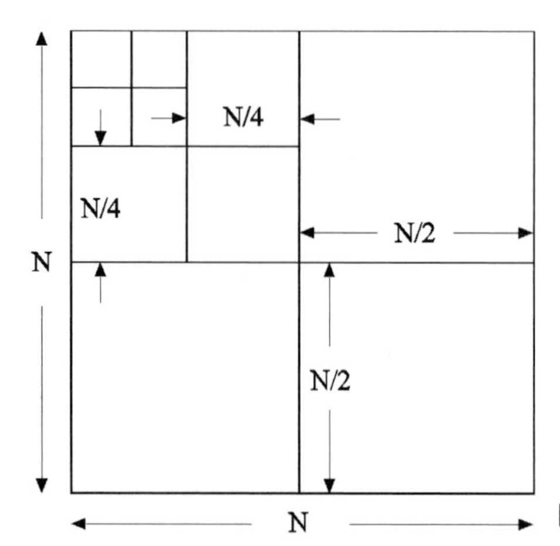

Figure 7.8 Hierarchical VQ.

A, B, C, and *D*. The interpolation errors in each block go through VQ. The reverse process at the decoder reconstructs the block.

Entropy-constrained VQ (Fig. 7.11) Apply VWL (variable word length) codes to the M codewords reflecting their probability distribution. The codebook is designed not only using the standard distortion criteria but also choosing the VWL codes so that the average bit rate is minimized.

Lapped VQ As in transform coding, at low bit rates block artifacts may arise when VQ is applied to image blocks. A technique called *lapped VQ* of images (Fig. 7.12) has been developed by Wu and Gersho to eliminate this [306]. The codevector dimension at the decoder is higher than that at the encoder. The area covered by each output codevector extends beyond the input block of pels into its adjacent neighborhood. This overlapping has resulted in improved perceptual image quality compared with ordinary VQ.

So far we have discussed many variations to the basic VQ approach. The list, of course, is far from complete. In the future many new variations will be developed. We will now focus on the hybrid schemes involving VQ.

7.5 Hybrid Systems

- Predictive VQ [296, 303, 318, 319, 320, 321, 322, 327]. As DPCM (Chapter 4) has been successfully applied to audio/image/video coding, it is only natural that the concept of prediction be extended to vectors, i.e., predictive VQ (PVQ) (Fig. 7.13) [322]. This scheme is strikingly similar to Fig. 4.1, i.e., replacing the scalars by vectors. Other ingredients such as entropy-constrained VQ (Fig. 7.11) can be incorporated in the PVQ.

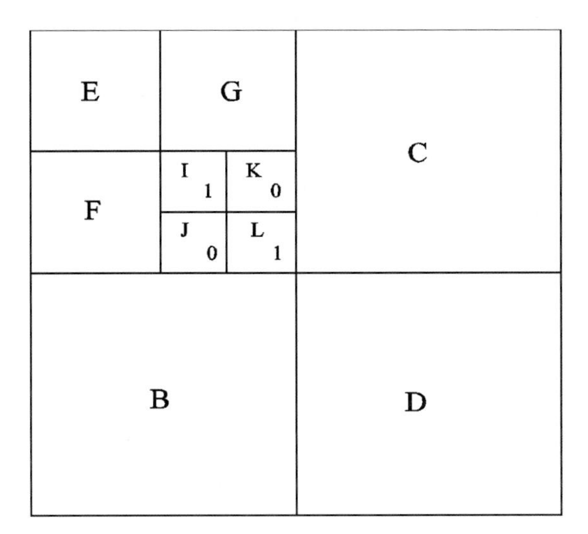

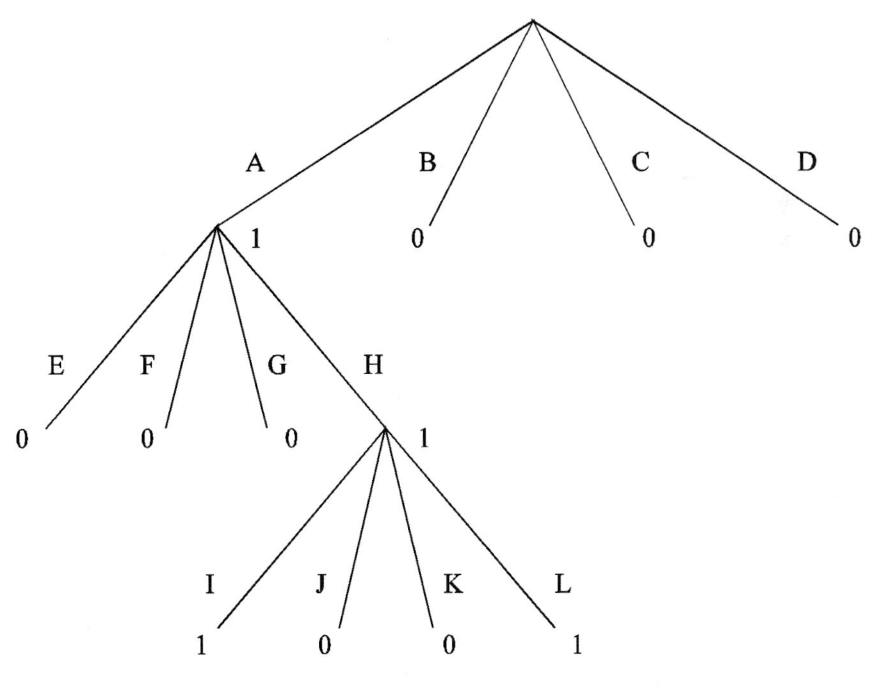

Figure 7.9 A typical quad-tree representation of an image.

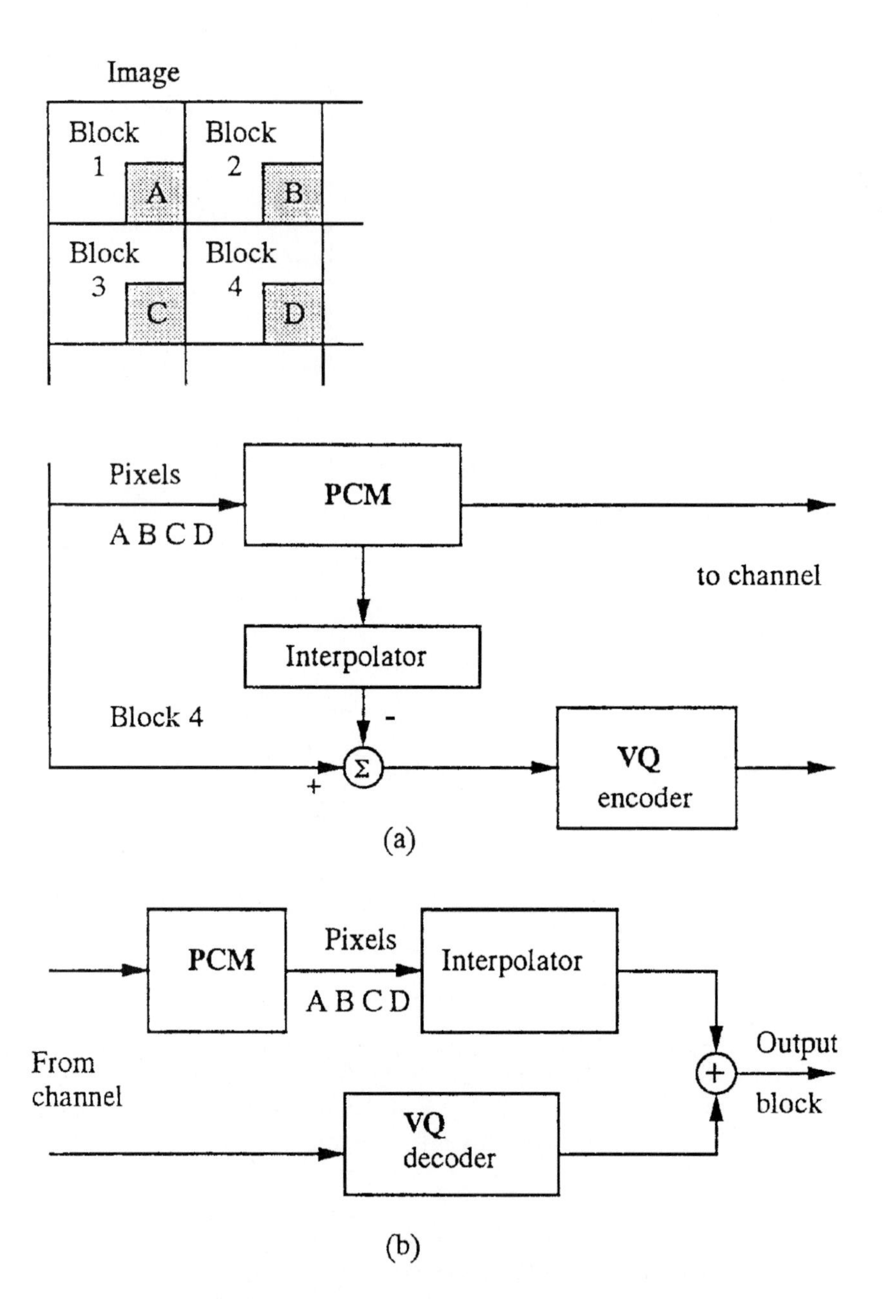

Figure 7.10 An interpolative vector quantization system. (a) Encoder; (b) decoder.

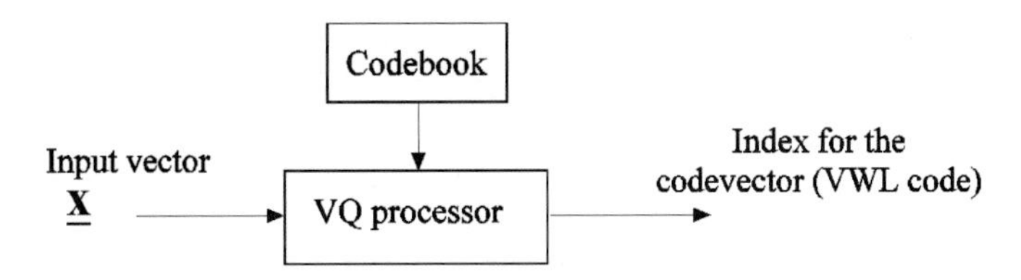

Figure 7.11 Entropy-constrained VQ.

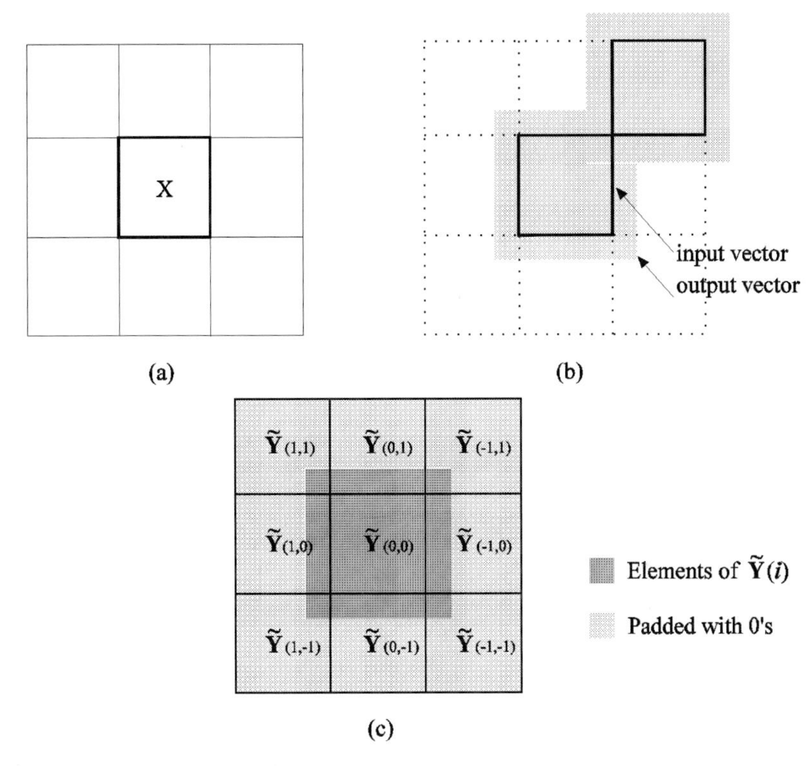

Figure 7.12 Lapped VQ. [306] (a) An example of a nine-block neighborhood of input vector, (b) block overlapping at the decoder, and (c) construction of partial codevectors for the nine-block neighborhood of (a). The index i is omitted in the figure for simplicity. [306]

- Transform VQ [324]. VQ can be applied to transform coefficients (Fig. 5.15). One approach is to combine groups of coefficients having similar variances as vectors that then undergo VQ (Fig. 7.14). As an example, coefficients in the three spectral bands are grouped as vectors and the corresponding VQ is then applied (VQ1, VQ2 and VQ3). In general, the DC coefficient is separately coded (SQ).

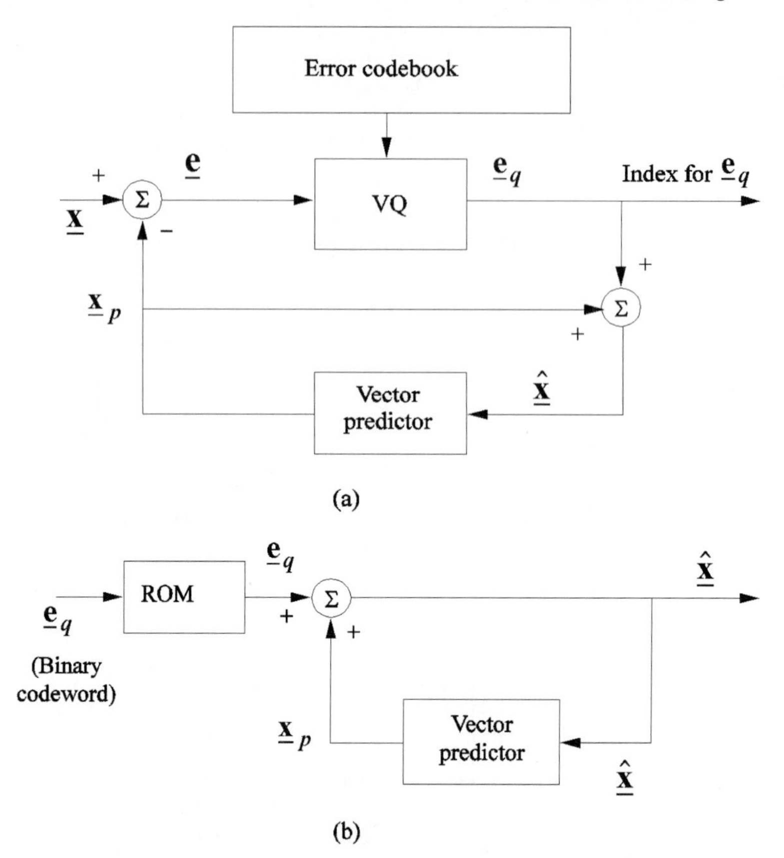

Figure 7.13 Predictive VQ [322]. (a) Encoder; (b) decoder. © 1985 IEEE.

By classifying the transform block into a finite number of classes, say 4, and appropriately selecting the vectors in each class (both the number of vectors and also the grouping), adaptivity can be added. This requires a large number of codebooks and the overhead identifying the category (class) of the block.

A simpler version of 2D-transform/VQ is separable 1D-transform/VQ, i.e., 1D-transform along rows (columns) of a block followed by VQ along columns (rows) of the semitransformed array.

VQ can also be applied to the temporal domain (Fig. 7.15). For example, the 2D-transform of each frame (on a block-by-block basis) is followed by VQ of the corresponding transform coefficients in successive frames. Memory constraints limit the number of frames.

In the MC hybrid (transform-DPCM) scheme (Fig. 9.8), VQ can be extended to transform coefficients of the prediction errors. The number of variations to the

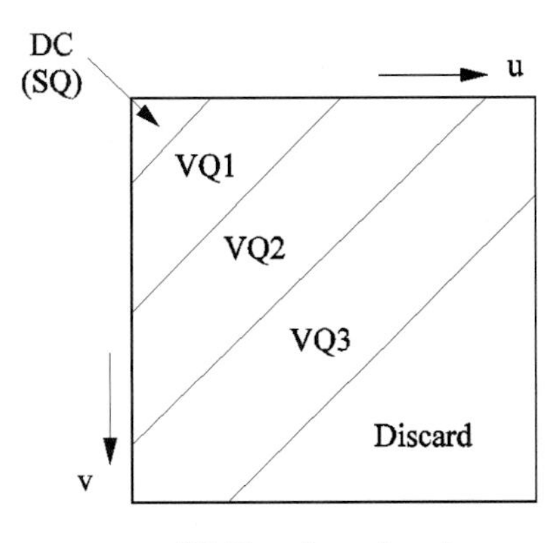

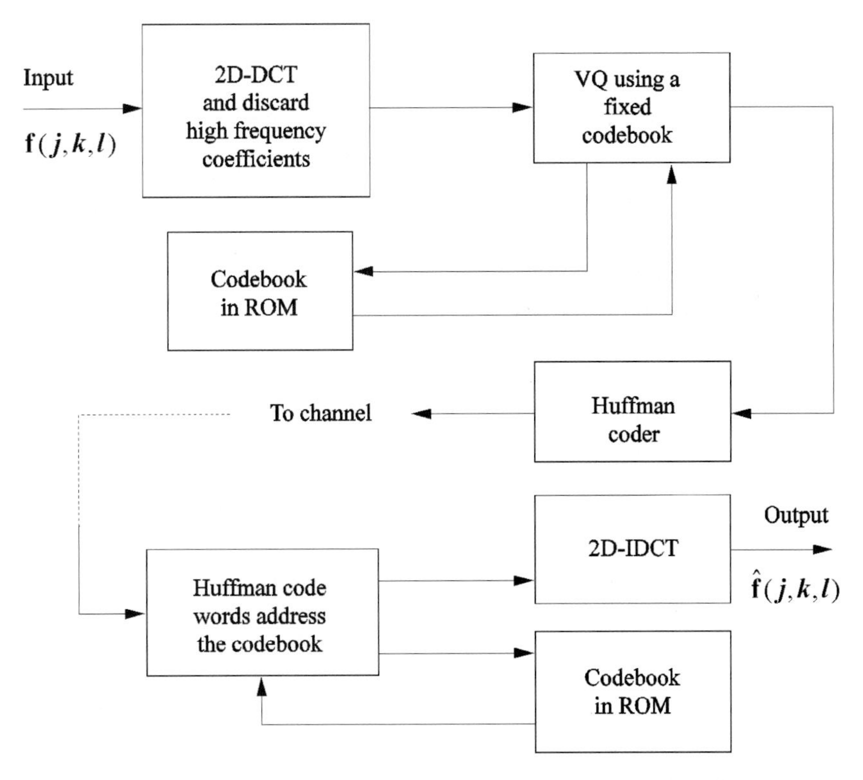

Figure 7.14 Two-dimensional transform (DCT)/VQ coding.

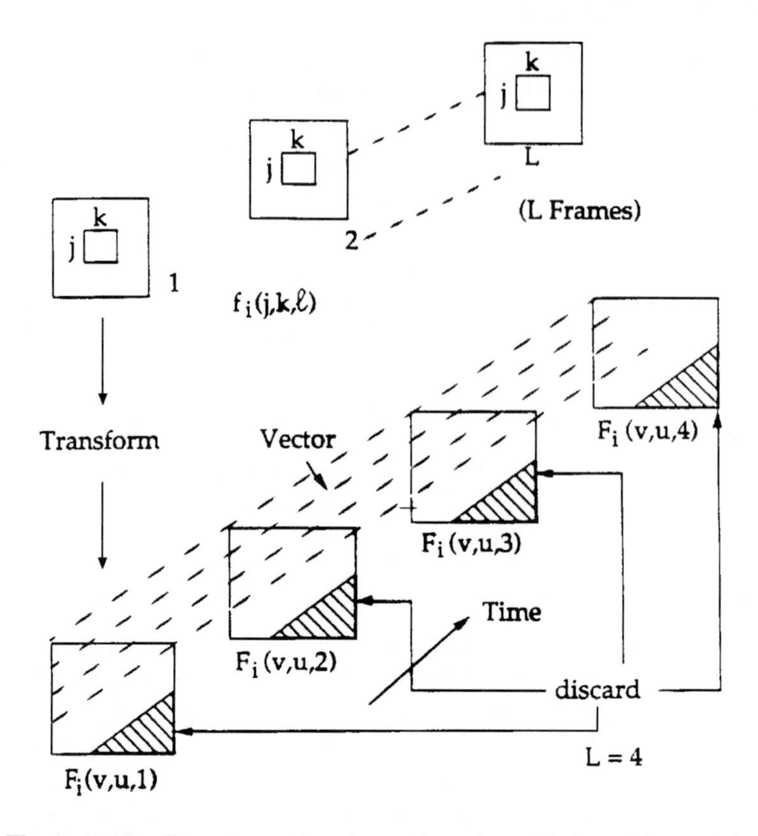

Figure 7.15 Two-dimensional transform (spatial domain) and VQ (temporal domain).

basic VQ, adaptivity, and hybrid VQ schemes is limitless and quite a few of them have been thoroughly researched, simulated and evaluated in the compression arena.

7.6 Introduction: Subband Coding

Subband coding (SBC) is another form of frequency decomposition; transform coding has been described earlier (Chapter 5). In SBC, a signal (either 1D or 2D) is decomposed into a number of equal- or unequal- frequency bands using filter banks that have been developed recently [345, 347, 350, 353, 354]. In fact, by using perfect reconstruction filter banks [345, 347, 354], the original signal going through the frequency decomposition-subsampling-interpolation-synthesis process can be fully recovered. The philosophy behind SBC is that coding techniques compatible with the frequency bands can be applied. Signal components in high-frequency bands can be either dropped out or coarsely quantized. This approach is used in MPEG-1 and 2 audio coding (Sections 10.3 and 11.3 respectively).

The audio coding is based on MUSICAM (Fig. 10.27), which involves decomposing the high-quality/large-bandwidth audio into 32 equal subbands. The MPEG audio encoder/decoder chips implement the subband analysis/systhesis operations as part of the overall compression algorithm. Similarly, the AC-2 and AC-3 audio codecs developed by Dolby Labs use filter banks (Fig. 13.30). The SBC can be extended to the 2D frequency domain [343, 344, 346, 348, 351, 352].

The HDTV codec (140 Mbps-SONET-ATM) proposed by Bellcore [359] utilizes 2D subband decomposition (unequal subbands). After evaluating other techniques, such as DCT and DPCM, Bellcore selected the subband approach, considering the performance-complexity criterion. Subband coding is also ideally suited for progressive image transmission (PIT) as bits related to lower bands can be transmitted first, followed by those related to the upper bands. Image buildup at the receiver follows this order. Another application is in asynchronous transfer mode (ATM) networks, assembling the bits related to different bands into different packets (also different priority levels). Our emphasis will be not on the theory of multirate filter banks (the book by Vaidyanathan [354] is an excellent resource) but on their applications in coding. One technique of subband decomposition (analysis filter bank) is shown in Fig. 10.32. Corresponding to this is the signal reconstruction (synthesis filter bank) described in [487]. The analysis filter bank by itself does not result in any compression. On the contrary, the input samples increase considerably. For example, consider the high-quality audio sampled at $f_s = 32$ KHz (Fig. 7.16).

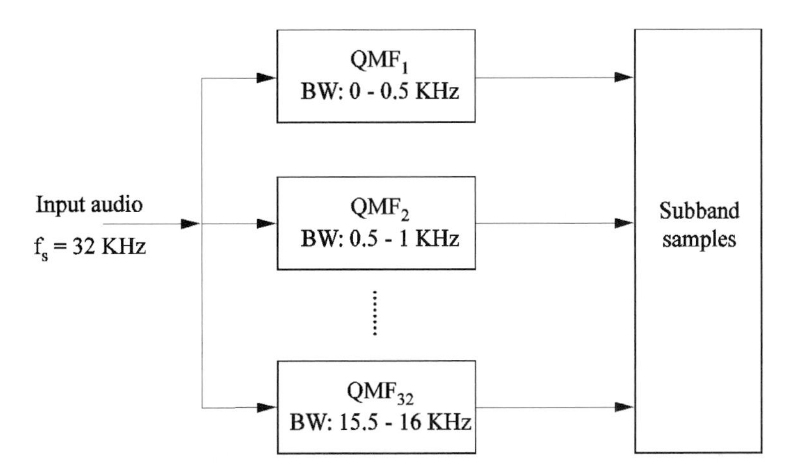

Figure 7.16 An analysis filter bank for subband decomposition (QMF: quadrature mirror filter).

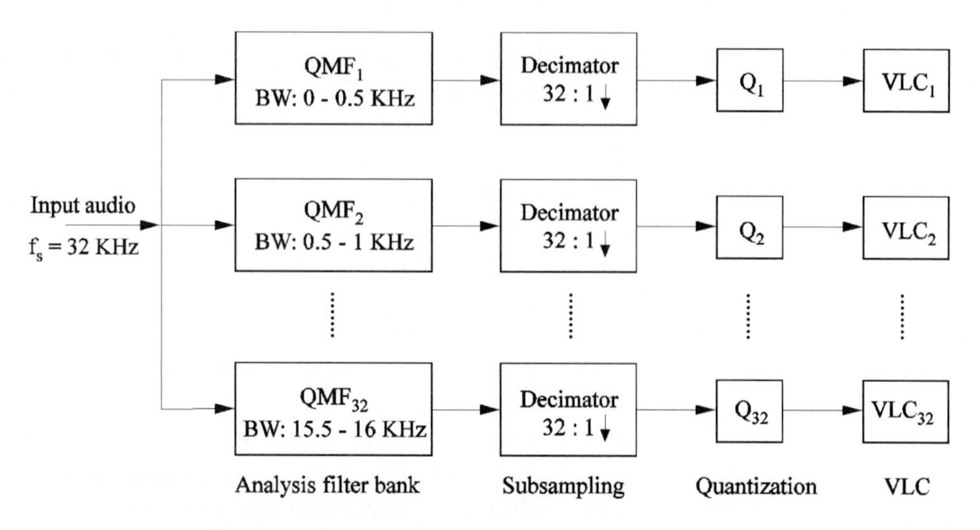

Figure 7.17 An example of a subband encoder.

As per MPEG-1 audio, the output of the analysis filter bank is 32 times (samples/sec) that of the input. Because the bandwidth of each filter is only 0.5 KHz, its output can be decimated by a factor of 32. The decimated samples from all the subbands are now equal to the input samples (Fig. 7.17). This is called critical sampling. By applying quantization (can be adaptive) and VLC (optional) to these decimated samples, compression is achieved. In audio coding, psychoacoustic masking effects (Section 10.3.3) are taken advantage of in reducing the bit rate

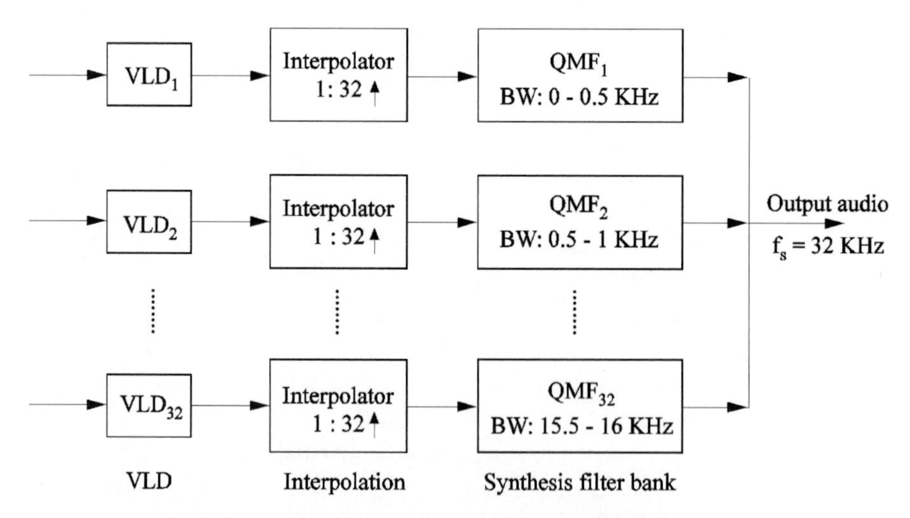

Figure 7.18 A subband decoder corresponding to the encoder shown in Fig. 7.17.

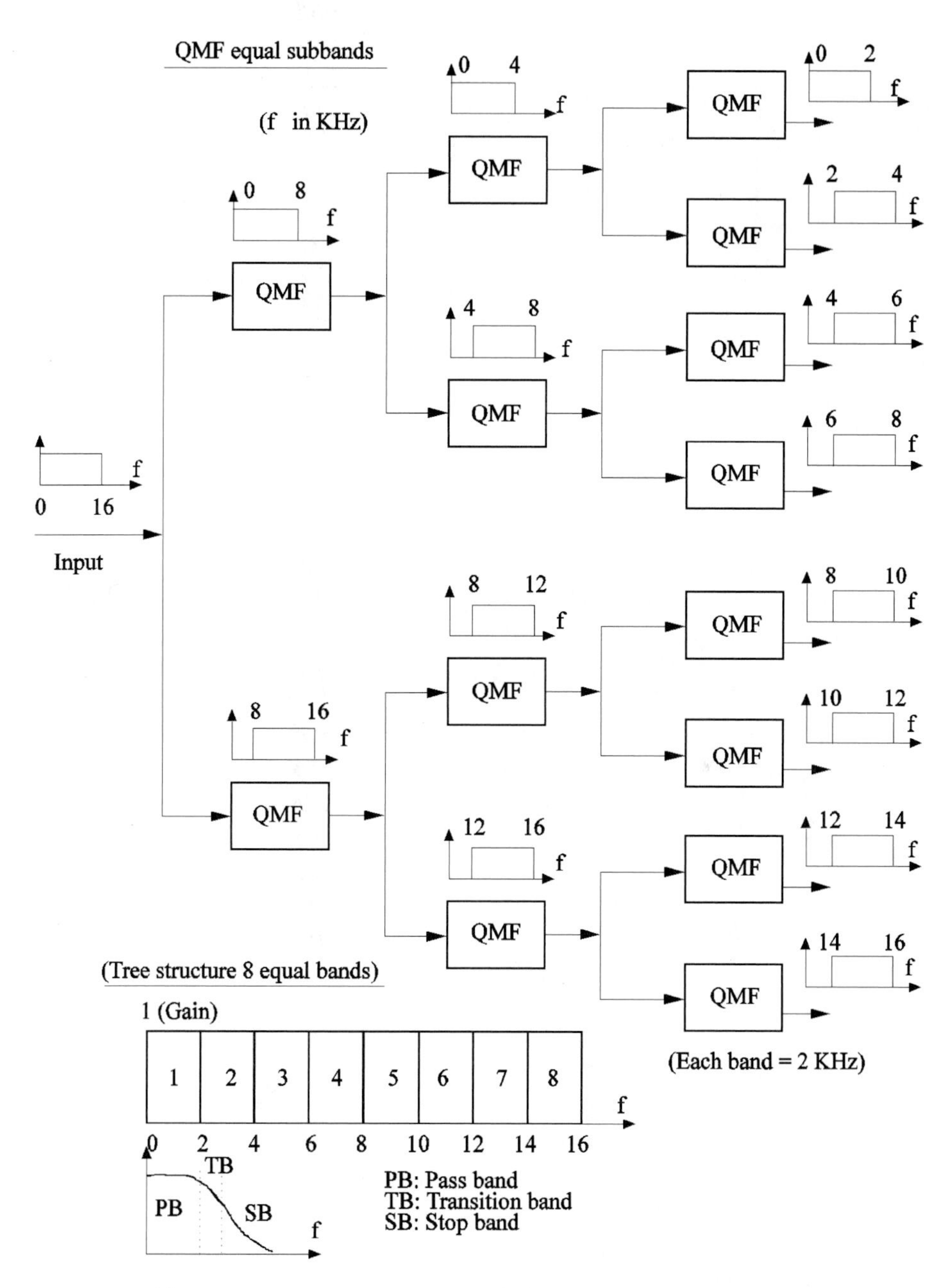

Figure 7.19 A tree structure for an analysis filter bank.

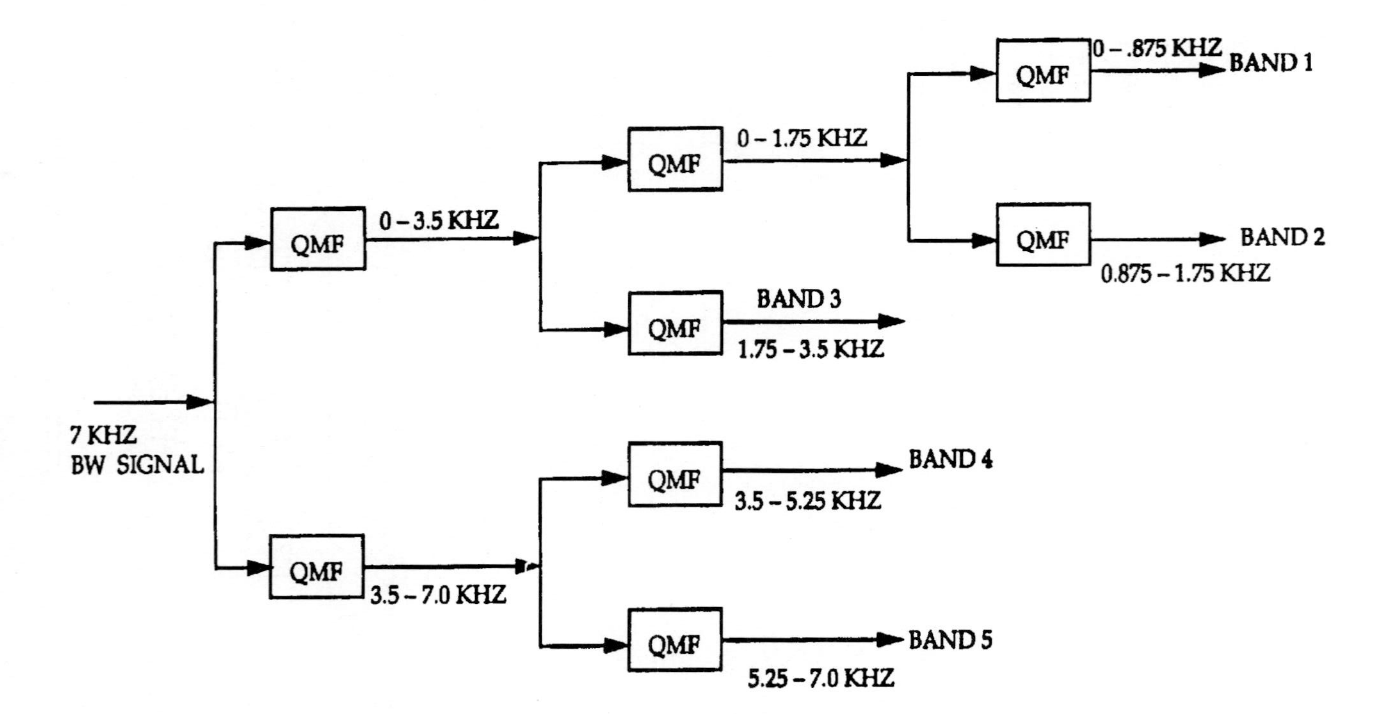

Figure 7.20 Original seven KHz wide-band audio split into five unequal bands [349]. © 1986 IEEE.

while maintaining the original audio quality. Similarly, psychovisual effects can be applied to subband video coding.

The decoder corresponding to the encoder (Fig. 7.17) is shown in Fig. 7.18. The decoder operations are inverse to those at the encoder and are in the reverse order. The encoder inherently is more complex as it has to implement modeling (psychoacoustic or psychovisual), adaptivity (bit allocation, dropping some subbands, etc.), and other features. The decoder, on the other hand, using the overhead bits, only has to track the particular modes and implement the corresponding operations.

Analysis filter bank A simple and straightforward tree structure can be used to split the input signal into equal frequency bands. An example of band decomposition (each band has a 2 KHz bandwidth) is shown in Fig. 7.19. Using separable filters, this decomposition can be easily extended to the 2D frequency domain. An example of unequal subbands is shown in Fig. 7.20. This has been applied in wideband audio coding, with the lowest band predictively coded [349].

7.7 Subband Image Coding

A popular approach to subband image coding [346] is to map the image into four equal subbands in the 2D frequency domain (Fig. 7.21). In general, a transform

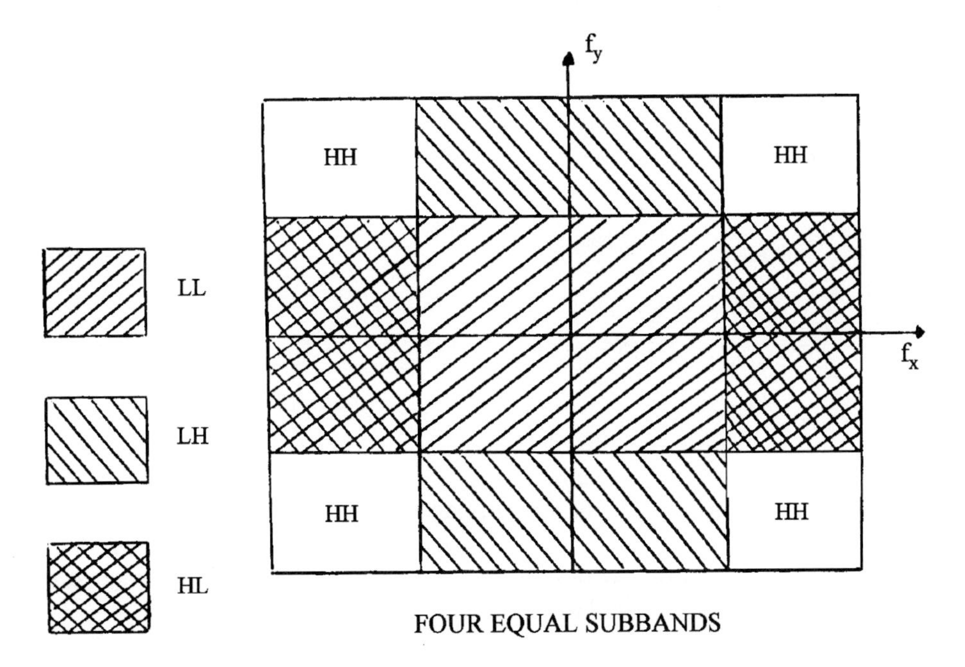

Figure 7.21 Mapping of an image into four equal subbands in the two dimensional frequency domain (L: low frequency, H: high frequency).

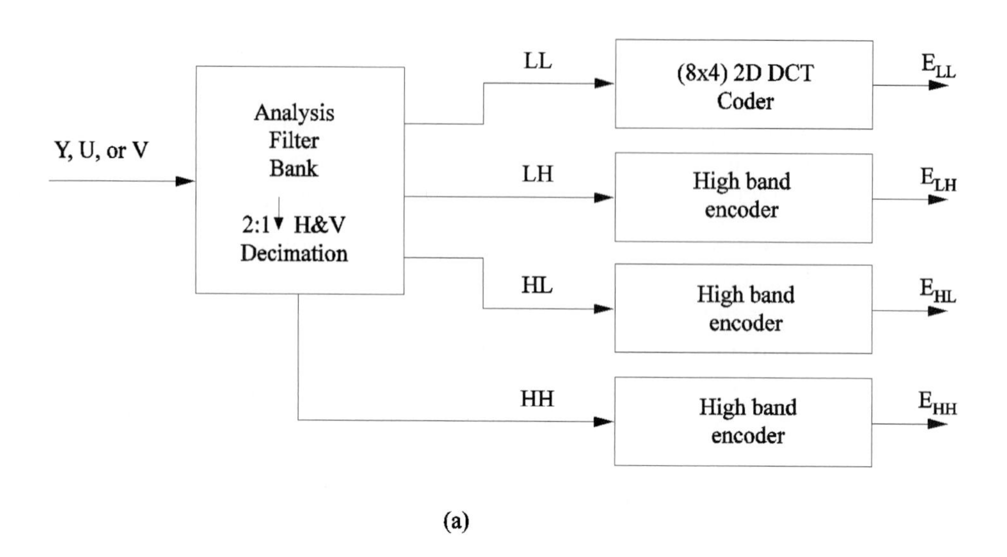

(a)

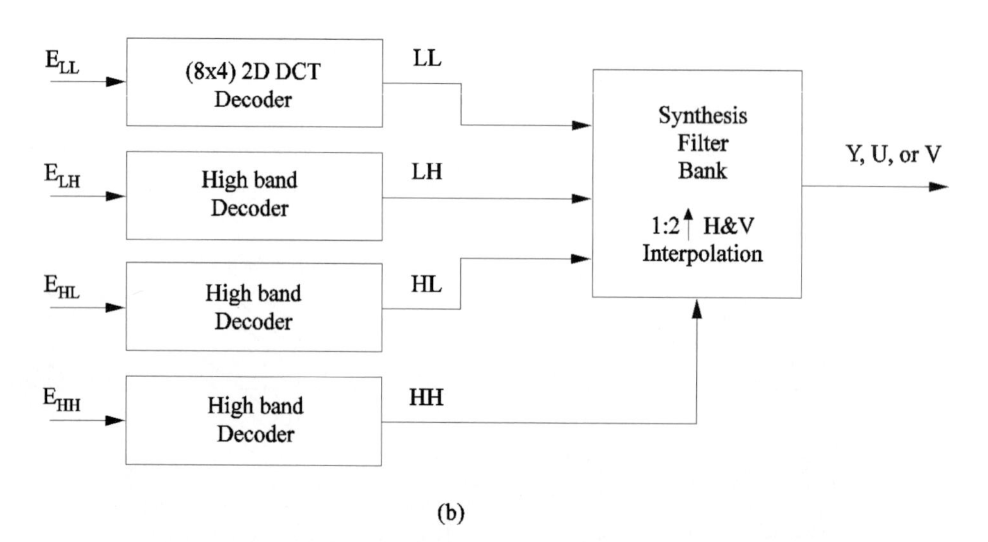

(b)

Figure 7.22 HDTV subband/DCT coding. (a) Subband/DCT encoder; (b) subband/DCT decoder. LL band is DCT coded. LH, HL, and HH bands are coarsely quantized and run-length coded [356]. © 1990 IEEE.

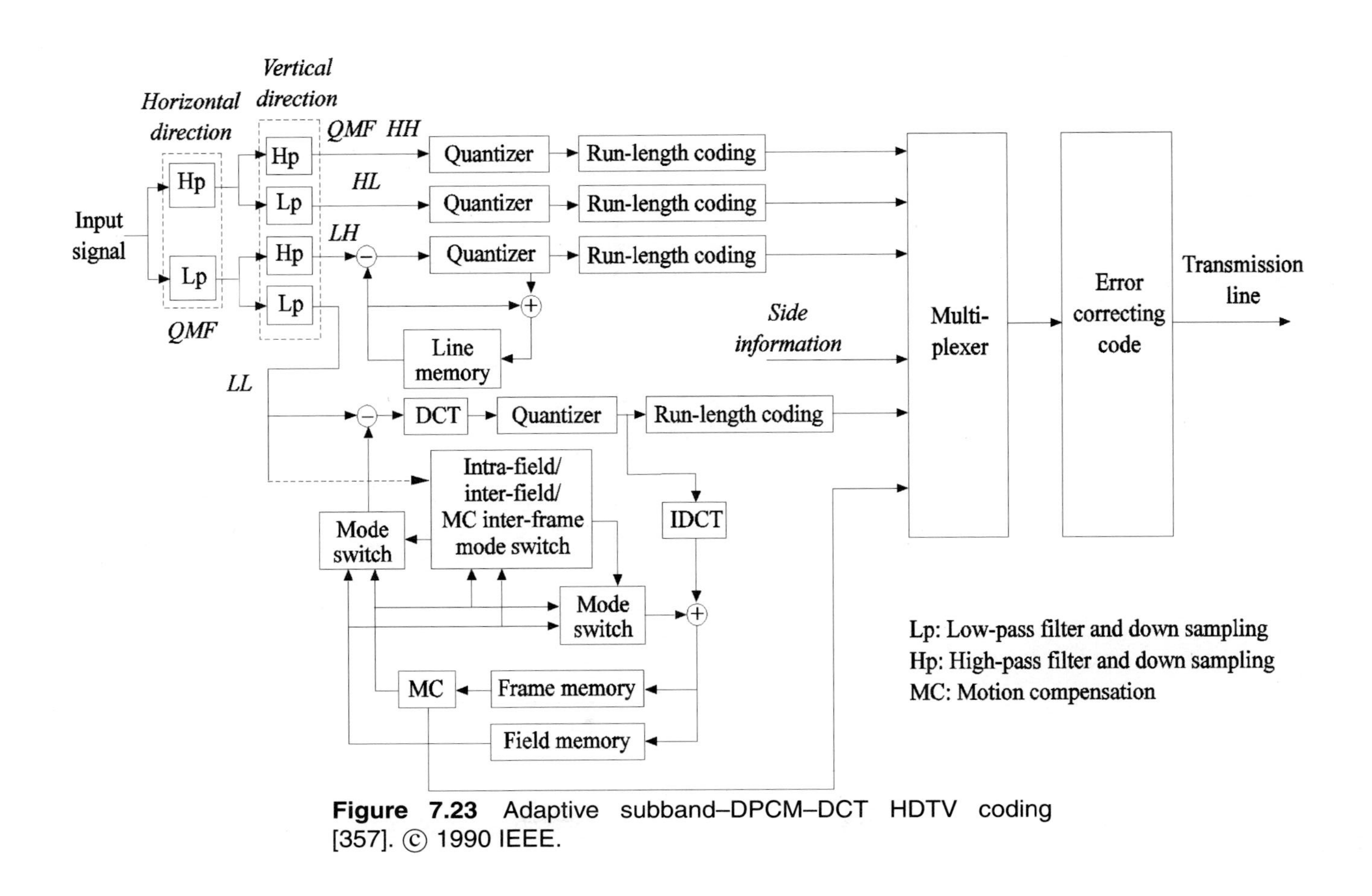

Figure 7.23 Adaptive subband–DPCM–DCT HDTV coding [357]. © 1990 IEEE.

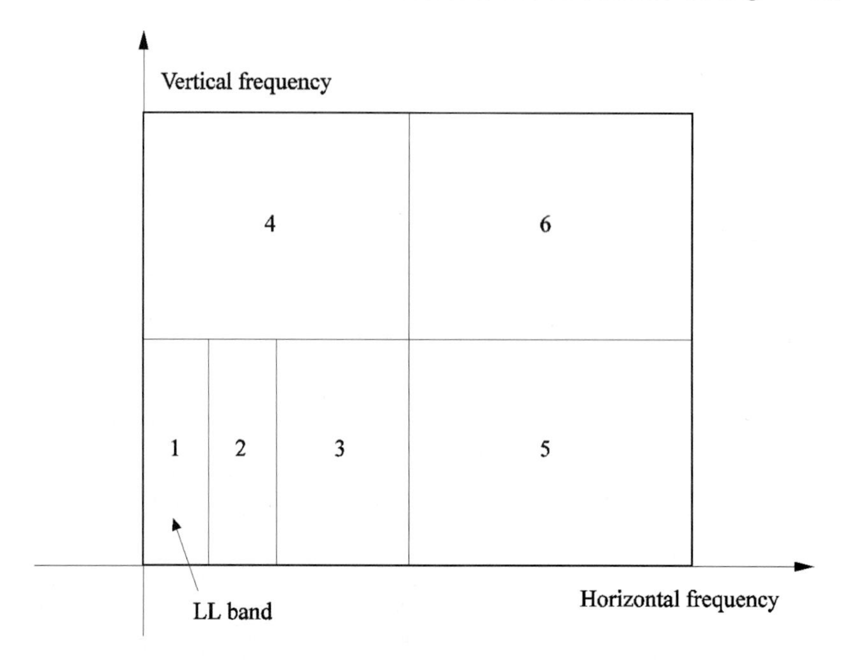

Figure 7.24 Unequal subband decomposition proposed by Bellcore for the ATM/SONET/H-4 B-ISDN HDTV project [359].

(such as DCT) or DPCM is applied to the lowest subband, followed by quantization and VLC (Fig. 7.22). The remaining subbands are coarsely quantized. Of course, VQ can also be extended to the subband samples [348]. Other variations include subband coding of interframe prediction errors (with or without MC) (Fig. 7.23) and subband coding of color components. In [357], the LH band is predictively coded, whereas the LL band goes through adaptive hybrid coding. The unequal decomposition explored by Bellcore [359] for HDTV coding is shown in Fig. 7.24. In this, the LL band (band 1) is predictively coded (previous pel predictor). The remaining bands are coarsely quantized and VWL coded.

Recently, subband coding has been extended to both spatio-temporal domains [360]. The 3D spatio-temporal subband decomposition is followed by the appropriate redundancy reduction techniques such as ADPCM and variations of VQ applied to the subbands.

Part II
International Standards for Image, Video, and Audio Coding

Chapter 8
JPEG Still-Picture Compression Algorithm

Summary

Since the mid 1980s, a joint ISO/CCITT committee group known as JPEG (Joint Photographic Experts Group) has been working to study an efficient coding scheme for continuous-tone still images. As a result, IS (International Standard) 10918 has been established by both international organizations (ISO/IEC and ITU). The standard may be applied to various fields such as image storage, color facsimile, newspaper photo transmission, desktop publishing, medical imaging, electronic digital cameras, and so forth. The JPEG standard provides several coding modes from basic to sophisticated according to application fields.

In this chapter we show the reader how the JPEG has worked for formulating the international standard, how the coding/decoding techniques have been developed, what procedures have to be done at the encoder and decoder, and how the algorithm can be implemented by VLSI chip sets. We illustrate the IS 10918-1 with examples (simulated images) followed by the latest developments in hardware. Finally, we also add JPEG extensions which allow adaptivity, transcoding, file format interchange, and other desirable features.

8.1 Introduction

The main obstacle for many applications of digital images, i.e., acquisition, data storage, printing, and display, is the huge amount of data required to represent an image directly. A digitized version of a single color picture at normal TV resolution contains on the order of one million bytes. Such an image needs to be compressed for storage or transmission. The actual compression ratio can vary from 100:1 to 2:1 depending on the specific application and encoder/decoder complexity. State-of-the-art techniques can compress typical images by a factor of 10 to 50 without significantly affecting the image quality, depending on the technique applied. For the applications of storage or transmission on the bandwidth-limited channels which are widespread in today's market, a standard image-compression method is needed to get more efficient use of media or channels and to enable interoperability of equipment developed by different manufacturers.

JPEG has recently been recognized as the most popular and efficient coding scheme for continuous-tone (multilevel) still images, both monochrome and color. At a moderate bit rate (less than 1bpp), JPEG can usually yield a satisfactory solution to most of the practical coding problems. In the middle of the 1980s, a standardization effort known by the acronym JPEG (Joint Photographic Experts Group) was initiated, with the objective of establishing the first international digital compression standard for still images. It was motivated by the ISO Working Group pursuing a progressive scheme at the ISDN rate (64 Kbps), but the word *joint* in JPEG refers to the collaboration between CCITT (now ITU-T) and ISO. Although JPEG convened officially as the ISO committee designated JTC1/SC2/WG10, the CCITT Study Group VIII also had a close informal relationship with JPEG. The International *de jure* standard JPEG has become both ISO/IEC 10918-1 [367] and ITU-T Recommendation T.81. A detailed history of the JPEG effort is described in this chapter and can also be found in [371, 372, 373, 374].

There are many international groups such as MPEG, ITU-T Study Group XV, and JBIG. One of the contributions of JPEG is that these groups have gained from JPEG's experiences. Unlike the traditional standardization efforts, JPEG has grown into a truly international research collaboration. It has also provided the industry with the incentive to develop hardware significantly.

To use the basic JPEG algorithm in various applications, the experts group has worked on JPEG extensions described in ISO/IEC 10918-3 (ITU-T Rec. T.84) [368]. Variable quantization for each block is necessary to transcode from any video compression format (e.g., MPEG). It also gives increased subjective quality and efficient utilization of channel bandwidth. The other functions, such as selective refinement and tiling, are also important in JPEG extensions, which will be discussed in Section 8.8. Compliance testing of JPEG efforts has been finalized as ISO/IEC 10918-2 (ITU-T Rec. T.82) [369].

From the image compression viewpoint, JPEG specifies one or more algorithms for each mode listed below. It gives a flexible and comprehensive encoding framework that could open a broad range of still-image applications. The four JPEG encoding modes are

1. Sequential encoding: The image is encoded in a raster scan fashion left-to-right/top-to-bottom, based on DCT. The baseline system belongs to this category.
2. Progressive encoding: The image is encoded in multiple scans at the same spatial resolution for applications in which the channel bandwidth is narrow and, hence, the transmission time may be long, but the viewer prefers to view the image contents in multiple coarse-to-fine stages.
3. Lossless encoding: The image is encoded to guarantee exact recovery of every source sample value. Irrelevant information is removed and compression efficiency is inherently lower than lossy methods.
4. Hierarchical encoding: The image is encoded at multiple spatial resolutions, so that lower-resolution images may be accessed and displayed without having to decompress the image at a higher spatial resolution.

Required coding techniques for the JPEG are mainly based on 2D (8×8) DCT, which was chosen for two reasons, compression capability and the availability of commercial hardware chips. JPEG chips are at present available in volume for about $50 to $100. A number of manufacturers produce multipurpose chips, not only for JPEG but also for video-compression standards (H.261 or MPEG; see Chapters 9 through 11). Typically, a 10 MHz chip can compress a full-page 24-bit-color, 300 dpi image from 25 MB to 1 MB in about 1 second, and the processing time continues to decrease rapidly. The fast implementation makes JPEG applicable to color facsimile, high-quality newspaper wire photos, desktop publishing, multimedia, graphic arts, medical imaging, electronic digital cameras, imaging scanners, and so forth.

8.2 History of JPEG

JPEG has kept close and informal collaboration between two committees, i.e., ISO/IEC JTC1/SC29/WG10 and the Special Rapporteur's committee Q.16 of CCITT SGVIII. The ISO Working Group has developed the main processes and ITU-T has provided the requirements that these processes must satisfy for applications such as facsimile, videotex, or audio graphic conferencing. We describe here the historical development of this standard. Readers may refer to [374, 375] for details.

- 1982: A photographic experts group was formed under ISO/TC97/SC2 WG8 — Coded Representation of Picture and Audio Information.
- 1986: The ISO Group joined the CCITT SGVIII SRG (Special Rapporteur's Group) on New Forms of Image Communication. JPEG was therefore established.
- Mar. 1987: Twelve proposals were registered as candidates for the compression algorithm. They were called Generalized Block Truncation Coding, Progressive Coding Scheme, Adaptive DCT, Component VQ, Quadtree Extension of Block Truncation Coding, Adaptive Discrete Cosine Transform, Progressive Recursive Binary Nesting, Adaptive Transform and Differential Entropy Coding, DCT with Low Block-to-Block Distortion, Block List Transform Coding of Images, Hierarchical Predictive Coding, and DPCM using Adaptive Binary Arithmetic Coding.
- Jan. 1988: Extensive evaluation of the various schemes had been done. JPEG reached a consensus that the adaptive DCT approach should be the basis for further refinement and development in both hardware and software.
- Sept. 1988: The Transform Technique Enhancement Group was formed within JPEG.
- Feb. 1989: Baseline capability based on DCT and Huffman coding was defined. This became the first draft of the JPEG Technical Specification (JPEG Revision 0). An editing committee rewrote the specification resulting in JPEG Revision 1.
- July 1989: JPEG Revisions 2 and 3 allowed higher precision for the DCT coefficients and source data. Revision 4 aggregated all the techniques.
- Oct. 1989: Amendments were made to allow for transcoding between the progressive and sequential DCT-based modes (JPEG Revision 5).
- Mar. 1990: A number of minor changes were adopted. The default Huffman table was abandoned.
- Apr. 1990: A plenary meeting of ISO/IEC JTC1/SC2 restructured WG8 and JPEG became the responsibility of WG10, Photographic Image Coding Group. JPEG Revision 6 included new probability estimation tables for arithmetic coding. A common arithmetic coder was made by coworking with JBIG, and JPEG Revision 7 was adopted.
- Aug. 1990: A final revision of the Draft Technical Specification (JPEG Revision 8) was completed.
- Feb. 1991: Committee Draft (CD) part 1 was submitted to ISO/IEC JTC1/SC2.
- June 1991: JPEG CD part 1 was approved.

- Oct. 1991: DIS document was submitted to the JTC1.
- Nov. 1991: A new study committee, SC29, was formed. Working Groups 9, 10, 11, 12, and 13 were transferred from SC2.
- Apr. 1992: CCITT SGVIII approved Recommendation T.81.
- July 1992: DIS was approved by the principal 22 members of ISO/IEC JTC1 and became IS 10918-1.
- Aug. 1994: Working Draft (WD) for JPEG extensions (ISO/IEC 10918-3, ITU-T Rec. T.84) [368] was developed.
- Sept. 1994: WD for compliance testing (ISO/IEC 10918-2, ITU-T Rec. T.83) [369] was developed.
- Nov. 1994: CD for ISO/IEC 10918-3 was finalized.

8.3 Goals and Directions

The JPEG standard is intended to compress still, continuous-tone, monochrome and color images. This may be achieved by one of the four main processing modes, i.e., sequential, progressive, lossless, or hierarchical (see Sections 8.4 through 8.7 for descriptions of these modes). Ambitiously low bit-rate targets have to be set for the lossy compression mode, however. Possible bit rates and qualities are [373]:

- $0.25 \sim 0.5$ bpp: moderate to good quality
- $0.50 \sim 0.75$ bpp: good to very good quality
- $0.75 \sim 1.5$ bpp: excellent images
- $1.50 \sim 2.0$ bpp: indistinguishable images (visually lossless)

The images chosen for the contest, i.e., testing and evaluation of the various algorithms, are based on the ITU-R BT.601 format (Table 2.1). Besides the four modes of functional flexibility, there are common required functions as follows:

- Both encoder and decoder could be **synchronized** for real-time applications such as videoconference.
- Both hardware and software should be **cost effective**.
- **Picture resolution** should be independent.
- Gray scale representation of **text images** should be handled well.
- There is no absolute **bit-rate** target.
- **2D-DCT** of 8×8 blocks is used for compression, in view of its superior performance and the availability of VLSI chips.

Table 8.1 Luminance quantization matrix Q_{uv} (example only) [367]

16	11	10	16	24	40	51	61
12	12	14	19	26	58	60	55
14	13	16	24	40	57	69	56
14	17	22	29	51	87	80	62
18	22	37	56	68	109	103	77
24	35	55	64	81	104	113	92
49	64	78	87	103	121	120	101
72	92	95	98	112	100	103	99

Source: © 1993 ITU-T.

Table 8.2 Chrominance quantization matrix Q_{uv} (example only) [367]

17	18	24	47	99	99	99	99
18	21	26	66	99	99	99	99
24	26	56	99	99	99	99	99
47	66	99	99	99	99	99	99
99	99	99	99	99	99	99	99
99	99	99	99	99	99	99	99
99	99	99	99	99	99	99	99
99	99	99	99	99	99	99	99

Source: © 1993 ITU-T.

- **DCT coefficients** are quantized by lookup tables (see Tables 8.1 and 8.2).
- DCT operations are restricted to 8- and 12-bit/sample (input pel) **precision**.
- The system is **extensible** in a number of ways. There are no bounds on the number of progressive or hierarchical stages.
- **Default quantization** tables are retained only as informative examples.

The standard gives much flexibility for improving performance. There have already been suggestions within the standard framework for possible addenda. One of the ideas is adaptive quantization. Since quantization noise is inevitable, reduction of this noise is important. Quantization noise can be reduced by as much as 30%, at the expense of additional computational burden. One way to improve the quality without any degradation is lossless coding. We can combine several entropy coding methods, develop statistical models, and utilize adaptive prediction algorithms. Different block sizes, wavelet, or fractal expansions may be considered as possible addenda, provided that backward compatibility (see Chapter 11) is guaranteed.

8.4 Sequential DCT-Based Mode

There are several variables for the sequential DCT-based mode of operation from a baseline system to an extended one. We illustrate the possible modes of sequential DCT operation in Fig. 8.1, which shows five possible combinations (arrowed paths indicating the combination of 8- or 12-bit sample precision, baseline or extended system, and Huffman or arithmetic coding). The encoder may select 8- or 12-bit precision data and choose Huffman or arithmetic coding. For each of these modes, the encoding and decoding processes are specified by means of specific functions for the procedures that comprise these coding modes. Detailed requirements are not defined for following the JPEG standard. It is only necessary that the codec complies with the functions and bitstream syntax provided by the standard. A standard confirming encoder needs to produce a valid bit stream according to the standard. In this case the bit stream is specified by JPEG.

We proceed with describing the baseline system functions followed by examples. For the extended system using the Huffman coding, the number of VLC tables may be increased up to four for DC and AC coefficients. Sequential arithmetic coding extensions give higher data compression than the Huffman coding, by about 10% [374].

8.4.1 Baseline system description

Any DCT-based decoder, if it embodies any DCT-based decoding process other than the baseline sequential process, shall also embody the baseline sequential decoding process. The block diagram depicted in Fig. 8.2 shows the overall functions of the baseline system.

We summarize the mode of operation that is applied for all DCT-based decoders as follows:

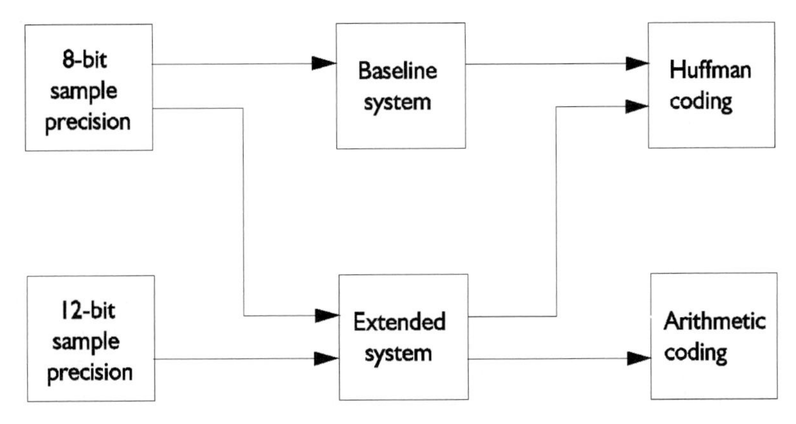

Figure 8.1 Possible extensions in sequential DCT mode of operation.

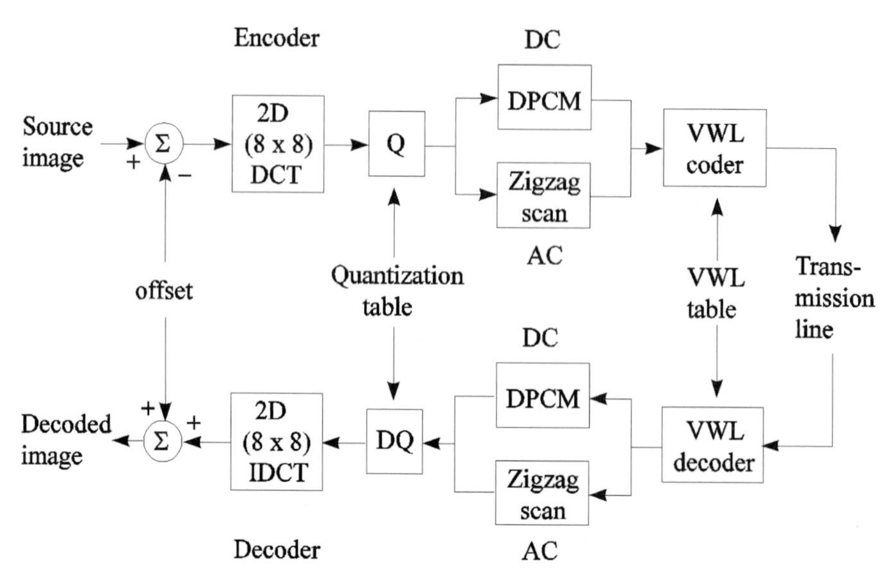

Figure 8.2 JPEG baseline encoder and decoder (Q: quantizer; DQ: dequantizer).

- DCT-based process
- Source image: 8-bit samples for each color component
- Sequential operation
- Huffman coding: 2 AC and 2 DC tables
- Decoders process scans with 1, 2, 3, and 4 components
- Interleaved and noninterleaved scans

Input source-image data are level-shifted to a signed two's complement representation by subtracting 2^{p-1}, where p is the precision parameter of the image intensity in bits. When $p = 8$ in baseline mode, the level shift is 128. An inverse level shift at the decoder (Fig. 8.2) restores the sample to the unsigned 8-bit representation. The level shifting does not affect AC coefficients or variances. It affects only the DC coefficient, shifting a neutral gray intensity to zero. Difference values between DC coefficients are also unaffected. The initial starting value at the beginning of the image and at restart interval may be set to zero.

The level-shifted image is partitioned into (8×8) blocks along a raster scan fashion for the 2D (8×8) DCT operations. DCT is the most popular block-transform method for compressing data in a lossy mode. Detailed algorithmic developments and properties of DCT are discussed in Chapter 5. The accuracy requirements for DCT and IDCT are not specified. They should, however, have sufficient accuracy to meet the application requirements.

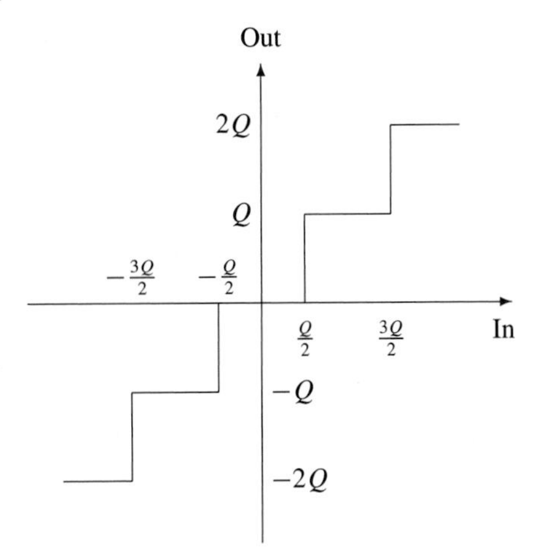

Figure 8.3 Uniform quantizer for the baseline mode. Q is the stepsize.

Quantizer designation is an important factor in a gray-level image coder. Bit-allocation algorithms also have a great role in transform coding, being closely connected with quantization. A number of quantizers and bit-allocation algorithms have been suggested [376, 377]. JPEG gives simple and easy quantization methods and suggests informative tables for DC and AC coefficients. Each of the 64 resulting DCT coefficients is quantized by a different uniform quantizer. The step sizes are based on a visibility threshold of 64-element quantization matrices for luminance and chrominance components. Quantized DCT coefficients, S_{quv}, are defined by the following equation:

$$S_{quv} = \text{Nearest integer}\left(\frac{S_{uv}}{Q_{uv}}\right) \tag{8.1}$$

where S_{uv} is the DCT coefficient and Q_{uv} is the quantization matrix element (see Tables 8.1 and 8.2). This quantization method is useful for simple and fast implementation, reducing complicated steps at the coder. Figure 8.3 shows uniform quantizer outputs versus the corresponding input values represented by multiples of level Q, given for each coefficient. In any case, the quantizer matrix elements are the same both at the encoder and at the decoder.

Tables 8.1 and 8.2 show the quantization matrices that are obtained empirically. These matrices yield fairly good performance. These quantization matrices need to be stored in the decoder. Inverse operation removes the normalized factor, as shown in Eq. (8.2). For the DC coefficient, reconstruction from differential value is preceded. Each AC coefficient is uniformly quantized. The step size Q_{uv} (Tables 8.1 and 8.2) is a function of the transform coefficient frequency:

$$R_{uv} = S_{quv} Q_{uv} \tag{8.2}$$

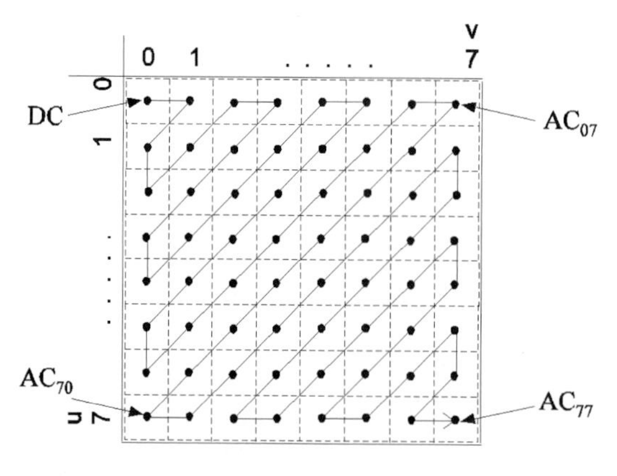

Figure 8.4 A zigzag scan of quantized DCT coefficients.

where R_{uv} is a dequantized DC or AC coefficient to be fed into the 2D (8×8) IDCT operation. It may be noted that Tables 8.1 and 8.2 are examples only. Any other custom-designed quantization matrices can be used as long as the receiver has knowledge of the same.

Since a blocking artifact mainly from the DC coefficient is sensitive to spatial frequency response by the human visual system, the DC coefficient is treated separately from the 63 AC coefficients. It is differentially coded by the following first-order prediction

$$\text{DIFF} = \text{DC}_i - \text{DC}_{i-1} \tag{8.3}$$

where DC_i and DC_{i-1} are the current (8 × 8) block and the previous (8 × 8) block DC coefficients, respectively.

The quantized 63 AC coefficients are formatted as per the zigzag scan shown in Fig. 8.4, in preparation for entropy coding. Along the zigzag scan the DCT coefficients represent increasing spatial frequencies and, in general, decreasing variances. Also, the HVS (human visual system) weighted quantization (Tables 8.1 and 8.2) results in many zero coefficients. An efficient VLC table can be developed that represents runs of zero coefficients along the zigzag scan followed by the size of the nonzero coefficient.

JPEG uses only Huffman coding for the baseline system. The encoder may employ two DC and two AC Huffman table lookups for luminance and chrominance DCT coefficients. It is suggested that all codes, for DC or AC, consist of a set of Huffman codes (maximum length 16 bits) followed by appended additional bits for representing the exact values.

Coding DC coefficients: The DIFF values as defined by Eq. (8.3) are classified into 12 categories (SSSS in Table 8.3) for 8-bit resolution. The dynamic range of DCT coefficients is 11 bits when source-image precision is 8 bits. Prediction based on the previous DC coefficient increases the prediction error by 1

Table 8.3 Difference categories for DC coding [367]

SSSS	DIFF values
0	0
1	$-1, 1$
2	$-3, -2, 2, 3$
3	$-7 \cdots -4, 4 \cdots 7$
4	$-15 \cdots -8, 8 \cdots 15$
5	$-31 \cdots -16, 16 \cdots 31$
6	$-63 \cdots -32, 32 \cdots 63$
7	$-127 \cdots -64, 64 \cdots 127$
8	$-255 \cdots -128, 128 \cdots 255$
9	$-511 \cdots -256, 256 \cdots 511$
10	$-1023 \cdots -512, 512 \cdots 1023$
11	$-2047 \cdots -1024, 1024 \cdots 2047$

Source: © 1993 ITU-T.

bit, resulting in 12 categories. Therefore, we have the difference categories written by two's complement expression. For the maximum source-image precision of 12 bits, 16 categories are defined in the same way.

The reader may generate Huffman tables for each category which has different probability distributions, or may employ informative JPEG Huffman tables as shown in Table 8.4 for luminance and chrominance components. Note that not a single code consists entirely of all 1 bits, since all-1s code would be difficult to decode.

In the case of SSSS = 0, i.e., the same value as the previous DC coefficient, additional bits are not required. For other categories, we need extra bits to express the exact value in the category, consisting of the sign and amplitude of the prediction error. When DIFF is positive, the sign bit is 1 and the SSSS low-order bits of DIFF are appended to the Huffman code. When DIFF is negative, the SSSS low-order bits of (DIFF − 1) are appended. The sign bit would be 0 and (DIFF − 1) operation implies one's complement representation to avoid all 1 bits of the two's complement operation. For example, the negative value −5 is represented as 010. This procedure for appending the additional bits is also applied to coding AC coefficients.

Coding AC coefficients

In general, because many of the AC coefficients become zero after quantization, runs of zeros along the zigzag scan are identified and compacted. Each nonzero AC coefficient is described by a composite R/S, where R is a 4-bit zero-run from the previous nonzero value and S represents the 10 (4-bit) categories, as shown in Table 8.5. If the zero-runs (the number of quantized zero coeffi-

Table 8.4 Huffman code table for luminance and chrominance DC difference [367]

	Luminance DC		Chrominance DC	
SSSS	Code length	Codeword	Code length	Codeword
0	2	00	2	00
1	3	010	2	01
2	3	011	2	10
3	3	100	3	110
4	3	101	4	1110
5	3	110	5	11110
6	4	1110	6	111110
7	5	11110	7	1111110
8	6	111110	8	11111110
9	7	1111110	9	111111110
10	8	11111110	10	1111111110
11	9	111111110	11	11111111110

Source: © 1993 ITU-T.

cients along the zigzag scan) are greater than 16, then R/S = x'F0' (15 zero-runs and 1 zero value) is coded by an 11-bit VLC as shown in Appendix C.1 and counted again. In addition, if all remaining coefficients along the zigzag scan are zero, a special value R/S = x'00' is coded as an EOB (end-of-block) code of 1010 (see Appendix C.1).

Huffman encoding and decoding tables are generated for each composite category R/S. We may use the informative tables that define the possible 162 codes in Appendix C.1 for luminance and in Appendix C.2 for chrominance. Here *Run* refers to the run length of the zero coefficients followed by the *Size* of the

Table 8.5 Categories assigned to AC coefficient values [367]

SSSS	AC coefficients
1	$-1, 1$
2	$-3, -2, 2, 3$
3	$-7 \cdots -4, 4 \cdots 7$
4	$-15 \cdots -8, 8 \cdots 15$
5	$-31 \cdots -16, 16 \cdots 31$
6	$-63 \cdots -32, 32 \cdots 63$
7	$-127 \cdots -64, 64 \cdots 127$
8	$-255 \cdots -128, 128 \cdots 255$
9	$-511 \cdots -256, 256 \cdots 511$
10	$-1023 \cdots -512, 512 \cdots 1023$

Source: © 1993 ITU-T.

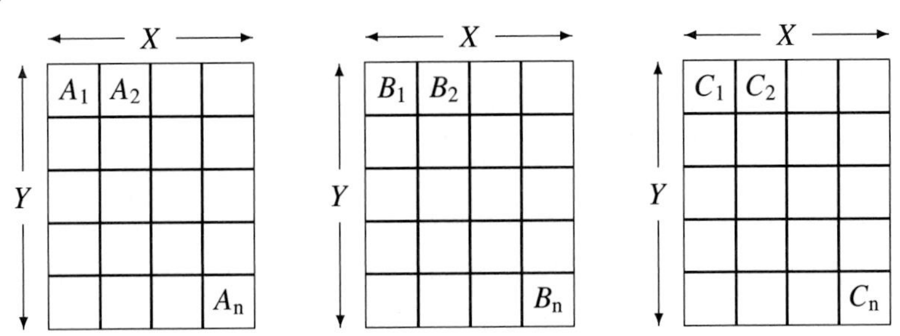

(a) Data unit encoding order, *noninterleaved*

$\underbrace{A_1, A_2, \cdots, A_n,}_{\text{Scan 1}}$ $\underbrace{B_1, B_2, \cdots, B_n,}_{\text{Scan 2}}$ $\underbrace{C_1, C_2, \cdots, C_n,}_{\text{Scan 3}}$

(b) Data unit encoding order, *interleaved*

$\underbrace{A_1, B_1, C_1, A_2, B_2, C_2, \cdots, A_n, B_n, C_n}_{\text{Scan 1}}$

Figure 8.5 Interleaved and noninterleaved encoding order.

nonzero coefficient along the zigzag scan (Fig. 8.4). A similar VLC table is also developed for the chrominance [367]. Each of the codes is uniquely defined and the composite R/S information is decoded at the receiver, using the same VLC tables. The format and rules for the additional bits are the same as those for coding the prediction error of the DC coefficient. The encoding process is illustrated by an example in the next section. We have described baseline encoding-decoding processes for the luminance component, Y. Color image consists of Y and two chrominance components. The decoder is able to decode up to four components, such as CMY_eK (cyan, magenta, yellow, and black) in color print image coding. The three components of the color image can be interleaved, or noninterleaved as shown in Fig. 8.5. Subject to the interleave or noninterleave status, the MCU (minimum coded unit) components are varied. If it is interleaved, the first MCU becomes $A_1\,B_1\,C_1$ (Fig. 8.5).

8.4.2 Baseline coding example

We present an example of a baseline system reflecting the coding techniques described earlier. The following data block (marked by a squared block) is selected from the LENA image (Fig. 8.6):

$$x_{ij} = \begin{pmatrix} 79 & 75 & 79 & 82 & 82 & 86 & 94 & 94 \\ 76 & 78 & 76 & 82 & 83 & 86 & 85 & 94 \\ 72 & 75 & 67 & 78 & 80 & 78 & 74 & 82 \\ 74 & 76 & 75 & 75 & 86 & 80 & 81 & 79 \\ 73 & 70 & 75 & 67 & 78 & 78 & 79 & 85 \\ 69 & 63 & 68 & 69 & 75 & 78 & 82 & 80 \\ 76 & 76 & 71 & 71 & 67 & 79 & 80 & 83 \\ 72 & 77 & 78 & 69 & 75 & 75 & 78 & 78 \end{pmatrix} \quad (8.4)$$

Note the region is a very bright area, the data are 128 level-shifted, and we choose a relatively flat area for an easy example. This block, transformed by 2D (8×8) DCT, is given by

$$S_{uv} = \begin{pmatrix} 619 & -29 & 8 & 2 & 1 & -3 & 0 & 1 \\ 22 & -6 & -4 & 0 & 7 & 0 & -2 & -3 \\ 11 & 0 & 5 & -4 & -3 & 4 & 0 & -3 \\ 2 & -10 & 5 & 0 & 0 & 7 & 3 & 2 \\ 6 & 2 & -1 & -1 & -3 & 0 & 0 & 8 \\ 1 & 2 & 1 & 2 & 0 & 2 & -2 & -2 \\ -8 & -2 & -4 & 1 & 2 & 1 & -1 & 1 \\ -3 & 1 & 5 & -2 & 1 & -1 & 1 & -3 \end{pmatrix} \quad (8.5)$$

The DC coefficient value 619 is eight times the average gray-level of the block. The actual DC level without level-shifting is obtained by adding 1024 to 619, resulting in 1643. It is seen that the energy in the low-frequency band is greater than in the high-frequency band. This means that the spatial block is dominated by the low-frequency components, i.e., there is a high correlation between pixels.

A quanitized version of the transformed block is achieved by applying the luminance quantization matrix (Table 8.1), resulting in

$$S_{quv} = \begin{pmatrix} 39 & -3 & 1 & 0 & 0 & 0 & 0 & 0 \\ 2 & -1 & 0 & 0 & 0 & 0 & 0 & 0 \\ 1 & 0 & 0 & 0 & 0 & 0 & 0 & 0 \\ 0 & -1 & 0 & 0 & 0 & 0 & 0 & 0 \\ 0 & 0 & 0 & 0 & 0 & 0 & 0 & 0 \\ 0 & 0 & 0 & 0 & 0 & 0 & 0 & 0 \\ 0 & 0 & 0 & 0 & 0 & 0 & 0 & 0 \\ 0 & 0 & 0 & 0 & 0 & 0 & 0 & 0 \end{pmatrix} \quad (8.6)$$

Note that each AC coefficient S_{uv} is uniformly quantized. The step size Q_{uv} is dependent on the corresponding coefficient S_{uv}. Table 8.1 shows that the low-frequency coefficients are finely quantized (smaller step sizes) whereas the

Figure 8.6 A selected (8×8) block (square on hat) in a LENA image.

high-frequency coefficients are coarsely quantized (larger step sizes). This process reflects the HVS.

There are only a few nonzero coefficients [see Eq. (8.6)]. Assuming the quantized DC coefficient of the previous block is 34, the prediction error is 5, which is entropy-coded. It belongs to category 3 in Table 8.4, and its Huffman code is 100. Additional bits of 101, representing the exact value of the prediction error in this category, are appended. The two-dimensional array is rearranged into a one-dimensional array based on the zigzag scan (see Fig. 8.4). It results in

$$\begin{pmatrix} 39 & -3 & 2 & 1 & -1 & 1 & 0 & 0 & 0 & 0 & 0 & -1 & \text{EOB} \end{pmatrix} \tag{8.7}$$

The first AC coefficient -3 belongs to category 0/2, which has codeword 01, in Appendix C.1. As previously discussed, the additional bits 00 are appended $(01 - 1 = 00)$. For the next coefficient we get the same category codeword 01 and extra bits 10. The last nonzero coefficient -1 belongs to category 5/1, which has codeword 1111010. This is followed by the EOB code 1010. The resulting bit stream is

$$(100101/0100/0110/001/000/001/11110100/1010) \tag{8.8}$$

A total of 35 bits are needed to transmit this block. Therefore the transmission bit rate is 0.55 bit/pel when we use 8-bit precision for the source image, and the compression ratio is about 15:1. The reconstructed block obtained by the operations VWL decoding, inverse quantization, and inverse DCT is given by

$$\hat{x}_{ij} = \begin{pmatrix} 74 & 75 & 77 & 80 & 85 & 91 & 95 & 98 \\ 77 & 77 & 78 & 79 & 82 & 86 & 89 & 91 \\ 78 & 77 & 77 & 77 & 78 & 81 & 83 & 84 \\ 74 & 74 & 74 & 74 & 76 & 78 & 81 & 82 \\ 69 & 69 & 70 & 72 & 75 & 78 & 82 & 84 \\ 68 & 68 & 69 & 71 & 75 & 79 & 82 & 85 \\ 73 & 73 & 72 & 73 & 75 & 77 & 80 & 81 \\ 78 & 77 & 76 & 75 & 74 & 75 & 76 & 77 \end{pmatrix} \tag{8.9}$$

The error block is given by

$$e_{ij} = \begin{pmatrix} 5 & 0 & 2 & 2 & -3 & -5 & -1 & -4 \\ -1 & 1 & -2 & 3 & 1 & 0 & -4 & 1 \\ -6 & -2 & -10 & 1 & 2 & -3 & -9 & -2 \\ 0 & 2 & 1 & 1 & 10 & 2 & 0 & -3 \\ 4 & 1 & 5 & -5 & 3 & 0 & -3 & 1 \\ 1 & -5 & -1 & -2 & 0 & -1 & 0 & -5 \\ 3 & 3 & -1 & -2 & -8 & 2 & 0 & 2 \\ -6 & 0 & 2 & -6 & 1 & 0 & 2 & 1 \end{pmatrix} \tag{8.10}$$

The errors due to the coarse quantization in the high-frequency area are greater in sharply changing areas. Since the selected block is a relatively flat area, the resulting normalized mean square error from Eq. (8.10) is 3.52 for the block, implying good performance. If we choose a block in the high-activity area, however, many coefficients become nonzero. The bit rate would be increased but greater errors due to larger quantization step sizes in high-frequency bands result in a blocking effect.

8.4.3 Simulation results

Using the coding procedures and informative tables, the resulting monochrome LENA image, which has (512×512) pixels, is presented in Fig. 8.7, at two average bit rates. The baseline system is quite efficient for most of the images, having moderately low complexity. The JPEG goal is to achieve moderate to good quality at 0.25 to 0.5 bpp [378]. We see in Fig. 8.7 that at 0.25 bpp, the lower bit rate, quality is quite degraded, but quality is very good at 0.5 bpp.

8.5 Progressive DCT-Based Process

For the sequential mode, (8×8) blocks from left to right and top to bottom are coded and decoded, i.e., FDCT, quantization, VWL coding, and corresponding inverse operations at the decoder. For the progressive mode, (8×8) blocks are also typically encoded in the same order as in the sequential mode, but in multiple scans through the image. As each transform block is quantized, its coefficients

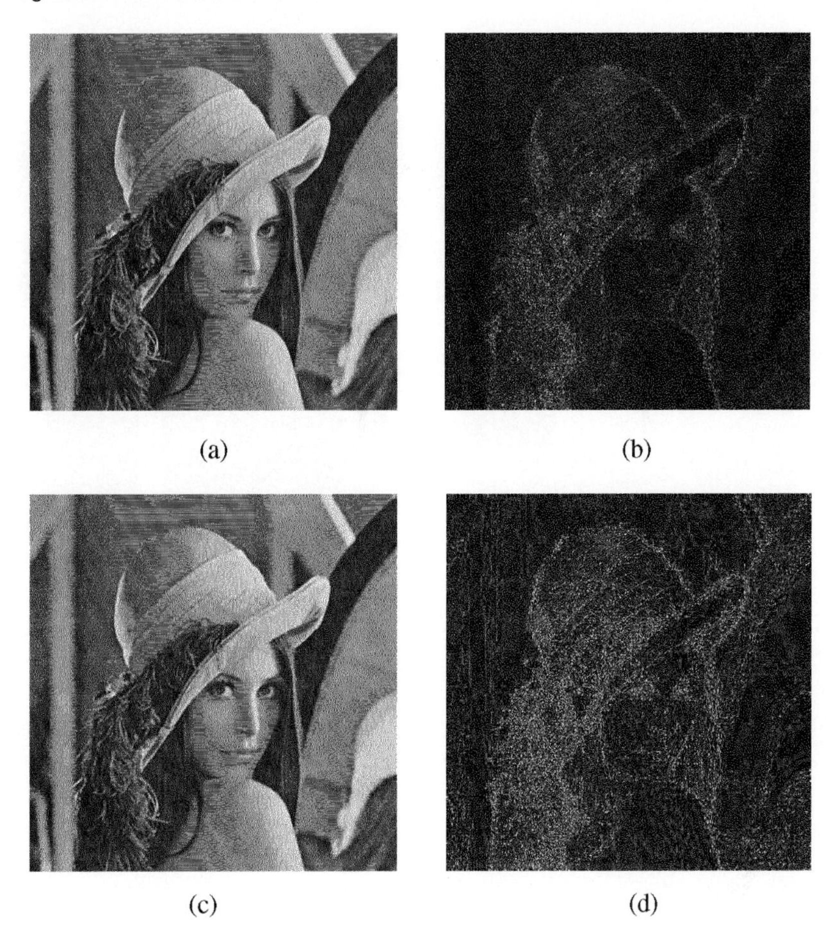

Figure 8.7 Reconstructed images and error images of JPEG baseline system [LENA.IMG]. (Image resolution is (512×512) and error images are magnified by a factor of 10.) (a) Image at 0.50 bpp; (b) errors of (a); (c) image at 0.25 bpp; (d) errors of (c).

are stored in the buffer. Progressive image transmission is receiving attention for applications in interactive image communications over restricted channel capacity. The least information necessary to represent each block is transmitted quickly with as few bits as possible. On the receiver's request, the image is progressively reconstructed in several stages until the required quality is achieved. Figure 8.8 shows how the progressive and sequential modes represent an image as a function of time at the decoder.

In the progressive mode, one can see a rough image during the first stage, while the sequential mode gives partial information of the full image. The JPEG

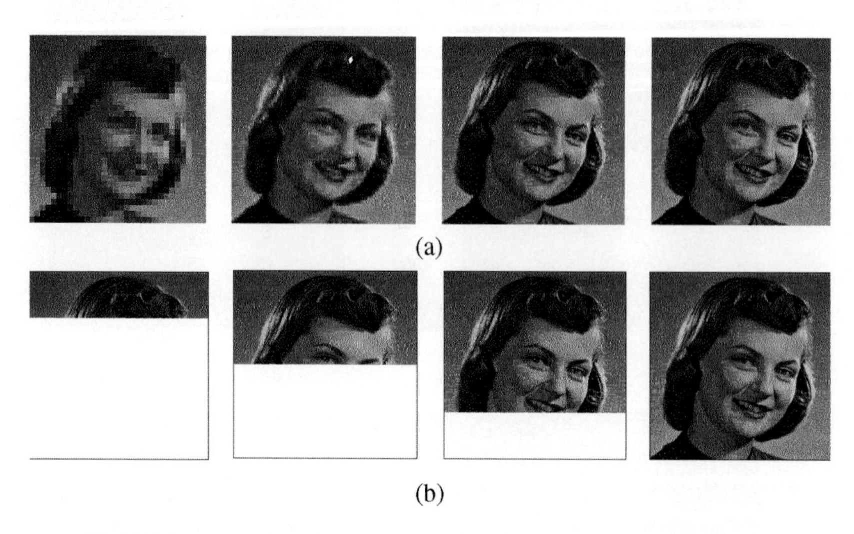

Figure 8.8 (a) Progressive and (b) sequential presentation of an image.

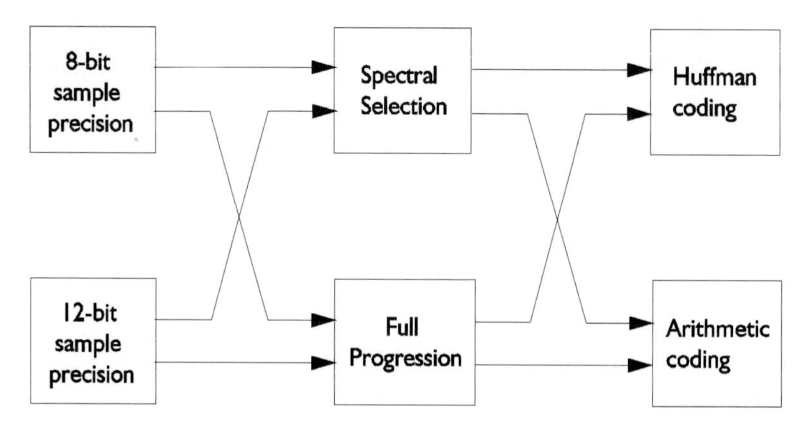

Figure 8.9 Possible extensions in the JPEG progressive mode of operation.

standard allows different kinds of operational modes with two input sample precisions, progressive methods, and variable coding algorithms, as shown in Fig. 8.9, where full progression denotes spectral selection within successive approximation.

There are two methods by which the quantized DCT coefficients in the buffer may be partially encoded and transmitted. After the 2D (8×8) DCT of

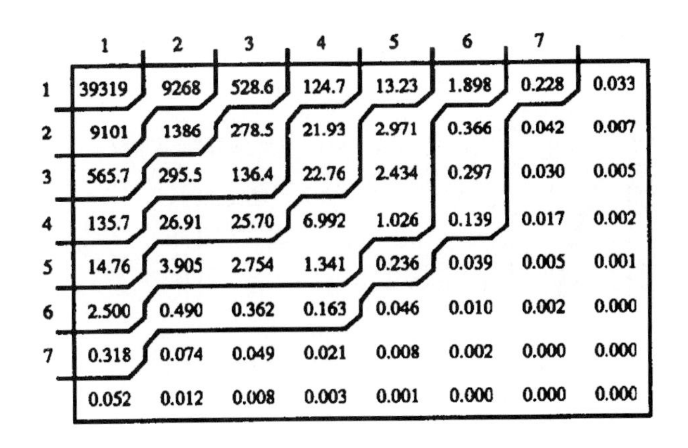

	1	2	3	4	5	6	7	
1	39319	9268	528.6	124.7	13.23	1.898	0.228	0.033
2	9101	1386	278.5	21.93	2.971	0.366	0.042	0.007
3	565.7	295.5	136.4	22.76	2.434	0.297	0.030	0.005
4	135.7	26.91	25.70	6.992	1.026	0.139	0.017	0.002
5	14.76	3.905	2.754	1.341	0.236	0.039	0.005	0.001
6	2.500	0.490	0.362	0.163	0.046	0.010	0.002	0.000
7	0.318	0.074	0.049	0.021	0.008	0.002	0.000	0.000
	0.052	0.012	0.008	0.003	0.001	0.000	0.000	0.000

Figure 8.10 Seven-stage spectral selection for progressive transmission of an (8×8) DCT block [379]. © 1993 SPIE.

the pels in the original image, a group of coefficients representing certain spectral properties (frequencies) are first quantized, coded, and transmitted, then followed by the remaining groups of coefficients in subsequent stages. This is the *spectral selection* method. Figure 8.10 shows an example of classifying spectral bands in an (8×8) block. The numbers represent the variances of the DCT coefficients. Those coefficients having a similar range of variances are sent in successive stages [379].

The DC coefficient implies the average value of the block. Therefore, a uniform gray-level block image is obtained with the DC coefficient. Similarly, the most significant bits of each DCT coefficient are the most important and thus a specified number of the most significant bits can be encoded first. In subsequent scans, the less significant bits are then encoded. This procedure is called successive approximation, that is, bit-plane encoding in the spectral domain. The two methods are compared in Fig. 8.11. Either procedure may be used separately, or they may be mixed in flexible combinations.

For a given bit rate in the intermediate stages, the absence of high-frequency bands typically leads to blocking effects, i.e., a less graceful image quality. Successive approximation increases complexity, but this system provides better image quality for a given bit rate. Combined transmission of both spectral selection and successive approximation may give a more graceful progression.

8.6 Lossless Process

Lossless technique implies that the encoding-decoding process results in a reconstructed image that is an exact replica of the original image, i.e., the gray levels of all the pels remain the same. The compression ratio in this case is very small, about 2:1. A lossless process can be achieved by predictive coding, which is not as efficient as DCT-based processes in compression capability.

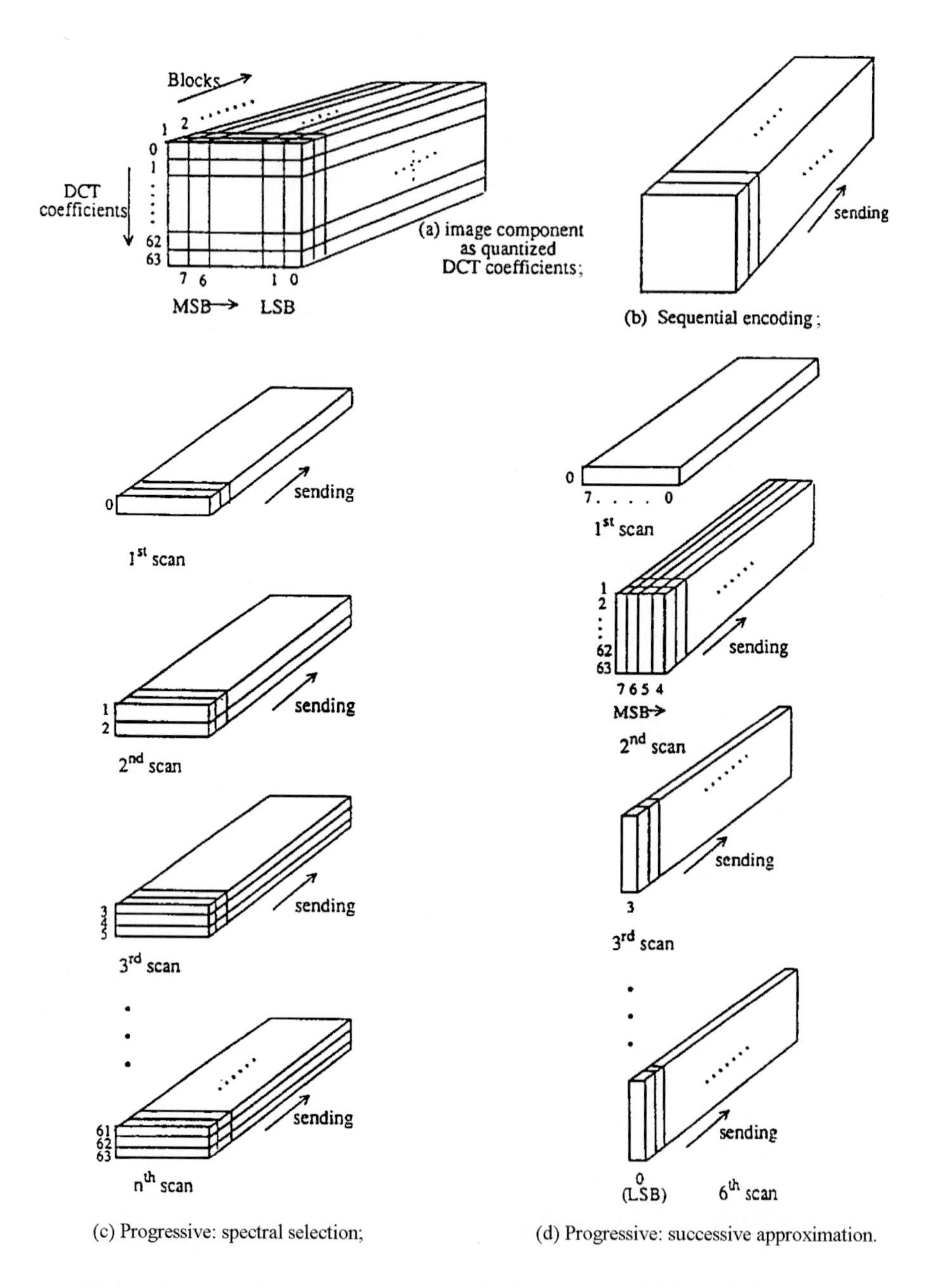

Figure 8.11 Spectral selection and successive approximation progressive processes [367]. © 1993 ITU-T.

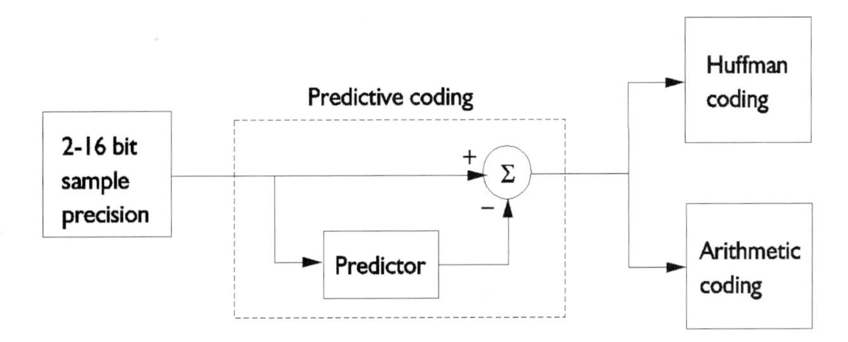

Figure 8.12 Possible extensions in lossless mode of operation.

The image, however, is totally reconstructable without any information loss. Functional descriptions of this mode are

- Predictive process (not DCT-based)
- Source image: N-bit samples ($2 \leq N \leq 16$)
- Sequential process
- Huffman or arithmetic coding: 4 DC tables
- Decoders process scans with 1, 2, 3, and 4 components
- Interleaved and noninterleaved scans

The main procedures for lossless encoding are predictive coding and entropy encoding, which are shown in Fig. 8.12. Source-image precision may be varied form 2 to 16 bits. The level shifting necessary in DCT mode is not needed here, and input data represented by unsigned integers are directly predicted. In lossless predictive coding there is no quantizer, as in DPCM. Lossless mode can remove redundant information only. Either Huffman coding or arithmetic coding may be used. The statistical model for the arithmetic coder is extended to a two-dimensional form. The categories in Table 8.3 for Huffman coding are extended up to 16-bit precision.

A predictor, which is one of the important factors in lossless mode, may be designed in one- or two-dimensional space. Figure 8.13 shows the neighboring causal pixels (a, b, c) used for prediction of the current pixel x.

The JPEG standard suggests eight different predictors, as shown in Table 8.6 (recognized by selection value). The selection value 0 means no prediction and is used for the hierarchical mode of operation. The others are three one-dimensional predictors and four two-dimensional predictors. P_x is the prediction value of x, and R_a, R_b, and R_c are the reconstructed values of pels a, b, and c, respectively.

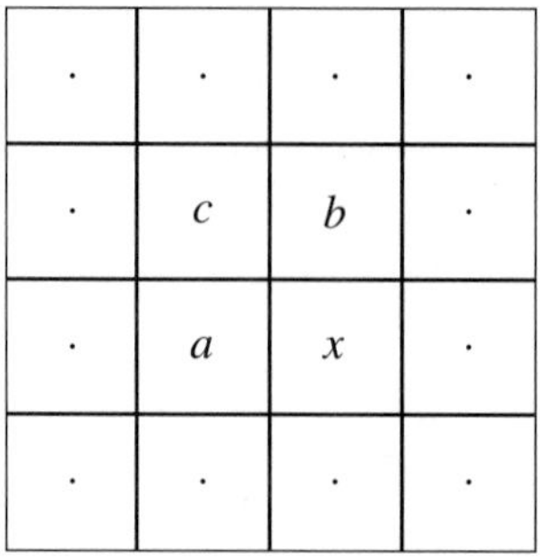

·	·	·	·
·	c	b	·
·	a	x	·
·	·	·	·

Figure 8.13 Causal pixels a, b, c, surrounding the pixel x under prediction.

Table 8.6 Predictors for lossless coding [367]

Selection value	Predictor design
0	no prediction
1	$P_x = R_a$
2	$P_x = R_b$
3	$P_x = R_c$
4	$P_x = R_a + R_b - R_c$
5	$P_x = R_a + ((R_b - R_c)/2)$
6	$P_x = R_b + ((R_a - R_c)/2)$
7	$P_x = (R_a + R_b)/2$

Source: © 1993 ITU-T.

The prediction error is calculated modulo 2^{16} and VWL coded using the same codes as those for the prediction error of the DC coefficient (see Table 8.4). The lossless mode allows 2- to 16-bit precision. Categories of Table 8.3 are extended as shown in Table 8.7.

A corresponding Huffman coding table may be defined using the entropy coding concept. Arithmetic coding can also be used for coding the prediction error [380, 381].

The draft recommendations established by the Joint Binary Image Group (JBIG) for progressive bilevel image compression are described in [400]. This specification can be effectively used for lossless coding of grayscale and color images by simply coding bit planes independently as if each bit plane were by itself a bilevel image. It is claimed that this simple approach can be more efficient than the JPEG lossless mode while at the same time providing for progressive buildup in both spatial and grayscale refinements.

Table 8.7 Difference categories for lossless Huffman coding [367]

SSSS	DIFF values
0	0
1	$-1, 1$
2	$-3, -2, 2, 3$
3	$-7 \cdots -4, 4 \cdots 7$
4	$-15 \cdots -8, 8 \cdots 15$
5	$-31 \cdots -16, 16 \cdots 31$
6	$-63 \cdots -32, 32 \cdots 63$
7	$-127 \cdots -64, 64 \cdots 127$
8	$-255 \cdots -128, 128 \cdots 255$
9	$-511 \cdots -256, 256 \cdots 511$
10	$-1023 \cdots -512, 512 \cdots 1023$
11	$-2047 \cdots -1024, 1024 \cdots 2047$
12	$-4095 \cdots -2048, 2048 \cdots 4095$
13	$-8191 \cdots -4096, 4096 \cdots 8191$
14	$-16383 \cdots -8192, 8192 \cdots 16383$
15	$-32767 \cdots -16384, 16384 \cdots 32767$
16	32768

Source: © 1993 ITU-T.

8.7 Hierarchical Process

Hierarchical processes may be implemented using any of the coding modes, i.e., lossy or lossless, sequential or progressive, and using Huffman or arithmetic coding as illustrated in Fig. 8.14. Any JPEG mode may be represented hierarchically with a number of spatial resolutions. In hierarchical mode, the source image is first successively down-sampled by a factor of two horizontally, or vertically, or both as shown in Fig. 8.15. The lowest resolution image is then compressed using any of the JPEG sequential or progressive algorithms. This image is then decompressed and bilinearly interpolated so that it matches the resolution of the next higher stage. This interpolated image is used as a prediction of the (down-sampled) source image and the resulting difference image is coded using a sequential or progressive algorithm.

The hierarchical mode is useful in applications in which a high-resolution image must be accessed by a low-resolution device that does not have the buffering or computational power to reconstruct the image to full size and then scale it down to the required size [382].

The basic functions in hierarchical processes are down-sampling (decimation) and upsampling (interpolation), which are shown in Figs. 8.16 and 8.17 as an example. Each intermediate stage, using the DCT-based process, is produced via FDCT and the relevant encoder. The reconstructed and upsampled version is

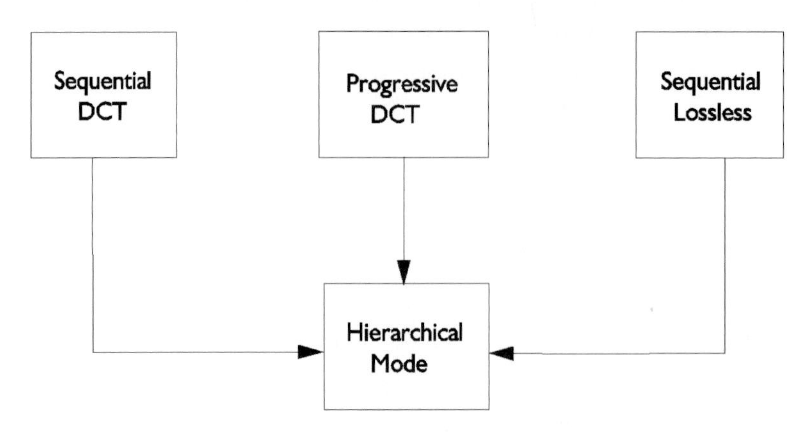

Figure 8.14 Possible hierarchical modes.

Figure 8.15 Down-sampling procedures horizontally (H) and vertically (V).

compared with the higher intermediate version. The down-sampling filter is to be compatible with the up-sampling filter. Filter coefficients are designed by users and Fig. 8.18 shows a low-pass filter (LPF) example that is suggested by JPEG IS. The same results are obtained by a separable or a composite filter. Another example is described in Fig. 8.19. The images shown in Fig. 8.20 are reconstructed from simple decimation and interpolation processes shown in Fig. 8.21. Note that only a part of the image is taken and magnified for easy detection. Although we used bilinear interpolation techniques, Fig. 8.20 implies that the simple structure produces blurred output images and we need well designed parameters for the LPF or interpolator.

In some stages of Figs. 8.16 and 8.17 a lossless DPCM or progressive coding technique may be used. Compressed data and the reconstructed image at the nth stage are obtained, denoted by $C(n)$ and $R(n)$, respectively.

To get a low-resolution image, a high-resolution image is decimated. When the $N \times N$ image is 2:1 horizontal and vertical decimated, the resulting image has $N/2 \times N/2$ pixels. An LPF is used for removing aliasing as a pre- and postfilter. A simple interpolation method is a bilinear one taking average values between two pixels and between four pixels for center pixel interpolation, as presented in Fig. 8.22.

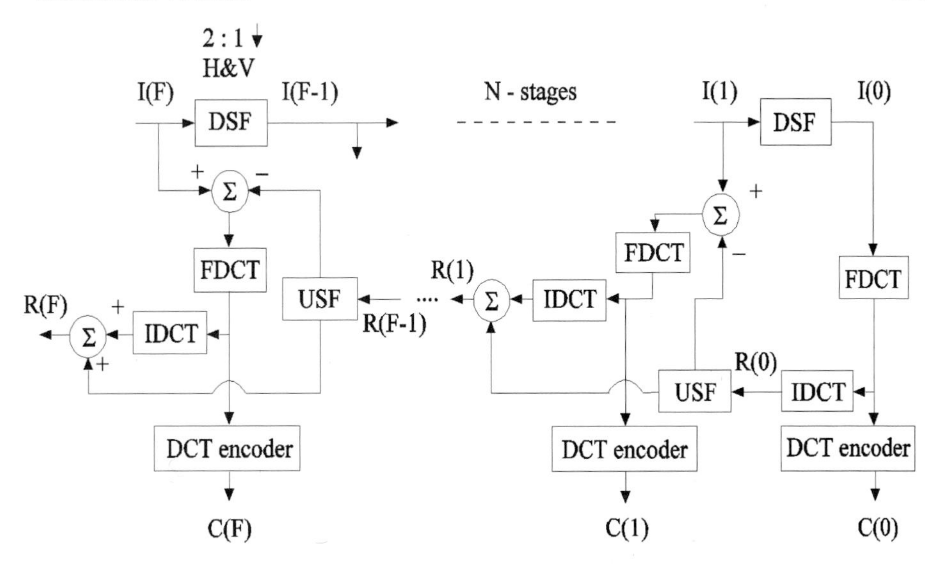

Figure 8.16 A hierarchical encoder example (DSF: downsampling filter, USF: upsampling filter).

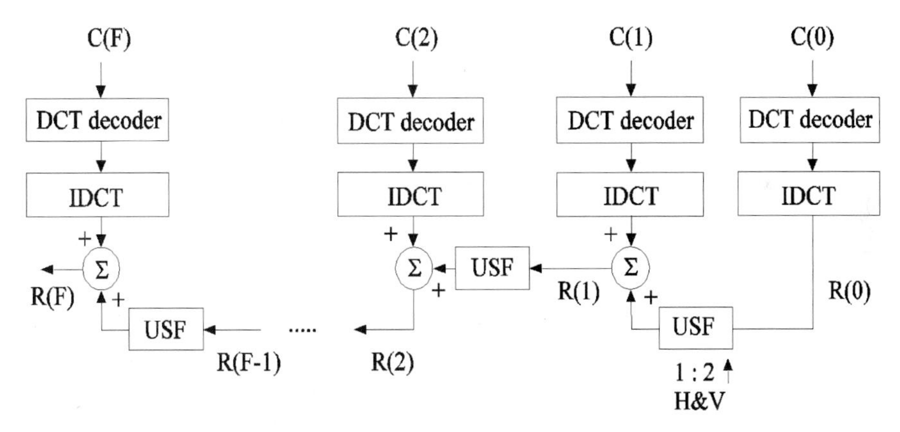

Figure 8.17 A hierarchical decoder example. (DSF: downsampling filter; USF: upsampling filter).

The hierarchical mode represents an image with multiple spatial resolutions. For example, one could provide 1024 × 1024, 512 × 512, 256 × 256, 128 × 128, and so on versions of the images, when we apply 2:1 down- and up-sampling. The higher-resolution images are coded by difference values from the next coarse images, and thus need much lower bit rates than they would if transmitted independently. However, the total number of bits will be greater than that needed to store just the highest-resolution image.

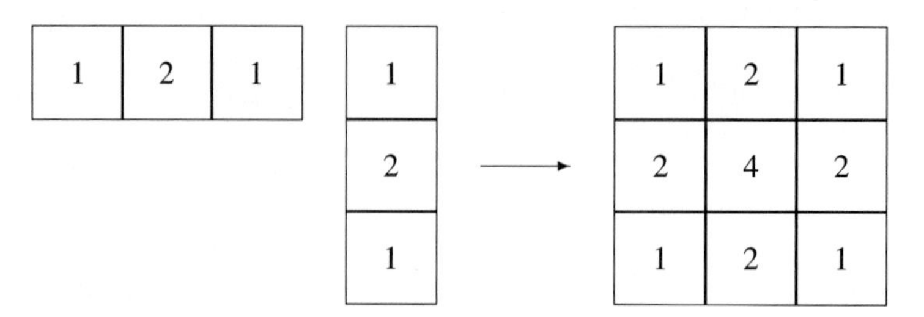

Figure 8.18 A separable 1D LPF and an equivalent 2D LPF (JPEG informative). The normalization factors for the 1D and 2D LPF are $\frac{1}{4}$ and $\frac{1}{16}$, respectively.

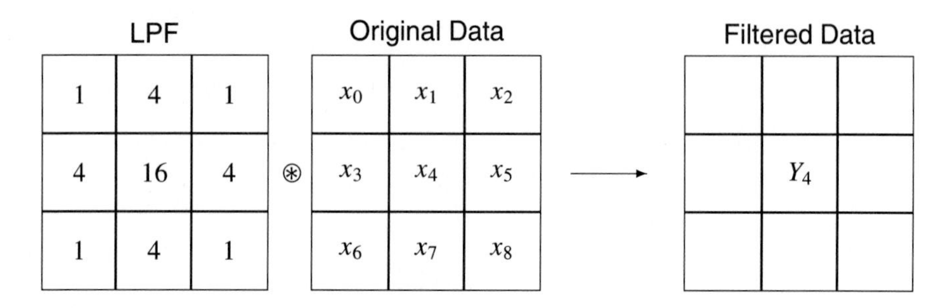

$$Y_4 = \frac{1}{36}(x_0 + 4(x_1 + x_3 + x_5 + x_7) + x_2 + x_6 + x_8 + 16x_4).$$

Figure 8.19 An example of a 2D LPF. (This LPF is also a separable filter. Y_4 is the filtered pixel.)

For the compression of image data, either DCT or a spatial-predictive version can be preselected, subject to the application purposes. Figure 8.23 shows two main paths and the corresponding resolution hierarchies. The DCT path may be used for applications that need high compression for a given level of visual distortion. Quantization of the DCT coefficients degrades the image quality, although it is possible to use lossless and progressive transmission (PT). The lossless path may be used for applications that need a simple progression with a truly lossless final stage.

8.8 JPEG Extensions

ISO/IEC 10918-1 (ITU-T Rec. T.81) [367] specifies requirements and implementation guidelines for continuous-tone still-image codec and for the coded representation of compressed image data, as discussed in the previous sec-

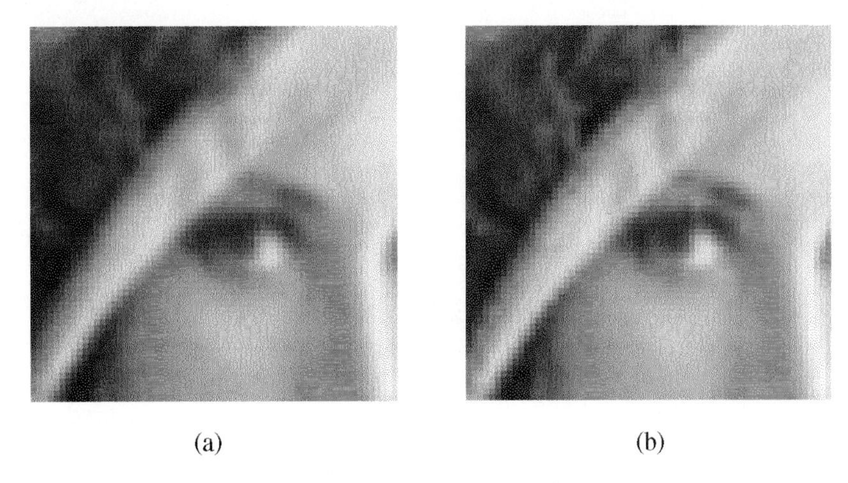

Figure 8.20 Output images from down- and up-sampling process, filtered by the separable filters (a) (1 2 1) and (b) (1 4 1).

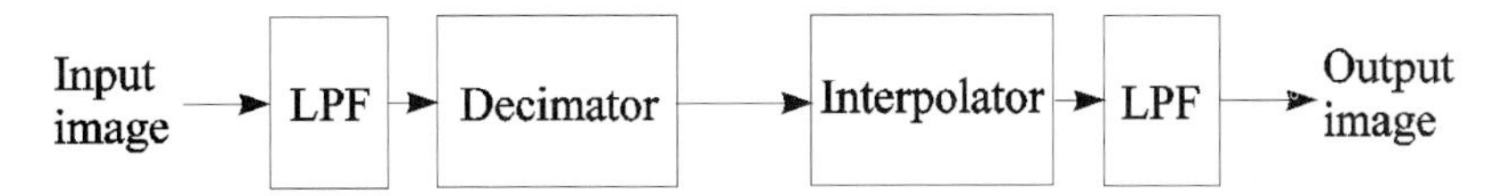

Figure 8.21 Typical decimation and interpolation processes.

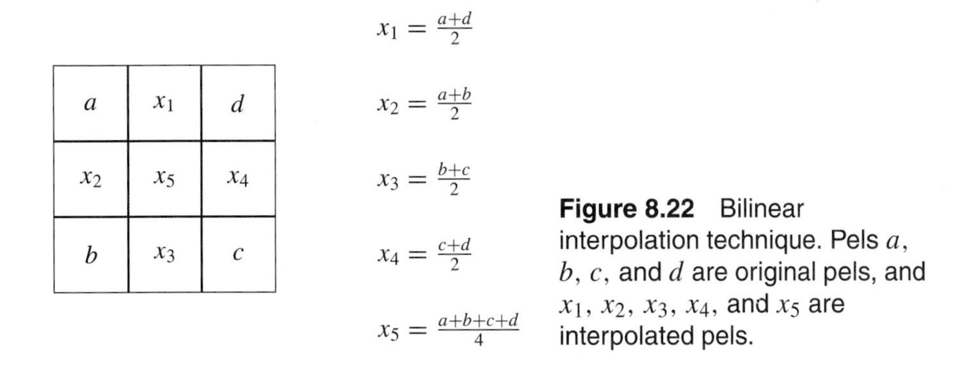

Figure 8.22 Bilinear interpolation technique. Pels a, b, c, and d are original pels, and x_1, x_2, x_3, x_4, and x_5 are interpolated pels.

tions. After these guidelines were established, the image-compression format *jpeg* has been utilized in applications such as computer graphics and image transmissions. It is required, however, to apply to a wide range of applications: large image (e.g., 1280×1280) processing, variable quantization on a block basis, extended modes from ISO/IEC 10918-1, and so forth. Hence, extensions for the JPEG algorithm have been established [368] and compliance testing for implementations has

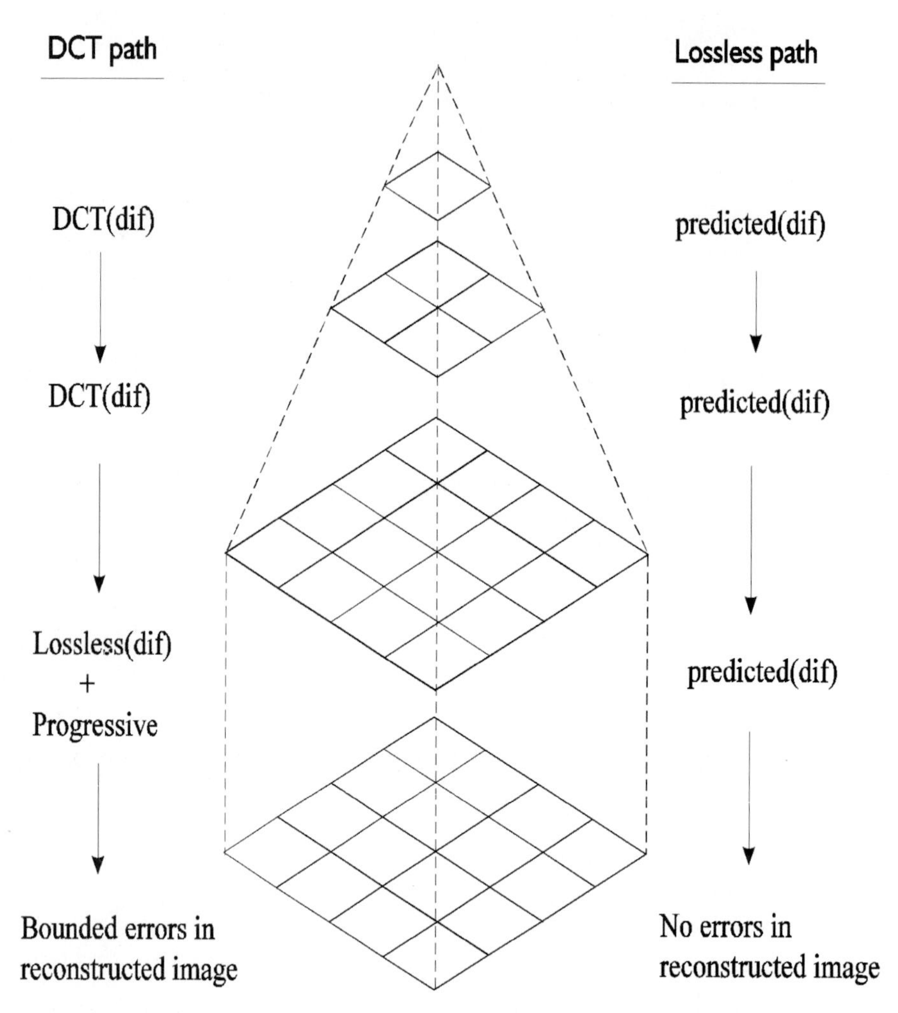

Figure 8.23 Hierarchical multi-resolutions and two possible paths [367]. © 1993 ITU-T.

been performed [369]. In this section, we describe the various extensions for general applications of the JPEG algorithm.

Variable quantization extension

ISO/IEC 10918-1 permits quantization tables to be redefined prior to the start of an image but does not allow modification of the tables within an image. Some applications may require capturing image information from a video sequence and preserving it in a digital form. For example, transcoding from an

Table 8.8 Quantizer scale code (5-bit) and corresponding quantizer scale factor (qs)

	Scale factor			Scale factor	
Scale code	Linear	Nonlinear	Scale code	Linear	Nonlinear
0	(forbidden)		16	32	24
1	2	1	17	34	28
2	4	2	18	36	32
3	6	3	19	38	36
4	8	4	20	40	40
5	10	5	21	42	44
6	12	6	22	44	48
7	14	7	23	46	52
8	16	8	24	48	56
9	18	10	25	50	64
10	20	12	26	52	72
11	22	14	27	54	80
12	24	16	28	56	88
13	26	18	29	58	96
14	28	20	30	60	104
15	30	22	31	62	112

MPEG-1 or MPEG-2 bitstream is useful in some cases. Since the quantization table in MPEG coding can be scaled in a macroblock, the variable quantization extension introduces a quantizer scale factor and provides a means for changing the quantization matrix values at the start of any (8×8) block. This extension provides the following advantages:

- Transcodability from other compression file formats such as MPEG
- The ability to compress an image to a fixed-size compressed image by changing the scale factor
- Increasing subjective quality by exploiting the masking properties of the human visual system

If it is needed to change a quantization matrix for a block, a symbol x'15' indicating quantizer change is added to the table of the DC coefficient differences. As a result, the tables used for 8-bit precision (Table 8.3) are extended to 13 symbols. The 4-bit change symbol is followed by a 5-bit scale code (1 through 31) and the corresponding quantizer step size is determined. Two different lookup tables, a linear table and a nonlinear table, as shown in Table 8.8, are allowed as in MPEG-2 (Fig. 11.19). To indicate which of the two tables is used, a marker code, called DQS (define quantizer select), can be included in the JPEG bitstream.

The quantizer scaling does not affect the quantization of the DC coefficient that is quantized by Eq. (8.1). Each of the resulting 63 AC coefficients S_{uv} is quantized by the following equation:

$$S_{quv} = \text{Nearest integer}\left(\frac{S_{uv} \times 16}{Q_{uv} \times qs}\right) \tag{8.11}$$

where qs denotes the quantizer scale factor defined by the linear or nonlinear table (Table 8.8). The value of qs is set to 16 at the start of each frame (i.e. not scaled from the quantization table), until it is modified by a subsequent occurrence of the qs change symbol, x'15'.

Using the variable quantization in a JPEG coder, the subjective and objective quality increases as shown in Fig. 8.24. At a fixed bit rate, the sensitive errors (block boundary) are moved to less sensitive regions where high-frequency details are dominant. This can be achieved by changing the scale factor according to the block activity (Fig. 14.3) and the human visual system [383].

Selective refinement and tiling extension

The selective refinement extension refers to selecting and defining a part of an image for further refinement. There are several types of selective refinement according to the mode of operation.

In all modes of operation (sequential, progressive, lossless, and hierarchical), any color component may be coded with fewer bits. Then the color component may be selected for further refinement, called *component selective refinement.* Images to be refined by this type of selective refinement are mixed grayscale and color.

In the DCT-based progressive mode of operation, only part of an image may be coded by *progressive selective refinement.* When it is applied to a scan that uses the spectral selection procedure, more DCT coefficients are transmitted for part of a frame. When it is applied to a scan that uses the successive approximation procedure, more bits are added to part of a frame.

In the hierarchical mode of operation, a particular region of interest in an image may be coded with greater detail than the remainder of the image by using *hierarchical selective refinement.* The location of the subpart of the image to be selectively refined is specified immediately prior to a differential frame within a hierarchical sequence. The size of the subpart is specified in the differential frame header. Any type of selective refinement is signaled by a corresponding marker segment as one of the tables/miscellaneous marker segments.

Tiling extension is used when an image is too large to be easily processed by either the compressor or the decompressor. A large image is divided into small subpart files so as to display a tile image on a given size screen and to

(a)

(b)

Figure 8.24 Reconstructed images from (a) a JPEG baseline system and (b) a variable quantization extension at the same bit rate. The blocking artifact is reduced in (b).

process image regions of interest. There are three types of tiling techniques, simple tiling, pyramidal tiling, and complex tiling, arranged in the order of increasing complexity.

Simple tiling is useful for dividing a large image into smaller rectangular tiles and for providing random access points in the middle of a compressed image. The tiles are of a fixed size and are nonoverlapping, with the exception of tiles that fall on the bottom and right of the image as shown in Fig. 8.25a. Tiles are coded sequentially from left to right and top to bottom. All the tiles have the

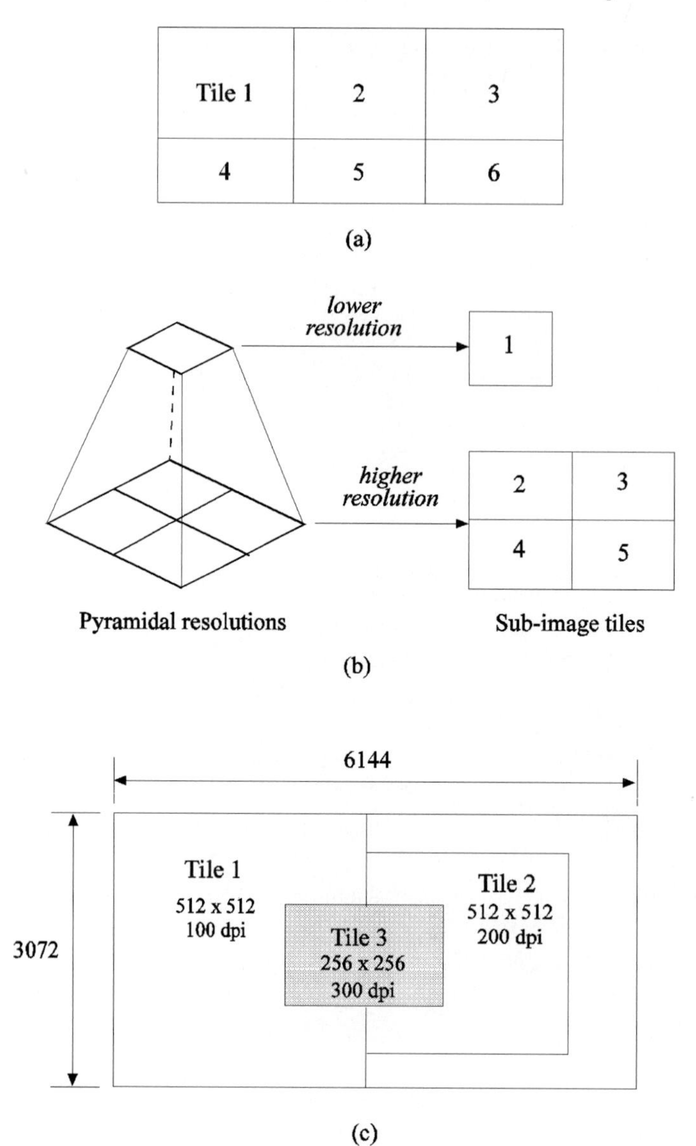

Figure 8.25 Examples of (a) simple tiling, (b) pyramidal tiling, and (c) complex tiling.

same compression ID, sampling frequency, and compression factor, whereas other compression parameters such as quantization table and VLC table may be changed for each tile.

Pyramidal tiling is used to provide multiple resolutions of an image as shown in Fig. 8.25b. Each resolution image can also be partitioned into subimage tiles.

Tiles in one resolution level may overlap those in other levels, but they do not overlap tiles in the same resolution image. The tiles are placed into the compressed image bitstream in increasing resolution order and raster order within a given resolution. A selected resolution image or tiles of the subimage can be displayed on a given screen.

Complex tiling is used to provide multiple resolutions on a single display image screen (higher-resolution tiles and lower-resolution tiles). A reference resolution is determined so that all the tiles that have different resolutions may be combined (displayed) without resampling of a tile. Tiles are placed into the compressed image data stream in order of the lowest-resolution to the highest-resolution level. For example, as shown in Fig. 8.25c, three subimages are displayed on a screen. They have different resolutions and sizes. Given one (512×512) image scanned at 100 dpi resolution, one (512×512) image at 200 dpi, and one (256×256) image at 300 dpi, reference resolution must be 600 dpi. The first and second subimages can be nonoverlapped, while the third can be overlapped. Hence, the complex tiling is a superset of simple tiling and pyramidal tiling.

Still-picture interchange file format

ISO/IEC 10918-1 (JPEG Part 1) does not define an image interchange file format that can be used by other applications. The increased interest in using JPEG data streams in applications such as MHEG multimedia has led to generating a still-picture interchange file format (SPIFF) conforming to JPEG Part 1 and JPEG extensions (Part 3). Thus the SPIFF extension provides for the interchange of compressed image files between various application environments. It is a complete coded image representation and includes all parameters necessary to reconstruct an image. It consists of file header, directories that contain information necessary to decode the image data, and interchange format data streams. The SPIFF allows other applications to encapsulate JPEG data within their own file formats such as JFIF, TIFF, and PostScript. Currently, ISO/IEC 10918-3 provides an example of how to transcode the file format JFIF into SPIFF with a sample program written in C code to carry out a simple translation.

8.9 JPEG Implementations

Hardware implementations of JPEG are now available from a number of sources. Since JPEG (baseline and extended, other than hierarchical) uses DCT, VLSI design techniques of the DCT have provided incentives for developing JPEG hardware [384, 398, 399, 402, 403]. Many manufacturers make JPEG encoder/decoder chip sets [385, 386, 387, 389, 391, 392]. Other companies provide board-level application sets on PCs or dedicated image-compression systems, using the chip sets. Some of the chip sets contain a subfunction to meet the JPEG algorithm, as well as moving picture compression standard, MPEG [487], or H.261 [393]. However, most vendors, until now, like to realize the baseline system of JPEG rather than its extension modes (progressive, hierarchical, lossless, or arithmetic coding). There are two reasons for this: first, they need dedicated faster hardware in progressive mode, and they need more bits than are required to store just the highest-resolution frame in hierarchical mode; and second, the ideal lossless or IDCT implementation is not possible because of round-off errors, and arithmetic coding offers 8 to 17% better compression, which may not be enough to justify patent fees for implementing these techniques [374]. Nevertheless, the chip sets can be applied to systems in a variety of fields, such as

- Computer and multimedia add-on boards
- Full motion video compression
- Digital still cameras and peripherals
- Video floppy disks
- Security and industrial systems
- Videophones and color fax machines
- Color printers and scanners
- Fixed bit-rate image transmission devices

For digital still-camera applications, several issues have to be solved, i.e., bit-rate control to guarantee the number of images in a memory card, high-speed compression techniques to meet fast contiguous shooting, and compatibility with other systems. Power dissipation is a major problem for portable applications operating at high speeds [395, 396]. The electronic digital cameras DS-505 and DS-515 developed by Fuji use JPEG data compression.

Obtaining constant bit rate by buffer status control, however, is not as important in storing applications as in transmitting JPEG still images. In this case, feedback buffer control is incorrect at scene changes that give output buffer status information to the quantizer. Feedforward control that gives scene change information to the quantizer gives improvement at scene changing instants [397].

A typical JPEG chip set consists of a DCT processor and an image coder/decoder, that work together to perform image compression and expansion (a two-chip solution) [392, 402]. The DCT processor implements both forward and inverse DCT computations on (8×8) blocks, while the coder/decoder implements quantization and variable length encoding operations for image compression and the corresponding inverse operations for decoding. The chip set is useful for reducing the large amounts of data required to store and/or transmit digital images.

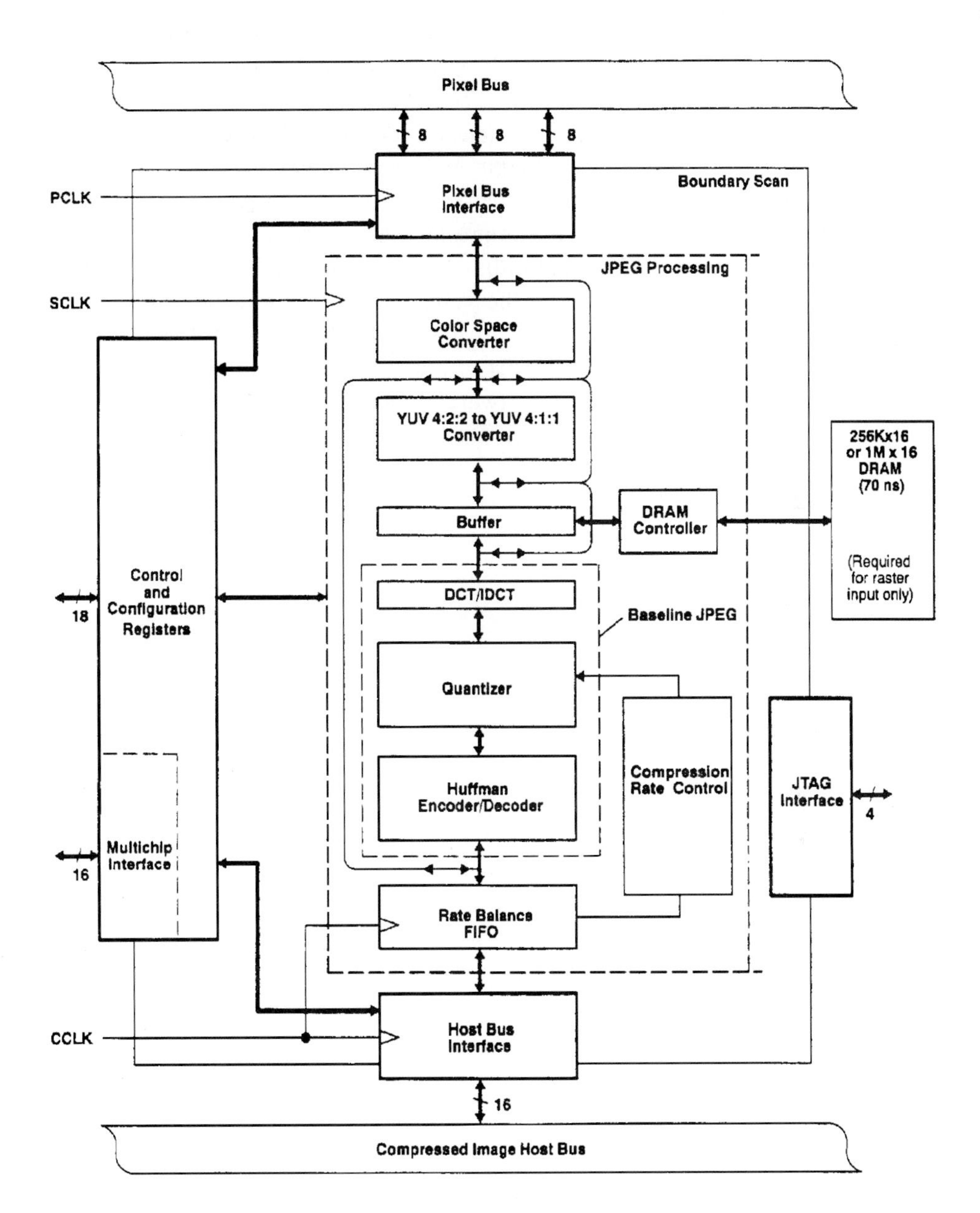

Figure 8.26 An example of a single-chip JPEG image-compression system [391]. © 1993 TI.

For example, the ZR36040 chip by Zoran Corp. reduces 69 Mbytes of storage data to 2.9 Mbytes, which is feasible on a PC hard disk [392].

The SN74ACT6341 JPEG engine with a 33 MHz clock rate by TI yields a processing rate of 16.285 Mpel/sec at 8 or 12 bpp precision [391]. This single chip can process DCT/IDCT and encoder/decoder operations. Compression rate can be controlled by an output buffer overflow. Figure 8.26 shows the JPEG system (baseline mode embedded) and peripheral interfaces. Such fast systems are capable of compression/decompression at real-time video rates. Therefore, some vendors use the JPEG algorithm for encoding full-motion video that is known as motion JPEG [370]. But the system may be useful only in limited areas, due to efficiency that is lower than that of MPEG and its lack of audio function. The reader may refer to the list of manufacturers who provide JPEG hardware in Appendix A. Also, JPEG software is available from several companies, such as the Picture Prowler from Xing Technology.

Chapter 9
ITU-T H.261 Video Coder

Summary

ITU-T Recommendation H.261 has been developed with the aim of video transmission at $p \times 64$ Kbps (p in the range of 1 to 30). Possible applications of H.261 include videophone and videoconferencing. The video coding algorithm must be able to operate in real time with limited delay and to compress moving images in the spatial and temporal domains, by using 2D (8×8) DCT and motion-compensated prediction, respectively.

We explain in this chapter the common intermediate format (CIF) that is a compromise based on the different television formats such as NTSC, PAL/SECAM, and ITU-R BT.601. A video-compression algorithm in the macroblock unit will be discussed. The compressed video bit stream is transmitted with a layered data structure and error correction coding. Finally, implementations by VLSI chips or system manufacturers and updated suggestions for the low bit-rate video will also be covered.

9.1 Introduction

Since the late 1980s, there has been much demand for visual telephony through the historical telephone lines, which have a narrow bandwidth. In view of the proliferation of PCs (personal computers), workstations, mainframes, terminals, digital networks, and communication lines, this is a natural trend. Increasing processing speed and storage and reduced costs have contributed to this proliferation.

Similar to the JPEG algorithm discussed in the previous chapter, another international standard for videophone and videoconferencing has been established. In JPEG, there is no process for time variables, such as field or frame processes, because JPEG deals with still images only. We indicated briefly the possibility that we can use the JPEG algorithm for motion video with some limitations, such as the temporal redundancy reduction problem.

The ITU-T (formerly CCITT) H-series standard [420, 449, 450], discussed in this chapter, is applicable to videophone or videoconferencing. It is similar to JPEG in that DCT is used as a main data compression tool in intraframe and the transform coefficients are coded by a VLC, e.g., a Huffman coder. But the primary difference between the two standards is the use of the motion-compensated temporal prediction adopted in H.261 (the video codec for audiovisual services at $p \times 64$ Kbps). The motion-compensated temporal prediction errors are mapped into the DCT domain (to remove the spatial correlation) followed by quantization and VLC [419].

Therefore, the overall structure is a hybrid system in the spatial and temporal domains. Hybrid denotes a system which involves more than one redundancy reduction technique, such as transform or predictive coding [432]. MPEG-1, developed for digital storage media (DSM) (see the following chapter for details), and H.261 are similar from a compression point of view, but MPEG provides greater flexibility at the expense of increased complexity for extended application purposes. Most of this complexity, however, is on the encoder side, as the objective is to make the decoder as simple (hence low-cost) as possible. Although H.261 specifies video coding, the audio coding is indirectly specified in ITU-T Recommendation H.320, narrow-band visual telephone systems and terminal equipment.

Brief history of the H.261 video coder

The standardization activities for low bit-rate video transmission can be summarized as follows:

- Dec. 1984: Specialists Group on Coding for Visual Telephony established (CCITT SG XV)
- Sept. 1985: Common intermediate format (CIF) adopted
- 1985–1988: The first step research for $n \times 384$ Kbps ($n = 1$ to 5)
- 1986: Coding concept $m \times 64$ Kbps ($m = 1, 2$) initiated
- 1988–1989: The second extension to $p \times 64$ Kbps ($p = 1$ to 30)
- Sept. 1988: A number of coding algorithms suggested
- Dec. 1988: Abandoned 4/9 CIF

- Mar. 1989: $p \times 64$ Kbps specification frozen
- June 1989: Minor modification to the algorithm (the final Reference Model 8 simulated)
- Sept.–Nov. 1989 : Hardware trials
- Dec. 1990: Standardization established (H-series)

Thus, recommendation H-series has been formulated by the CCITT SGXV Specialists Group on Coding for Visual Telephony. There are five related recommendations in the H-series developed by CCITT SGXV:

1. H.221: Frame structure for a 64 to 1920 Kbps channel in audiovisual teleservices
2. H.230: Frame synchronous control and indication signals for audiovisual systems
3. H.242: System for establishing communication between audiovisual terminals using digital channels up to 2 Mbps
4. H.261: Video codec for audiovisual services at $p \times 64$ Kbps
5. H.320: Narrow-band visual telephone systems and terminal equipment

In this chapter, we focus on Recommendation H.261 among these series. To show specific properties of H.261 more efficiently, we compare the three typical coding standards, H.261, JPEG, and MPEG-1, in Table 9.1. MPEG-1 is discussed in Chapter 10. Here we can compare common modules and different characteristics [474]. Table 9.2 shows the common video formats used in the standards.

Table 9.1 Brief comparison of H.261, JPEG, and MPEG-1

		JPEG (Baseline)	H.261	MPEG-1
(8×8) DCT		Yes	Yes	Yes
Zigzag scan		Yes	Yes	Yes
2D run-length coding		Yes	Yes	Yes
Quantizer		Default tables	32 step sizes	INTRA: similar to JPEG INTER: similar to H.261
Motion estimation		No	Forward	Forward/backward
MV accuracy		No	Integer pel accuracy	Half pel accuracy
Rate buffer control		No	Yes	Yes
	TCOEFF	Downloadable	Fixed	Superset of H.261
VLC	MBA, CBP	No	Fixed	Same as H.261
	MV, MTYPE	No	Fixed	Superset of H.261

Table 9.2 Digital video formats specified by ITU-R 601, JPEG (J), H.261 (H), and MPEG-1 (M) (I: interlaced, P: progressive)

Image format	Resolution	YC_bC_r	Scan type	Mbyte/s	Use
ITU-R 601 (NTSC)	720×480	4:2:2	I	20.7	J
ITU-R 601 (PAL)	720×576	4:2:2	I	20.7	J
Square pixel (NTSC)	640×480	4:2:2	I	18.4	J
Square pixel (PAL)	768×576	4:2:2	I	22.1	J
ITU-R SIF (NTSC)	352×240	4:2:0	P	3.8	M, J
ITU-R SIF (PAL)	352×288	4:2:0	P	3.8	M, J
Square pixel SIF (NTSC)	320×240	4:2:0	P	3.5	M, J
Square pixel SIF (PAL)	384×288	4:2:0	P	4.1	M, J
CIF (NTSC & PAL)	352×288	4:2:0	P	4.6	H
QCIF (NTSC & PAL)	176×144	4:2:0	P	1.1	H

The data rate of the ITU-R BT.601 format is 20.7 Mbyte/sec in both the NTSC and PAL systems. The square pixel systems retain the 4:3 aspect ratio that is applicable to conventional television monitors. The CIF (common intermediate format) and the QCIF (quarter CIF) provide the same resolutions in both the NTSC and PAL systems, while the SIF (source input format) video provides equivalent data rates, as discussed in Section 10.2.2.

The use of 2D (8×8) DCT to remove intraframe correlation, zigzag order to scan the transform coefficients, and run-length coding for zero-valued coefficients after quantization is common to all three standards. In H.261, motion estimation is applied to video sequence to improve the prediction between successive frames, but it is not as complex as in MPEG, which is designed for a number of interactive applications. This is because application fields of H.261 are limited to videophone and videoconference, which have relatively slow-moving objects. For the purpose of simplex or duplex communication, different from JPEG, the transmission rate has to be controlled in the range of $p \times 64$ Kbps, where p is up to 30. Naturally, data-compressed image transmission is more sensitive to channel errors. Error resilience including synchronization and concealment technique is required in the transmission coder shown in Fig. 9.1.

To permit a single recommendation for the different video formats, such as the 625-line (PAL or SECAM) and 525-line (NTSC) formats, common intermediate format (CIF) was developed [409, 413]. In CIF, the luminance (Y) sampling structure is 352 pels per line, 288 lines per picture, in an orthogonal arrangement, since the source coder works in the (8×8) block dimension.

The H.261 coder operates under a hierarchical data layer structure. The four layers are Picture layer, Group of Block (GOB) layer, Macroblock (MB) layer, and the fundamental (8×8) block layer. The layers are multiplexed for transmission

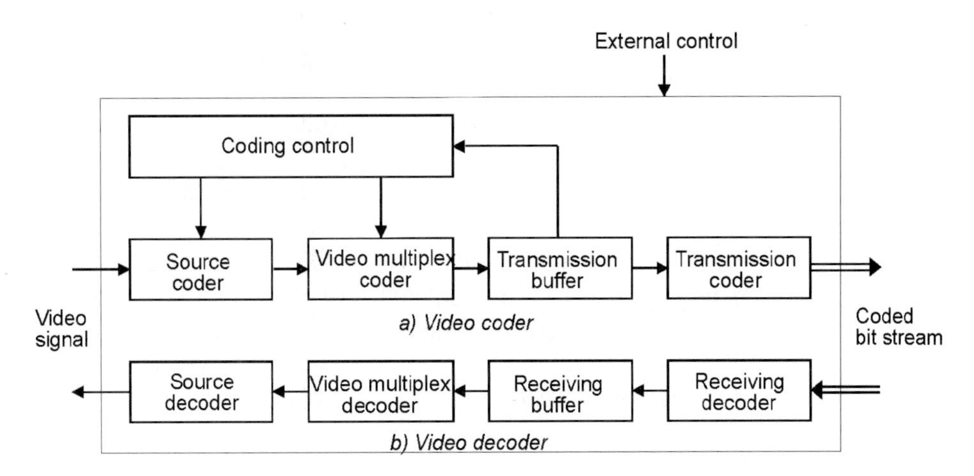

Figure 9.1 A block diagram of the H.261 codec [450]. © 1990 ITU-T.

in series. Each layer is composed of data and corresponding header information bits, which will be discussed later.

The standard does not specify all the parameters or exact components for each function. The following operations are not subject to Recommendation H.261:

1. The conversion to CIF from the video source such as NTSC, PAL, SECAM, ITU-R 601, etc., and vice versa.
2. The decoding of the BCH (511,493) error correction code.
3. The use of intra or inter mode.
4. Motion estimation in the encoder. (One MV (motion vector) per macroblock may be transmitted.)
5. The use of the loop filter in the encoder.
6. The arithmetic process for computing the FDCT.
7. The control of the video data rate.
8. Any pre- or postprocessing.

In the following sections we describe video data formats and hierarchical structure, source coding algorithms including motion estimation and prediction, channel coding problems, hardware implementation methods, and suggestions for improving the codec performance.

9.2 Hierarchical Multiplex Format

Hierarchy denotes several layers in the overall coding process. The basic block is composed of (8 × 8) pixels, the same as in JPEG. The next higher level in the hierarchy is a macroblock, which is used for motion estimation. Another reason to use a hierarchical structure is that the loss of a whole frame can be avoided by the number of hierarchical synchronizations. Certain regions that have uncorrectable errors can be duplicated or interpolated. We describe the frame format called common intermediate format (CIF) applicable to H.261.

9.2.1 Common Intermediate Format

For the video communication services, the video format has to be based on the existing broadcasting standards: NTSC, PAL, and SECAM. Visual telephony, which usually has slow moving objects (head and shoulders configuration), does not, in general, need a large bandwidth. Also, the viewer can tolerate the visual quality, which can be less than that of the TV broadcasting level. Moreover, cost reduction is required for a wide range of applications. Thus, spatial resolution may be reduced, and the CCITT SGXV Specialists Group has adopted the Common Intermediate Format (CIF) and Quarter CIF (QCIF) considering the possibility of information exchange and sharing the burden between 525/60 and 625/50 systems, as described in Fig. 9.2 [411]. The parameters for the two formats (CIF and QCIF) are shown in Fig. 9.3 [418]. For the luminance component, Y, 360 pels per line were adopted, just a 2:1 decimation from the 720 active pels per line used in both the TV systems (ITU-R 601). Two parameters are adopted based on sharing the horizontal resolution burdens: one parameter, 288 active lines, is influenced by the 625/50 (PAL/SECAM) system which has 576 active lines (twice the CIF resolution), and the other one is a frame rate of 29.97, which is based on the 525/60 (NTSC) system [409]. It may be emphasized that for both CIF and QCIF configurations, the frames are noninterlaced (progressive).

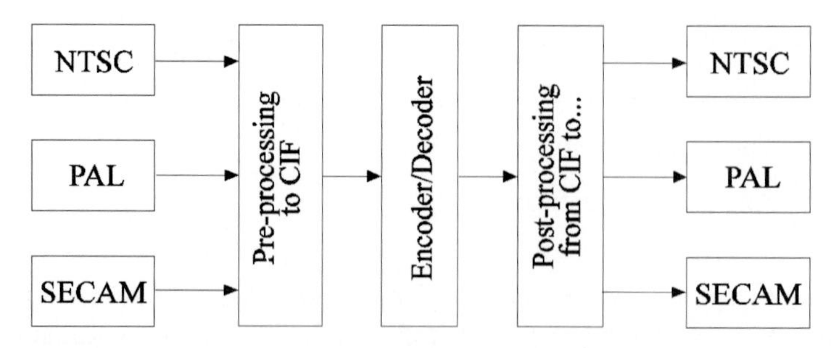

Figure 9.2 Information exchange via the common intermediate format (CIF).

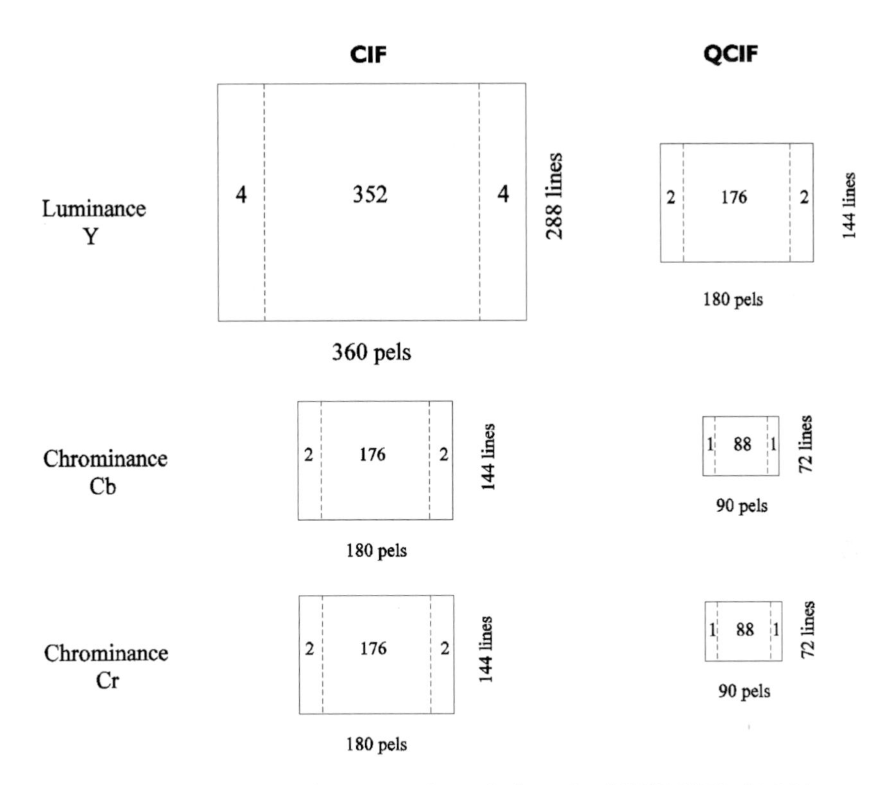

Figure 9.3 Picture format and resolutions for CIF/QCIF in H.261.

The processing of images is block based, and the basic block size is (8×8) pixels. Because motion estimation is based on (16×16) luminance, Y blocks, vertical stripes of 4 pels on either side of the Y-component, are cropped out, resulting in 352 pels/line and 288 lines/picture resolution. This is called the significant pel area (SPA). Henceforth, for the CIF/QCIF we will use the term picture rather than frame to stress that it is noninterlaced. The picture resolution of the SPA is now integer divisible by the (16×16) block. The corresponding SPA for the QCIF and the chrominance components of the CIF are shown in Fig. 9.3. Roughly speaking, CIF still retains the 4:3 TV screen aspect ratio.

In Table 9.3, CIF/QCIF and the two systems (NTSC and PAL) are compared. Note that parameters of the two systems are based on ITU-R BT.601, and CIF/QCIF are noninterlaced (progressive) formats, whereas the TV systems have interlaced formats. The color components have half the spatial resolutions (both horizontal and vertical) of luminance as shown in Fig. 9.3. Thus, the color components (one C_b pel and one C_r pel) are located in the middle of four luminance pixels [431], as illustrated in Fig. 9.4.

Table 9.3 Parameters of the CIF and TV systems

		CIF	QCIF	NTSC (525/30)	PAL (625/25)
Active pixels/line	Y	360	180	720	720
	C_b, C_r	180	90	360	360
Active lines/picture	Y	288	144	480	576
	C_b, C_r	144	72	480	576
Pictures/sec		29.97	29.97	29.97	25

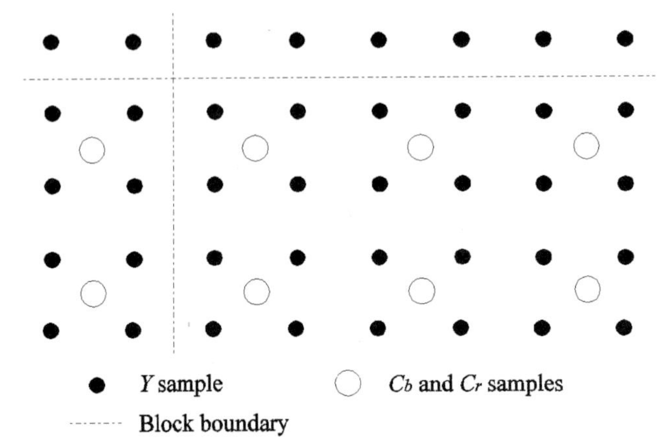

Figure 9.4 Positioning of luminance and chrominance pixels (4:2:0) in H.261.

In addition, further compression may be obtained by decreasing the picture rate, since H.261 allows 0, 1, 2, 3 or more nontransmitted pictures between transmitted ones at the decoder side. The latter may be reasonable, as objects in the videophone move in general quite slowly. The decoder should be able to decode a bit stream with any number of nontransmitted pictures according to its decoding capability. Encoders need not maintain a constant picture rate such as 10 Hz, but cannot transmit $p_0, p_3, p_5, p_8, \ldots$, if the decoder's capability is up to 10 Hz (two or more nontransmitted pictures between transmitted ones) because there is only one transmitted picture between p_3 and p_5, but $p_0, p_3, p_8, p_{12}, \ldots$ is valid. The bit rate for transmitting CIF at 15 pictures/sec is 18.3 Mbps in uncompressed form, still requiring a large compression ratio (from 10:1 to 300:1 in the range of $p \times 64$ Kbps). This is based on a spatial resolution (352×288) for Y and (176×144) for C_b and C_r, 15 fps, and 8-bit PCM (see Fig. 9.3). Note that spatial or temporal resolution reduction causes aliasing problems. Picture qualities are degraded and should be compensated by the appropriate pre- and postfilters.

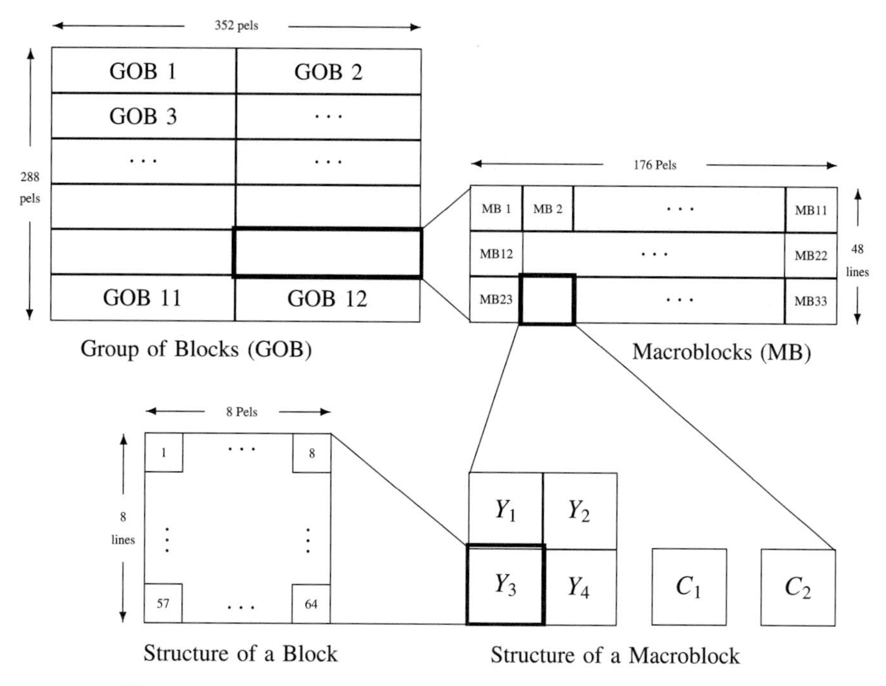

Figure 9.5 Hierarchical block structure in a CIF image.

9.2.2 Video multiplex arrangement

We shall discuss the video data structure of H.261 before describing coding techniques. The standard defines a consistent structure so that the decoder may decode the received bit stream without any ambiguity. The structure that has four layers of block-based data is shown in Fig. 9.5. The four layers and corresponding luminance (Y) pels in each layer are [450]

1. Picture layer: (352×288) pels (1584 basic blocks)
2. Group of block layer: (176×48) pels (132 basic blocks)
3. Macroblock layer: (16×16) pels (4 basic blocks)
4. Block layer: (8×8) pels (basic block)

In QCIF mode, the lower three layers are the same as in the CIF structure. Only the pels comprising the picture layer are a quarter of normal CIF, i.e., (176×144) pels. Thus, we have only 3 GOBs of luminance in QCIF.

Although this hierarchical concept has been introduced in early stages of the standard activities, parameter decisions have been the main subject [408]. Each layer consists of header information and data for the following layer. The header includes several kinds of information, as shown in Fig. 9.6.

Picture and GOB layers include start codes (PSC and GBSC) so that the decoder may synchronize again, under transmission error conditions. The others are relevant information, such as quantizer, motion vector, address, group of block number, and picture type. Acronyms used in Fig. 9.6 and necessary information bits are summarized as follows:

- PSC: Picture Start Code (20 bits, 0000 0000 0000 0001 0000)
- TR: Temporal Reference (5-bit FLC) is to indicate the source CIF picture number for decoder's use (how to use it is outside the scope of H.261). It is not for informing the picture rate.
- PTYPE: Picture Type (6-bit FLC). This informs the split screen indicator on/off, document camera indicator on/off, freeze picture release on/off, and source format CIF (1) or QCIF (0).
- PEI: Picture Extra Insertion
- PSPARE: Spare bits for future use (0, 8, 16, . . . bits)
- GBSC: GOB Start Code (16 bits, 0000 0000 0000 0001)
- GN: Group Number representing 12 GOBs (4-bit FLC)
- GQUANT: Group Quantizer information (5-bit FLC). This indicates one of the 31 quantizers to be used in a GOB until overridden by any subsequent MQUANT information.
- MBA: MB Address, indicating its position within a GOB (up to 11-bit VLC). MBA is the difference between the absolute address of the macroblock and the last transmitted macroblock in a GOB.
- MTYPE: MB type information as shown in Table 9.4 (VLC table)
- MQUANT: MB Quantizer information (5-bit FLC)
- MVD: Motion Vector Data (up to 11 bits, 32 VLCs)
- CBP: Coded Block Pattern (up to 9 bits, 63 VLCs)
- TCOEFF: Transform Coefficient (zigzag scanning and 8-bit FLC or up to 13 bits, 66 different VLCs)

In Table 9.4, CBP is not transmitted in intra mode and MVD-only mode. It implies that the macroblock has to be coded and transmitted in the two modes. If the CBP signal is defined, only the selected blocks in a macroblock are coded and the remaining blocks are reproduced from the previous picture. Figure 9.7 shows explicitly the codeword length of the possible 63 patterns for the four luminance and two color blocks. The most probable (the shortest length) pattern is

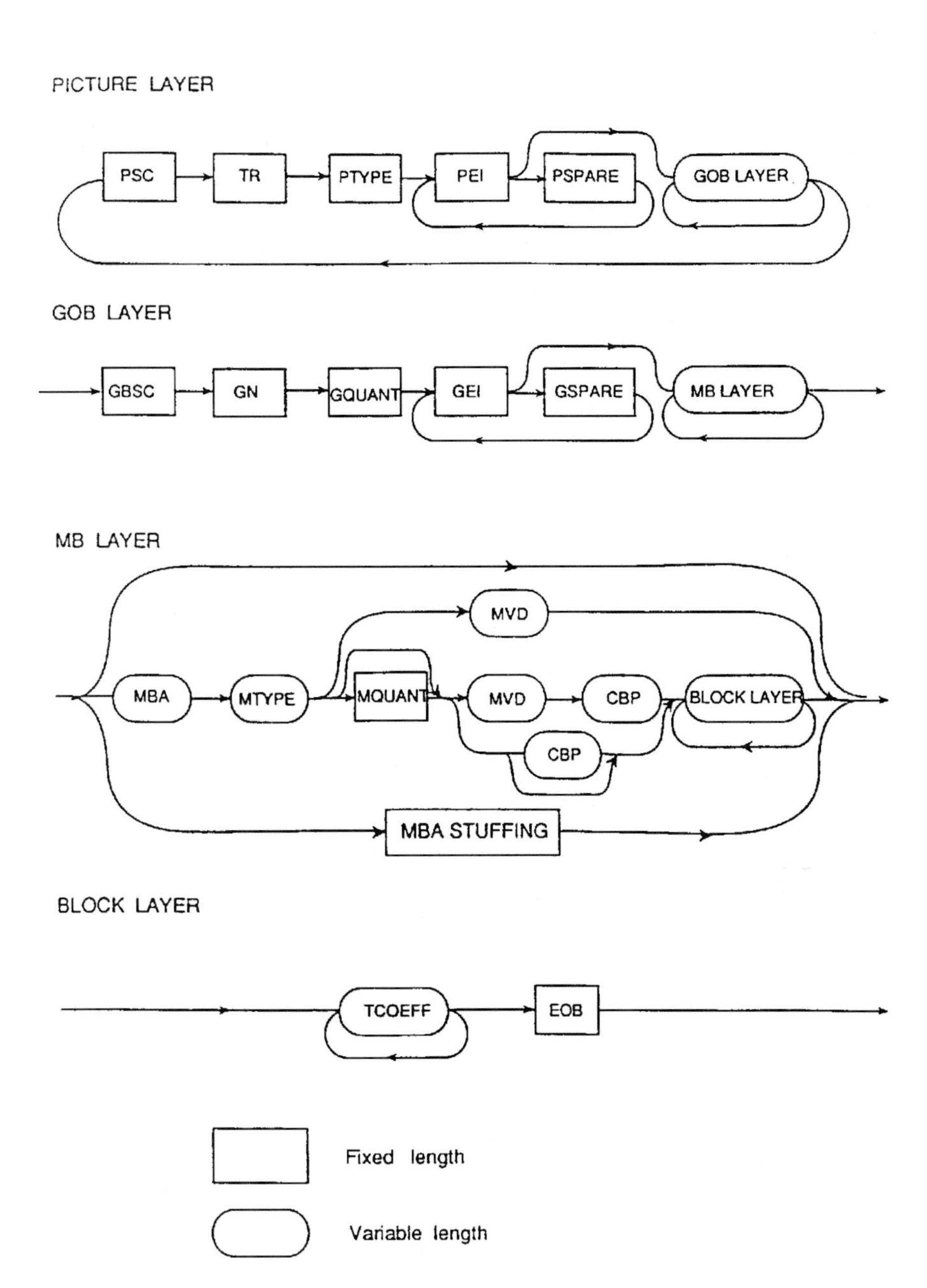

Figure 9.6 A syntax diagram for the hierarchical video multiplexing. Each layer is composed of possible headers and a corresponding low-level layer [450]. © 1990 ITU-T.

Table 9.4 VLC table for macroblock type (MTYPE) [450]

	MQUANT	MVD	CBP	TCOEFF	VLC
INTRA				o	0001
INTRA	o			o	0000 001
INTER			o	o	1
INTER	o		o	o	00001
INTER+MC		o			0000 0000 1
INTER+MC		o	o	o	0000 0001
INTER+MC	o	o	o	o	0000 0000 01
INTER+MC+LF		o			001
INTER+MC+LF		o	o	o	01
INTER+MC+LF	o	o	o	o	0000 01

Note 1: o means that the item is present in the MB.

Note 2: It is possible to apply the loop filter in a nonmotion compensated MB by declaring it as MC + FIL but with a zero MV.

Table 9.5 VLC table for macroblock addressing (MBA) [450]

MBA	Codeword	MBA	Codeword
MBA stuffing	0000 0001 111	17	0000 0101 10
Start code	0000 0000 0000 0001	18	0000 0101 01
1	1	19	0000 0101 00
2	011	20	0000 0100 11
3	010	21	0000 0100 10
4	0011	22	0000 0100 011
5	0010	23	0000 0100 010
6	0001 1	24	0000 0100 001
7	0001 0	25	0000 0100 000
8	0000 111	26	0000 0011 111
9	0000 110	27	0000 0011 110
10	0000 1011	28	0000 0011 101
11	0000 1010	29	0000 0011 100
12	0000 1001	30	0000 0011 011
13	0000 1000	31	0000 0011 010
14	0000 0111	32	0000 0011 001
15	0000 0110	33	0000 0011 000
16	0000 0101 11		

	63 Patterns codeword length				
		C_b C_r			
Y_0 Y_1 Y_2 Y_3	(0)	-	5	5	6
	(4)	4	7	7	8
	(8)	4	7	7	8
	(16)	4	7	7	8
	(32)	4	7	7	8
	(12)	5	8	8	8
	(48)	5	8	8	8
	(20)	5	8	8	8
	(40)	5	8	8	8
	(24)	6	8	8	9
	(36)	6	8	8	9
	(28)	5	8	8	9
	(44)	5	8	8	9
	(52)	5	8	8	9
	(56)	5	8	8	9
	(60)	3	5	5	6

■ Coded block('1') □ Non-coded block('0')

Figure 9.7 Coded block patterns (CBP), coded or not, for macroblock transmission [450]. © 1990 ITU-T.

four luminance block coding (code length 3). Figures in parentheses are decimal numbers, D_n, derived by

$$D_n = 32Y_0 + 16Y_1 + 8Y_2 + 4Y_3 + 2C_b + C_r \tag{9.1}$$

which implies 63 patterns and Y_0, Y_1, Y_2, Y_3, C_b, and C_r can be 0 or 1. The actual VLC code can be found in Table 10.15. Note that all-zero, i.e., non-transmission of a macroblock, is not defined here, as this can be indicated by macroblock addressing (MBA).

MBA is only included in transmitted macroblocks. The transmitted order of macroblocks is as shown in Fig. 9.5. MBA for the first transmitted macroblock in a GOB is the absolute address shown in Table 9.5. For the subsequent macroblocks, the difference between the absolute address of the macroblock and the last transmitted macroblock is coded by using the VLC table. If the bit stream buffer is underflow, the 11-bit MBA stuffing code is transmitted.

9.3 Source Coding Algorithms

The aim of this section is to describe the compression algorithm so that the information can be transmitted through the narrow bandwidth channel of videophone line. The coding algorithms play the great role of evaluating picture quality of commercial application systems (videophone or videoconference). The compression ratio should be more than 300:1 for the CIF image when we transmit on the DS_0 (64 Kbps) channel. For this high compression ratio, temporal redundancy reduction techniques have to be employed as well as spatial techniques.

Figure 9.8 shows an encoding diagram based on H.261. Two main compression techniques in intra/inter mode are shown. First, (8×8) DCT is used for intra picture correlation reduction. Second, motion-estimation and compensation technique may be used for interpicture correlation reduction. During the activities for standardization, several replacements have been considered, although the main structure (DCT and MC/DPCM) has been kept. Control signals for MC/no MC, intra/inter, and loop filter (LF) on/off are to be determined using specific algorithms by the designer. Buffer status is used to select the quantizer step sizes [432]. Quantized coefficients may be classified into several groups [411, 420]. The location of the loop filter was simulated at various positions in the feedback loop [414]. In addition to these experiments carried out before finalizing the standards, various modifications and enhancements to the basic H.261 algorithm have been suggested. These features will be discussed later.

It needs to be clarified here that not all the functions shown in this figure are part of H.261. The following items are optional:

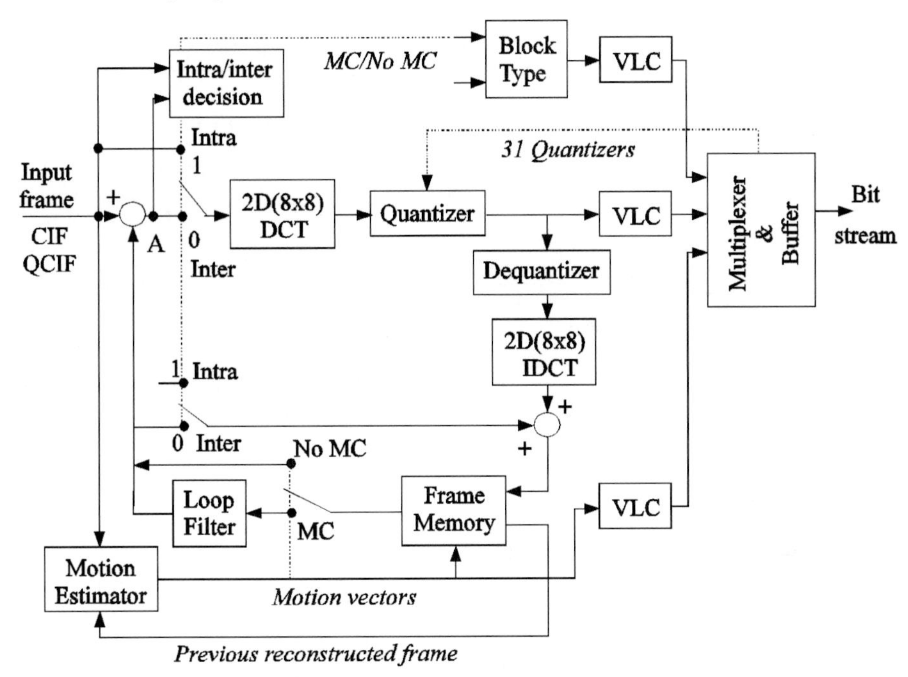

Figure 9.8 A video encoder based on H.261 [450]. © 1990 ITU-T.

1. The FEC decoder (encoder has a BCH (511, 493) FEC)
2. Motion compensation in the encoder. The decoder, however, must accept one motion vector per macroblock.
3. Loop filter
4. The arithmetic process for computing the 2D (8×8) DCT

As in all the standards, the decision-making processes (intra or inter, MC or no MC, loop filter on or off, buffer control of quantizers, etc.) are left to the designer.

In general, the four mode decisions, intra or inter, MC or no MC, loop-filtering or not, and coded or not-coded, have to be included in the header MTYPE (Table 9.4). Simple algorithms can be applied for the mode decisions. First, intra or inter mode may be decided, for example, on the basis of the energies of the luminance prediction error (Fig. 9.9). If the energy is abnormally large, it implies scene change or fast motion, and inter-picture prediction (even with motion estimation) can be ineffective. In this case, we need full accuracy information of the DC coefficient for the next inter prediction. If the energy is too small, however, it implies that there is no significant change between consecutive pictures and we do not need to transmit the macroblock. In inter mode, *A* in Fig. 9.8, is the displaced block difference (DBD) when MC is used, and it is the block difference

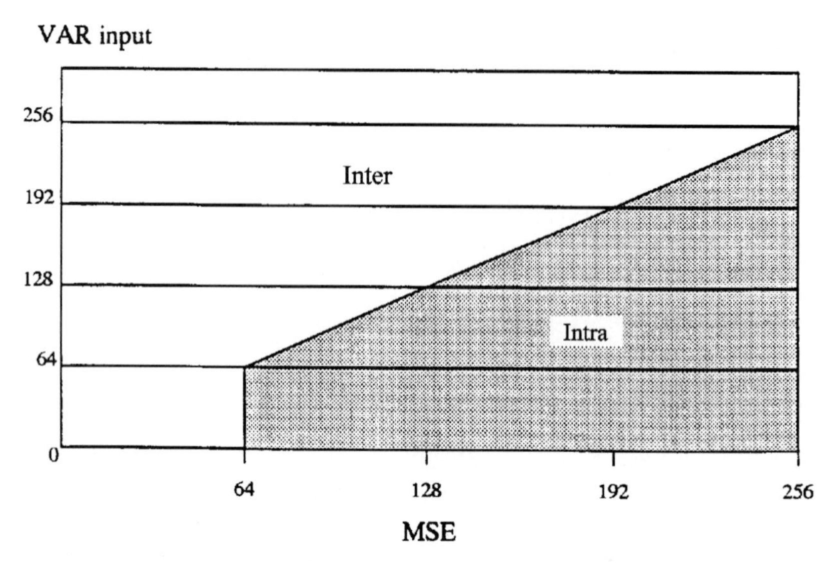

Figure 9.9 An intra/inter mode decision curve, as an example only [447]. © 1990 ITU-T.

(BD) when no MC is used. The decoder can reconstruct the macroblock from the previous picture.

Figure 9.9 shows the two decision areas for intra or inter mode, which was simulated during the standardization, but not included in the final version [447]. MSE in Fig. 9.9 is based on a (16×16) Y block i.e.,

$$\text{MSE} = \frac{1}{256} \sum_{m=0}^{15} \sum_{n=0}^{15} (x_o(m, n) - x_{mc}(m, n))^2 \tag{9.2}$$

where $x_o(m, n)$ denotes original macroblock and $x_{mc}(m, n)$ denotes the motion-compensated macroblock. "VAR input" is the variance of the input macroblock. Generally speaking, from Fig. 9.9, intra area has smaller variance and greater MSE. This curve is empirically optimized. Inter mode includes the boundary line. Although the algorithm is not mandatory, many manufacturers listed in Appendix A usually employ this approach.

Second, the motion compensation is performed on the macroblock with a maximum displacement of ± 15 pels/picture allowed in each dimension. The previous frame is stored in the frame memory so that each macroblock can be motion compensated. The decoder finds the best matching macroblock from the previous frame by using the MV and replaces the current MB in this mode.

Third, loop filtering (optional) is normally activated to remove the blocking artifacts associated with the motion compensation. The (3×3) loop filter (Fig. 8.18) processes the data in each (8 × 8) block, without overlap between the blocks. However, the MC/No MC switch does not always control the loop filter shown in Fig. 9.8. Even in MC mode, the LF may not be activated as shown in Table 9.4. The loop filter is optional at the encoder but is mandatory for the decoder implementation.

The last decision is done for transform coefficients coded or not-coded indicated by CBP (Table 9.4). For example, it is not necessary to code and transmit a macroblock that has the same contents as the previous one. In intra mode, all blocks have to be coded and CBP information is not required. In case of no nonzero transform coefficients (TCOEFF) or only motion vector data (MVD), CBP also has no meaning.

Transform coefficients are not the only output from the encoder; there is also some control information, such as flags for intra or inter, MB transmitted or not, quantizer indication, quantizing index, motion vector, and loop filter on or off.

Similar to the conventional DPCM encoder, H.261 decodes and reconstructs the image in the encoder. The output of the IDCT represents the reconstructed prediction error in inter mode, whereas in intra mode it represents the reconstructed picture. This output is added to the current frame prediction (normally done by motion compensation) to obtain the decoded frame, which is then stored in the frame memory. The decoder is much simpler than the encoder because all the decision-making processes (intra or inter, MC or no MC, loop filtering or not, etc.) are implemented in the encoder. Only the overhead indicating these options is sent to the decoder.

The coefficients and control information are coded by VLC or FLC, subject to their properties. The coded bit stream is counted to control the constant $p \times 64$ Kbps rate. The buffer status signal is fed back to the quantizer where the step sizes are changed. For correcting channel errors, BCH(511, 493) code, which can correct up to two bit errors in every block of 511 bits, is mandatory in the encoder and optional in the decoder.

Figure 9.10 depicts the H.261 decoder block diagram. The compressed data is buffered and variable-length decoded. The remaining part of the decoder is similar to the back end of the encoder. The only difference is that motion estimation is no longer necessary. The motion vector and other side information are obtained directly from the variable-length decoder. Detailed descriptions for every mode in the encoder and decoder are provided in the following subsections.

9.3.1 DCT/IDCT in a loop

The 2D (8 × 8) DCT used for spatial data compression in H.261 as well as in JPEG and MPEG, is a good compromise between coding efficiency and hardware complexity. Input data to the 2D (8 × 8) DCT (Fig. 9.8) are prediction errors in any of the decision modes other than intra mode. The system is similar to a typical closed-loop DPCM coder with forward transform before quantization

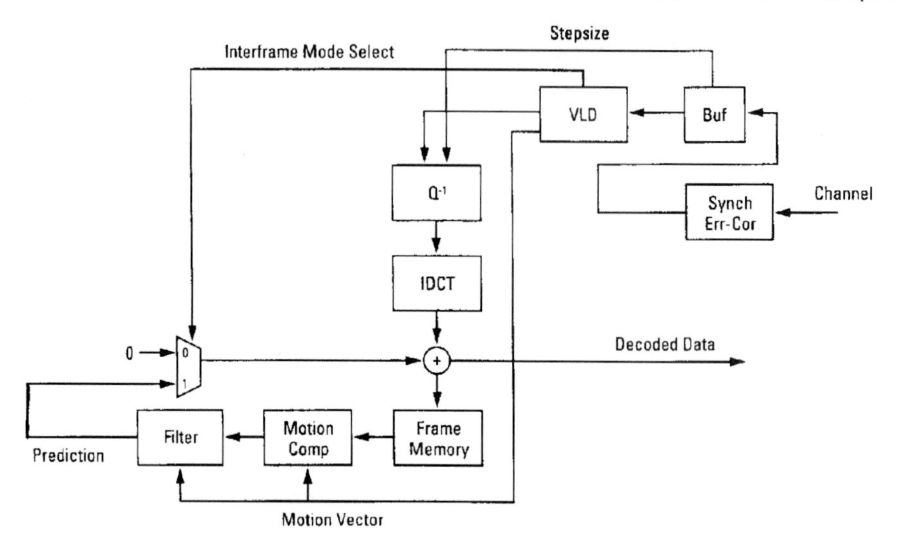

Figure 9.10 An H.261 video decoder example.

and inverse transform after inverse quantization in the loop. The basic reason for inverse transform is that prediction (motion compensation) is performed with the previous reconstructed image. Therefore forward and inverse transforms coexist in the prediction loop.

DCT is close to the optimal transform for the first-order Markov image in energy compaction and consequently in decorrelating a signal. Also, as it is orthogonal, IDCT is straightforward. Besides, fast algorithm for FDCT implies fast algorithm for IDCT. If the two IDCTs in the encoder and decoder, one for DPCM loop and the other for reconstruction of the transmitted image, are implemented with different accuracies, the difference between the two IDCT outputs can accumulate, leading to degraded reconstructed images. This is DCT/IDCT mismatch error [448]. The errors are accumulated due to the property of predictive coding (Chapter 4). The DCT matrix that consists of irrational number operations requires full-precision arithmetic. If the IDCT output is in the vicinity of the boundary value of the round-off operation, i.e., $g + 0.5$, where g is an arbitrary pel value, any two slightly different IDCT operations would cause substantial distortions due to the round-off operations and the accumulative nature.

The mismatch errors obtained from the test simulation of FDCT/IDCT should not exceed the following limitations. (This error is defined against a reference IDCT having at least 64-bit floating point accuracy, and it is calculated over 10,000 blocks.)

1. For any pel, the peak error shall not exceed 1 in magnitude.
2. For any pel, the mean square error shall not exceed 0.06.

3. Overall, the mean square error shall not exceed 0.02.
4. For any pel, the mean error shall not exceed 0.015 in magnitude.
5. Overall, the mean error shall not exceed 0.0015 in magnitude.
6. For all-zero input, the IDCT shall generate all-zero output.

For the simulation, random integer pixel data are generated in the range $-L$ to $+H$ according to the following random number generator in C language. Then (8×8) blocks are the data sets. More than 10,000 blocks of data are simulated with 64-bit floating-point accuracy.

```
/* L and H must be long, i.e., 32 bits */
long rand(L,H)
long L,H;
{
static  long randx = 1;   /* long in 32 bits */
static  double z = (double) 0x7fffffff;

long  i,j;
double  x;  /* double is 64 bits */
randx = (randx * 1103515245) + 12345;
i = randx & 0x7ffffffe; /* keep 30 bits */
x = ((double)i)/z;   /* range 0 to 0.99999... */
x *= (L+H+1);   /* range 0 to < L+H+1 */
j = x;
return(j-L);   /* range -L to H */
}
```

In general implementation, the accumulated mismatch error is removed by a forced updating process giving a new intra mode. A macroblock should be updated (intra mode) at least once for every 132 times it is transmitted.

9.3.2 Quantization and VLC

Two types of quantizers are applied for quantizing DCT coefficients in the H.261 encoder/decoder. The intra DC coefficient is uniformly quantized with a step size of 8 and no dead zone. Each of the other 31 quantizers for AC and inter DC coefficients is nearly uniform but with a central dead zone around zero, viz. nearly uniform quantizer with dead zone $= 2Th$ (Fig. 9.11). The step size Q is an even integer in the range of 2 to 62, which represents 31 quantizers. Since many AC coefficients have near-zero levels, the midtread quantizer with zero as one of the output values is used. H.261 specifies only the reconstruction levels (see Table 9.6). How to perform quantization is left to the designer.

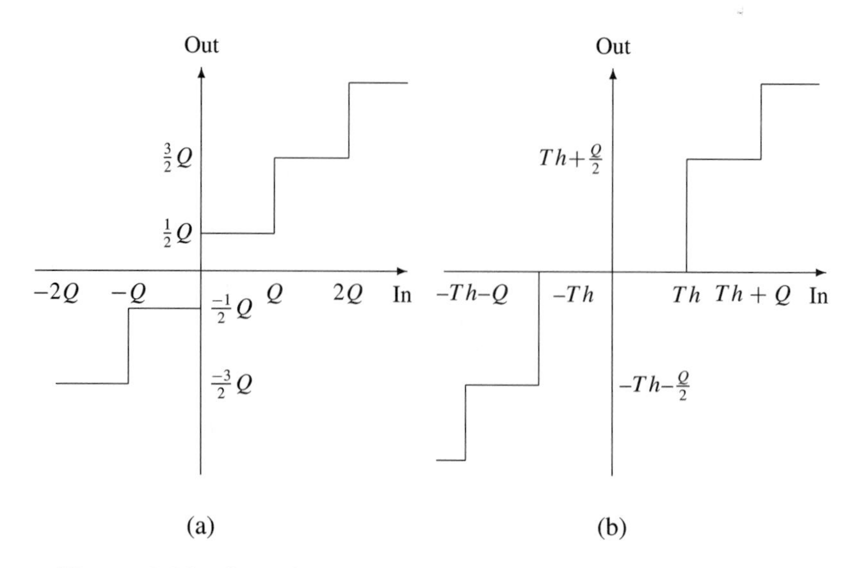

Figure 9.11 Quantizers in H.261: (a) uniform quantizer for the intra DC coefficient only (see Table 9.8), and (b) nearly uniform midtread quantizer for the inter DC and all AC coefficients (dead zone $= 2Th$; the step size Q can be changed from MB to MB in increments of 2 from 2 to 62 adaptively).

For the coefficients intra AC or inter DC/AC, a nearly uniform midtread quantizer is employed. The input between $-Th$ and $+Th$ (Th is the threshold) is quantized to level zero. Except for the dead zone, the step size Q is uniform. All the coefficients in a macroblock except for the intra DC coefficients go through the same quantizer.

The quantization step size represents the distance between possible values of the quantized signal. By varying the step size, the amount of information used to describe a particular pixel or block of pixels can be changed. Larger step sizes result in less information being required, while accuracy is reduced in the representation. Smaller step sizes result in better quality but also in an increase in the amount of information to be transmitted.

In the H.261 coder, the length of the coded bit stream is dependent on the image properties, complexity, motion, or scene changes. The easy way to control the output bit rate may be found in quantizer step sizes. The Reference Model 8 [451] inspects buffer status after coding 11 macroblocks. The buffer size is on the order of 64 Kbits. If the buffer status is full, there is no room for accepting residual information in the frame. One should increase step size by the feedback control signal in Fig. 9.8. This is called the liquid level control model [414]. If the buffer status is zero, the encoder can use the fine quantizer that requires more bits. Rate control is totally left to the designer.

Table 9.6 Reconstruction levels for all coefficients other than the intra DC [450]

	Q_n								
Q_l	1	2	3	4	...	9	...	30	31
−127	−255	−509	−765	−1019	...	−2048	...	−2048	−2048
−126	−253	−505	−759	−1011	...	−2048	...	−2048	−2048
⋮	⋮	⋮	⋮	⋮	...	⋮	...	⋮	⋮
−3	−7	−13	−21	−27	...	−63	...	−209	−217
−2	−5	−9	−15	−19	...	−45	...	−149	−155
−1	−3	−5	−9	−11	...	−27	...	−89	−93
0	0	0	0	0	...	0	...	0	0
1	3	5	9	11	...	27	...	89	93
2	5	9	15	19	...	45	...	149	155
3	7	13	21	27	...	63	...	209	217
⋮	⋮	⋮	⋮	⋮	...	⋮	...	⋮	⋮
56	113	225	339	451	...	1017	...	2047	2047
57	115	229	345	459	...	1035	...	2047	2047
⋮	⋮	⋮	⋮	⋮	...	⋮	...	⋮	⋮
126	253	505	759	1011	...	2047	...	2047	2047
127	255	509	765	1019	...	2047	...	2047	2047

Source: © 1990 ITU-T.

Table 9.7 Dynamic range of data at each coding stage

	Precision (bpp)	Dynamic range
Input	8	0 ~ 255
DPCM	9	−255 ~ 255
FDCT	12	−2047 ~ 2047
Q_l	8	−127 ~ 127

The related headers in the hierarchical structure are GQUANT and MQUANT. GQUANT has initial information of quantizer step size. The step size is the same for all coefficients within a macroblock, but can be changed for each macroblock that has MQUANT information. GQUANT may be replaced by MQUANT in the macroblock layer, and the MQUANT may be replaced by another possible MQUANT.

In the H.261 encoder, the input image has 8-bit gray levels. Dynamic range of the levels is changed by the encoding techniques (DPCM, quantizer, and DCT). We summarize the sample precisions and dynamic ranges in Table 9.7. Q_l represents the quantizing index (see Table 9.6).

Decision levels for the transform coefficients are not defined in H.261, but reconstruction levels, $\hat{S}_{uv}$, for the quantized levels Q_l are obtained by the following formulas:

- case (Q_n is odd)

  if ($Q_l > 0$)
  $$\hat{S}_{uv} = Q_n \times (2 \times Q_l + 1)$$
  if ($Q_l < 0$)
  $$\hat{S}_{uv} = Q_n \times (2 \times Q_l - 1)$$
- case (Q_n is even)

  if ($Q_l > 0$)
  $$\hat{S}_{uv} = Q_n \times (2 \times Q_l + 1) - 1$$
  if ($Q_l < 0$)
  $$\hat{S}_{uv} = Q_n \times (2 \times Q_l - 1) + 1$$
- case ($Q_l = 0$)
  $$\hat{S}_{uv} = 0$$

where Q_n and Q_l are quantizer numbers from 1 through 31 and quantized levels from -127 through 127, respectively. The above equations yield the reconstruction levels as shown in Table 9.6. Quantizer numbers from 1 to 31 are transmitted by either GQUANT in GOB layer or MQUANT in the MB layer. All reconstruction levels are symmetrical except for the clipped 2047/$-$2048. Note that the dead zone around zero level becomes larger as Q_n increases.

Example 9.1 *Given the* Q_n *and* Q_l *as (a)* (31, 5) *and (b)* (10, 127)*, respectively, then the reconstruction values are*

(a) $\hat{S}_{uv}(31, 5) = 31 \times (2 \times 5 + 1) = 341$

(b) $\hat{S}_{uv}(10, 127) = 10 \times (2 \times 127 + 1) - 1 = 2549 \Rightarrow 2047.$

Since the reconstruction levels are in the range of -2048 *to* 2047*, the levels that are out of this range have to be clipped.*

The transform coefficients are quantized in the range of -127 to 127 for AC and 1 to 254 for intra DC (level 255 forbidden) and sequentially scanned by zigzag order as shown in Fig. 8.4. The intra DC coefficient is uniformly quantized with a step size of 8 with no dead zone (Fig. 9.11 a). The resulting values are represented with 8-bit FLC as shown in Table 9.8.

All AC and inter DC coefficients are quantized by VLC with *RUN* and *LEVEL* (Q_l) as shown in Appendix C.3. This table is for the most frequent combination of zero-run and level of the following nonzero coefficient. H.261 defines relatively smaller levels in the VLC table than those in MPEG (Appendix

Table 9.8 Reconstruction levels for the intra DC coefficient (the decoded value corresponding to decimal n is $8n$, except decimal 255 gives 1024 [450])

FLC	Decimal	$\hat{S}_{uv}$ of intra DC
0000 0001	1	8
0000 0010	2	16
⋮	⋮	⋮
0111 1111	127	1016
1111 1111	255	1024
1000 0001	129	1032
⋮	⋮	⋮
1111 1110	254	2032

Source: © 1990 ITU-T.

C.4 and C.5), since it mainly treats slow-moving video. They are also comparable with the VLC table for the luminance component of JPEG shown in Appendix C.1.

The remaining combinations of (RUN, Q_l) are encoded with a 20-bit word, consisting of 6 bits *ESCAPE*, 6 bits *RUN* and 8 bits Q_l. The quantizer level zero is not necessary to denote a code in both cases, since it only increases *RUN*. The other 8-bit FLCs are two's-complement ordered from –127 to 127.

9.3.3 Motion estimation and compensation

Although the motion estimation technique can play a key role for the output video quality in inter mode, H.261 does not specify how to estimate the motion vector. It does, however, specify how to use the MV for motion compensation both at the encoder and decoder, when the MB type is inter + MC or inter + MC + LF (Table 9.4). Only the motion vector data (MVD) can be transmitted via up to 11-bit VLC (Table 9.9). MVD is the difference between the MV of the current macroblock and the MV of the preceding macroblock. MVD consists of VLC for the horizontal component MV_x followed by VLC for the vertical component MV_y. However, for the following situations, the encoder regards the MV of the preceding macroblock as zero.

- When evaluating MVD for macroblocks 1, 12, and 23 (left side in a group of macroblocks as shown in Fig. 9.5).
- When the MBA does not represent the difference of 1 (macroblock increment is greater than 1).
- When the MTYPE of the previous macroblock is not MC.

The MV consists of a pair of horizontal and vertical components in the two-dimensional spatial domain.

Table 9.9 shows that the MVD zero, i.e., relatively no motion, is the most probable, and each code represents two possible values. The decoder, however, uniquely decodes the correct MVD, due to the characteristic of maximum displacement ± 15 pels/lines per picture, i.e., one of the two values would lead to the displacement beyond this range, which would be invalid.

Example 9.2 *Let $MV(n)$ and $MV(n-1)$ denote the motion vectors of the present and previous macroblocks, respectively. Given $MV_x(n-1) = -10$ pels/picture, find the possible MVD range and show how to find the correct one from VLC in the decoder. Here MV_x represents the horizontal component of MV.*

$$\begin{aligned} MVD &= MV_x(n) - MV_x(n-1) \qquad (9.3) \\ &= 15 - (-10) = 25 \; pels/picture, \;\; if \;\; MV_x(n) = 15 \\ &= -15 - (-10) = -5 \; pels/picture, \;\; if \;\; MV_x(n) = -15 \end{aligned}$$

Therefore, the permissible range of MVD is -5 to 25. The MVDs in this range do not appear in the same row of Table 9.9. When the decoder detects this range, a corresponding MVD can be obtained from the VLC.

Though the motion compensation is optional in the encoder, motion estimation and compensation result in an efficient inter mode compression. Thus, most H.261-based codecs employ this technique with frame memory. The basic requisites are first, that both the horizontal and vertical components of the motion vector have integer values not exceeding ± 15 pels/picture, second, that the ME is based on the 16×16 luminance block in the macroblock, and third, that the MV for the (8×8) color-difference blocks is derived by halving the MV of luminance and truncatin to integer value. For example, if the MV is $(-5, -6)$, then the MV for chrominance is $(-2, -3)$.

Interpretation of a moving object may include three-dimensional analysis. Currently, most of motion estimations are performed in the two-dimensional domain, i.e., between successive frames. Thus, the expression *displacement estimation* is frequently used in this field. Displacement can be measured by an error or matching criterion between consecutive frames.

Denoting $F_n(j,k)$ and $F_{n-1}(j,k)$ as the pel intensities of the current frame n and previous frame $n-1$, respectively (row j and column k), the frame difference is

$$FD_{n,n-1}(j,k) = F_n(j,k) - F_{n-1}(j,k) \qquad (9.4)$$

The block difference BD is defined as

$$BD = \sum_{MB} \left| FD_{n,n-1}(j,k) \right| \tag{9.5}$$

where $\sum_{MB}$ implies summation over the (16 × 16) Y block. Similarly, the displaced frame difference is

$$DFD_{n,n-1}(j,k) = F_n(j,k) - F_{n-1}(j + MV_x, k + MV_y) \tag{9.6}$$

where MV_x and MV_y are the horizontal and vertical components of the MV of the (16 × 16) Y block. The displaced block difference DBD is then

$$DBD = \sum_{MB} \left| DFD_{n,n-1}(j,k) \right| \tag{9.7}$$

The criterion to determine MC/no MC can be decided by the relationship of BD and DBD as shown in Fig. 9.12, which is experimentally optimized. Note that the no MC area includes the boundary line.

Obtaining the exact motion vector is a formidable task in terms of complexity and efficiency. Block-matching techniques that estimate the motion of a block in the present frame based on a search area in the previous frame are in general applied in video coding. The size of the search area implicitly limits the range of the motion estimation. Brute force full search, three-step hierarchical search, logarithmic search, and conjugate directional search are typical block matching algorithms [507] (see Section 6.3). H.261, however, does not specify the motion estimation technique. As we discussed in Chapter 6, the optimum algorithm for motion estimation is still under study.

9.3.4 Loop filtering

The introduction of a low pass filter after motion compensation (MC) could have the following advantages:

- Reduction of high-frequency artifacts, like mosquito effects, introduced by MC
- Reduction of quantization noise in the feedback loop

The loop filter activation (filter on) could be controlled by the motion vector. The noise due to MC plays a large role in the quality impairment. Therefore, loop filtering can be tied with the MC operation, i.e., if the motion vector is nonzero, then the filter is on. No separate overhead for the filter on/off beyond what is

Table 9.9 VLC for motion vector data (MVD) [450]

MVD	VLC code	MVD	VLC code
–16 , 16	0000 0011 001	0	1
–15 , 17	0000 0011 011	1	010
–14 , 18	0000 0011 101	2 , –30	0010
–13 , 19	0000 0011 111	3 , –29	0001 0
–12 , 20	0000 0100 001	4 , –28	0000 110
–11 , 21	0000 0100 011	5 , –27	0000 1010
–10 , 22	0000 0100 11	6 , –26	0000 1000
–9 , 23	0000 0100 01	7 , –25	0000 0110
–8 , 24	0000 0101 11	8 , –24	0000 0101 10
–7 , 25	0000 0111	9 , –23	0000 0101 00
–6 , 26	0000 1001	10 , –22	0000 0100 10
–5 , 27	0000 1011	11 , –21	0000 0100 010
–4 , 28	0000 111	12 , –20	0000 0100 000
–3 , 29	0001 1	13 , –19	0000 0011 110
–2 , 30	0011	14 , –18	0000 0011 100
–1	011	15 , –17	0000 0011 010

Source: © 1990 ITU-T.

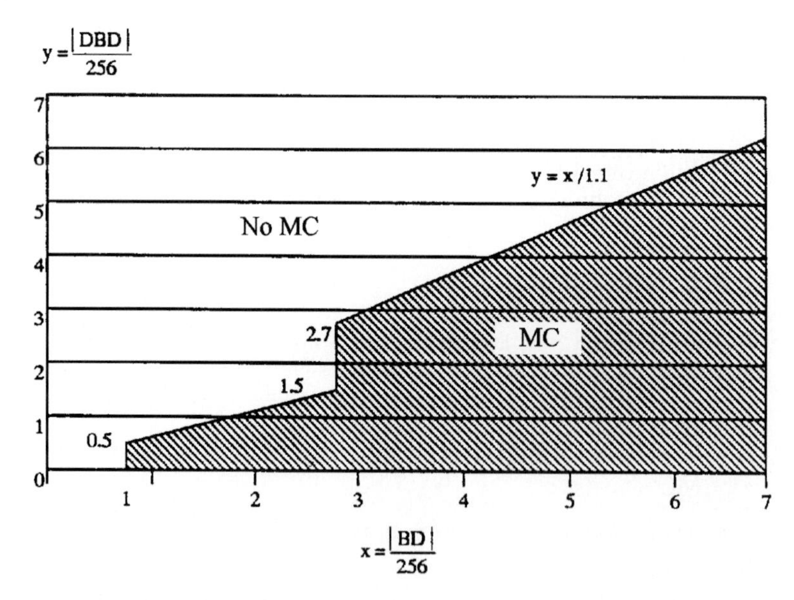

Figure 9.12 MC/no MC mode decision curve, as an example only [447]. © 1990 ITU-T.

shown in Table 9.4 is then needed [443]. The MC and the LF, however, can be basically independent of each other, as stated earlier.

No special loop filter coefficients are defined in H.261, but the same LPF we discussed in JPEG (Fig. 8.18) is recommended. The filter is applied to both luminance and chrominance in (8×8) blocks. It is nonrecursive but separable into one-dimensional horizontal and vertical functions. At block edges where one of the filter taps would fall outside the block, the one-dimensional filter coefficients are (0, 1, 0), which means no filtering. Then the filter becomes (4, 8, 4) for pixels on the block edges and it becomes (16) for pixels on the block corners. During the standardization, three different filter coefficients were considered [447].

1. For pixels inside the block

$$\frac{1}{16}\begin{bmatrix} 1 & 2 & 1 \\ 2 & 4 & 2 \\ 1 & 2 & 1 \end{bmatrix}$$

2. For pixels on the block edges

$$\frac{1}{16}\begin{bmatrix} 3 & 1 \\ 6 & 2 \\ 3 & 1 \end{bmatrix}$$

3. For pixels on the block corner

$$\frac{1}{16}\begin{bmatrix} 9 & 3 \\ 3 & 1 \end{bmatrix}$$

For the best location of the LPF in the H.261 encoder, various simulations have been performed as shown in Fig. 9.13. In addition to the filtering in the encoder, possible locations in the decoder are also considered. Preprocessing (h_{pre}) of the source material improves overall coding results [452]. Filtering the transform coefficients (H_7), reconstructed coefficients (h_4), predicted block (h_5), and local output (h_3, h_n) also has shown significant improvement of the subjective picture quality of a codec working at 64 Kbps [414, 446]. Since motion compensation in the frame memory may cause visible artifacts, filtering in the loop after the MC gives more efficient suppression of the mosquito effect than does noise filtering as postprocessing.

The (1, 2, 1) filter used in the motion compensation loop is the simplest and is in a convenient form for hardware implementation. It does not always match the different statistical characteristics of the motion-compensated signals, however. The optimum filter that minimizes the prediction error can be designed in terms

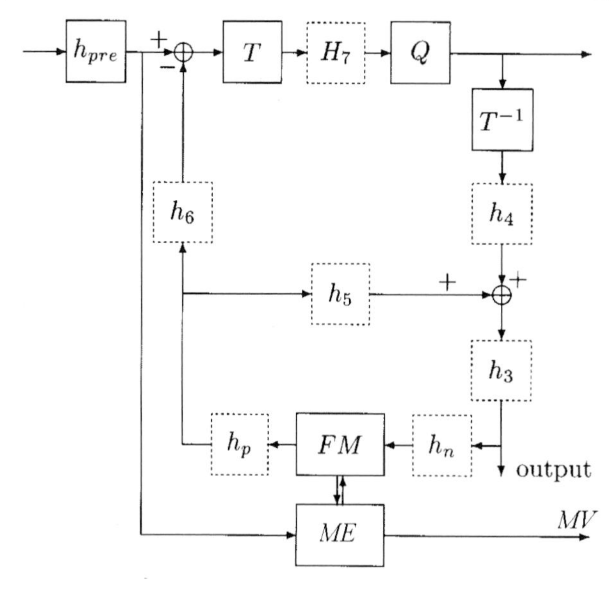

Figure 9.13 Possible locations of the loop filter that were considered in the H.261 encoder [446]. © 1990 ITU-T.

of a correlation parameter and an inaccuracy parameter using the Wiener filter concept [751].

9.4 Transmission of H.261 Video

This section includes error correcting code before transmission and interface to ISDN network with the other related standards. The synchronization and check bits are added to the bit stream to allow error correction in the decoder. The error correcting code is BCH (511, 493), which can correct up to two-bit errors in every block of 511 bits. The code includes 18 redundant bits (parity bits) for each 493 data bits (Fig. 9.14).

The generator polynomial is

$$g(x) = (x^9 + x^4 + 1)(x^9 + x^6 + x^4 + x^3 + 1) \tag{9.8}$$

Implementation of the polynomial will be discussed in the following section.

A synchronization bit is added to each codeword to determine the BCH codeword boundary in the group of eight data frames. The fill indicator (F_i) can be set to zero in the encoder. Then, all bits are filled with ones to provide video data on every valid clock cycle.

For the purpose of transmitting the bit stream as per H.261 for videophone application, the transmission coder includes the other necessary information, i.e., audio, data, and control signal as depicted in Fig. 9.15. The audio signal is processed by separate standards or by proprietary algorithms. G.722 deals with

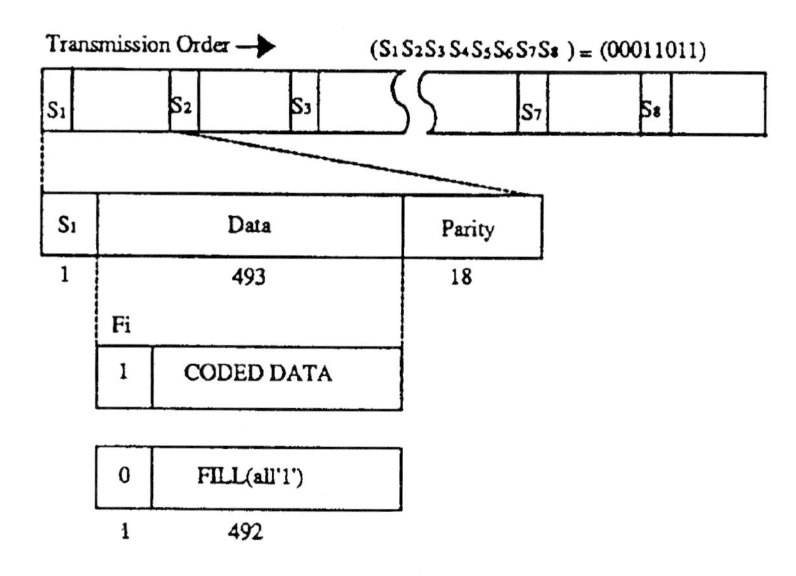

Figure 9.14 Transmission format of a group of error correcting codes [450]. © 1990 ITU-T.

ADPCM in broad band (50 Hz ~ 7.0 KHz) [532] and G.711 uses A-law or μ-law in narrow band (300 Hz ~ 3.4 KHz) [531]. Pre- and postprocessing are needed for interfacing between the region-specific video systems, PAL or NTSC, and the CIF or QCIF.

The videophone on ISDN channel, so-called H.320 videophone, involves several Recommendations, i.e., H.221, H.242, H.320, H.230, G.711, G.722, and I.400, (Fig. 9.16), system aspects of audiovisual terminals and systems. MUX/DMUX is used to multiplex/demultiplex coded video, coded audio, and data signal, corresponding to a bit rate of $p \times 64$ Kbps, with a frame structure specified in H.221 [423]. Network interface complies with Recommendation I.400 series, typically D_0 (16 Kbps), D_1 (64 Kbps), B (64 Kbps), H_0 (384 Kbps), and so forth. System control is used to control network access, establishing communication by end-to-end signaling.

A schematic diagram of H.320 that covers the technical requirements for narrow-band telephone services defined in H.200/AV.120 series recommendations (channel rates not exceeding 1920 Kbps) is shown in Fig. 9.17. The communication modes of H.320 visual telephones are listed in Table 9.10.

The videophone transmission can be established on two commercial networks, PSTN (non-ISDN) and ISDN. PSTN and ISDN denote public switched telephone network and integrated services digital network, respectively. Interworking problems, however, should be solved between any two different videophone systems (Fig. 9.18) [442]. The transmission channel of PSTN is a 3.1 KHz ana-

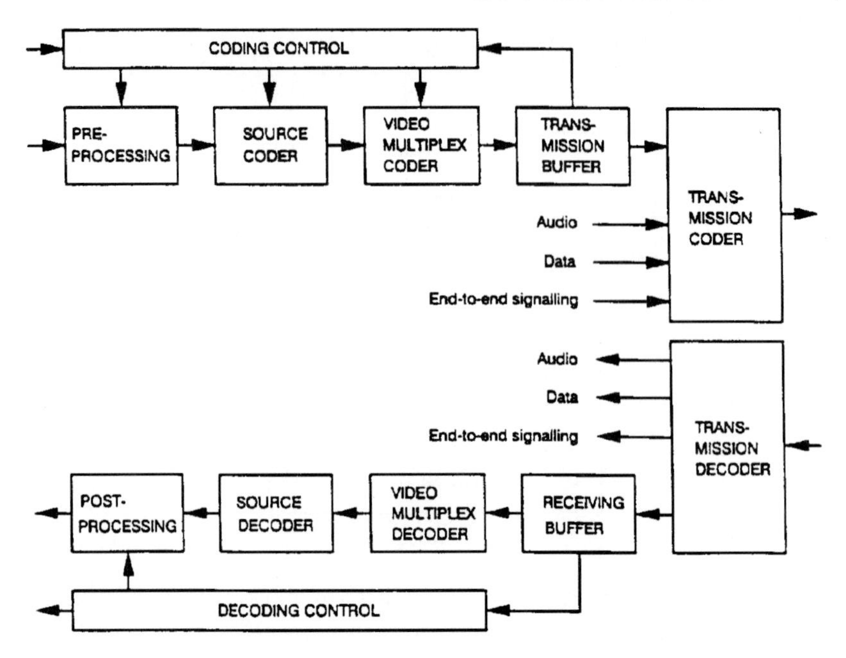

Figure 9.15 A block diagram for video/audio interfaces [450]. © 1990 ITU-T.

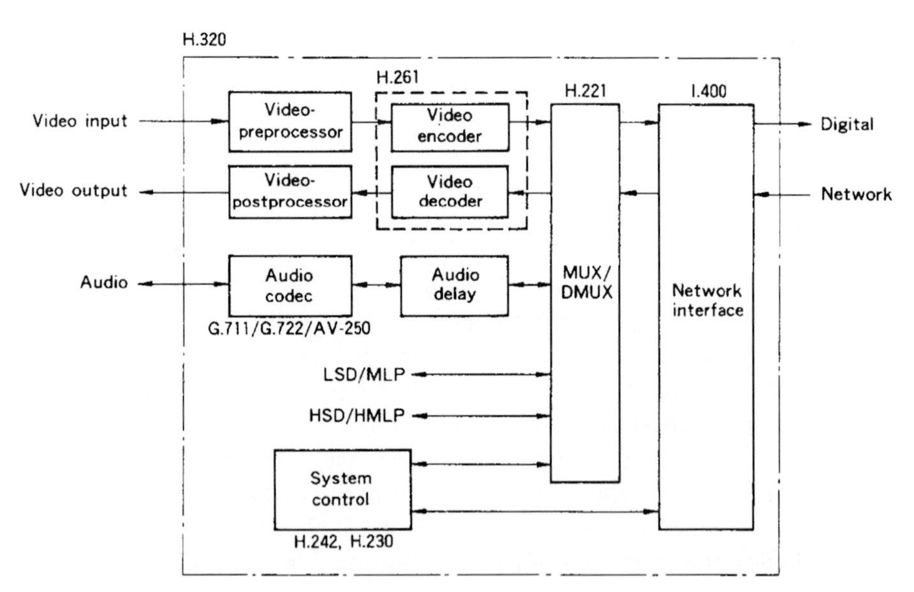

Figure 9.16 A functional block diagram of an ISDN H.320 videophone. (LSD: low-speed data; HSD: high-speed data; MLP: multilayer protocol; HMLP: high-speed multilayer protocol) [423]. © 1991 NEC.

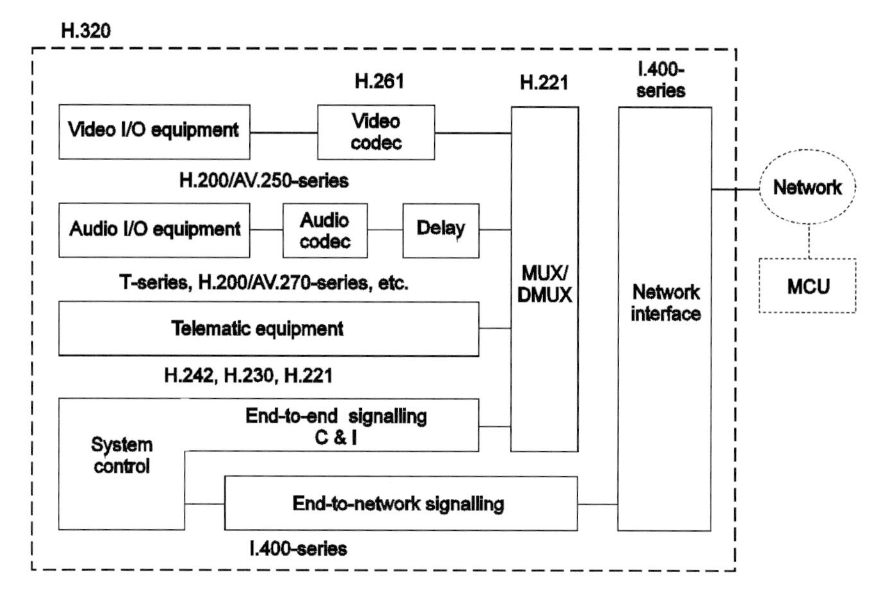

Figure 9.17 A generic visual telephone system [450]. Video I/O equipment includes cameras, monitors, and video processing units (split-screen and other capabilities). Audio I/O equipment includes microphones-loudspeakers and audio processing units (echo cancellation and other capabilities). Telematic equipment includes visual aids such as electronic blackboard, and still-picture transceiver. Delay in the audio path compensates for video codec delay to maintain lip synchronization. © 1990 ITU-T.

log channel with modem, while an ISDN B-channel is a 64 Kbps digital channel. From this basic difference, some kind of interworking functions, including audio transcoding, end-to-end signaling, and mode switching, are needed between the two videophone systems. With the advent of new transmission media, H.261 has to work with both of them. We can consider layered coding in ATM transmission [429] and can reconstruct the low-resolution image from a high-resolution transmitted one that has forward/backward compatibility [445].

9.5 H.261 Implementations

Since the H.261 was established in 1990, the low bit-rate audiovisual communication has been realized, owing to the development of low-cost and real-time implementation techniques [410, 430]. A number of manufacturers or vendors who have developed software, VLSI chip sets, add-on boards, and video-

Table 9.10 Communication modes of visual telephone [450]

<table>
<tr><th colspan="2" rowspan="2">Visual Telephone Mode (Note 1)</th><th rowspan="2">Channel Rate (Kbit/s)</th><th rowspan="2">ISDN Channel 1 (Note 2)</th><th colspan="2">ISDN Interface</th><th colspan="2">Coding</th></tr>
<tr><th>Basic</th><th>Primary Rate</th><th>Audio</th><th>Video</th></tr>
<tr><td rowspan="2">a</td><td>a0</td><td rowspan="2">64</td><td rowspan="2">B</td><td rowspan="5"></td><td rowspan="16">applicable</td><td>G.711</td><td>not applicable</td></tr>
<tr><td>a1</td><td>H.200/AV.254</td><td rowspan="15">H.261</td></tr>
<tr><td rowspan="3">b</td><td>b1</td><td rowspan="3">128</td><td rowspan="3">2B</td><td>G.711</td></tr>
<tr><td>b2</td><td>G.722</td></tr>
<tr><td>b3</td><td>H.200/AV.254
/253
(Note 1)</td></tr>
<tr><td colspan="2">c</td><td>198</td><td>3B</td><td rowspan="11">not applicable</td><td rowspan="11">G.722</td></tr>
<tr><td colspan="2">d</td><td>256</td><td>4B</td></tr>
<tr><td colspan="2">e</td><td>320</td><td>5B</td></tr>
<tr><td colspan="2">f</td><td>384</td><td>6B</td></tr>
<tr><td colspan="2">g</td><td>384</td><td>H0</td></tr>
<tr><td colspan="2">h</td><td>768</td><td>2H0</td></tr>
<tr><td colspan="2">i</td><td>1152</td><td>3H0</td></tr>
<tr><td colspan="2">j</td><td>1536</td><td>3H0</td></tr>
<tr><td colspan="2">k</td><td>1536</td><td>H11</td></tr>
<tr><td colspan="2">l</td><td>1920</td><td>5H0</td></tr>
<tr><td colspan="2">m</td><td>1920</td><td>H12</td></tr>
</table>

Note 1: (Audio coding of mode b3) In addition to H.200/AV.254, higher quality audio coding such as H.200/AV.253 may be used for this mode.

Note 2: For multiple channels of B/H0, all channels are synchronized at the terminal according to Section 2.7/H.221

Source: © 1990 ITU-T.

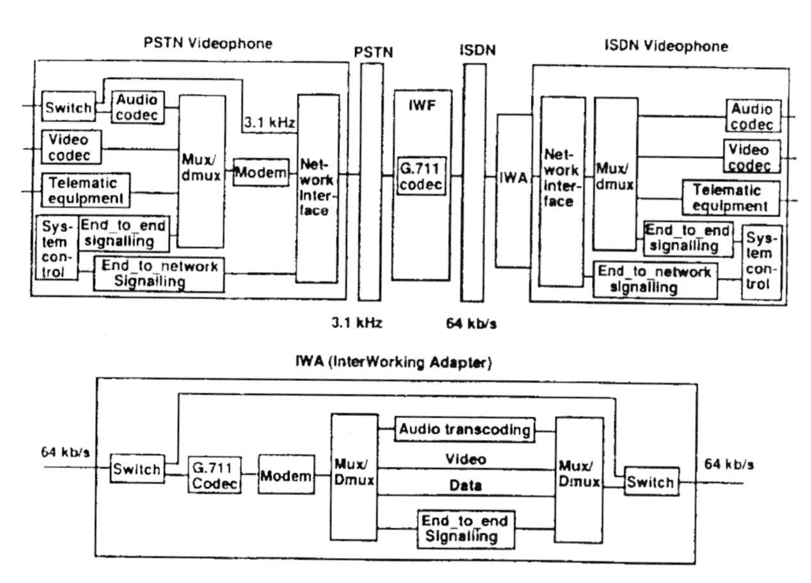

Figure 9.18 Interworking example between a PSTN videophone and an ISDN H.320 terminal [442]. © 1993 IEEE.

phone systems are shown in Appendix A. We now discuss implementation algorithms for the H.261 coding techniques.

Even before H.261, low bit-rate motion video codecs were studied [453, 412]. These include a hybrid coder with motion compensation, transform, and vector quantization.

For real-time operation, hardware design was developed by means of 1-μm CMOS technology. DCT has been implemented by a number of manufacturers already [384]. Later, the DCT chip set had to comply with the IDCT mismatch criterion specified in H.261 [437]. The remaining coding functions were implemented in turn, with the clock rate 10 to 20 MHz [425, 428]. To have cost-effective flexibility, the encoder is partitioned into several functional modules that utilize the state-of-the-art VLSI technology. Therefore, typical systems are composed of pre- and postprocessing chip, source coding chip, VLC/VLD chip set, and so forth. The chip sets can be implemented on a single board, still requiring peripheral and interface operations [426].

Software-based implementation shows slow moving pictures up to five frames/sec [438]. Such a success is possible by means of a massively parallel computer. As shown in Table 9.11, the most time-consuming operations are quantization and motion estimation. That quantization may not be executed by fast algorithm is the main reason for long processing time. Since the H.261 uses 32 quantizers, quantization would be more complicated.

Table 9.11 Timing of individual operations for an H.261 source coder simulator [438]

Function	Clock time (ms)
Frame data distribution	6
DCT	13
IDCT	14
Zooming (for motion estimation)	10
Motion estimation	49
Zigzag scanning	1
Zigzag descanning	1
Forward quantization loop	64
Inverse quantization	4
Variable length coding (VLC)	8
Loop filtering	4
Frame total	174

Source: © 1992 IEEE.

A combination of dedicated DCT or loop filter circuit and vector pipelining processor can realize the H.261 with 15 frames/sec, 64 Kbps in CIF mode. In [434], performance of 2.0 GOPS at a clock rate of 60 MHz is achieved. CMOS process has also been developed up to 0.6 μm process. Motion estimation and quantization take two-thirds of the overall processing.

An all-ASIC system can be designed with 7 to 11 chip sets that can implement separate functions [439, 436]. Figure 9.19 shows that the system is input from NTSC video and allows external parameters. DCT/IDCT uses the same module in the encoder and decoder (Fig. 9.20). The general properties that are found in most of the chip sets are summarized as follows:

- *Prefiltering:* Conversion of the NTSC signal to a CIF or QCIF [409, 413] needs temporal prefiltering to remove the flicker noise. A three-tap filter may be used for the conversion. Additional filtering is needed in the spatial domain.
- *DCT/IDCT:* Input video is stored in frame memory and the interframe difference is transformed by DCT. The DCT/IDCT mismatch problem needs to be solved.
- *Quantizer/dequantizer:* The quantizer step size can be controlled either by the control signal or by external parameters. Dequantization and IDCT are performed for temporal prediction.
- *Loop filter:* Loop filtering and motion estimation are controlled by one decision parameter. A cache memory of the search area for motion estimation is installed.

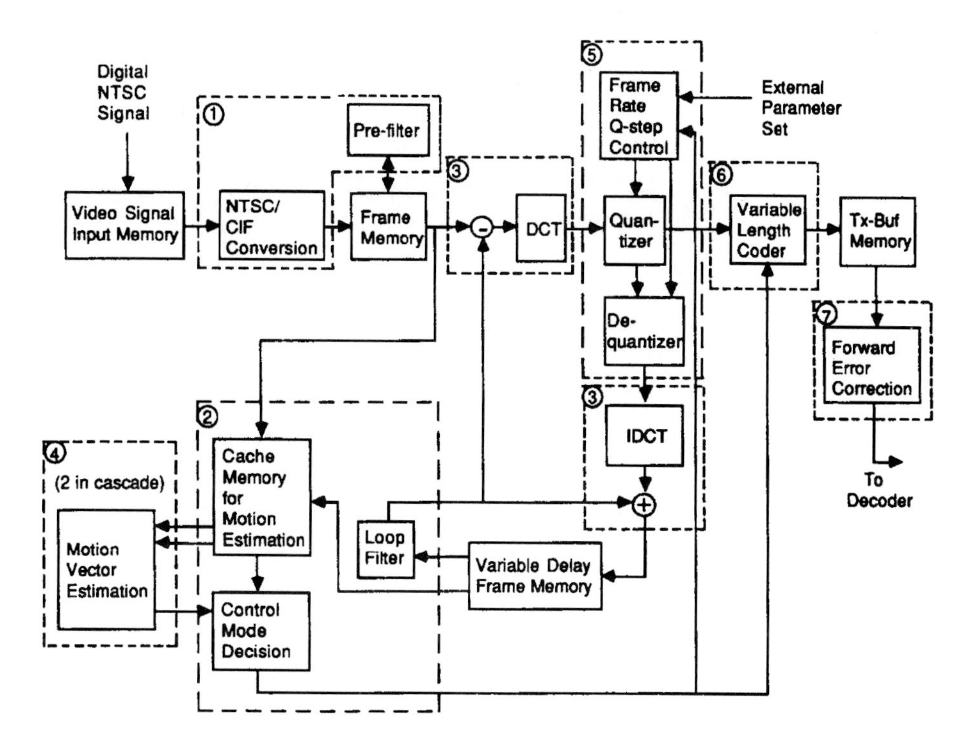

Figure 9.19 A functional partition of an ASIC-based encoder [436]. © 1992 IEEE.

- *Motion estimation:* In general, a full-search or three-step-search block-matching algorithm (Fig. 6.10) is used for the motion estimation. The search area can go up to (46 × 46), since H.261 allows maximum displacement of ±15 pels(lines)/frame. The macroblock is input sequentially and processed in parallel.
- *VLC/VLD:* Headers that need VLC (MVD, MTYPE, CBP) and transform coefficients (TCOEFF) are coded/decoded. These are separately coded and multiplexed/demultiplexed in the bit stream.
- *Forward error control:* BCH (511, 493) is implemented by an 18-tap shift register in the encoder, separated by EX-OR gates at positions corresponding to the nonzero terms of the generator polynomial. After all the message bits are in, the registers contain the check bits that are then transmitted in serial.
- *Postfiltering:* It is possible either that the postfilter is positioned before format conversion to NTSC for the reduction of the blocking effect or mosquito noise, or that it is located with the format converter for the reduction of aliasing noise in the interpolation process.

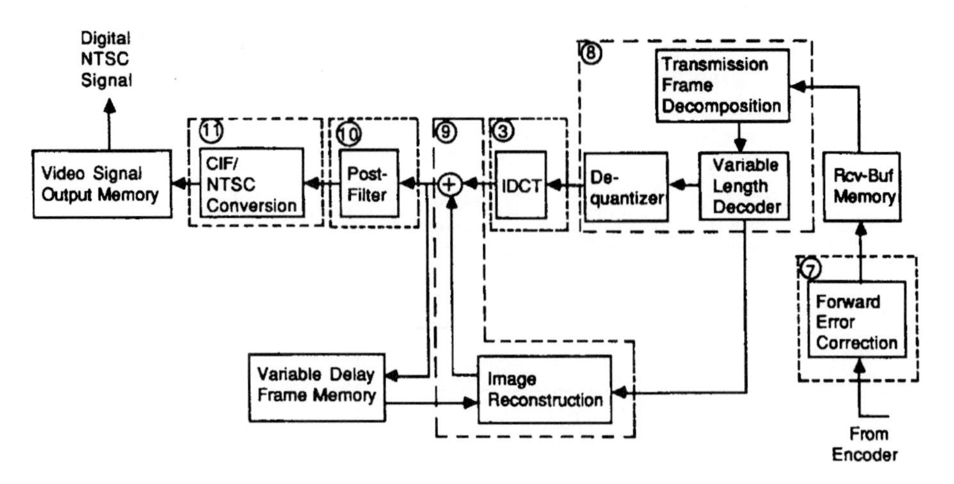

Figure 9.20 A functional partition of an ASIC-based decoder [436]. © 1992 IEEE.

The implementation example is composed of several functional chip sets. As new techniques have been developed in the hardware design, however, fast and simple implementations have become possible. Several chip sets are combined into one or two chips. Moreover, the chip sets are designed to be programmable and thus provide high performance, cost-effective solutions not only to H.261 but also to JPEG or MPEG requirements [393, 390, 391]. The improved motion video chip sets will be discussed further in the following chapters.

9.6 Additional Techniques

The coding algorithms specified in H.261 have influenced other standardization activities, MPEG, HDTV, CMTT, and so forth. In MPEG-1, designed for digital storage media (DSM) [416], more complicated motion estimation and HVS weighting are introduced. Lin et al. [433] suggest random ordering of macroblocks rather than sequential one-in-a-frame. Since the random distribution can be analyzed by a deterministic model, any visible artifacts at low bit rates can be reduced by randomizing them. Quantizer needs are to be changed, adopting random properties, such as reducing step sizes or Laplacian modeling. In this section, suggestions that are not compatible to H.261 are mentioned as new directions.

Instead of DCT in the coder, subband coding (SBC) with vector quantization (VQ) can be introduced [421, 424]. Motion compensation is desirable because it reduces temporal redundancy. It should improve performance at the expense of

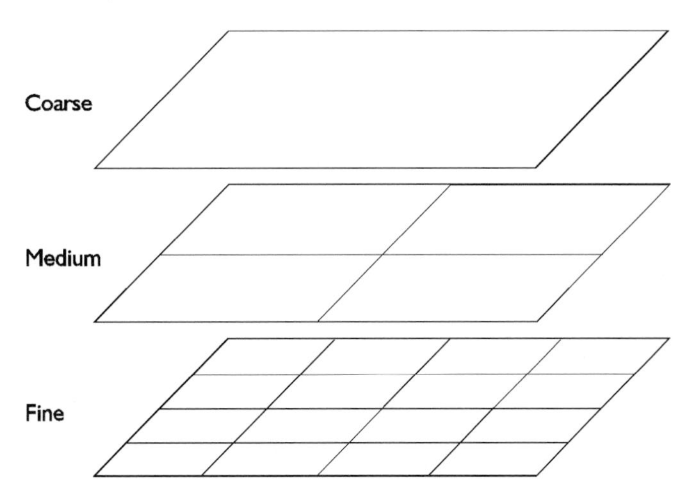

Figure 9.21 A variable block size search algorithm for motion estimation. The next stage becomes coarse or fine search, subject to motion statistics.

computational complexity. Using short kernel filters [404] for subband decomposition may simplify the analysis and synthesis filters. Wavelet transform has also been suggested, using a space-variant wavelet packet decomposition for energy compaction in the spatial domain [444].

For motion compensation, variable block sizes (multigrid) of the search area can reduce block artifacts [435]. The multigrid search especially affects subband coding as block-based motion estimation may introduce annoying block effects. The coder searches in the initial block area and then goes to a smaller or a larger block determined by certain rules. The motion vectors are interactively estimated and refined at different scales, as shown in Fig. 9.21. Although the algorithm gives a lower prediction error for a given amount of overhead motion field information, more high-frequency noises that do not imply true motion may occur. Phase correlation may be used as a powerful frequency domain technique that is insensitive to noise [440].

Assuming a low bit-rate transmission such as 64 Kbps in H.261, directional edge-oriented, object-oriented, and model-based coding have been suggested. The edge-oriented method is similar to subband coding in view of spatial filtering, and to edge detection using the edge information. Both high- and low-frequency components are coded by DPCM [415].

An object-oriented coder analyzes input image and extracts three parameters: motion, shape, and color information [755]. The analysis is based on source models of moving two-dimensional or three-dimensional objects and coded by an object-dependent parameter coding. Compression ratio could be increased, when the decoder synthesizes the image using these three parameters only. The model-based approach also relies on the object information. The objects are segmented from the background. Initial segmentation is obtained if the image is divided into changed and unchanged regions. Moving objects are detected and modeled as significant information to be transmitted [422, 441]. The procedures might be hierarchical ordering. The decoder receives the output image by means of the hierarchical model parameters. However, to obtain such a model description for a real scene, criteria and algorithms are necessary for performing the segmentation, analysis, modeling, and synthesis. Often such a whole three-dimensional modeling is difficult to develop. Therefore, simple and stable algorithms have to be developed. Further details are described in Chapter 12, including the very low–bit-rate coders such as waveform-based, object-oriented, model-based, fractal and morphological-based coders, and ITU-T Recommendation H.263.

Chapter 10
MPEG-1 Audiovisual Coder for Digital Storage Media

Summary

The MPEG committee has been active since 1988, with the aim of achieving plausible video and audio recording and transmission at about 1.5 Mbps.This standard as a committee draft was approved by ISO IEC/JTC1 SC29 in 1991. The draft, CD 11172, "Coded representation of picture, audio, and multimedia/hypermedia information," was prepared by SC29/WG11, known as MPEG (Moving Picture Experts Group). The draft became the international standard IS 11172 in November 1992. The data rate is motivated from compact disc (CD), which is usually recorded with audio information with high quality. It is only natural that consumers like to use CD-ROM as a digital storage medium for storage and retrieval of video and audio. Thus, the interactive systems on CD-ROM allow the development of many new applications involving video, sound, images, text, and graphics. The standard can be implemented by a personal computer, workstation or dedicated hardware/chip sets. Both the video and audio information have to be integrated in the MPEG standard so as to provide interactive services.

In this chapter, we discuss the first phase of MPEG activities (MPEG-1) and audiovisual coding algorithms resulting in higher quality and greater flexibility than in the H.261-based video coder (Chapter 9). Forward and backward prediction techniques for video coding and perceptual audio coding using the spectral and temporal masking effects of the human ear will be covered. We also include hardware implementation algorithms and suggestions for improvements from various researchers.

10.1 Introduction

MPEG is another group of experts that meet under ISO-IEC/JTC1 to recommend standards for digital video and audio encoding and decoding. This group was formed in 1988, and it has developed a committee draft CD 11172, "Coding of moving pictures and associated audio for digital storage media at up to about 1.5 Mbps," in December 1991 [487]. The draft is composed of five parts: system configurations, video coding, audio coding, compliance testing, and software for MPEG-1 coding. Parts 1 to 3 attained the IS status in November 1992 (Part 4 in Nov. 1994).

MPEG does not specify an encoding process. It specifies the syntax and semantics of the bit stream and the signal processing in the decoder. The detailed compression algorithms, however, are up to the individual designers. The video and audio parts address specific coding of full-motion video at about 1.15 Mbps and audio at rates (in Kbps) ranging from 32 to 448 for layer I, 32 to 384 for layer II, and 32 to 320 for layer III. The total bit rate 1.406 Mbps is nearly the bit rate of a CD, 1.412 Mbps. The system part specifies a system coding layer for synchronization and multiplexing of the compressed audio and video bit streams [497]. MPEG is thus the first technical group in charge of studying jointly the coding of the two media, i.e., video and audio. Figure 10.1 illustrates the interactive operations with digital storage media. CD and laser-disc players using MPEG-1 video/audio decoders have already come into the market. These are mass-consumer electronics and have a wide spectrum of applications (entertainment, education, training, medicine, sports, karaoke video, etc.).

Brief history of MPEG-1

The MPEG committee started with a tight schedule and a great deal of work has been done by the members to arrive at an international standard (IS). The audio schedule was a little different from the video schedule, as shown below. The brief history of the committee's efforts can be summarized as follows:

- Oct. 1988: The group was formed in ISO IEC/JTC1/SC2/WG8.
- June 1989: The preregistration deadline passed.
- Sept. 1989: The proposal registration passed. Fourteen different proposals were presented.
- Oct. 1989: Subjective tests (video) were performed [485].
- Mar. 1990: The video algorithm was defined as Simulation Model (SM) 1.
- Apr. 1990: ISO IEC/JTC1/SC29 WG11 became responsible for the group.
- July 1990: Subjective tests (audio) were performed [514].

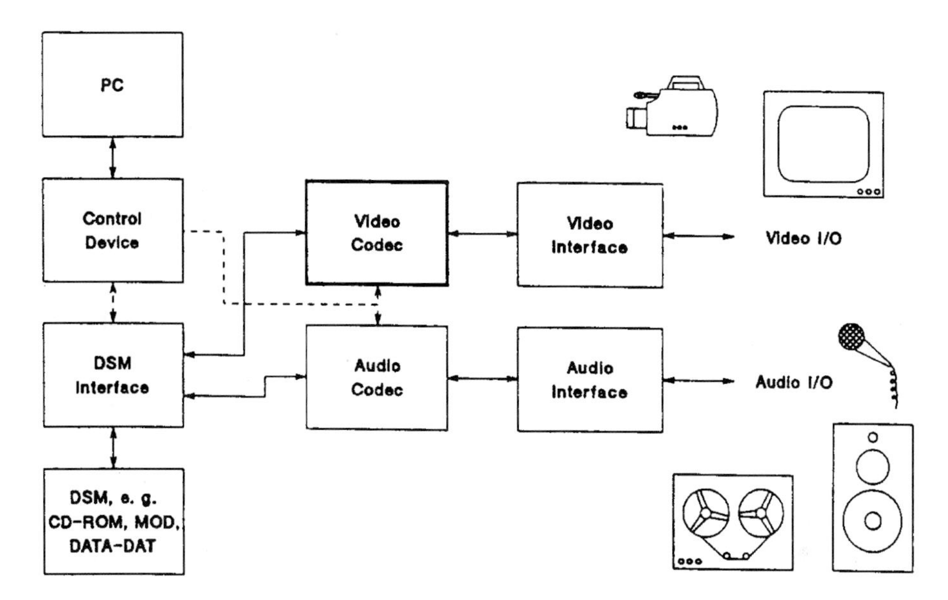

Figure 10.1 A block diagram of an interactive video/audio codec in a digital storage media environment [416]. © 1990 Signal Processing.

- July 1990: The final Simulation Model (SM 3) was adopted [488].
- Sept. 1990: Agreement was reached on a draft proposal.
- Dec. 1991: CD 11172 was adopted [487].
- Nov. 1992: DIS 11172 was issued.
- Nov. 1992: International Standard (IS) 11172 was established (Part 1, Systems; Part 2, Video; Part 3, Audio).
- Nov. 1994: IS 11172-4 (compliance testing) and Draft Technical Report (DTR) 11172-5 (software for MPEG-1 coding) were completed.

Many storage media and telecommunication channels are well suited to the compressed video at the rate of 1 to 1.5 Mbps. Standards for the provision of interactive moving pictures using inexpensive digital storage media are considered to be an essential item for the development of the related industry. CD-ROM is certainly one of the most important storage media because of its large capacity and low cost. Digital audio tape (DAT) is a recordable medium, but its sequential nature is a major drawback when random access is required. The Winchester-type computer disk can be used with both recordability and random access possibility, but it is expensive. The writable optical disk can be widely used due to its recordability, random access, portability, and low price. Telecommunication links include buses, LAN, and other transmission channels. The MPEG coder should allow a throughput around the rate of these media.

The objective of the MPEG standard is to define a source coding algorithm with a large degree of flexibility that can be used in many different applications. Interactivity between these applications is an important role in the standard. Essentially no constraints on algorithm complexity are defined in the encoder. It is allowed to exploit the benefits of non–real-time processing. The decoder algorithm should be made simple and inexpensive by using a few VLSI chips. A brief description of the procedure adopted to test and evaluate the various proposals is given below [502].

- *Motion video decoding (playback)*: Moving images in the four basic operating modes, forward and reverse playback at normal and fast speed (8 to 10 times the normal speed).
- *Wideband audio decoding (playback)*: Audio sequences sampled at 32, 44.1, and 48 KHz can be decoded in monophonic (up to 128 Kbps) or stereophonic (up to 256 Kbps) mode.
- *Random access*: Fast access and reconstruction (forward/backward) of an arbitrary video frame from the bit stream requires periodic breaks in interframe coding (single frame video accessibility).
- *High-resolution still frame*: Encode certain frames at very high resolution for reconstruction and display on demand.
- *Audiovisual synchronization*: The video signal should be accurately synchronizable to an associated audio signal.
- *Error robustness*: The source coding scheme should be robust to any remaining uncorrectable errors.
- *Zoom capability*: Zooming image (camera movement) causes motion estimation error by using a block-based local estimator. A global estimation technique is required.
- *Direct transcoding of H.261, JPEG, or consumer standard audio*: This means that the MPEG decoder is able to decode the bit streams exactly, describing other video services such as JPEG and H.261.
- *Cost tradeoffs*: The decoder is implementable with a small number of VLSI chips (VLSI implementability) and hence with a lower cost.

With the interaction of video and audio, many application fields have developed. Asynchronous applications of digital video are electronic publishing, education and training, travel guidance, video text, point of sale, electronic games, entertainment movies, video karaoke, and so forth [484–496]. Synchronous applications include electronic publishing, video mail, video telephone, and videoconferencing.

The MPEG algorithm is based heavily on the H.261 video coder (see Chapter 9) developed for videoconferencing. The basic encoder of the MPEG is also composed of temporal and spatial redundancy reduction techniques. Because the H.261 coder works well for slow moving video, this type of coding alone does not yield enough compression necessary for obtaining good quality when all the frames are transmitted and the video includes a large amount of motion data, even if the data rate is in the mid range of p for the $p \times 64$ Kbps coder. The encoder should yield a higher compression ratio with video quality comparable to that of a VCR, and audio quality comparable to that of a CD.

For the videophone implementations as described in Chapter 9, the H.261 video coder can be interfaced with one of the CCITT digital audio standards. Recommendation G.711 is based upon μ-law or A-law transformation for the 3.2 KHz bandlimited signal [531, 537]. Recommendation G.722 is based upon ADPCM technique for the 7 KHz signal [527, 528].

The MPEG audio coding standard, however, calls for an audio quality comparable to that of a compact disc, which has a full 20 KHz bandwidth [517]. The digital representation of a stereo sound signal requires a sampling frequency of 48 KHz and 16-bit resolution, which amounts to 2×768 Kbps. The data rate of a compact disc sampled at 44.1 KHz is 2×706 Kbps. The standard is aimed at storing both audio and video on the compact disc, which has limited capacity. Therefore, the data rate of the audio signal is reduced by about 6:1 (256 Kbps) to 24:1 (64 Kbps) from the original 2×768 Kbps stereo sound. The possible application fields of the MPEG audio coding are

- Radio sound-program emission (terrestrial and satellite, stationary and mobile)
- Television sound emission (terrestrial TV, conventional satellite TV, EDTV, and HDTV)
- Contribution links (music, commentary, and reporting circuits)
- Distribution links
- Production (tapeless studio, editing system, and postprocessing)
- Storage (studio and consumer)

The main requirements for the digital audio coding in MPEG-1 standardization are as follows:

1. Bit rate: Monophonic audio (64, 96, 128, and 192 Kbps); stereo audio (128, 192, 256, and 384 Kbps)

2. Sampling rate: 32 KHz (FM broadcasting), 44.1 KHz (CD), and 48 KHz (DAT)
3. Input resolution: 16-bit uniform
4. Quality: 128 and 192 Kbps, contribution (studio) quality; 64 and 96 Kbps, distribution quality
5. Access unit length: < 100 ms (the smallest part of the encoded bit stream that can be decoded by itself)
6. Addressable unit: $< 1/30$ ms (the maximum distance between closest entry points in the encoded bit stream)
7. Total system delay: < 80 ms at a bit rate of 2×128 Kbps and a sampling rate of 48 KHz

In 1989, 14 proposals were submitted for establishing a digital audio coding standard. The proposals were clustered into four groups according to similarities [521] and are listed below:

1. ASPEC (Audio Spectral Perceptual Entropy Coding): transform coding with overlapping blocks
2. ATAC (Adaptive Transform Aliasing Cancellation): transform coding with nonoverlapping blocks
3. SB/DPCM (Subband Coding and DPCM): subband coding with ≤ 8 subbands
4. MUSICAM (Masking-pattern Universal Subband Integrated Coding and Multiplexing): subband coding with > 8 subbands

Based on evaluation of these four algorithms at the Swedish Broadcasting Corporation, Stockholm, combination of ASPEC as a representative of transform coders [534] and MUSICAM as a representative of subband coders [524] was proposed for further study in 1990. The most important performances are the sound quality, random access, decoder complexity, and so forth.

In the following sections, overall MPEG video and audio algorithms are discussed. The reader can observe additional and more efficient algorithms than the H.261 algorithm described in the last chapter. The algorithms will be emphasized here. This chapter includes two main parts: video and audio. General algorithms to implement the MPEG-1 algorithm on hardware or software are covered. Suggestions for further improvements are also mentioned.

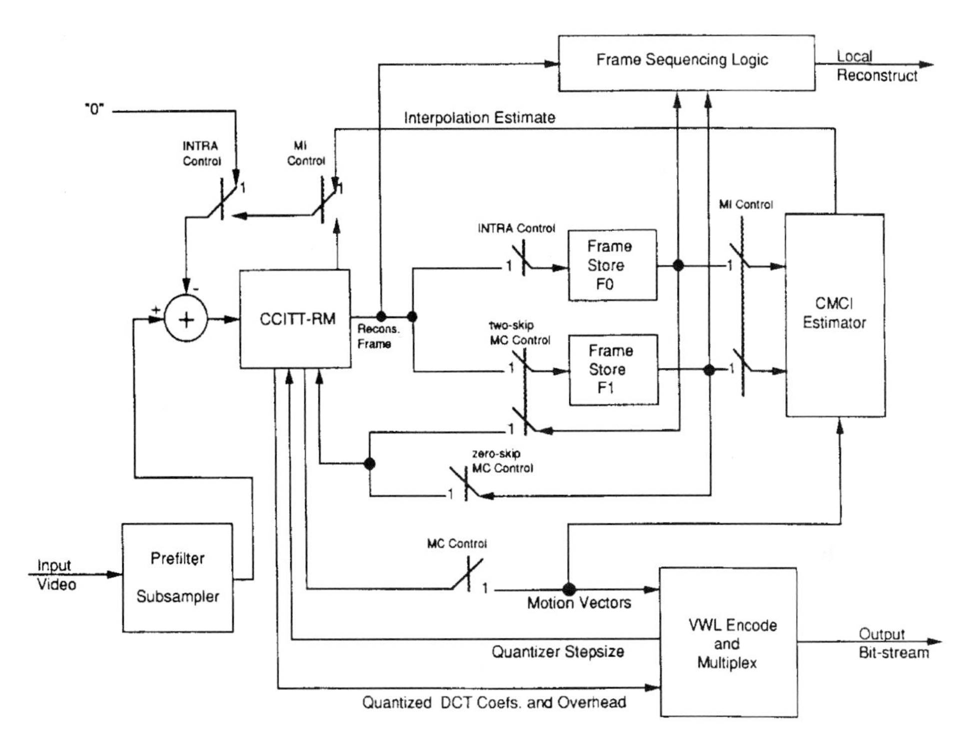

Figure 10.2 The CMCI–MPEG encoder proposed by Puri et al. [476]. © 1990 Signal Processing.

10.2 Video Coding

10.2.1 Extended coding algorithms

To meet the special requirements of the digital storage media as stated earlier, additional techniques beyond the H.261 video coder have been incorporated in the MPEG-1 coder. Some of the extended systems are described here.

Puri and Aravind employ another type of interframe processing called conditional motion-compensated interpolation [476, 477]. A motion vector appropriate to the interpolated frame is obtained by integer motion vector interpolation resulting in half-pel motion compensation. Every fifteenth frame is regarded as an intraframe. Predictive and interpolative frames are defined between two intraframes. Figure 10.2 shows a block diagram for the possible MPEG coder by using CCITT–RM (H.261) and a conditional motion-compensated interpolation (CMCI) estimator. The error signal is encoded using the H.261 algorithm. The CMCI estimator controls motion interpolations from two reconstructed frames stored in frame memories, F_0 and F_1. The same estimator can be used in the decoder of Fig. 10.3, using the transmitted motion vectors.

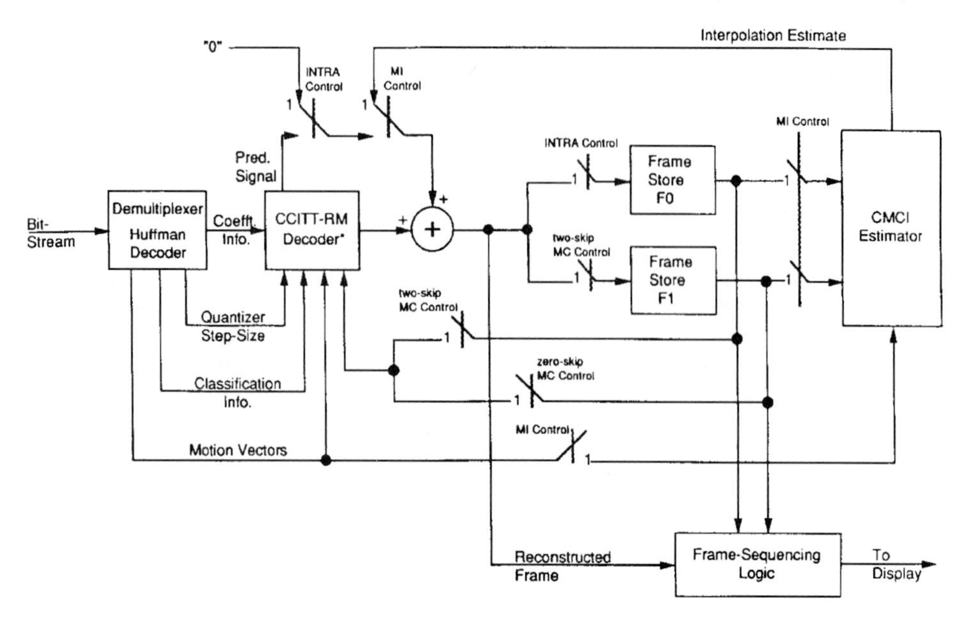

Figure 10.3 CMCI–MPEG decoder proposed by Puri et al. [476]. © 1990 Signal Processing.

The interpolation can be done from two or more neighboring frames. For example, in Fig. 10.4, the motion field for the interpolated frame F_1 can be obtained by scaling the available motion fields, e.g., d_{10} from F_0 and d_{13} from F_3.

Gonzales et al. [478, 480] suggest the video coder with arithmetic coding, which is more efficient than integer length Huffman coding. Complicated techniques such as nonlinear loop filtering and AC correction are also proposed. Median filtering is applied for motion-compensated blocks but certain pixels are restored at the output of the filter, to avoid boundary blurring effects, by using luminance characteristics of the HVS (human visual system). Some of the AC coefficients are predicted from considering (3×3) neighboring DC values and are corrected to avoid blocking effects.

Frame interpolation in a group of frames is also utilized in the proposed coder (Fig. 10.5) regarding every 16th frame as an intraframe. The rest of the algorithm is similar to the H.261 coder. Some AC coefficients which are significant for picture quality are corrected in the decoder only, as shown in Fig. 10.6. AC correction is a technique for modifying the first few low-frequency coefficients, if the dequantized value is zero. This technique reduces the block artifacts.

Haghiri and Denoyelle [481] introduced efficient pre- and postprocessing techniques to the H.261 coder. Motion-adaptive subsampling and interpolation are performed in the encoder and decoder, respectively. The motion-adaptive

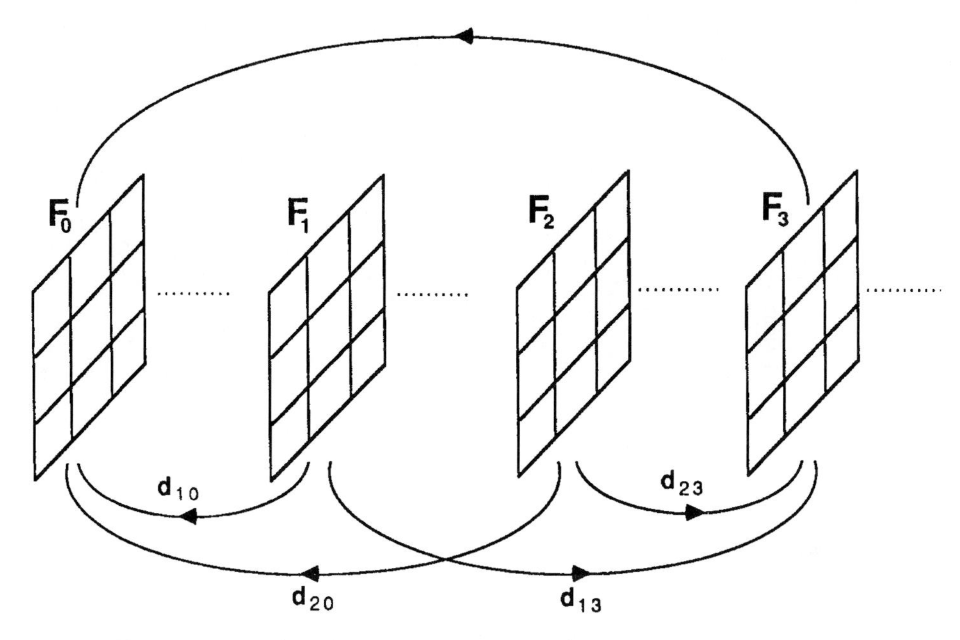

Figure 10.4 Two frame interpolations (F_1 and F_2) from two predicted frames (F_0 and F_3) [477]. © 1990 SPIE.

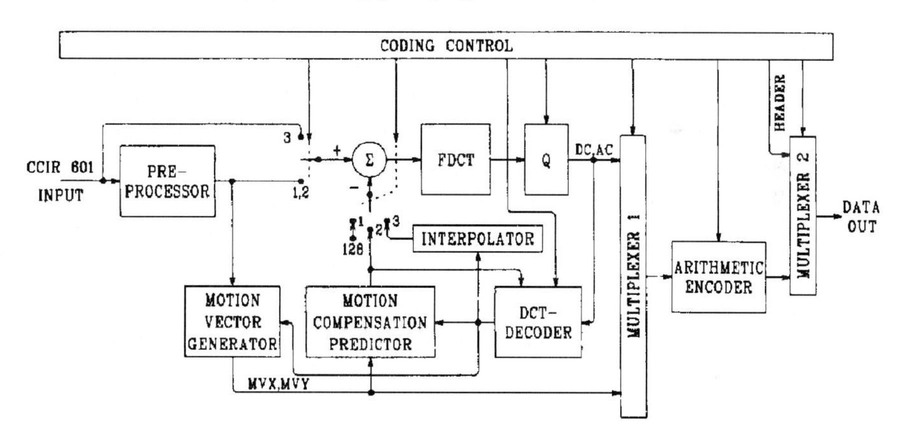

Figure 10.5 The encoder block diagram proposed by Gonzales et al. [480]. © 1990 Signal Processing.

subsampling is to reduce the sampling rate of input images by a factor of two without losing any visual information. In Fig. 10.7, the preprocessing algorithm is described. When objects move, one frame may be skipped and replaced by the motion vector and mode data.

Kamikura and Watanabe [499] introduce I, B, and P frame structure (see Fig. 10.8) adopted in the MPEG video coder. To reduce processing time in

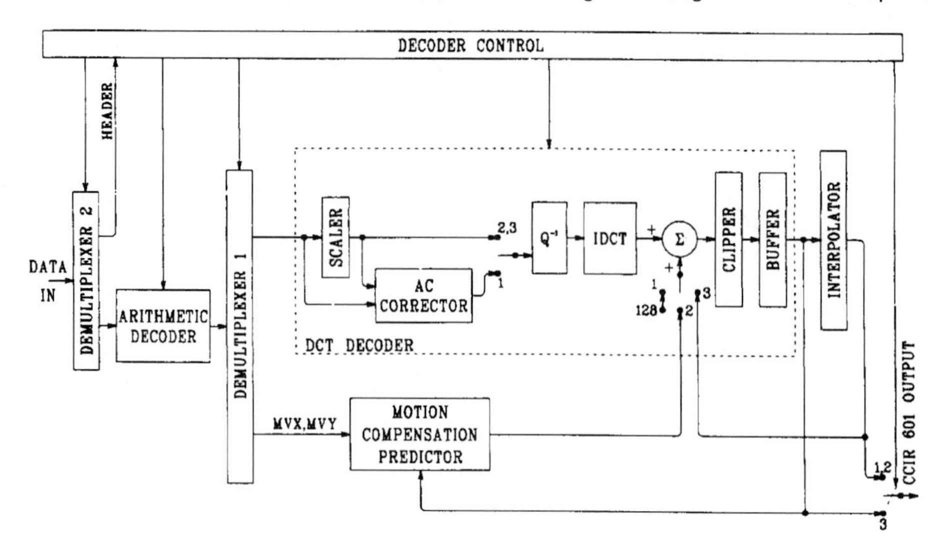

Figure 10.6 The decoder block diagram proposed by Gonzales et al. [480]. © 1990 Signal Processing.

intraframe coding, two layer quantization techniques are used as shown in Fig. 10.9. Global motion estimation is applied for improving coding efficiency by compensating for camera movement and zoom/pan operation.

Nagata et al. [479] also suggest a group of frames (GOF) concept. (The cycle of intraframe coding is once per 10 frames.) To increase the compression ratio and decrease the coding noise sensitivity, transform coefficients are weighted by the HVS weighting function in the coder suggested by Herpel, Hepper, and Westerkamp [416] and Pereira et al. [482] (Fig. 10.10). The illumination correction proposed by Hürtgen et al. [475] can be applied for shadows created by moving or rotating objects, nonuniform illuminations, changes in the reflectance, and changes in the global illumination.

A global motion compensation scheme is required in the video coder because of the zoom or pan in image sequences due to camera movement. The global motion estimation can be obtained from the conventional local estimator with estimation of the three parameters (zoom, horizontal, and vertical pan parameters) [506]. It was originally proposed to MPEG but was not adopted.

The three-step scheme [499] detects local motion using a block-matching technique first, and then the three parameters are obtained. This can be global estimation in that the following frame prediction represents true motion compensation. The zoom parameter can be determined as a focal length extension of the camera used. Displaced (enlarged) coordinates $P_2(X_2, Y_2)$ in the two-dimensional plane in Fig. 10.11 can be described by

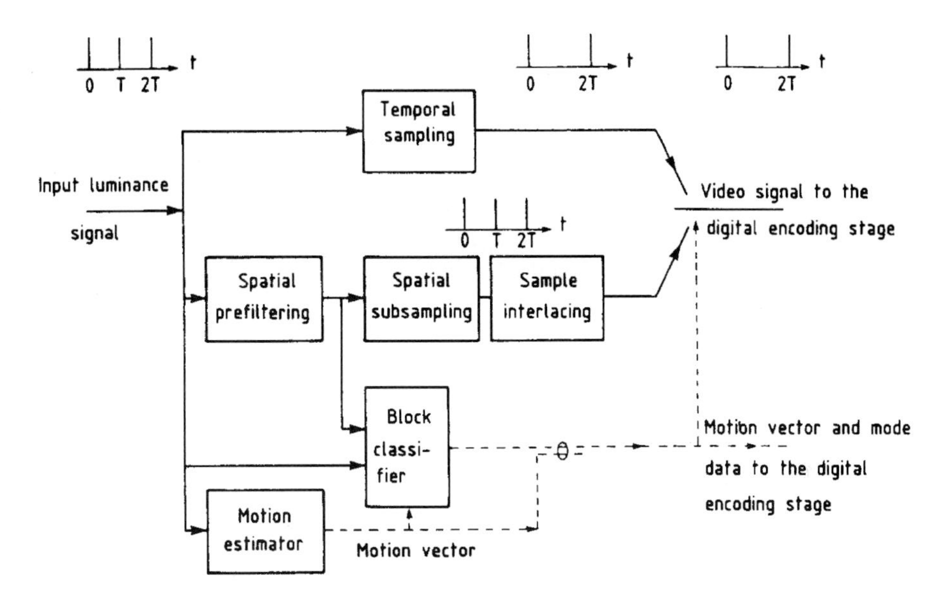

Figure 10.7 The processing stage for luminance signal proposed by Haghiri and Denoyelle [481]. © 1990 Signal Processing.

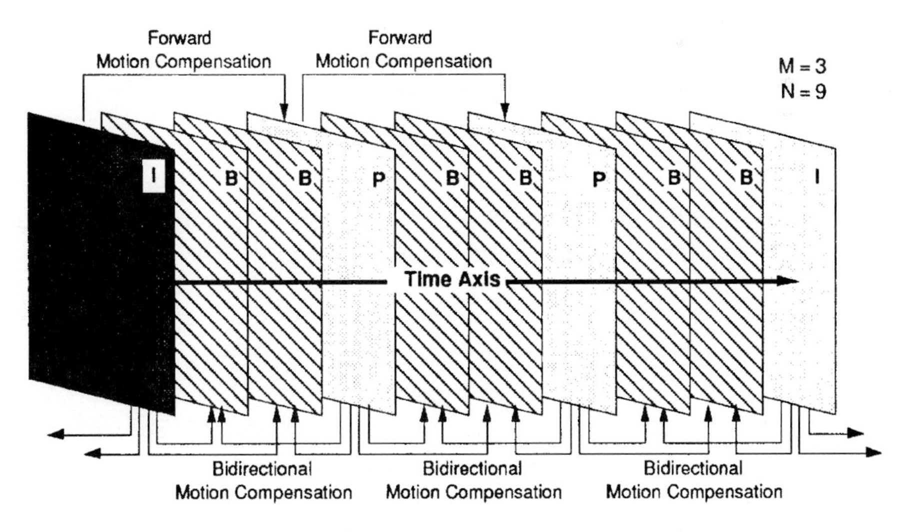

Figure 10.8 An example of temporal picture structure (I, B, and P pictures).

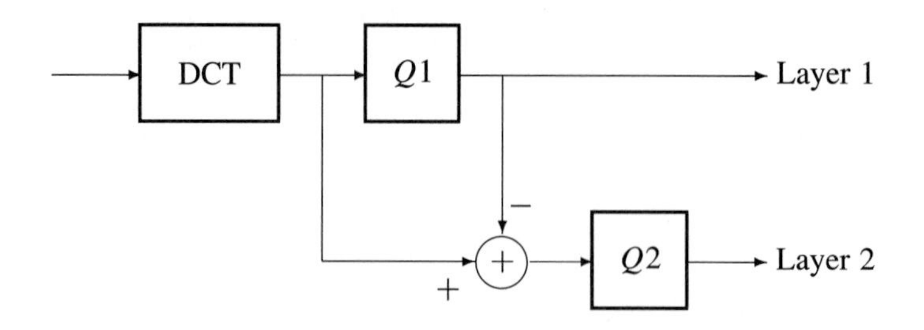

Figure 10.9 The hierarchical intraframe coding algorithm proposed by Kamikura and Watanabe [499]. © 1990 SPIE.

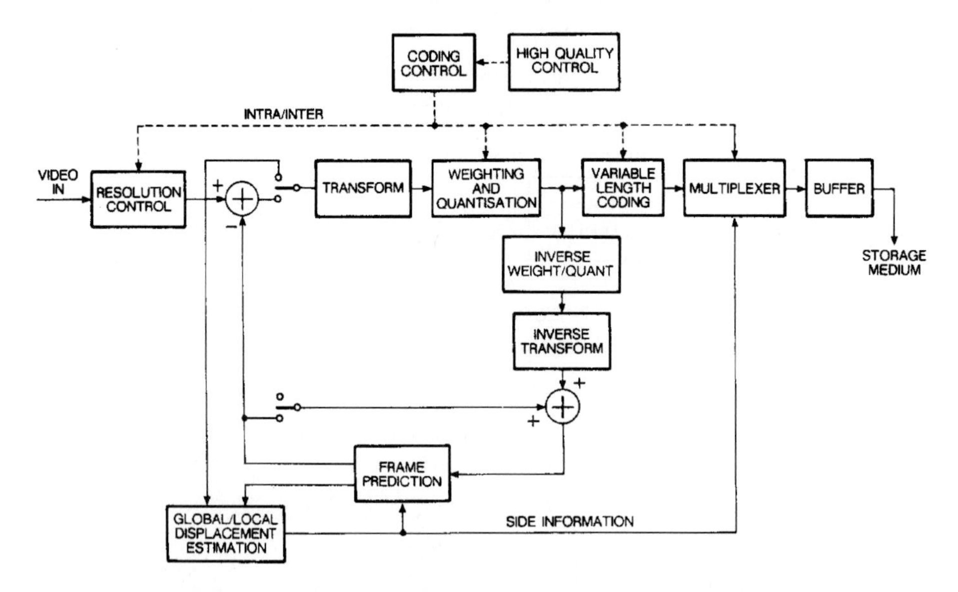

Figure 10.10 The encoder block diagram proposed by Pereira et al. [482]. © 1990 Signal Processing.

$$P_2(X_2, Y_2) = (a_1 X_1, a_1 Y_1) \tag{10.1}$$

where only one parameter, a_1, is to be determined [475], while pan operation can be regarded as a rotation around the x-axis or y-axis [501]. The two rotation parameters can be easily obtained by using the conventional image deformation algorithm.

In the case of bidirectionally predicted pictures (B pictures), the prediction is based on forward (previous I or P picture), backward (future I or P picture) or both motion estimations (Fig. 10.8). More accurate motion vectors are required in the coder when motion interpolation is introduced (Fig. 10.4) [477]. Half-

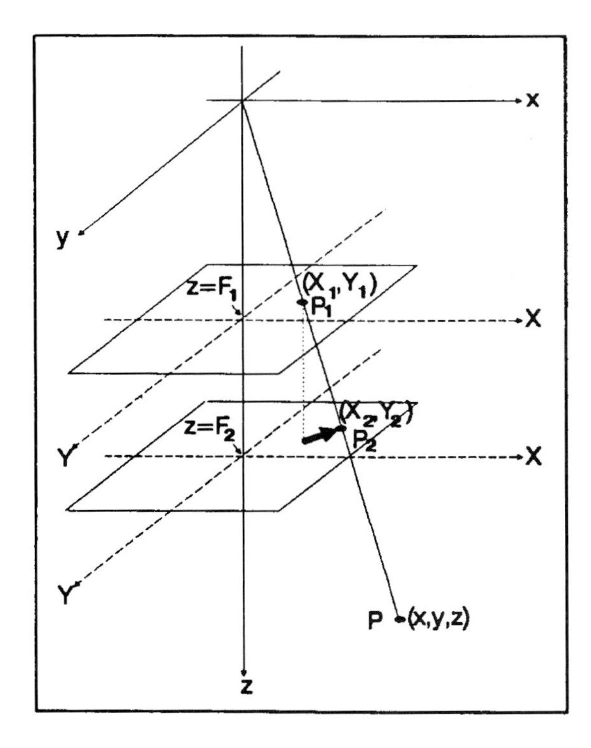

Figure 10.11 Modeling of camera zoom operation [475]. © 1990 SPIE.

pixel accuracy for motion estimation is a good trade-off between compression efficiency and computational complexity or increased overhead. Since a simple linear interpolation is efficiently used in digital image processing [508], the half-pixel resolution for motion estimation is based upon the interpolation algorithm as described in Fig. 8.22.

For the purpose of motion estimation, two steps are defined, i.e., a block-matching algorithm and a search range decision (Section 6.3). Mean absolute difference and mean square error criteria can be used in seeking a match between past and present pictures. The best matching algorithm can be adopted by the designer. Logarithmic search (Fig. 6.7), three-step search (Fig. 6.10), and many other simplified searches reduce complexity and obtain good local minima [507]. Once the best integer pel match has been found by either full search (exhaustive) or any fast algorithm, half-integer pel displacements are evaluated with the neighboring eight pels, as shown in Fig. 10.12. A further half-pel search with 8 half-pels is performed around the best integer matching pel (x, y). If one of the half-pel interpolated error criteria is lower than that of the integer resolution motion vector, then the corresponding half-pel resolution motion vector and the associated error are used in the MC/No MC decision.

The extended coding algorithms described so far have contributed to the fundamental algorithms in MPEG-1 video compression. Obtaining very high

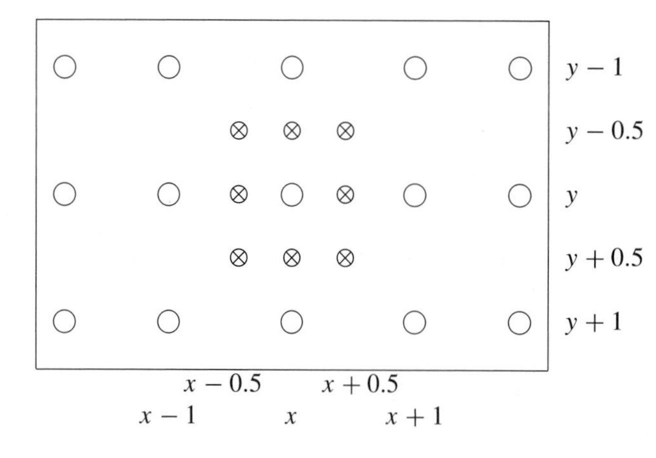

Figure 10.12 Illustration of half-pel displacements (○: integer-pel, ⊗: half-pel).

compression demands both intraframe and interframe coding. The two techniques are DCT-based transform coding for the reduction of spatial redundancy and block-based motion compensation for the reduction of temporal redundancy, as shown in Fig. 10.13. Let us briefly compare the encoder block diagram to the H.261 coder shown in Fig. 9.8. DCT, quantizer, VLC, multiplexer, and buffering blocks can be found in both coders. The quantizer is also controlled by the buffer status. Input procedures are not the same. The MPEG-1 coder receives input pictures that have more flexible resolutions. As defined in Table 10.1, the conventional television (NTSC or PAL) pictures or any other machine-generated pictures can be processed, once they satisfy the constrained parameters.

The MPEG standard does not specify an encoding process. It only specifies the syntax and semantics of the bit stream and signal processing in the decoder. Hence, many options are left open to the encoders to trade-off cost and speed against picture quality and coding efficiency.

The input video signal is followed by preprocessing and format conversion to select an appropriate window, resolution, and format before going through the encoder. If B pictures are used, the input group of pictures are reordered (Table 10.2) so that reference pictures (I and P) are preceded, because B pictures can be coded based on forward and/or backward motion estimations. There are also a number of other different modes for coding the P and B pictures. Further details are discussed below.

In Fig. 10.13, motion vectors used for motion compensation are obtained using intra (I) and predicted (P) pictures. In H.261, the motion vectors are obtained between input picture and the reconstructed picture based on minimizing the prediction error. Motion-compensated interpolation is a key feature of the MPEG video coder. Motion estimation is performed in the predefined group of pictures and the obtained motion vectors and mode decisions are used for motion compensation and interpolation in the decoder as shown in Fig. 10.14.

Table 10.1 Summary of the constrained parameters in MPEG-1 [487]

Parameter	Constraint
Horizontal picture size	$\leq$ 768 pels
Vertical picture size	$\leq$ 576 lines
Picture area	$\leq$ 396 macroblocks
Pixel rate	$\leq 396 \times 25$ macroblocks/s
Picture rate	$\leq$ 30Hz
Motion vector range	$\leq \pm 64$ pels (using half-pel vectors)
Input buffer size (VBV model)	$\leq$ 327,680 bits
Bit rate	$\leq$ 1,856,000 bits/s (constant)

Source: © 1993 ISO/IEC.

The MPEG standard specifically defines four types of pictures: intra (I), predicted (P), bidirectional (B) (Fig. 10.8), and DC (D) picture. These four types can be described as follows:

- An **I picture** uses only transform coding (Fig. 10.13) and provides a random access point into the compressed video data. In an I picture, all the (8×8) blocks are coded using DCT, quantization, and VLC. I pictures can be used for predicting P and B pictures.
- A **P picture** is coded using motion compensated prediction from a previous I or P picture. This technique is called forward prediction from I/P to P as shown in Fig. 10.8. This mode is similar to interframe coding in the H.261 coder. P pictures can accumulate coding errors. P pictures go through the feedback loop and can be used for predicting P and B pictures.
- A **B picture** is coded using both a past and/or future picture as a reference. Thus it is called bidirectional prediction, as shown in Fig. 10.8. The errors are not accumulated, since the B picture is never used as a reference. B pictures can be coded using forward or backward (or both) motion compensation.
- A **D picture** is a special case of intra in which only the DC coefficient of each (8×8) block is coded. D pictures provide simple and fast forward mode but yield limited image quality.

Motion compensated interpolation has many advantages. In the group of pictures, B pictures can be predicted based on bidirectional motion compensation. This improves random access and reduces the effect of MC prediction errors, as the B pictures do not go through the feedback loop. They are not used for predicting B or P pictures. Bidirectional prediction can cope with the uncovered area, since an area just uncovered is not predictable from the past reference. For example, if the camera is panned from right to left, the new picture segment is coming from the left edge of the picture. Forward motion compensation, however,

Table 10.2 Example of groups of pictures and ordering. (The coded bit stream is transmitted in decoding order and the decoder converts it into display order)

(a) Display	I	B	P	B	P								
	0	1	2	3	4								
Decoding	I	P	B	P	B								
	0	2	1	4	3								
(b) Display	B	B	I	B	B	P	B	B	P	B	B	P	
	0	1	2	3	4	5	6	7	8	9	10	11	
Decoding	I	B	B	P	B	B	P	B	B	P	B	B	
	2	0	1	5	3	4	8	6	7	11	9	10	
(c) Display	B	I	B	B	B	B	P	B	I	B	B	I	I
	0	1	2	3	4	5	6	7	8	9	10	11	12
Decoding	I	B	P	B	B	B	B	I	B	I	B	B	I
	1	0	6	2	3	4	5	8	7	11	9	10	12

cannot completely predict this changed area. The moving (changed) area can be efficiently predicted from the future reference picture using backward motion estimation. The effect of noise can be decreased by averaging between the past and the future pictures using both motion estimations, i.e., forward and backward. If the number of consecutive B pictures increases within a group of pictures, however, the correlation between the references decreases and interpolation errors become larger. Therefore, tradeoff between reasonable quality and complexity is determined by the designer. Many proposals mentioned earlier show good results by using two or at most three consecutive B pictures with 10 to 15 pictures in a group of pictures.

The MPEG is intentionally specified such that the decoder is much simpler than the encoder. As the MPEG is designed mainly for digital storage media, such as the storage and interactive display of motion video and audio on a CD, a simplified decoder assures this as an affordable (low cost) mass market consumer electronics item. The decoding process is defined by the MPEG-1 standard. But architecture for decoding the bit stream is not specified. The decoder reconstructs the data elements in the bit stream according to the defined syntax. The decoding process is in the reverse order from the encoding process, i.e., (8×8) basic block, macroblock, slice, and picture. The macroblock type and motion vectors are used to construct a prediction based on past and/or future reference pictures. Finally, it decodes the (8×8) coefficients by an inverse DCT and the result is added to

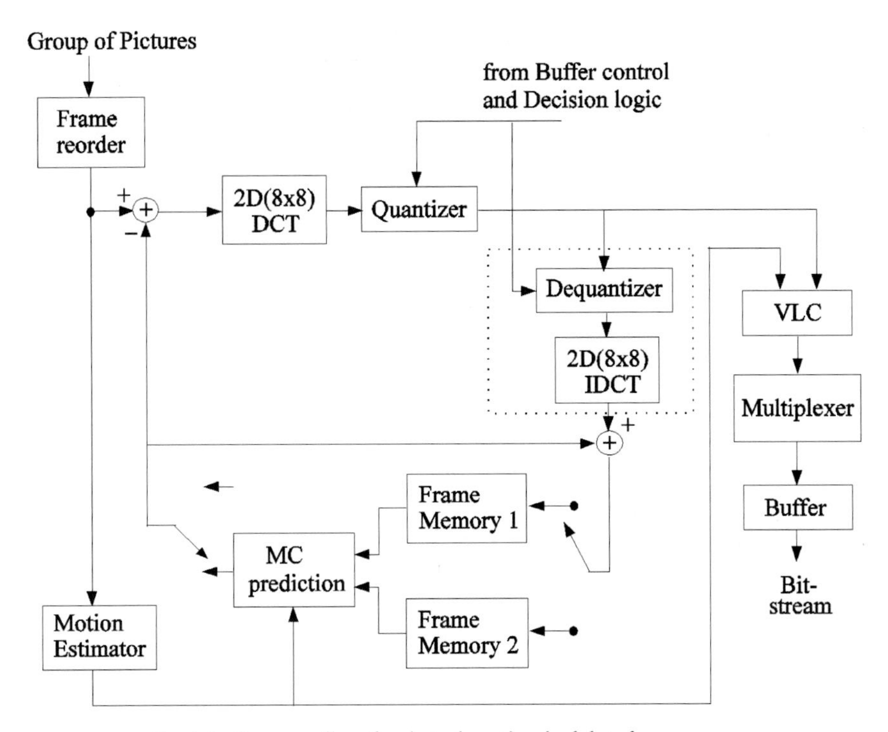

Figure 10.13 A simplified encoder block diagram of the MPEG-1 video coder. For intraframe coding, the dotted portion is deleted.

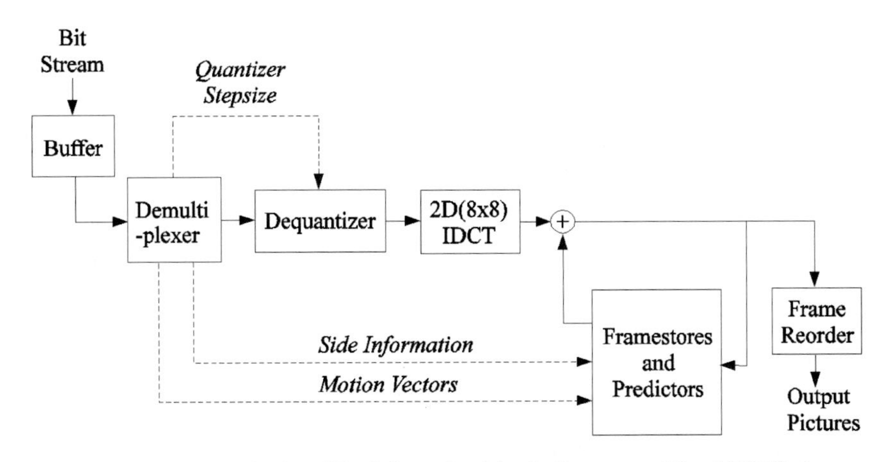

Figure 10.14 A simplified decoder block diagram of the MPEG-1 video coder.

the predicted block resulting in the reconstructed pictures. If the picture is an I or P picture, it is stored as a reference picture. Decoded pictures are reordered in display order and postprocessed before displaying as shown in Table 10.2.

10.2.2 MPEG-1 video structure

For the purpose of very high compression and good video quality, the MPEG-1 standard defines interpolated (B), intra (I), and predicted (P) pictures as stated earlier. The layered video structure concept is also extended from the H.261 structure. The input format conversion to SIF (source input format) is discussed here.

Source input formats

The source information may exist with various formats, satisfying the constrained parameters listed in Table 10.1. But the optimum parameters for the picture rate and resolutions (horizontal and vertical) are determined in view of the best perceived quality at a given data rate. For example, SIF-525, which has 360H×240V pels at a 30 Hz picture rate, and SIF-625, which has 360H×288V pels at 25 Hz, represent the same information of about 2.6 Mpels/sec, as shown in Table 10.3.

Let us give an example from ITU-R 601 video. Popular source resolutions are specified in this format. It consists of two components: luminance (Y) and chrominance(C_b, C_r). MPEG-1 requires that the color-difference signals be subsampled with respect to the luminance by 2:1 in both vertical and horizontal directions. This procedure is shown in Fig. 10.15, where each chrominance component is filtered horizontally and vertically and then decimated giving one-half resolution of even or odd luminance fields. Both components are split by

Table 10.3 Conversion of two source formats to the MPEG SIF (for SPA (significant pel area) the temporal rate is in pictures (noninterlaced) per second)

		(NTSC) 525/30	(PAL) 625/25
Luminance	ITU-R	720 × 480	720 × 576
(Y)	SIF	360 × 240	360 × 288
(pels/picture)	SPA	352 × 240	352 × 288
Chrominance	ITU-R	360 × 480	360 × 576
(C_b, C_r)	SIF	180 × 120	180 × 144
(pels/picture)	SPA	176 × 120	176 × 144
Frame rate (Hz)		30	25

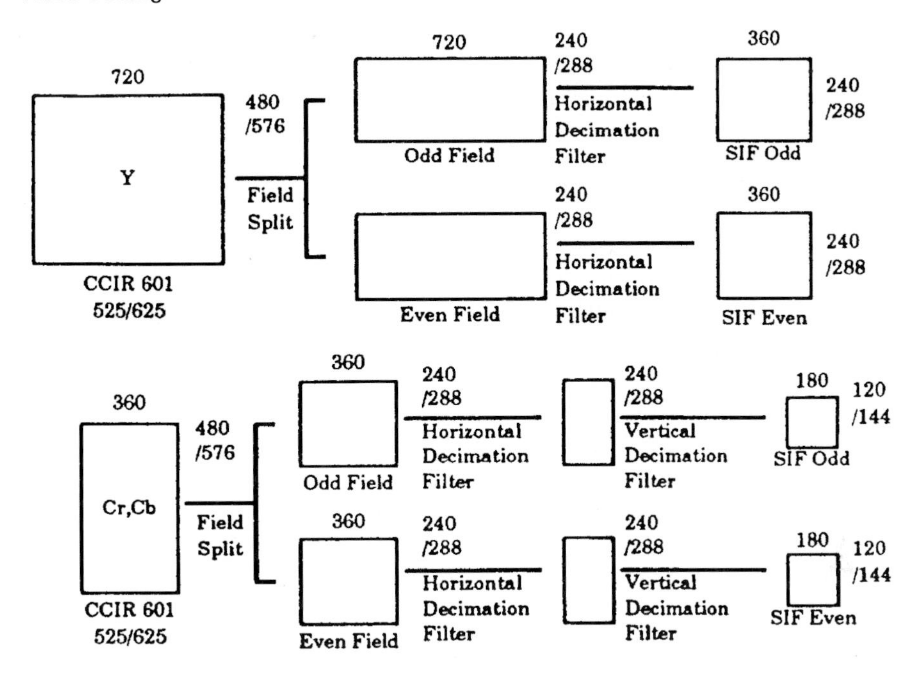

Figure 10.15 A block diagram for conversion from ITU-R 601 into SIF Odd and SIF Even.

fields to achieve the noninterlaced SIF. The luminance and chrominance samples are positioned as shown in Fig. 9.4.

Decimation filters are used for the conversion of ITU-R 601 into SIF, whereas interpolation filters are used for the reconversion (postprocessing) in the decoder, as shown in Fig. 10.16. MPEG-1 defines only the codec features for the SIF picture and as usual pre- and postprocessing are left to the designer. Reconversion to ITU-R 601 at the decoder is performed in the reverse order of the conversion process from ITU-R 601 at the encoder. Hence, vertical filtering is implemented before horizontal filtering in the postprocessing. Field splitting in the preprocessing corresponds to vertical downsampling in the noninterlaced picture. Therefore, the chrominance part of SIF should be up-sampled once horizontally and twice vertically.

The nonlinear filter is designed for pre- and postprocessing with respect to the filtering points: first, the filtering reduces the aliasing effect; second, it gives a smooth transmission region and reduces the visibility of false contours, and it retains image sharpness. Considering these constraints, the standard discusses decimation and interpolation filter coefficients as an example only. In Fig. 10.17 the filters are illustrated and are applied to the input pels as shown in Example 10.1. The notation // denotes integer division with rounding to the nearest integer away from zero.

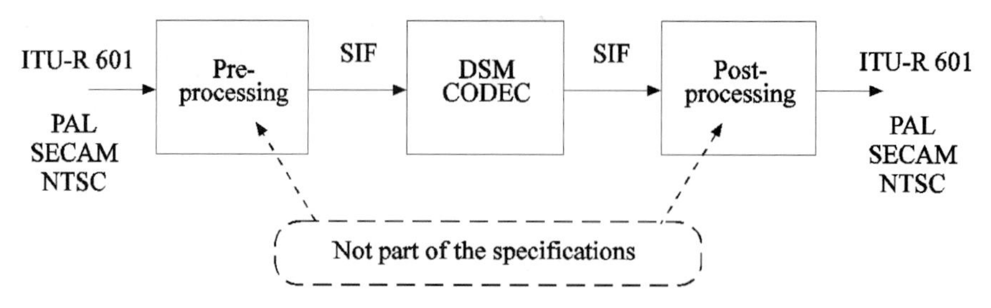

Figure 10.16 Pre- and postprocessing at the encoder and decoder.

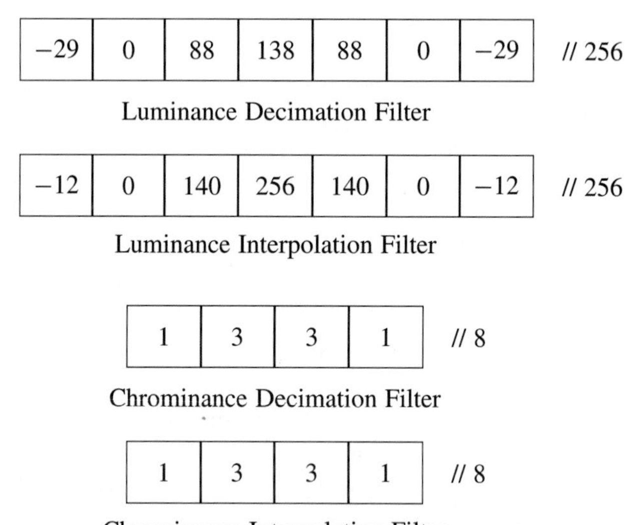

Figure 10.17 Decimation and interpolation filters for conversion between ITU-R 601 and SIF pictures. (These are examples only.)

Example 10.1 *Image data samples are excerpted from a picture. Then the* 10 *pels are decimated by using the filter (Fig. 10.17), replicating the last pel at the both ends of pel-array. Filtered outputs are rounded up to the nearest integers. The results are shown in the table below:*

Input pels	x_0	x_1	x_2	x_3	x_4	x_5	x_6	x_7	x_8	x_9
	80	76	79	75	79	82	82	86	94	85
Decimated pels		$\hat{x}_1$		$\hat{x}_3$		$\hat{x}_5$		$\hat{x}_7$		$\hat{x}_9$
		78		76		80		88		88
Interpolated pels	a_0	$\hat{x}_1$	a_2	$\hat{x}_3$	a_4	$\hat{x}_5$	a_6	$\hat{x}_7$	a_8	$\hat{x}_9$
	78	78	77	76	78	80	84	88	89	89

Table 10.4 Functional comparison of six layers

Layers of syntax	Functions
Sequence layer	One or more groups of pictures
Group of pictures layer	Random access into the sequence
Picture layer	Primary coding unit
Slice layer	Resynchronization unit
Macroblock layer	Motion compensation unit
Block layer	DCT unit

For example,

$$\hat{x}_5 = (138x_5 + 88(x_4 + x_6) - 29(x_2 + x_8))//256$$

$$a_4 = \left(140(\hat{x}_3 + \hat{x}_5) - 12(\hat{x}_1 + \hat{x}_7)\right)//256$$

MPEG bit stream hierarchy

The layered structure in the MPEG bit stream is as follows: sequence, group of pictures (GOP), picture, slice, macroblock, and block layer. Each layer consists of the appropriate header and following lower layers in a manner similar to that of the H.261 coder. The layered structure supports flexibility and efficiency in the coder/decoder [484]. Coding processes can be logically distinct and layers can be decoded systematically. Each layer supports a specific function as described in Table 10.4. The MPEG-1 layered structure is shown in Fig. 10.18. A number of blocks or pels in a layer are defined by an input format identified in the sequence header.

(a) The **sequence layer** consists of a sequence header, one or more groups of pictures, and an end-of-sequence code. A sequence header consists of several entities as given below. Not all items are shown.

- Sequence header code: beginning header ('00 00 01 B3')
- Horizontal and vertical sizes: 12 bits each
- Pel aspect ratio: height/width ratio (4 bits), e.g., VGA (1.0000), ITU-R 601/625 (0.9375), ITU-R 601/525 (1.1250)
- Picture rate: 4 bits (23.976, 24, 25, 29.97, 30, 50, 59.94, and 60 pictures/sec)
- Bit rate: 18-bit integer that represents multiples of the unit rate (400 bps), giving the bit rate of the data channel (constant or variable)
- VBV buffer size: 10-bit integer that represents multiples of the unit buffer (16384 bits), defining the size of the video buffering verifier (VBV)
- Intra and nonintra quantizer matrix: (8×64) bits default tables

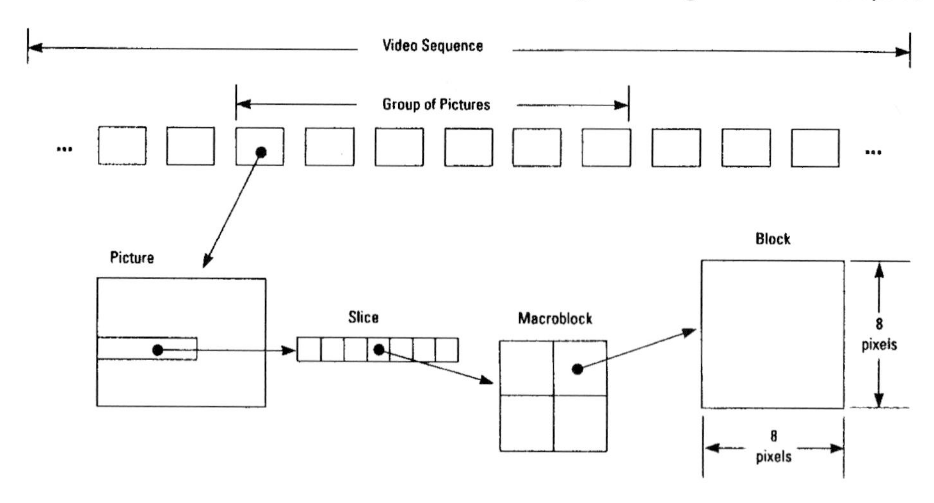

Figure 10.18 The layered MPEG-1 video structure. (Besides the (16×16) Y block, an MB has a C_b and a C_r (8×8) block.)

(b) The **group-of-pictures layer (GOP)** is a set of pictures that are in a continuous display order. It begins with an I or a B picture and ends with an I or a P picture. The smallest size is a single I picture and the largest is not specified. The group-of-pictures header starts with the start code (x '00 00 01 B8'). A time code of 25 bits refers to the first picture in the group, as a unit of hours, minutes, and seconds. A closed GOP code represents closed prediction in the group only or open prediction that requires decoded pictures of the previous group for motion compensation.

(c) The **picture layer** is a primary coding unit that consists of the luminance and two chrominance components. The layer starts with a picture header that begins with a start code (x '00 00 01 00'). Some of the other header entities are:

- Temporal reference: A 10-bit number that denotes a serial number of a picture in display order, where the first picture in a group of pictures is numbered as zero.
- Picture coding type: A 3-bit index defining I, P, B, and D picture types.
- VBV delay: A 16-bit number defining the time in units of 1/90000 second. It is used by the decoder in conjunction with the bit rate in the sequence header.
- Forward/backward frame code: A 3-bit number representing the maximum size of the forward/backward motion vectors up to 64 frames, respectively.

(d) The **slice layer** is important in the handling of errors. The decoder can skip the corrupted slice and go to the start of the next slice, if the bit stream is corrupted by noise. The number of slices in a picture can range from one to the number of macroblocks in a picture, depending upon error environments. The principal entities in the slice header are a slice start code that consists of the code x '00 00 01' and an 8-bit vertical position code in a picture, and a quantizer scale index in the range that can be changed by the next slice or the macroblock layer.

(e) The **macroblock layer** is composed of a (16×16) luminance block and the corresponding chrominance blocks as in the H.261 coder (Fig. 9.5). The header contains information as follows:

- MB stuffing: An 11-bit (0000 0001 111) string that is included only when the transmission buffer underflow occurs. The macroblock escape code (0000 0001 000) is used when the MBA increment is larger than 33.
- MBA increment: A 1- to 11-bit VLC implying the increment value from 1 to 33. The VLC is basically the same as that of H.261 (Table 9.5).
- MB types: I, P, and B pictures have their own MB types.
- Quantizer scale, motion vector, and coded block pattern (CBP) are also defined in the header.

(f) The **block layer** is composed of (8×8) pels that are transformed by 2D (8×8) DCT. The coded block layer contains the size of the DC coefficient, the DC difference, the AC coefficients, and an end-of-block (EOB) code. The EOB signifies that all DCT coefficients along the scan (zigzag) beyond the EOB code are zero.

An MB is a skipped MB when its $\text{MV} = 0$ and all the quantized DCT coefficients are zero. These MB occur when the MBA is larger than 1.

- In I pictures, all MB are coded, i.e., there are no skipped MB. The MBA increment within a slice is always one.
- In P pictures, a skipped MB is an MB with a zero MV and no DCT coefficients. The decoder copies the skipped MB from the previous picture into the current one.
- In B pictures, a skipped MB is the same MB type (forward, backward, or both MV) as the previous MB, i.e., the differential motion vector (DMV) is zero and all quantized DCT coefficients are zero.

To guarantee the constant bit rate, the buffer status is checked after each subsequent picture interval. The video buffering verifier (VBV) is a hypothetical decoder that is conceptually connected to the output buffer of an encoder. In general, the VBV status is used for changing the quantizer scale to be larger or smaller to avoid both overflow and underflow problems. After each subsequent

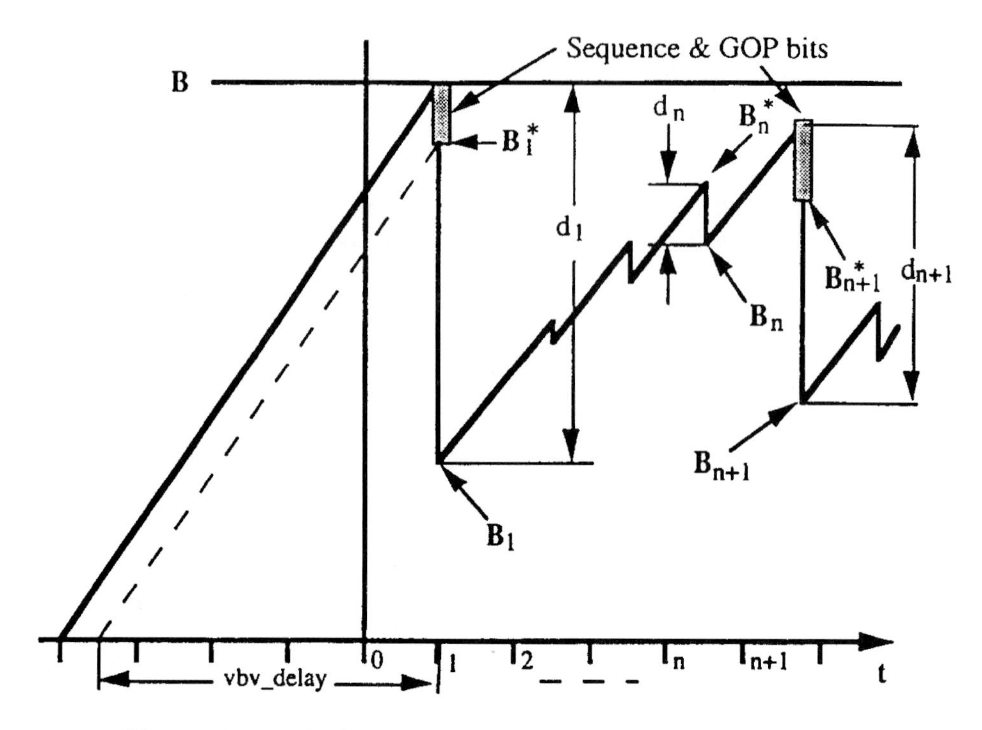

Figure 10.19 Buffer occupancy in the video buffering verifier (VBV) [487]. © 1993 ISO/IEC.

picture interval, all the buffer contents are instantaneously removed. Sequence header and group-of-picture layer data elements that immediately precede a picture are removed at the same time as the picture is coded. The VBV is examined immediately before and after removing any data. The number of bits for the $(n + 1)$th coded picture, d_{n+1}, must be in the range given by

$$B_n + \frac{2R}{P_r} - B < d_{n+1} \le B_n + \frac{R}{P_r} \qquad (10.2)$$

where B_n is the buffer occupancy just after time t_n, R is the bit rate and B is the minimum buffer size in bits required to decode the sequence, which is given in the sequence header, P_r is the number of pictures per second, and n is the time index that the nth coded picture is removed from the VBV buffer. Figure 10.19 illustrates the changing buffer occupancies picture by picture. The actual buffer status, however, can be calculated at any macroblock in a picture and compared with the nominal fullness that is obtained by assuming uniform distribution among all the macroblocks in the picture. According to the comparison, the quantizer scale can be larger or smaller.

Example 10.2 *Assume that each picture is comprised of* 10 *slices in a* 30 *Hz system and* 12 *slices in a* 25 *Hz system and the target number of bits for a slice are given in the following table. N is the spacing between two successive I frames and M is the spacing between anchor frames (I or P).*

	Number of bits per slice			
	30 Hz		25 Hz	
Picture type	(a) $M = 2$ $N = 15$	(b) $M = 3$ $N = 15$	(c) $M = 5$ $N = 6$	(d) $M = 3$ $N = 12$
I	12,700	20,000	15,000	16,000
P	5,600	6,400		6,400
B	800	1,200	1,600	1,350

Then total bit rate in each case becomes 1.15 *Mbps as follows:*

(a) $(12,700 + 7 \times 5600 + 7 \times 800) \times 10 \times (30/15)$
$I\,B\,P\,B\,P\,B\,P\,B\,P\,B\,P\,B\,P\,B\,P$

(b) $(20,000 + 4 \times 6400 + 10 \times 1200) \times 10 \times (30/15)$
$I\,B\,B\,P\,B\,B\,P\,B\,B\,P\,B\,B\,P\,B\,B$

(c) $(15,000 + 5 \times 1600) \times 12 \times (25/6)$
$I\,B\,B\,B\,B\,B$

(d) $(16,000 + 3 \times 6400 + 8 \times 1350) \times 12 \times (25/12)$
$I\,B\,B\,P\,B\,B\,P\,B\,B\,P\,B\,B$

10.2.3 Coding I pictures

By deleting motion estimation and motion-compensated interframe prediction, the block diagram shown in Fig. 10.13 can be reduced to the coding of I pictures, i.e., intraframe coding. This procedure is similar to that of JPEG or the intra mode in the H.261 coder (Fig. 9.8). If the picture header indicates the 3-bit code (001) for the picture coding type, the picture is coded in intra mode. The decisions for the length of slices and quantizer scaling are not subject to the standard. All the motion information in the MB header is disregarded. But two types of quantizers are defined (Table 10.5): one is the default type (intra–*d*), which is indicated in the sequence and slice layer, and another is the new quantizer scale (intra–*q*), which is generally defined by the buffer status.

Table 10.5 Macroblock type VLC for I pictures [487]

	VLC code	MB-quant	MB-intra
Intra–*d*	1	0	1
Intra–*q*	01	1	1

Source: © 1993 ISO/IEC

For coding the I picture, the corresponding headers in each layer are intra quantizer matrix in the sequence header, picture coding type and VBV delay in the picture header, quantizer scale in the slice header, and macroblock type and quantizer scale in the MB header.

If the MB type is intra–*q* (Table 10.5), then the MB header contains a 5-bit integer that defines the quantizer scale. The scale can be changed between 1 and 31 inclusive (the scale zero is forbidden). If the MB type is intra–*d*, no quantizer scale is transmitted and the decoder uses the previously set value.

DCT block coding

A block that contains (8×8) pels is transformed by using 2D DCT. The DCT coefficients are arranged in zigzag order in the order of decreasing spatial frequency (Fig. 5.18a). The energy of the block, in general, is concentrated in the low-frequency coefficients. DCT coefficient variances decrease along the zigzag scan (Fig. 3.8). In addition, the visual perception is less sensitive to the high-frequency coefficients. Therefore, more bits are assigned to the low-frequency portion. The variance distribution is well matched to the zigzag scanning order (Fig. 8.4). These properties are based on the block coding adopted in most of the standards including JPEG and H.261.

The 2D (8×8) DCT can be implemented in a separable fashion. One-dimensional DCT is performed along the horizontal direction followed by the 1D DCT along the vertical direction (Fig. 5.8). The dynamic range of pel values is from 0 to 255 (8 bits). For 11-bit DCT precision, DC coefficients range from 0 to 2047 (11 bits) and AC coefficients range from -1024 to 1023. (See Table 9.7 for 12-bit DCT precision.)

Each coefficient produced by the 2D (8×8) DCT is quantized with a uniform quantizer. The default quantizer matrix is defined as shown in Table 10.6, but it can be changed by the next sequence header. If the 1-bit flag load-intra-quantizer matrix in the sequence header is set to zero, then the coder uses the default quantization table. The table is similar to that of the JPEG (Table 8.1). For spatial resolutions close to 350 lines and 250 pels/line, the default table gives good results and normally there is no need to redefine the intra quantization matrix. In case of significantly different spatial resolution, some other quantization matrix may be

Table 10.6 Default quantization matrix Q_{uv} for (8×8) DCT coefficients:(a) intra and (b) nonintra [487]

(a)								(b)							
8	16	19	22	26	27	29	34	16	16	16	16	16	16	16	16
16	16	22	24	27	29	34	37	16	16	16	16	16	16	16	16
19	22	26	27	29	34	34	38	16	16	16	16	16	16	16	16
22	22	26	27	29	34	37	40	16	16	16	16	16	16	16	16
22	26	27	29	32	35	40	48	16	16	16	16	16	16	16	16
26	27	29	32	35	40	48	58	16	16	16	16	16	16	16	16
26	27	29	34	38	46	56	69	16	16	16	16	16	16	16	16
27	29	35	38	46	56	69	83	16	16	16	16	16	16	16	16

Source: © 1993 ISO/IEC.

applied. If the flag is set to 1, (8×8) 8-bit integers are transmitted in the zigzag order. This transmission may be a great burden, i.e., as much as 512 bits. But the new matrix can be used until the next occurrence of a sequence header.

DC coding The DC coefficient is quantized by a uniform quantizer that has a step size of 8 (the same as in the H.261 coder) as the visual perception is highly sensitive to the DC coefficient. Even when a new quantization matrix is applied, DC quantization is not changed. If the DC coefficient is divided by 8, the dynamic range is within the range of 0 to 255. The quantized DC coefficient is obtained by dividing the DC coefficient by 8 and rounding to the nearest integer. Since the DC coefficient of the present block is correlated with the DC coefficient of the preceding block, the difference between the two quantized DC coefficients is coded using Tables 10.7 and 10.8.

The quantized DC coefficient is coded losslessly by the DPCM technique as in JPEG. In this I picture mode, the basic processing units are macroblocks that are numbered as in Fig. 10.20. The DC coefficient of the last block (Block 3) in the previous MB (MB1) is the prediction for the DC coefficient of the first block (Block 6) in the current MB (MB2). The DC coefficient of each chrominance block is the prediction for the DC coefficient of the adjacent chrominance block along the raster scan. At the beginning of each slice, predictions for DC coefficients for luminance and chrominance blocks are reset to 1024 (128×8). The differential DC values are categorized according to their absolute values as shown in Table 10.7. The size information is transmitted using a VLC code. The size also defines the number of additional bits required to represent the value as shown in Table 10.8. Thus, the size VLC is followed by the additional bits for the actual value of the prediction error. For example, a luminance differential DC value of 10 is coded as 110/1010. A size of zero requires no additional bits.

Table 10.7 VLC code for the differential DC value in I picture [487]

Differential DC (absolute value)	Size	VLC code	
		Luminance	Chrominance
0	0	100	00
1	1	00	01
2 ~ 3	2	01	10
4 ~ 7	3	101	110
8 ~ 15	4	110	1110
16 ~ 31	5	1110	1111 0
31 ~ 63	6	1111 0	1111 10
64 ~ 127	7	1111 10	1111 110
128 ~ 255	8	1111 110	1111 1110

Source: © 1993 ISO/IEC

Table 10.8 Additional size code following the differential DC code [487]

Differential DC	Size	Additional code
−255 ~ −128	8	00000000 ~ 01111111
−127 ~ −64	7	0000000 ~ 0111111
−63 ~ −32	6	000000 ~ 011111
−31 ~ −16	5	00000 ~ 01111
−15 ~ −8	4	0000 ~ 0111
−7 ~ −4	3	000 ~ 011
−3 ~ −2	2	00 ~ 01
−1	1	0
0	0	
1	1	1
2 ~ 3	2	10 ~ 11
4 ~ 7	3	100 ~ 111
8 ~ 15	4	1000 ~ 1111
16 ~ 31	5	10000 ~ 11111
32 ~ 63	6	100000 ~ 111111
64 ~ 127	7	1000000 ~ 1111111
128 ~ 255	8	10000000 ~ 11111111

Source: © 1993 ISO/IEC

AC coding AC coefficients are independently coded, since they are not correlated. In the decoder, the quantized DC coefficient is multiplied by 8 before applying the inverse DCT. AC coefficients are quantized using the intra quantization matrix. The quantizer step size is derived from the quantization matrix and the quantizer scale. The basic rule for the quantization is that each AC coefficient is divided by the corresponding element in the quantization matrix (Table 10.6).

Furthermore, the division factor can be made larger or smaller by introducing the quantizer scale factor 1 thru 31 and constant value 8 given by the designer. The quantized AC coefficients, S_{quv}, can be obtained by an example as follows:

$$S_{quv} = 8 \times S_{uv}//(qs \times Q_{uv}) \tag{10.3}$$

where Q_{uv} is the corresponding element in the quantization matrix (Table 10.6), S_{uv} denotes the transform coefficient, and qs denotes the quantizer scale which is in the range of 1 to 31. For MPEG-2 quantization, the corresponding quantizer step size is defined by using a linear or nonlinear curve as shown in Fig. 11.19. The quantized coefficients are clipped to the range of -255, to $+255$, and inverse procedures are performed in the decoder to obtain the reconstructed AC coefficients.

Example 10.3 *Let $S_{uv} = 56$, $qs = 2$, and $Q_{uv} = 19$, as an example. Then the quantized coefficient S_{quv} is*

$$S_{quv} = (8 \times 56)//(2 \times 19) = 12 \tag{10.4}$$

In the decoder, the reconstruction procedure is

$$\hat{S}_{uv} = (2 \times S_{quv} \times qs \times Q_{uv})/16 \tag{10.5}$$

If $(\hat{S}_{uv}\ \&\ 1) = 0$,

$$\hat{S}_{uv} = \hat{S}_{uv} - Sign\,(S_{uv})$$

where & denotes logical AND *operation and*

$$Sign(x) = \begin{array}{ll} 1 & x > 0 \\ 0 & x = 0 \\ -1 & x < 0 \end{array}$$

By this equation the reconstructed value is

$$\hat{S}_{uv} = (2 \times 12 \times 2 \times 19)/16 = 57$$

The quantization error between the coder and decoder is $57 - 56 = 1$.

Coding of AC coefficients in I pictures is similar to that in the H.261 coder, i.e., a run length and level technique is used. This is a two-dimensional code

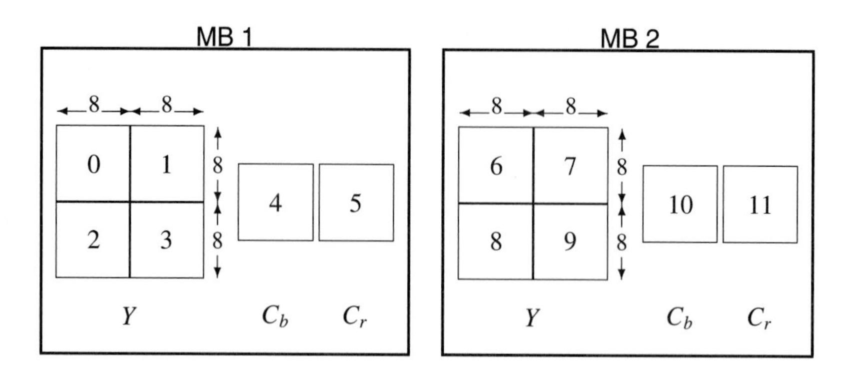

Figure 10.20 The raster scan ordering of the coded blocks.

indicating the run length of zero coefficients preceding the level of the nonzero coefficient. The run length is measured along the zigzag scan. An extended VLC table from that of H.261, as shown in Appendix C.3, is defined for MPEG-1 coding. Example 10.4 shows coding of I pictures explicitly with a sample block used in Chapter 8. If the combination of run length and level does not exist in the table, a special code is generated by the following rule: escape code (0000 001), run length (6-bit FLC), and level (8-bit FLC up to ± 127, 16-bit FLC up to ± 255, the levels zero and -256 are forbidden) as shown in Table 10.9. The table is also used for both P and B pictures. We compare the coding algorithms for DC and AC coefficients in I, P, and B pictures in Table 10.10.

Example 10.4 *Given the quantized transform coefficients we discussed in Section 8.4.2 [Eq. (8.6)] find the resulting MPEG-1 bit stream. The JPEG bit stream for the block is given in Eq. (8.8). Now the coefficients can be coded by using the I picture luminance coding technique. The differential value* 5 *corresponds to the VLC code* 101 *(Table 10.7) and additional size bits are also* 101 *(Table 10.8). The first AC coefficient,* -3*, has a run length of zero and its VLC code is* 001011*, from the table in Appendix C.4. The last bit* 1 *implies a negative sign. The resulting bit stream for the block is given by*

$$(101101/001011/01000/110/111/110/0001111/10)$$

The same number of bits are produced as in the JPEG baseline example described in Section 8.4.2.

10.2.4 Coding P pictures

Each P picture is composed of one or more slices that are, in turn, composed of a number of macroblocks. Coding P pictures is much more complex than coding I pictures, since the former, in general, involves motion estimation and

Table 10.9 Level codes for special coding of quantized AC coefficients [487]

Level	Word length	VLC code
−256		Forbidden
−255	16	1000 0000 0000 0001
−254	16	1000 0000 0000 0010
⋮		⋮
−129	16	1000 0000 0111 1111
−128	16	1000 0000 1000 0000
−127	8	1000 0001
−126	8	1000 0010
⋮		⋮
−5	8	1111 1011
−4	8	1111 1100
−3	8	1111 1101
−2	8	1111 1110
−1	8	1111 1111
0		Forbidden
1	8	0000 0001
2	8	0000 0010
3	8	0000 0011
4	8	0000 0100
5	8	0000 0101
⋮		⋮
126	8	0111 1110
127	8	0111 1111
128	16	0000 0000 1000 0000
129	16	0000 0000 1000 0001
⋮		⋮
254	16	0000 0000 1111 1110
255	16	0000 0000 1111 1111

Source: © 1993 ISO/IEC

compensation. The coding procedures are similar to those of the H.261 coder, which involves interframe motion-compensated hybrid coding. The picture type P is defined in the picture coding type header of the picture layer. Then the header is set to 010 for a P picture, i.e., predictive interframe coding.

The (8×8) blocks representing the interframe motion-compensated prediction errors of the macroblock are transformed by 2D-DCT giving an array of (8×8) transform coefficients. Then the DCT coefficients are quantized to produce a set of quantized coefficients. The quantized coefficients are then variable-length coded before transmitting. We go on to discuss MB type decisions to support every possible case in the MB layer and the (8×8) block DCT coding.

Table 10.10 Comparison of VLC coding for DCTcoefficients in MPEG-1

		Intra	Nonintra
I	DC	Difference (Size VLC/additional)	×
	AC	Run/Level (VLC code)	×
P, B	DC	Difference[1] (Size VLC/additional)	Run/Level[2] (VLC code)
	AC	Run/Level (VLC code)	Run/Level (VLC code)

Notes: (1) The predicted value is set to 128, unless the previous block was intra coded.
(Refer to Tables 10.7 and 10.8.)

(2) Since the motion compensated prediction errors are transformed, there is no spatial prediction of the DC term. (Refer to table in Appendix C.4.)

Table 10.11 Macroblock type VLC for P pictures [487]

Type	VLC	Intra	Motion forward	Coded pattern	Quant
pred–*mc*	1		1	1	
pred–*c*	01			1	
pred–*m*	001		1		
intra–*d*	0001 1	1			
pred–*mcq*	0001 0		1	1	1
pred–*cq*	0000 1			1	1
intra–*q*	0000 01	1			1
skipped					

Source: © 1993 ISO/IEC.

Macroblock types in P pictures

Slices are divided into macroblocks for P pictures in the same way as for I or B pictures. The major difference from I pictures is the introduction of motion compensation. The MB stuffing may occur in case of buffer underflow. The MBA increment in the I picture is restricted to one, whereas it may be larger for P pictures. For example, if the MBA indicates 2, then there is one skipped MB. The decoder copies the corresponding MB from the previous picture as the best representation of the skipped MB, i.e., predicted MB with zero motion vector.

There are eight types of macroblocks in P pictures as shown in Table 10.11. The acronyms for the eight types are also indicated in Fig. 10.21. In Table 10.11 the marking 1 implies that both row and column elements are matched, i.e., the type pred–*mc* transmits the forward motion vector and coded pattern. The abbreviations for MB type imply:

- *pred* (predictive, nonintra mode)
- *m* (motion compensated, forward MV transmitted)
- *c* (at least one block in the MB is coded and transmitted)
- *d* (default quantizer is used)
- *q* (quantizer scale is changed)
- *skipped macroblocks* (no VLC code, since they are notified by having the MBA increment skipped over them)

An encoder has the difficult task of choosing the different types of macroblocks. To provide the best trade-off, a simple and less expensive sequential series of decisions is illustrated in Fig. 10.21.

The **MC/no MC decision** can be made by using the same algorithm shown in Fig. 9.12 for the H.261 coder. The encoder has an option as to whether to transmit motion vectors or not. If the MV is zero due to MC decision, then some bits can be saved by not transmitting it. The definitions for $|BD|$ and $|DBD|$ are given in Eqs. 9.5 and 9.7, respectively. The curious shape of the boundary between the two regions in Fig. 9.12 is because any movement of the background caused by the drag-along effect of nearly moving objects is very objectionable. If the macroblock type is pred–*m*, pred–*mc*, pred–*mq*, then the prediction errors of the horizontal and vertical components of the forward motion vectors are transmitted in succession. In P pictures, the differential motion vector as discussed in Section 9.3 (Table 9.9) is coded by using the VLC code shown in Table 10.12. The reason is that the motion vector of a MB tends to be well correlated with the motion vector of the previous MB. In a panning image, for example, all motion vectors would be roughly the same. At the start of each slice and at each intra coded MB the predicted motion vector is set to zero.

Two entities in the picture header are defined for the coding of motion vectors in P pictures. One is the full-pel-forward-vector, which indicates integer-pel precision if set to 1, and half-pel precision if set to 0, as shown in Table 10.13. Another parameter is the forward-f-code, which is an unsigned integer taking values 1 through 7 (3 bits). An encoder chooses one of these ranges. The range must be constrained to take place within the boundaries of the decoded reference picture. Any motion vectors outside the picture are invalid. The MPEG-1 constrained parameters define the vector range as less than ± 64 pels using half-pel precision, i.e., the maximum forward-f-code is 4.

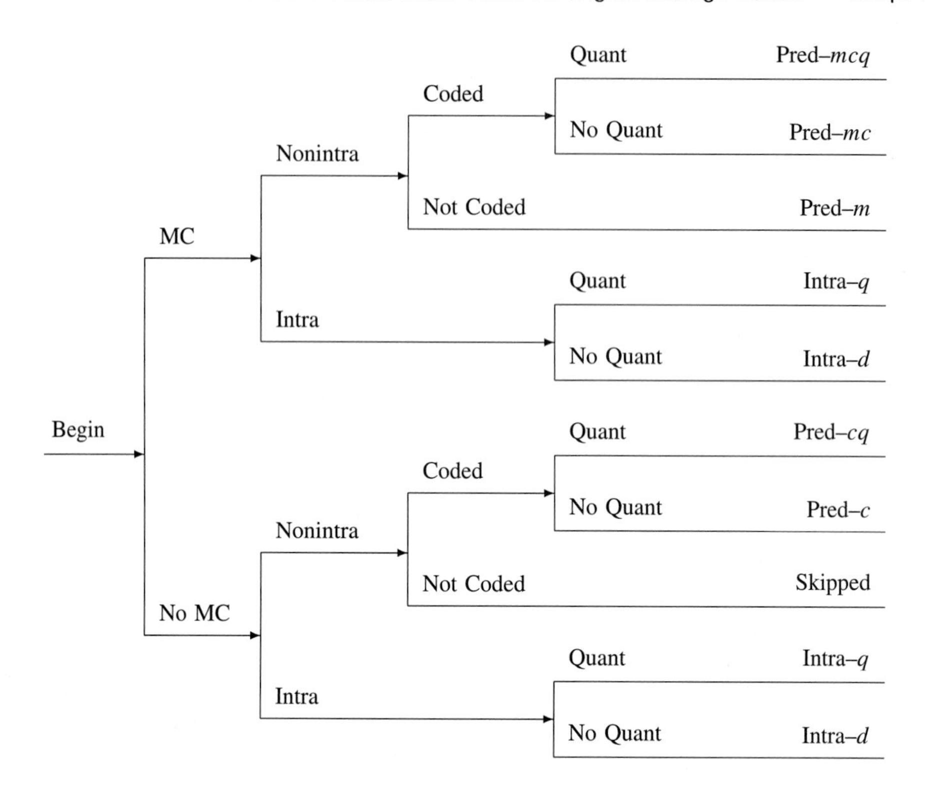

Figure 10.21 The macroblock type decision procedure in P pictures. m: motion compensated, c: at least one block coded, q: quantizer changed, d: default quantizer used, pred: predictive (forward) mode, MC: motion compensation mode.

The forward-f-code is related to the number of bits appended to the VLC codes listed in Table 10.12. The forward-f-size is obtained by subtracting 1 from the forward-f-code. Hence, the signed quotient in the differential motion vector is coded by the VLC code and the unsigned remainder (forward-f-size bits) coded by FLC is followed. The differential motion vector must be kept within the range defined in Table 10.13 by adding or subtracting a modulus that is described in Table 10.14. In Example 10.5, the procedure for coding the motion vectors in P pictures is illustrated.

Example 10.5 *Assume that a slice has the following motion vectors for the eight successive macroblocks, defined as integer-pel resolution, and that the forward-f-code is* 2 *(Table 10.13):*

$$3 \ 8 \ 15 \ 31 \ 19 \ -15 \ -20 \ 5$$

Table 10.12 VLC for differential motion vectors in MPEG-1 coder [487]

VLC code	DMV	VLC code	DMV
0000 0011 001	−16	010	1
0000 0011 011	−15	0010	2
0000 0011 101	−14	0001 0	3
0000 0011 111	−13	0000 110	4
0000 0100 001	−12	0000 1010	5
0000 0100 011	−11	0000 1000	6
0000 0101 11	−10	0000 0110	7
0000 0101 01	−9	0000 0101 10	8
0000 0101 11	−8	0000 0101 00	9
0000 0111	−7	0000 0100 10	10
0000 1001	−6	0000 0100 010	11
0000 1011	−5	0000 0100 000	12
0000 11	−4	0000 0011 110	13
0001 1	−3	0000 0011 100	14
0011	−2	0000 0011 010	15
011	−1	0000 0011 000	16
1	0		

Source: © 1993 ISO/IEC.

Table 10.13 Range of motion vectors in half-pel and integer-pel resolutions [487]

Forward-f-code or	Motion vector range	
backward-f-code	Full-pel = 0	Full-pel = 1
1	−8 ~ 75	−16 ~ 15
2	−16 ~ 15.5	−32 ~ 31
3	−32 ~ 31.5	−64 ~ 63
4	−64 ~ 63.5	−128 ~ 127
5	−128 ~ 127.5	−256 ~ 255
6	−256 ~ 255.5	−512 ~ 511
7	−512 ~ 511.5	−1024 ~ 1023

Source: © 1993 ISO/IEC.

Since the initial prediction is zero, the differential motion vectors are obtained as

$$3\ 5\ 7\ 16\ -12\ -34\ -5\ -25$$

We now apply the modulus value of 64 *in Table 10.14, to limit the range of differential MV to* −32 *through* 31, *and the results are*

$$3\ 5\ 7\ 16\ -12\ 30\ -5\ 25$$

Table 10.14 Modulus for motion vectors

Forward-f-code or backward-f-code	Modulus
1	32
2	64
3	128
4	256
5	512
6	1024
7	2048

After dividing by the forward-f value of 2, the resulting code words are

0101/00101/000101/00000101110/000010010/

000000110100/00111/000001000001/

The **intra/nonintra coding decision** can also be made from Fig. 9.9. As in all the standards, the actual decision-making process is not defined. After the encoder has determined the best MC prediction, it is in a position to decide whether to use it or disregard it and code the MB as intra. The obvious way to do this is to code the block as intra and compare it with the total number of bits needed when coded as MC with the same quantization scale. The mode giving the fewer bits may be selected. This mode decision is, however, quite complex. To provide a simple algorithm, extensive experiments resulted in the decision curve shown in Fig. 9.9. It can be seen that if the variance is small, then the MB can be coded as intra.

The **coded/not coded decision** is a result of quantization. If all the quantized DCT coefficients of an (8×8) block are zero then the block need not be coded. The coded block pattern (CBP) shows how many and which of the blocks (each of size 8×8) are coded in the MB. The VLC table is shown in Table 10.15, which is also used for the CBP in the H.261 coder (Fig. 9.7). The decimal numbers for the VLC table can be derived from Eq. (9.1).

The **quant/no quant decision** is made as to whether the quantizer scale is to be changed or not in the MB. This is generally based on local image content and on the buffer fullness assessment of the model decoder. Changing the quantizer scale improves the picture quality while keeping a constant bit rate in the picture layer and prevents the buffer overflow and/or underflow. If the same quantizer scale is maintained over a picture, the total mean square error of the coded picture may be close to minimum for a given number of coding bits. The visual appearance of most pictures can be improved, however, by varying the quantizer scale for each MB, making it smaller in smooth areas and larger in busy areas. The decision as

Table 10.15 VLC table for coded block pattern (CBP) [487]

CBP	VLC code	CBP	VLC code	CBP	VLC code
60	111	5	0010 111	51	0001 0010
4	1101	9	0010 110	23	0001 0001
8	1100	17	0010 101	43	0001 0000
16	1011	33	0010 100	25	0000 1111
32	1010	6	0010 011	37	0000 1110
12	1001 1	10	0010 011	26	0000 1101
48	1001 0	18	0010 001	38	0000 1100
20	1000 1	34	0010 000	29	0000 1011
40	1000 0	7	0001 1111	45	0000 1010
28	0111 1	11	0001 1110	53	0000 1001
44	0111 0	19	0001 1101	57	0000 1000
52	0110 1	35	0001 1100	30	0000 0111
56	0110 0	13	0001 1011	46	0000 0110
1	0101 1	49	0001 1010	54	0000 0101
61	0101 0	21	0001 1001	58	0000 0100
2	0100 1	41	0001 1000	31	0000 0011 1
62	0100 0	14	0001 0111	47	0000 0011 0
24	0011 11	50	0001 0110	55	0000 0010 1
36	0011 10	22	0001 0101	59	0000 0010 0
3	0011 01	42	0001 0100	27	0000 0001 1
63	0011 00	15	0001 0011	39	0000 0001 0

Source: © 1993 ISO/IEC

to smooth and busy areas can be arrived at by computing the activities in the MB. The criterion can be based on variance, AC energy, the sum of the deviations of the pel intensities from their average value, and so forth.

Coding (8×8) blocks

Only the DCT coefficients of the basic (8×8) blocks indicated by the CBP in an MB are coded. The DCT coefficients of the remaining basic (8×8) blocks in the MB are quantized as zeros. Different coding techniques are applied for intra or nonintra mode. The MC/no MC decision is not a matter of block DCT coding. If the MB is intra (in both MC and no MC mode), then the MB is transformed, quantized, and coded in the same way as the MB in I pictures. If the MB is nonintra, the subsequent decision is coded/not coded.

Statistics of DC and AC coefficients vary in different pictures or modes. Coding AC coefficients for intra blocks is similar to coding the I, P, and B pictures. Prediction of the DC coefficient differs, however. The DC predicted values are all set to 128 for intra blocks in P and B pictures, unless the previous block was intra coded. The main difference in coding the DCT coefficients of I and P or B pictures is that the transform coefficients represent differences between pel values (because of the nonintra mode) rather than representing the pel values themselves. Therefore, adjacent DC coefficients are not correlated and there is no spatial prediction of the DC term.

Coded blocks are quantized by using the previously established quantizer or updated quantizer scale. Intra macroblocks in P and B pictures are quantized using the same method as in the I picture, but nonintra macroblocks in P and B pictures are quantized using the quantizer scale and the nonintra uniform quantization matrix (Table 10.6b). For both MB, Eq. (10.3) is applied. The reconstruction formula is derived by inverting the quantizer equation.

Example 10.6 *Assume that the quantization scale* qs *is* 10 *for the nonintra coefficient and the default matrix is used for the quantization* ($Q_{uv} = 16$). *Find the quantized value and the reconstructed value* $\hat{S}_{uv}$ *for the transform coefficient value* S_{uv} *of* 21.

From Eq. (10.3) the quantized value is

$$S_{quv} = 8 \times 21//(10 \times 16) = 21//20 \Rightarrow 1 \tag{10.6}$$

For reconstruction it is

$$\hat{S}_{uv} = (((2 \times S_{quv}) + Sign(S_{quv})) \times qs \times Q_{uv})/16 \tag{10.7}$$

If $((\hat{S}_{uv}\ \ \&\ \ 1) = 0)$,

$$\hat{S}_{uv} = \hat{S}_{uv} - Sign(\hat{S}_{uv})$$

where & denotes the logical AND operation. Following this formula gives us

$$\hat{S}_{uv} = (((2 \times 1) + 1) \times 10 \times 16)/16 = 30$$

Here $(\hat{S}_{uv}\ \ \&\ \ 1) = 0$, *hence,*

$$\hat{S}_{uv} = 30 - 1 = 29$$

is the reconstructed value, at the receiver.

10.2.5 Coding B pictures

The major difference between coding B and P pictures is that three types of motion-compensated macroblocks can be predicted for the former, i.e., forward, backward, and bidirectionally interpolated (Fig. 10.8). There are two estimated motion vectors, forward and backward. Each motion vector is coded by the prediction of the same type. Both motion vectors are set to zero at the start of each

Table 10.16 Macroblock type VLC for B pictures [487]

Type	VLC	Intra	Motion forward	Motion backward	Coded pattern	quant
pred–i	10		1	1		
pred–ic	11		1	1	1	
pred–b	010			1		
pred–bc	011			1	1	
pred–f	0010		1			
pred–fc	0011		1		1	
intra–d	0001 1	1				
pred–icq	0001 0		1	1	1	1
pred–fcq	0000 11		1		1	1
pred–bcq	0000 10			1	1	1
intra–q	0000 01	1				1
skipped						

Source: © 1993 ISO/IEC.

slice and at each intra MB. The ranges of the motion vectors are defined by the full-pel-backward-vector as well as the full-pel-forward-vector. Backward-f-code is also defined in a fashion similar to that in P pictures. Differential motion vector and modulus operation are processed separately. The same example, i.e., Example 10.5, can be applied for B pictures.

The encoder does not store the decoded B pictures in the frame memory because they are not used for motion estimation and compensation. However, the B pictures are coded by more complex techniques. The encoder makes more decisions, such as for dividing the picture into slices, for determining motion vectors, for deciding forward, backward, or interpolated motion compensation or coding as intra, and for setting the quantizer scale.

Macroblock types in B pictures

The MB header in a B picture defines two MB types as shown in Table 10.16, where f denotes forward, b denotes backward, and i denotes interpolated.

There are three ways of coding the motion vectors. If only a forward MV is present, then the motion-compensated MB is predicted from a previous I or P picture. This is same as in a P picture where MC is based on forward MV only. If only a backward MV is present, then the motion-compensated MB is predicted from a future I or P picture. If both forward and backward motion vectors are present, such as MB type pred-i, then motion-compensated macroblocks are constructed from both previous I or P and future I or P pictures, using forward and backward MV, respectively, and the result is averaged to form the interpolated motion-compensated macroblock.

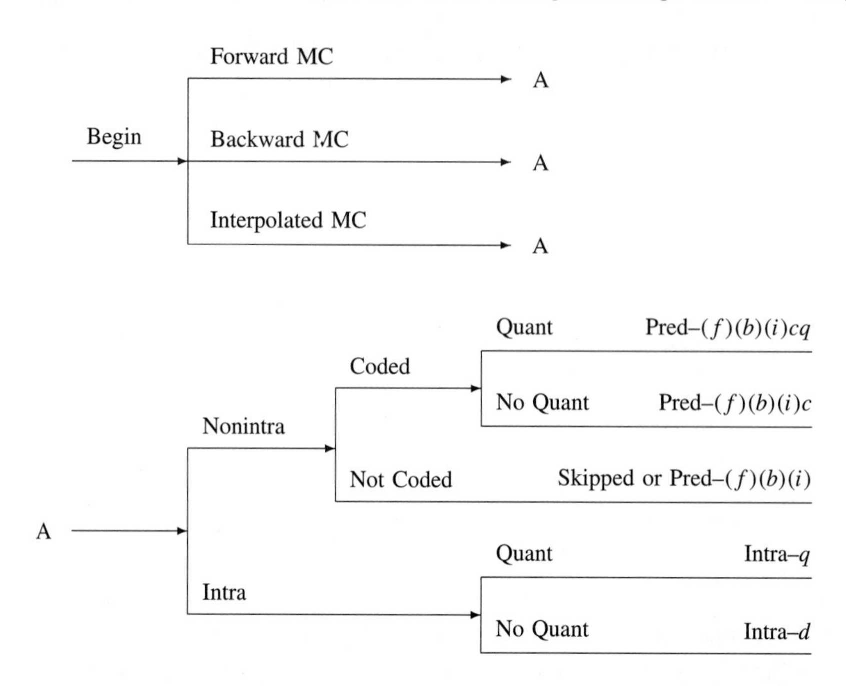

Figure 10.22 Macroblock type decision procedure in B pictures (f: forward; b: backward; i: interpolative).

Skipped macroblocks in B pictures have the same motion vectors and same MB types as the previous MB, whereas skipped macroblocks in P pictures have motion vectors equal to zero. The skipped MB in B picture, like P pictures, has no VLC.

The sequential mode decision procedures described in Fig. 10.22 lead to the MB type selection. The encoder at first determines the motion compensation mode, i.e., forward, backward, or interpolative. The MC/no MC decision is not necessary for B pictures. (All macroblocks are regarded as motion compensated unless they are coded as intra.) Even the skipped MB has the MV and the MB type. Each motion compensation mode used in P pictures (Fig. 10.21) can be one of the three modes (Fig. 10.22) as follows:

pred–*m* → pred–*f*, pred–*b*, pred–*i*

pred–*mc* → pred–*fc*, pred–*bc*, pred–*ic*

pred–*mcq* → pred–*fcq*, pred–*bcq*, pred–*icq*

The second decision is intra/nonintra coding. Although the MB is dealt with one of the three MC modes in the first step, the MB is evaluated again to select intra/nonintra mode. The same decision criterion as in P pictures can be used for the possible modes, i.e., the MB type intra or MC interframe. For the former, the MB is always coded as the intra MB in I picture. For the latter, the MC prediction error is checked to see if it is large enough to be coded using the DCT. The last step is to decide whether the quantizer scale is satisfactory for the quantization of DCT coefficients.

The selection of a motion compensation mode for a MB can be based on the MSE minimization of the luminance difference (16×16 block) as shown in Fig. 9.12. The encoder determines the best motion-compensated macroblock based on both for forward MV and for backward MV. Finally, it averages the two motion-compensated macroblocks to produce the interpolated macroblock. The encoder then selects a predicted MB which has the smallest mean square difference from the current MB. The actual selection criterion is outside the domain of the MPEG-1 standard.

As stated earlier, the motion vectors in B pictures have two directions, forward and backward. If the MV is forward, then the prediction errors of the horizontal and vertical components of the forward motion vectors are transmitted in succession as in P pictures. If the MV is backward, then the prediction errors of the horizontal and vertical components of the backward motion vectors are transmitted in succession. If both types are present, then the prediction errors of the four components representing the forward MV and backward MV are transmitted in succession.

Blocks of (8×8) pels are transformed into arrays of (8×8) coefficients using the 2D-DCT that is common for all pictures (I, P, B). Quantization and coding of the DCT coefficients in B pictures are the same as in P pictures.

In Table 10.10, VLC coding methods for the three picture types are listed. The differential DC value in I pictures is coded by using VLC for size and FLC for the actual value as in JPEG (Section 8.4). Nonintra mode in I pictures does not exist. The DC coefficient coding in P and B pictures is the same as in I pictures but for the prediction value setting. One VLC table is used for coding the AC coefficients of the three picture modes and the DC coefficients of P and B pictures.

10.2.6 Coding D pictures

D pictures are intended to be used for fast visible search modes. In this mode, the low-frequency information they contain is sufficient for the user to locate the desired frame.

D pictures are defined by picture coding type (100) in the picture header. D pictures are coded as the DC coefficients of the (8×8) blocks. Therefore, the picture is composed of the average intensity levels of the (8×8) blocks. D pictures are not part of the constrained parameter bit stream. They have a macroblock type that only indicates intra coding in the MB as shown in Table 10.17.

Table 10.17 Macroblock type VLC for D pictures

VLC code	Macroblock intra
1	1

There is a 1-bit header denoting the end of the macroblock, which is set to 1 when D pictures exist. Hence all procedures in D pictures are simple: macroblock type, end of macroblock, and coding only DC coefficients.

10.3 Audio Coding

The standardization efforts for digital audio coding have been initiated by the ISO/IEC (MPEG), as mentioned earlier. The prerequisites for the audio coder include bandwidth, quality, and multiplexing. First, the input signal is wideband (full 20 KHz). According to the Nyquist theorem, the sampling rate is at least 40 KHz. For example, the audio data on a compact disc is sampled at 44.1 KHz that requires an uncompressed data rate of about 1.4 Mbps for stereo sound (2 channels) with 16-bit resolution. The second requirement in this application is high quality comparable to the original CD quality. Furthermore, since the audio signals originate from multiple sources (more than one channel), the encoder must embrace them. The MPEG-1 bit stream consists of audio and video signals that are synchronized at the decoder. This will allow lip synchronization. For this purpose, video encoding is implemented with minimum delay. The decoder should be simple so that the simplest (hence the cheapest) decoder is available for decoding-only customers, i.e., consumers.

Digital audio coding techniques can be classified into two groups: time domain and frequency domain processing. The time domain technique can be implemented with low complexity, but it requires more than 10 bits/sample for maintaining high quality. Most of the known techniques belong to the frequency domain. With this kind of algorithm, high quality can be achieved at bit rates of 3 bits/sample or above. Subband coding and transform coding are used for mapping the audio signal into the frequency domain. These techniques yield certain compression ratios due to their frequency distribution capabilities.

To achieve a higher compression ratio, all coding methods rely heavily on human audio perception. This is called psychoacoustic masking, in which the masked signal must be below the masking threshold (masker). The noise masking phenomenon has been observed through a variety of psychoacoustic experiments [542]. This masking occurs whenever the presence of a strong audio signal makes a spectral neighborhood of weaker signals imperceptible.

For the purposes of the CD 11172-3, the following definitions apply for audio coding [487].

Bark: The unit of critical band rate.

Critical band: The part of the spectral domain that corresponds to a width of one Bark.

Critical band rate: A psychoacoustic measure in the spectral domain that corresponds to the frequency of the human ear.

Frame: A part of the audio signal that corresponds to a fixed number of audio PCM samples.

Granules (layer II): 96 subband samples, 3 consecutive subband samples for all 32 subbands that are considered together before quantization.

Granules (layer III): 576 (18×32) frequency lines that carry their own side information.

Intensity stereo: A method of exploiting stereo irrelevance or redundancy in stereophonic audio programs based on retaining at high frequencies only the energy envelope of the right and left channels.

Joint stereo coding: Any method that exploits stereophonic irrelevance or stereophonic redundancy.

Joint stereo mode: A mode of the audio coding algorithm using joint stereo coding.

Masking threshold: a function in frequency and time below which an audio signal cannot be perceived by the human auditory system.

Masking: Property of the human auditory system by which an audio signal cannot be perceived in the presence of another audio signal.

MS (middle/side) stereo: A method of exploiting stereo irrelevance or redundancy in stereophonic audio programs based on coding the sum and difference signal instead of the left and right channels.

Nontonal components: A noiselike component of an audio signal.

Psychoacoustic model: A mathematical model of the masking behavior of the human auditory system.

Tonal component: A sinusoidlike component of an audio signal.

10.3.1 Overview of MPEG-1 audio coding

The MPEG audio algorithm is a psychoacoustic algorithm that provides signal-to-mask ratio (SMR) for bits or noise allocation and bit stream formatting as shown in Fig. 10.23.

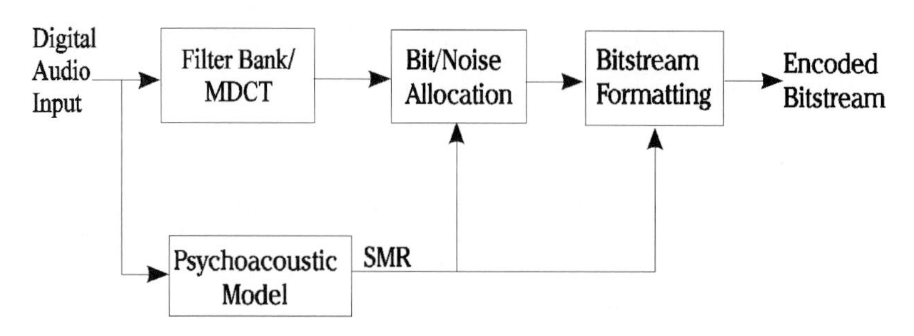

Figure 10.23 A block diagram of the MPEG-1 audio encoder

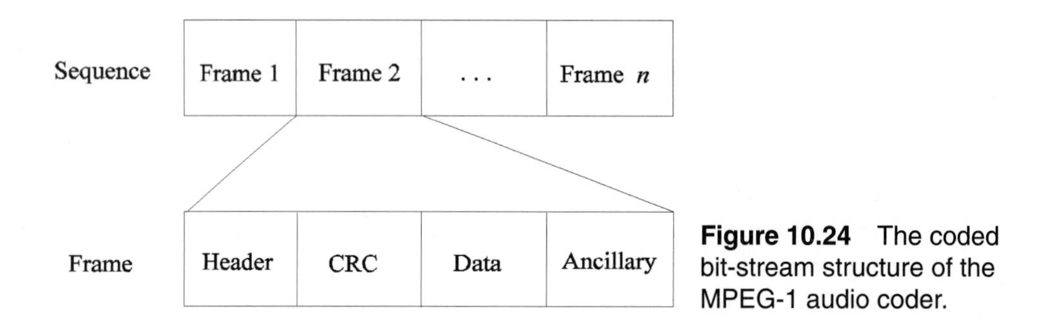

Figure 10.24 The coded bit-stream structure of the MPEG-1 audio coder.

The coded audio bit stream consists of separate frames that include header (32 bits), error check (16 bits), audio data, and ancillary data as shown in Fig. 10.24. The header contains the synchronization and status information of the audio frame. Some of the header contents are:

- **Syncword** (12 bits)
- **Layer code** (2 bits), representing layers I, II, and III
- **Bit-rate index** (4 bits): The all-zero value indicates the free format condition. The index is referred to a table which is different for each layer. Fifteen bit rates in the range of 32 Kbps up to 448 Kbps are listed in Table 10.18 for layers I, II, and III.
- **Sampling frequency** (2 bits), representing 48, 44.1, and 32 KHz sampling rate.
- **Padding bit**, indicating the number of slots, is N or $N + 1$. The bit stream is adjusted by the unit of slot in the frame.
- **Mode** (2 bits): stereo, joint stereo, dual or single channel

The MPEG audio standard includes three different layers for corresponding applications, with increasing encoder complexity and performance. The layers

Table 10.18 Bit rates available in MPEG-1 audio coding [487]

Index	Bit rate (Kbps)		
	Layer I	Layer II	Layer III
0000	free format	free format	free format
0001	32	32	32
0010	64	48	40
0011	96	56	48
0100	128	64	56
0101	160	80	64
0110	192	96	80
0111	224	112	96
1000	256	128	112
1001	288	160	128
1010	320	192	160
1011	352	224	192
1100	384	256	224
1101	416	320	256
1110	448	384	320

Source: © 1993 ISO/IEC.

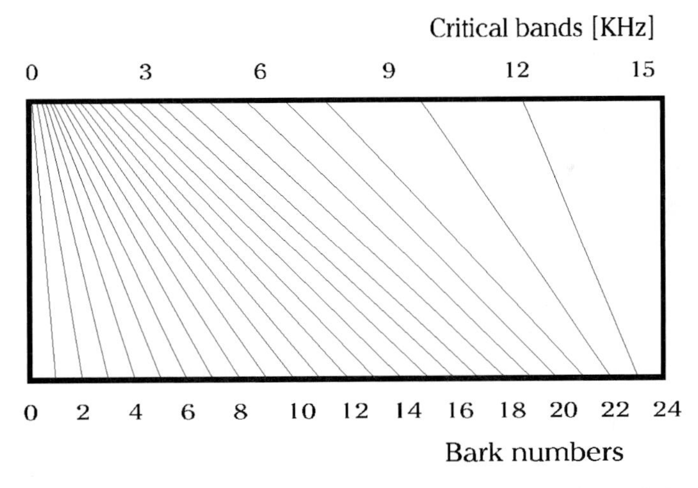

Figure 10.25 Nonlinear critical bands measured by Scharf showing 24 bands up to 15.5 KHz [542]. © 1970 Academic Press.

have upward compatibility, i.e., an audio layer N decoder is able to decode bitstream data that have been encoded in layer N and all layers below it.

Layer I forms the most basic algorithm. The filter bank divides the audio signal into 32 constant-width (each band has the same frequency range) frequency bands. The filters are relatively simple and easy to design. But the 32 subbands do not accurately conform to the ear's critical bands illustrated in Fig. 10.25. The encoder calculates the masking threshold in each subband by using the critical bands. Even without quantization, the filter bank and its inverse do not result in lossless reconstruction. Fortunately, the error introduced by the filter bank is not large and is imperceptible. The encoder formats the 32 subbands each consisting of 12 samples into a frame (384 samples). Layer I contains a psychoacoustic model to determine adaptive bit allocation, and quantization using block companding and formatting. Application fields using this layer include digital home recording on tapes or discs.

Layer II is an enhancement that uses some elements found in layer I. The improved techniques are additional coding of bit allocation, scale factors, and different frame structures. The encoder forms larger groups of $3 \times 12 \times 32$ samples into a frame (1152 samples), i.e., 12 samples/block, 3 blocks, and 32 subbands. Only one type of bit allocation and up to three scale factors are provided for the 3 blocks each composed of 12 samples. A scale factor is used for each block of 12 samples. Application fields include consumer and professional studio-like broadcasting, recording, telecommunications, multimedia, and audio workstations.

Layer III provides the highest compression ratio but at the cost of greater encoder/decoder complexity. This layer consists of a hybrid filter bank (subband filter bank and MDCT). Additional frequency resolution is obtained by using the MDCT. Layer III specifies two different MDCT block lengths: a long block of 36 samples or a short block of 12 samples. The short block length improves the time resolution to cope with transients. This layer provides nonuniform quantization, entropy coding and noise allocation instead of bit allocation. Application fields include telecommunications over narrow band ISDN and audio coding at very low bit rates. The detailed description of the three layers is provided in the following sections.

The **joint stereo coding** technique can be used for stereo redundancy reduction. The MPEG audio standard is available for possible audio sources from single-channel stereo or dual-channel. Stereo and dual-channel signals require twice the channel bandwidth to transmit the two channel signals if we code them separately. For the purpose of decreasing the bit rate or increasing the audio quality, the stereo or dual-channel signal can be coded by using intensity stereo mode or middle/side (MS) stereo coding. All layers provide the intensity stereo mode, and layer III provides the MS stereo coding.

The basic idea for intensity stereo coding is that, instead of transmitting separate left and right subband samples, a single summed signal is transmitted with scale factors for both the left (L) and right (R) channels. The decoder decodes the two channels based on the scale factors. The frequency spectra of the decoded stereo signals are the same but the magnitudes (intensities) are different.

In MS stereo mode, middle (sum of left and right) and side (difference of left and right) values are transmitted instead of left (L) and right (R) channel values. The value of middle (M) is transmitted in the left channel and the value of side (S)

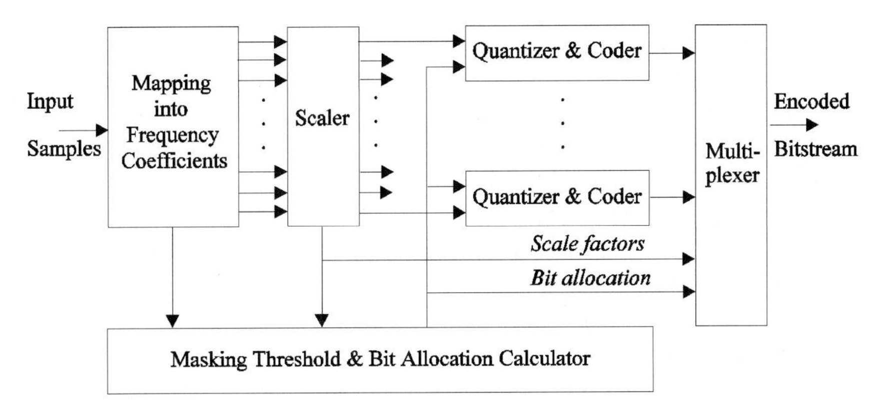

Figure 10.26 Common algorithms proposed by ASPEC and MUSICAM [521]. © 1990 IEEE.

is transmitted in the right channel. Thus L and R channels can be reconstructed using

$$L = (M + S)/\sqrt{2} \qquad R = (M - S)/\sqrt{2} \tag{10.8}$$

10.3.2 Proposed algorithms

Among the 14 proposals originally considered for MPEG audio, MUSICAM and ASPEC have shown remarkable results, and they have slightly different performances in subjective and objective tests [521]. Thus the two groups (MUSICAM and ASPEC) collaborated and prepared a draft standard combining the most efficient components of the two algorithms. The ASPEC algorithm uses MDCT with overlapping blocks and dynamic windowing. The MUSICAM algorithm uses a subband analysis filter bank with 32 equally spaced subbands to map the input samples into frequency components [520, 529]. Both algorithms exploit psychoacoustic masking effects to control quantization steps or bit allocations [525, 518].

The developments in MPEG audio compared to ITU-T audio compression standards (Rec. G.711 and G.722) are frequency domain operations, scale factors to control quantizers, and utilization of the psychoacoustic model.

Figure 10.26 shows the main algorithms that are common to the ASPEC and MUSICAM proposals. Audio samples are mapped into the frequency domain either by a transform or by a subband filter bank. The audio coefficients in the frequency domain are normalized by a scale factor that is determined from the masking threshold of psychoacoustic responses. The scale factor and bit allocation information are transmitted as overheads. These common functions have become basic components in the MPEG-1 and MPEG-2 audio coding. Before we discuss the three layers in the MPEG audio, a technical comparison between MUSICAM and ASPEC follows.

The subband coding system MUSICAM relies on a psychoacoustic model established from the critical bands and 32 subbands. Spectrum analysis techniques to decide masking thresholds and frequency-dependent bit allocation are used in order to match the quantizing noise to human perception. Each subband is quantized with a bit allocation based on the masking thresholds. Figure 10.27 shows a block diagram of the encoder and decoder [524].

A polyphase filter bank that has a low computational complexity and linear phase besides perfect reconstruction property is used for the subband filtering. If the input audio signal is sampled at 48 KHz and is divided into 32 equal subbands then each band ranges 0.75 KHz. The subband signals are divided into *digital frames* of 12 successive audio samples with a duration of 8 ms. The sampling interval in each band is $\frac{2}{3}$ ms.

The masking threshold is computed from an estimate of the short-time power spectral density (PSD) of the signal by means of the fast Fourier transform (FFT). The constant bandwidth of the subbands is not coincident with the critical bands. This problem can be solved in the frequency domain. This computation is repeated every 24 ms [529]. After computation of the masked noise powers, bits are allocated to the quantizers minimizing the sum of noise-to-mask ratios (NMR).

Scale factors can be calculated by using adaptive quantization with forward technique (or block companding) [543], in such a way that the samples are in the range $[-1, 1]$. The scale factors are coded and transmitted as side information: Statistical evaluations of the scale factors show strong redundancy which can be eliminated by using an appropriate coding technique. The scale factors and the bit allocation information are coded and integrated with the audio signal, and multiplexed before transmission. Table 10.19 shows some features of MUSICAM.

The ASPEC algorithm also relies heavily on psychoacoustic masking as in MUSICAM. For mapping into the frequency domain, however, MDCT is used. The block diagram of the encoder as shown in Fig. 10.28 consists of four main functional blocks. Digital audio samples are windowed for the purpose of time domain aliasing cancellation (TDAC). The MDCT together with the downsampling maps $2N$ audio samples into N frequency coefficients. Two different block lengths N are defined: the 256-sample block is used for coding at 96 Kbps and at 128 Kbps per mono channel, while the 512-sample block is selected for 32 Kbps and 64 Kbps transmission rates. Details of the ASPEC algorithm are described in [534].

The computation of the masking threshold is performed with the following procedures. First, the tonality of the signal energy is calculated on a line-by-line basis in the frequency domain (magnitude and phase analysis). Second, the energy in each critical band is calculated by using the tonality. This energy gives the unspread threshold, which does not consider the energy spreading in the upward direction. The spread threshold is computed by a modified spread function. The final step is computing the perceptual entropy that is used to estimate the number

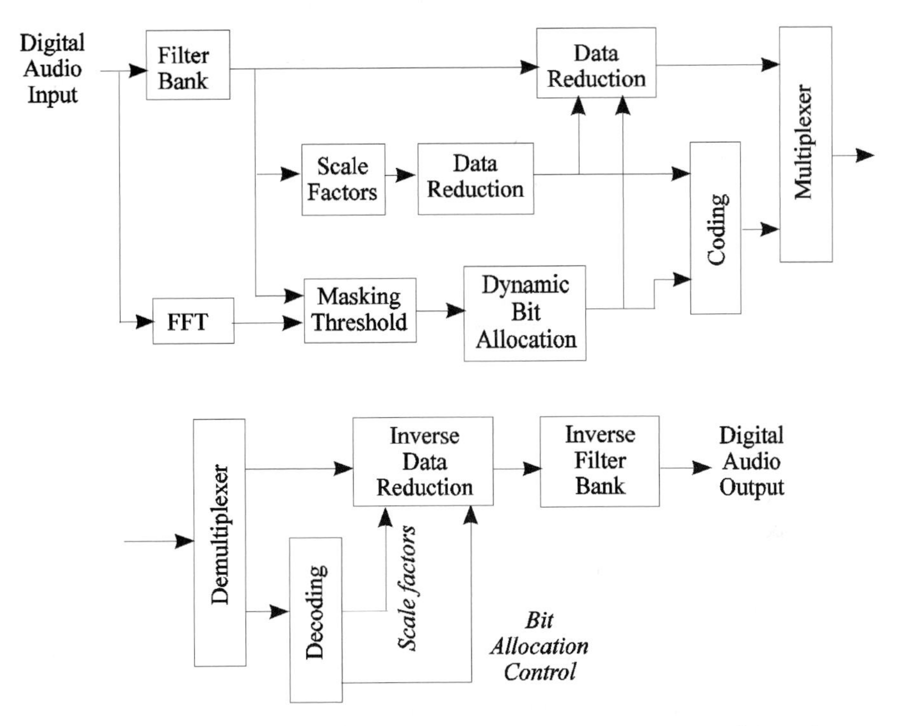

Figure 10.27 A block diagram of the MUSICAM encoder (above) and decoder (below) [524]. © 1990 IEEE.

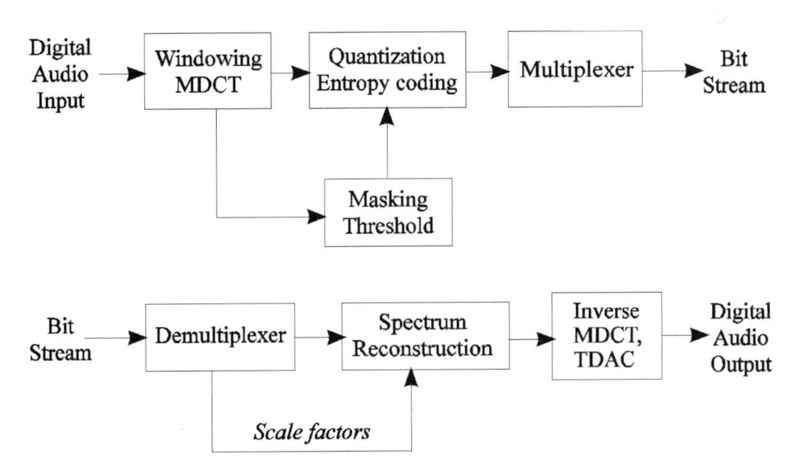

Figure 10.28 A block diagram of the ASPEC encoder (above) and decoder (below).

Table 10.19 Technical features of the MUSICAM algorithm

Sampling frequency	32, 44.1, 48 KHz
Bit rate	64, 96, 128, 192 Kbps
Delay time	Encoder 24 ms Decoder 6 ms

Table 10.20 Technical features of the ASPEC algorithm

Sampling frequency	32, 44.1, 48 KHz
Bit rate	32, 64, 96, 128 Kbps
Block length	256, 512 samples

of bits needed for the current signal block. The quantized data are coded by the Huffman coder. Scale factors and the other side information are multiplexed into the ASPEC bit stream. The technical features can be used for a brief comparison of ASPEC with MUSICAM (Tables 10.19 and 10.20).

10.3.3 Psychoacoustic weighting

Audio samples are, in general, mapped into the frequency domain by using a transform (MDCT) or a subband filter bank, which suits well the psychoacoustic phenomena of the human ear. Simultaneous masking is the phenomenon in which a weak signal is made inaudible (masked) by a simultaneously occurring stronger signal. Similar psychovisual masking effects occur in the human eye. An interesting fact in pschoacoustic masking is, however, that the masking occurs in each critical band [542]. The critical band represents the bandwidth at which subjective responses change rather abruptly. The bandwidth of the critical bands varies from 100 Hz at low frequencies to about $0.2\times$ actual frequency for frequencies above 500Hz. The loudness of a band of noise at a constant sound pressure remains constant in the critical band. The corresponding unit for the critical band is the bark shown in Fig. 10.25, which illustrates the 24 bands measured by Scharf [542] with the highest frequency of 15.5 KHz.

The threshold in quiet segments (absolute threshold) masked by a stronger signal can be measured as a function of frequency. Figure 10.29 shows masking thresholds of narrow band noise maskers centered at frequencies $f_m = 250$ Hz, $f_m = 1$ KHz, and $f_m = 4$ KHz. Targets with levels below the threshold (bottom curve) are masked and are in quiet regions. In the frequency band of about 2 KHz to 5 KHz, the SPL (sound pressure level) of the threshold curve is lower than in other frequencies, implying that this is the most sensitive frequency region.

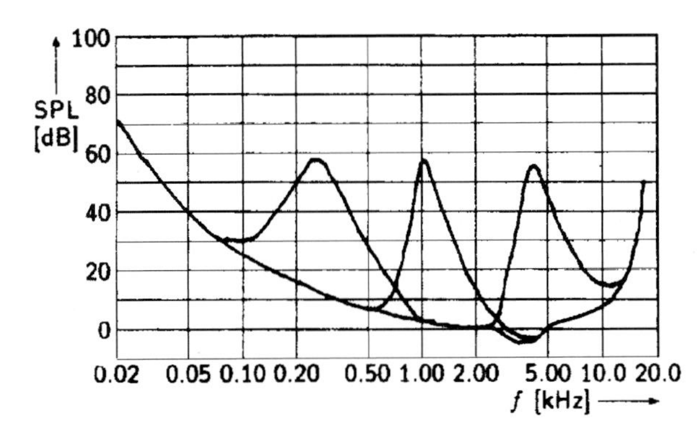

Figure 10.29 Masking thresholds of narrow band noise signals centered at frequencies, 250 Hz (left), 1 KHz (middle), and 4 KHz (right) for tonal targets [529]. © 1992 IEEE.

A signal masked by a stronger one is comparable to the quantization noise in the audio coding. The masking function gives allowable distortion for each critical band. Furthermore, interband masking is possible, and it is formulated as a spreading function [547]. The frequency characteristics of the human ear have been applied to high-quality audio coding. It is the key component of the MPEG audio encoder. A number of coding algorithms initially proposed to the MPEG audio group also utilize the psychoacoustic model. The algorithms are OCF [520] (Optimum Coding in the Frequency Domain), PXFM [518, 525] (Perceptual Transform Coder), ATC [538, 541, 546] (Adaptive Transform Coding), MUSICAM [524], and ASPEC [516, 545].

The common strategy of utilizing the masking effect is that an encoder analyzes the input audio signal and determines how much quantization noise can be masked and hence can be disregarded. The encoder uses this information to decide how best to represent the audio signal with the limited number of bits. Minimizing the bit rate under the masking constraint results in optimized audio coding [529]. Although the final decision from the psychoacoustic analysis is bit allocation and quantization, perceptual entropy [526] in the subband or signal-to-mask ratio [516] is generally computed as a tool for making this decision.

The MPEG audio standard supports two psychoacoustic models that are applied to the corresponding layers. Model I is valid for layers I and II. It can also be used for layer III, however. The psychoacoustic model in layer III is based on model II. The two models are implemented with complicated steps as described below.

Psychoacoustic model I

- The FFT in parallel with the subband filtering compensates for the lack of spectral selectivity obtained at low frequencies by the filter bank. The transform length for FFT is 512 samples for layer I and 1024 samples for layer II.
- The sound pressure level (SPL) in each subband is computed.

- The threshold in quiet regions, also called the absolute threshold, is also provided. An example is shown in Fig. 10.30, which is analogous to the lower curve in Fig. 10.29.
- The tonal and the nontonal components are extracted from the FFT power spectrum, since they influence the masking threshold in the critical band. For calculating the global masking threshold, we first have to determine the local maxima from neighboring spectral lines.
- Decimation is a procedure that is used to reduce the number of maskers that are considered for the determination of the global masking threshold. Only the tonal or nontonal components greater than the absolute threshold are considered. Two or more components that are smaller than the highest power within a distance of less than 0.5 Bark are removed from the list of tonal components.
- Individual masking thresholds of both tonal and nontonal components are defined by adding the masking index and the masking function to the masking component (tonal or nontonal). Both index and masking function are provided in the standard as formal equations.
- The global masking threshold LT_G for a frequency sample is derived by assuming the powers corresponding to the individual masking threshold (tonal LT_{tm} and nontonal LT_{nm}) and the threshold in quiet LT_q.

$$LT_G(i) = 10\log_{10}\left[10^{LT_q(i)/10} + \sum_{j=1}^{m} 10^{LT_{tm}(j,i)/10} + \sum_{j=1}^{n} 10^{LT_{nm}(j,i)/10}\right] \text{dB} \qquad (10.9)$$

- The minimum global masking threshold $LT_{min}(n)$ in subband n is used for determining the signal-to-mask ratio (SMR), which is given by

$$\text{SMR}_{sb}(n) = L_{sb}(n) - LT_{min}(n) \text{ dB} \qquad (10.10)$$

where $L_{sb}(n)$ is the signal component in subband n and the SMR is computed for every subband n. Bit allocation in layers I and II is based on the SMR.

Psychoacoustic model II

Psychoacoustic model I can be used for all layers in the coder. Layer III, however, is based on model II. More constraints are included in this model. The improvements and adaptation for layer III can be described as:

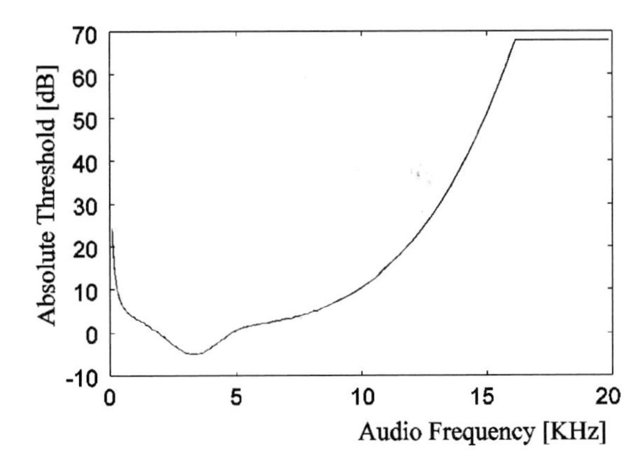

Figure 10.30 The absolute threshold for the psychoacoustic model I at a sampling rate of 48 KHz in layer I.

- The size of FFT and the Hann window can be varied. In practice, layer III computes the model twice in parallel with FFTs of 192 samples (short block) and 576 samples (long block).
- The spreading function is considered between neighboring critical bands. The function is based on the fact that a sound stimulus leaves a trace of aftereffects that die out gradually (forward masking) and the masking threshold can be changed by the previous stimulus (backward masking) [516, 544].
- The final energy threshold of audibility is obtained through the convolution of the spreading energy and the partitioned energy of the corresponding original power via FFT.
- The SMR is calculated by the ratio of the energy in the scale factor band, $epart_n$, and the noise level in the scale factor band, $npart_n$. The SMR_n to be sent to the coder is given by

$$\mathrm{SMR}_n = 10\log_{10}(epart_n/npart_n)\ \mathrm{dB} \tag{10.11}$$

where n denotes the index of the coder partition.

10.3.4 Coding layer I

Audio data can be coded by different algorithms subject to the application purposes. Layer I is the basic algorithm, which contains only three kinds of components as shown in Fig. 10.31. Bit allocation indicates the number of bits used to code the 12 samples in a subband. The all-zero code indicates that no samples are transmitted. Inverse quantized samples are multiplied by the scale factors in the decoder.

Data (Layer I)	Bit Allocation (4 bits)	Scale Factors (6 bits)	Samples (2~15 bits)

Figure 10.31 The data bitstream structure of layer I.

The coding steps in layer I are very close to the MUSICAM described in Fig. 10.27 using the filter bank for transforming the audio into the frequency domain. The filter bank splits the full-band signal with sampling frequency f_s into 32 equally spaced subbands with sampling frequencies $f_s/32$. The flow chart describing this process is given in Fig. 10.32. This kind of analysis filter bank is also valid for layer II.

Subband filtering: Thirty-two audio samples are input to the lower part of the 32-sample right-shifted data vector, X, of 512 elements. For the time domain aliasing cancellation, it is windowed by the window coefficients, C_i, listed in CD 11172-3 (see Table 3–C.1) [487]. The windowed data vector is subsampled at every 64th sample and the 8 samples are summed, resulting in a 64-coefficient vector, Y. It is filtered by the 32 subband matrixing. The coefficients for the matrix are generated by

$$M_{ik} = \cos\left((2i+1)(k-16)\frac{\pi}{64}\right) \qquad (10.12)$$

$$i = 0, 1, \ldots, 31, \quad k = 0, 1, \ldots, 63$$

Thus the 32 input samples are represented as the 32 subband filter coefficients.

Psychoacoustic modeling: Either model I or model II can be used for the determination of the psychoacoustic parameters. However, model I is sufficient for layer I, which requires the FFT of 512 samples. The SMR is determined from the psychoacoustic model.

Scale factor: The maximum of the absolute values of 12 samples in a subband is determined. The next largest value can be found in the lookup table and the index is coded as a scale factor for the 12 samples (6 bits, MSB first). The scale factor is transmitted only if a nonzero number of bits has been allocated to the subband.

Bit allocation: The basic concept in the bit allocation procedure is minimization of the total noise-to-mask ratio (NMR) over the frame with the constraint that the number of bits used does not exceed the number of bits available for the frame, B_f, which can be formulated as

$$B_f = \frac{Bit\ rate}{f_s} \times 384 \text{ bits/frame} \qquad (10.13)$$

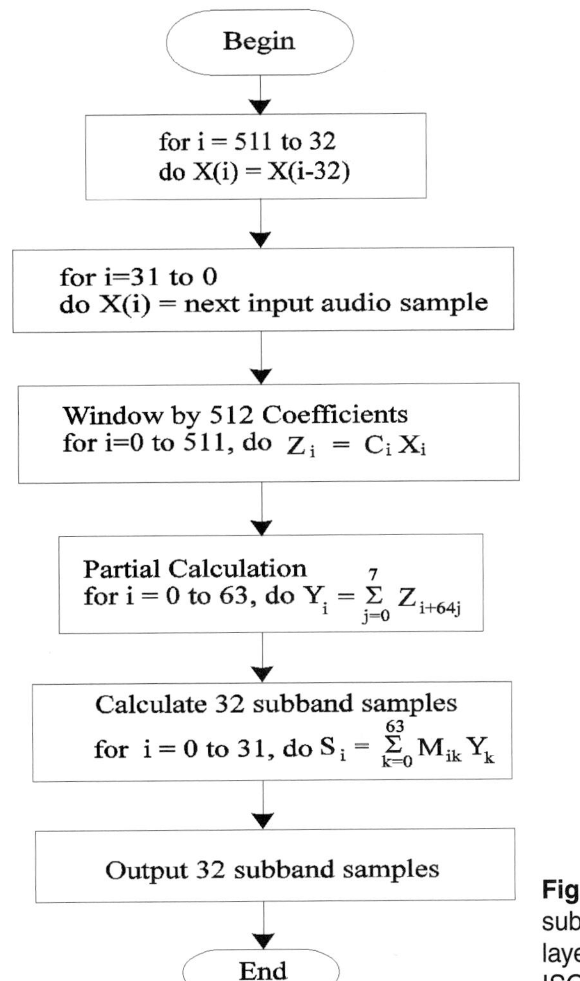

Figure 10.32 An analysis subband filter flow chart for layers I and II [487]. © 1993 ISO/IEC.

where f_s denotes the sampling rate and 384 samples are processed in the frame. Note that the frame consists of header, error check, data and ancillary data (Fig. 10.24). Furthermore, the total number of bits available for coding the audio represents the bit allocation among the subbands, scale factors, audio samples, and ancillary data.

The bit allocation procedure is an iterative process that starts with zero bit allocation. First the mask-to-noise ratio (MNR) is obtained by

$$\mathrm{MNR} = \mathrm{SNR} - \mathrm{SMR}\ \ \mathrm{dB} \tag{10.14}$$

where the SNR can be found in Table 10.21 and the SMR is provided from the psychoacoustic model. The MNR indicates the discrepancy between

Table 10.21 Bit allocation and corresponding SNR in layer I [487]

Bits	Code	Number of steps	SNR (dB)
0	0000	0	0.00
2	0001	3	7.00
3	0010	7	16.00
4	0011	15	25.28
5	0100	31	31.59
6	0101	63	37.75
7	0110	127	43.84
8	0111	255	49.89
9	1000	511	55.93
10	1001	1023	61.96
11	1010	2047	67.98
12	1011	4095	74.01
13	1100	8191	80.03
14	1101	16383	86.05
15	1110	32767	92.01
invalid	1111		

Source: © 1993 ISO/IEC.

waveform error and perceptual measurement. The audio samples can be further compressed as much as the MNR. Therefore, the minimum MNR of all subbands is determined at each iteration loop. The iterative procedure is repeated until the MNR is minimized and the number of bits used for the four components becomes closer to the total number of bits available, B_{tav}. The marginal bits B_{mg}, computed in the iteration loop, can be described by

$$B_{mg} = B_{tav} - (bbal + bscf + bspl + banc) \tag{10.15}$$

where *bbal*, *bscf*, *bspl*, and *banc* denote the number of bits for allocation, scale factor, samples, and ancillary data, respectively.

Quantization and encoding: The subband samples are quantized by a uniform quantizer with a symmetric zero representation. Each of the subband samples S_i is normalized by dividing its value by the scale factor scf and quantized using the following formula:

$$S_{qi} = \left(A \left(\frac{S_i}{scf} \right) + B \right) \Big|_N \tag{10.16}$$

where A and B are constants (Table 10.22), S_{qi} denotes quantized sample in the subband, scf is scale factor, and N represents the necessary number of bits to encode the number of steps in Table 10.22. To avoid the all-1 representation of the code, the most significant bit (MSB) of S_{qi} is inverted before formatting the bit stream.

Table 10.22 Quantization coefficients in layers I and II [487]

Number of steps	A	B
3	0.750000000	–0.250000000
7	0.875000000	–0.125000000
15	0.937500000	–0.062500000
31	0.968750000	–0.031250000
63	0.984375000	–0.015625000
127	0.992187500	–0.007812500
255	0.996093750	–0.003906250
511	0.998046875	–0.001953125
1023	0.999023438	–0.000976563
2047	0.999511719	–0.000488281
4095	0.999755859	–0.000244141
8191	0.999877930	–0.000122070
16383	0.999938965	–0.000061035
32767	0.999969482	–0.000030518

Source: © 1993 ISO/IEC.

Bit-stream formatting: The encoded subband information is multiplexed in a frame unit (Fig. 10.24), which becomes the main part of the standard. Only the bit-stream formatting is performed as a final procedure with no additional coding. A frame is composed of an integer number of slots to adjust the mean bit rate. In layer I a slot equals 32 bits, while in layers II and III a slot equals 8 bits. Therefore, the number of slots in a frame is obtained by dividing the total number of bits in Eq. (10.13) by the number of bits in a slot, i.e., $B_f/32$. If the sampling frequency is 44.1 KHz, the number of slots ($4N$) is not an integer, as shown in Example 10.7. In such cases the frame needs to be adjusted by adding bits (padding). This means that the number of slots can be N or $N + 1$.

Example 10.7 *Padding is performed to adjust the average length of an audio frame in time to the duration of the corresponding PCM samples. The following table shows three sampling frequencies f_s, corresponding frame rates and sizes, and the number of slots, assuming the bit rate of* 64 *Kbps.*

f_s (KHz)	Frame rate (frames/sec)	Frame size (ms)	Number of slots
48	125	8	16
44.1	114.84	8.70 ...	17.41 $\Rightarrow$ 18
32	83.33 ...	12	24

Data (Layer II)	Bit Allocation (2 ~ 4 bits)	Scale factor Select Information (2 bits)	Scale Factor (6 bits)	Samples (2 ~ 16 bits)	Ancillary

Figure 10.33 The data bit-stream structure of layer II.

10.3.5 Coding layer II

Layer II follows basically the same rules for coding and decoding as discussed in layer I. The main difference is that layer II introduces correlation between subbands. Layer II contains information for 1152 samples ($3 \times 12 \times 32$). The bit stream in layer II consists of scale factor–select information as well as bit allocation, scale factor, and data samples (Fig. 10.33).

Psychoacoustic models I and II can be used for bit allocation in layer II, provided that the FFT shiftlength is 1152 samples in model I and 576 samples in model II. Either model provides the SMR for every subband.

Coding scale factors: The same analysis and synthesis filters as those applied in layer I can be used for layer II, and the technique for calculating the scale factor is also the same. In layer II, however, a frame corresponds to 36 (3×12) subband samples (12 granules as shown in Fig. 10.34) and contains three scale factors per subband (one scale factor per 12 consecutive samples). The two differences are obtained from the three scale factors as follows:

$$\begin{aligned} Dscf1 &= scf2 - scf1 \\ Dscf2 &= scf3 - scf2 \end{aligned} \tag{10.17}$$

Each scale factor difference (*Dscf*1 or *Dscf*2) is classified into one of the five classes as follows:

$$\begin{aligned} \text{Class} &= 1 && \text{if} \quad Dscf \le -3 \\ &= 2 && -3 < Dscf < 0 \\ &= 3 && Dscf = 0 \\ &= 4 && 0 < Dscf < 3 \\ &= 5 && Dscf \ge 3 \end{aligned}$$

The two classified scale factors correspond to the transmission pattern that implies the three scale factors before classification. The transmission pattern is correlated between the three components as shown in Table 10.23. Therefore, redundancy in the pattern can be reduced at the expense of coding the scale

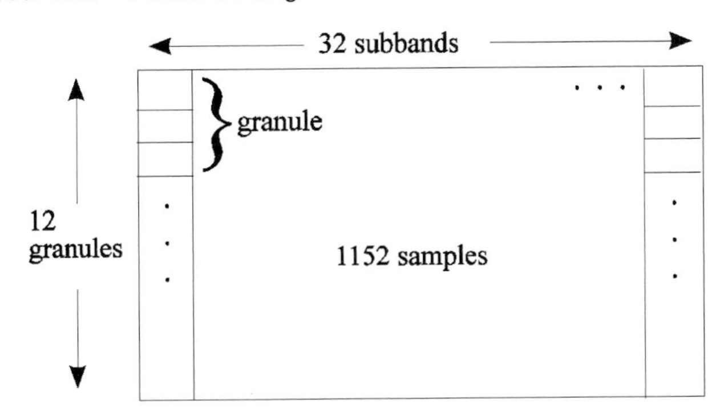

Figure 10.34 Structure of layer II subband samples.

factor–select information. The numbers 1, 2, 3, or 4 in the column of transmission pattern represent first, second, or third scale factor or the largest one among them, respectively. Example 10.8 shows some cases of coding and decoding the scale factors, *scf*, and scale factor–select information, *scfsi*. The *scfsi* represents the number and the position of the *scf* in each subband. Table 10.24 illustrates the coding procedure of the *scfsi*.

Example 10.8 *As stated earlier, the* 36 *samples in the subband within a frame are divided into three equal parts having three scale factors. Assume that the scale factors A, B, and C are produced in a subband. They can be classified for example as* (1, 1), (1, 3), *and* (3, 2) *in Table 10.23. Each scale factor is coded by a* 6*-bit integer. The table below shows that transmission bits are reduced if there is redundancy between the scale factors. In the decoder the three scale factors are reconstructed by the rule shown in Table 10.24 and they are multiplied by the samples after dequantization.*

Class	Transmitting scale factor	scfsi	Decoded scale factor
(1, 1)	ABC (18 bits)	00	ABC
(1, 3)	AB (12 bits)	11	ABB
(3, 2)	A (6 bits)	10	AAA

Bit allocation: The SMR from the psychoacoustic model is used for obtaining the MNR and the iterative operation is performed as shown in layer I, including the scale factor–select information.

Quantization and encoding: The same algorithm used for the quantizer in layer I is applied with the corresponding table for quantization coefficients. Given the number of steps from bit allocation, grouping by granule is determined as shown in Table 10.25. If grouping is required, three consecutive samples are coded

Table 10.23 Layer II scale factor transmission patterns [487]

Class 1	Class 2	Transmission pattern	Scale factor–select information
1	1	1 2 3	0
1	2	1 2 2	3
1	3	1 2 2	3
1	4	1 3 3	3
1	5	1 2 3	0
2	1	1 1 3	1
2	2	1 1 1	2
2	3	1 1 1	2
2	4	4 4 4	2
2	5	1 1 3	1
3	1	1 1 1	2
3	2	1 1 1	2
3	3	1 1 1	2
3	4	3 3 3	2
3	5	1 1 3	1
4	1	2 2 2	2
4	2	2 2 2	2
4	3	2 2 2	2
4	4	3 3 3	2
4	5	1 2 3	0
5	1	1 2 3	0
5	2	1 2 2	3
5	3	1 2 2	3
5	4	1 3 3	3
5	5	1 2 3	0

Source: © 1993 ISO/IEC.

Table 10.24 Coding of scale factor–select information

scfsi	Number of coded scale factors	Decoding scale factors
0 (00)	3	scf1, scf2, scf3
1 (01)	2	1st ⇒ scf1 and scf2 2nd ⇒ scf3
2 (10)	1	one for all
3 (11)	2	1st ⇒ scf1 2nd ⇒ scf2 and scf3

Table 10.25 Quantization/dequantization classes in layer II

Number of steps	C	D	Grouping	Samples/code	Bits/code
3	1.3333333333	0.5000000000	yes	3	5
5	1.6000000000	0.5000000000	yes	3	7
7	1.1428571428	0.2500000000	no	1	3
9	1.7777777777	0.5000000000	yes	3	10
15	1.0666666666	0.1250000000	no	1	4
31	1.0322580645	0.0625000000	no	1	5
63	1.0158730158	0.0312500000	no	1	6
127	1.0078740157	0.0156250000	no	1	7
255	1.0039215686	0.0078125000	no	1	8
511	1.0019569471	0.0039062500	no	1	9
1023	1.0009775171	0.0019531250	no	1	10
2047	1.0004885197	0.0009765625	no	1	11
4095	1.0002442002	0.0004882812	no	1	12
8191	1.0001220852	0.0002441406	no	1	13
16383	1.0000610388	0.0001220703	no	1	14
32767	1.0000305185	0.0000610351	no	1	15
65535	1.0000152590	0.0000305175	no	1	16

as one code word. There are three cases, i.e., the numbers of steps for grouping are 3, 5, and 9. The combined code has to be separated in the decoder. The following algorithm will supply the three separate codes $s(0)$, $s(1)$, and $s(2)$.

For $i = 0$ to 2

$s(i) =$ (code) **MOD** (no. of steps)

code = (code) // (no. of steps)

where **MOD** and // denote modulo and integer division, respectively.

The three codes obtained by the separation algorithm are decoded and dequantized by using the coefficients C and D shown in Table 10.25, which are applied in the formula

$$\hat{S}(i) = C(S''(i) + D) \tag{10.18}$$

where $S''(i)$ is the fractional number inverted to get the two's compliment value and $\hat{S}(i)$ is the dequantized value that has to be multiplied by the scale factor before applying the synthesis filtering.

Bit-stream formatting: The structure of the bit stream in layer II is shown in Fig. 10.33. Except for the length of a slot being 8 bits, the remaining algorithms for padding or formatting the bit stream are the same as in layer I.

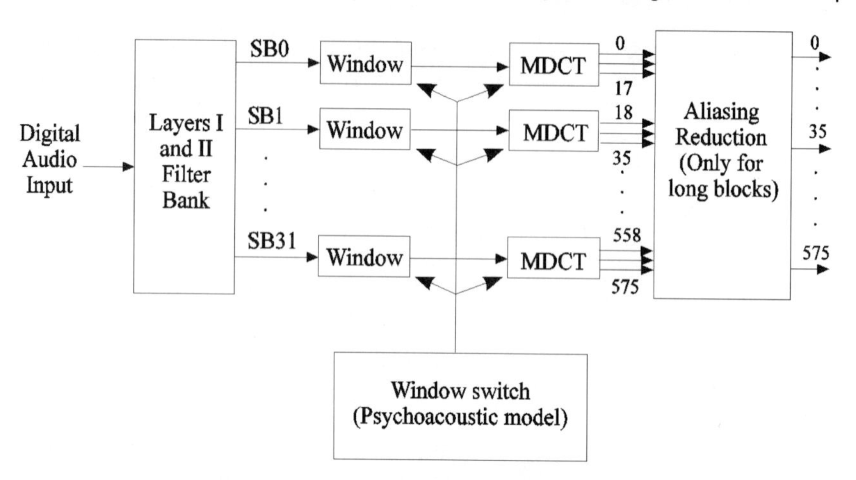

Figure 10.35 A hybrid filter bank for MPEG audio layer III.

10.3.6 Coding layer III

Coding layer III is a much more sophisticated technique than coding layers I and II. Additional frequency resolution is provided by the use of a hybrid filter bank. The nonuniform quantizer is applied by means of a power law. Entropy coding (Huffman coding) is introduced for coding data values. Furthermore, iteration loops for the psychoacoustic modeling and bit allocation are also elaborated.

Figure 10.35 shows a block diagram of a hybrid filter bank (the equal-bandwidth filter bank used for layers I/II and MDCT). The filter bank outputs are transformed to provide better spectral resolution. A number of samples in each subband are multiplied by the time domain window functions. Layer III specifies two different MDCT block lengths — a long block of 36 samples or a short block of 12 samples — due to the trade-off between time and frequency resolutions. The short block length improves the time resolution to cope with transients (abruptly changing portions).

The corresponding windows are defined for the long-block and the short-block MDCT. Figure 10.36 shows the window responses. Normal long block (block type 0) is given by

$$h(k) = x(k) \sin\left(\frac{\pi}{N}\left(k + \frac{1}{2}\right)\right) \tag{10.19}$$

$$k = 0, 1, \ldots, 35, \qquad N = 36$$

which is depicted by the notations o and x in Fig. 10.36 forming a sine function. The same equation with $N = 12$ is applied for the short-block (block type 2)

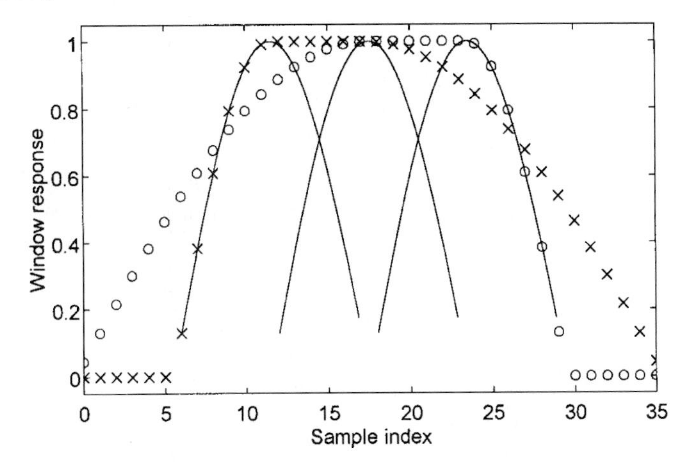

Figure 10.36 Long-block (36 samples) and short-block (12 samples) window responses (o: long-to-short; x: short-to-long; solid line: short block).

windowing. The block of 36 samples is divided into three overlapping blocks as shown in Fig. 10.36. Each of the three small blocks is windowed separately.

Switching between long and short blocks is not instantaneous. Transient windowing [short-to-long (block type 3) and long-to-short (block type 1)] functions are shown in this figure. The decision whether the filter bank should be switched to short or long windows is derived from the masking threshold obtained by the estimate of the psychoacoustic entropy (PE). If the value of PE is greater than a certain level ($PE \geq 1800$), then the block needs to be shorter. Hence, block activities can be considered for efficient coding.

The major enhancements and distinctions of layer III over layers I and II are summarized as follows.

MDCT: The following equation is used for obtaining $N/2$ coefficients S_i from N input samples x_k. For short blocks N is 12, and for long blocks N is 36. The analytical expressions for the forward MDCT and inverse MDCT are

$$S_i = \sum_{k=0}^{N-1} x_k \cos\left(\frac{\pi}{2N}\left(2k+1+\frac{N}{2}\right)(2i+1)\right) \tag{10.20}$$

$$i = 0, 1, \ldots, \frac{N}{2}-1, \qquad N = 12 \text{ or } 36$$

$$x_k = \sum_{i=0}^{\frac{N}{2}-1} S_i \cos\left(\frac{\pi}{2N}\left(2k+1+\frac{N}{2}\right)(2i+1)\right) \tag{10.21}$$

$$k = 0, 1, \ldots, N-1$$

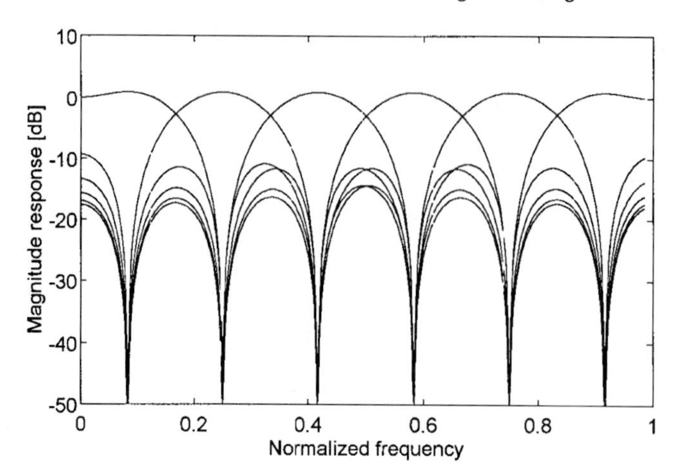

Figure 10.37 The frequency spectrum of an MDCT resulting in $N/2$ coefficients from N samples. (For this example, $N = 12$.)

Figure 10.37 shows the frequency response of an MDCT that subdivides sample frequencies into 6 bands, when $N = 12$. Hence, the MDCT used in layer III does not need to decimate after the transform, whereas the ASPEC (Section 10.3.2) obtained $N/2$ coefficients by decimation.

Aliasing reduction: The calculation of aliasing reduction in the encoder is performed as in the decoder. Only the long blocks are input to the aliasing reduction procedure. The MDCT results in 18 coefficients from 36 input samples in each subband. Between the two sets of 18 coefficients the butterfly operation is performed as shown in Fig. 10.38, where i represents the distance from the last line of the previous block or the first line of the current block. Eight butterflies are defined with different weighting factors cs_i, ca_i that are given by

$$cs_i = \frac{1}{\sqrt{1 + c_i^2}} \qquad ca_i = \frac{c_i}{\sqrt{1 + c_i^2}} \tag{10.22}$$

where the eight c_is are defined as −0.6, −0.535, −0.33, −0.185, −0.095, −0.041, −0.0142, −0.0037.

Noise allocation: The bit allocation process used in layers I and II only allocates the available bits and approximates the amount of noise caused by quantization. The layer III encoder introduces the noise allocation iteration loop. The inner loop quantizes the input vector and increases the quantizer step size until the output vector can be coded with the available number of bits. The outer loop calculates the distortion in each scale factor band so that a set of frequency lines is scaled by one scale factor. If the allowed distortion is exceeded, the scale factor band (critical band) is amplified and the inner loop is again activated.

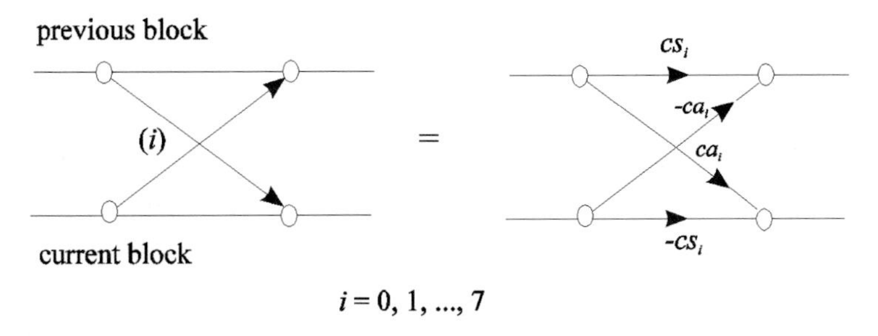

Figure 10.38 The aliasing butterfly for the layer III encoder and decoder.

Nonuniform quantization: The nonuniform quantizer is designed by a power law that raises input samples to the 3/4 power to provide more consistent SNR over the range of the quantizer. The decoder calculates the 4/3 power of the Huffman decoder output samples.

Entropy coding: The huffman coder is used for entropy coding. The lookup tables are provided for using the encoder and the decoder.

Bit reservoir: The coded data bit stream does not necessarily fit into a fixed-length frame. The slots are still used for adjusting the bit rate. Hence, bits are saved in the reservoir when fewer bits are used to code one granule. A granule in layer III consists of 18 samples in 32 subbands. If bits are saved for a frame, the remaining space in the frame can be used for the next frame data. Hence, the encoder can donate bits to or borrow bits from the reservoir when appropriate. The number of bytes in the bit reservoir is given by side information in the data bit stream.

10.4 MPEG-1 Implementations

The MPEG-1 standard supports video and audio coding/decoding. It does not, however, specify the encoding process. The standard ISO 11172 part 1 specifies the syntax and semantics of the system layer coding of combined coded audio, video, and ancillary data. Video and audio components are synchronized by the system encoder and decoder, which are illustrated in Fig. 10.39. The standard is intended mainly for digital storage media such as the compact disc (interactive system of digital consumer recording), which gives the benefits of a flawless reproduction or multiple copying [530].

The synchronization is performed using system clock reference (SCR), decoding time stamp (DTS), and presentation time stamp (PTS). The system time clock (STC) is operated using the SCR transmitted. The system layer shown in Fig. 10.40 includes a pack layer and a packet layer. The pack layer starts with the

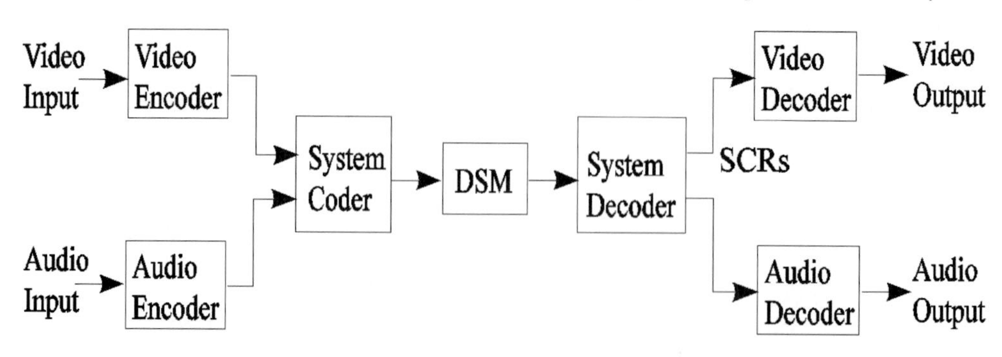

Figure 10.39 System realization of MPEG-1 video and audio (SCR: system clock reference).

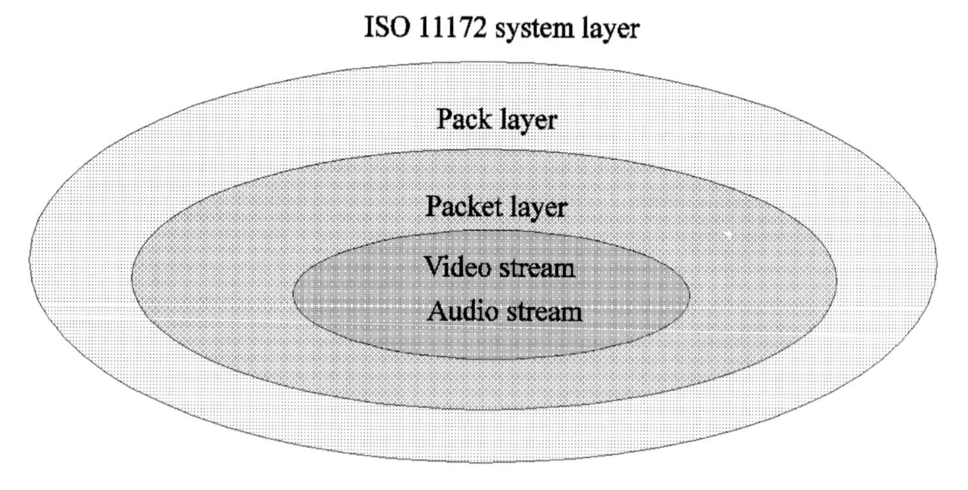

Figure 10.40 Layered implementation of MPEG-1 system.

pack start code. The SCRs are included in the pack header. A pack consists of one or more packets. The video and audio stream are multiplexed in the packet layer. In practical implementations, the video is coded with a bit rate of about 1.15 Mbps, while the audio is coded at 128 or 192 Kbps. Thus, the video bandwidth is six or nine times broader than that of audio. In the packet layer the video and audio stream are multiplexed. Considering the bandwidth ratio, one audio packet is interleaved with every six or seven video packets.

Some designers and chip makers are opting for dedicated functions and algorithm-specific chips, while others are selecting general-purpose programmable DSP approaches. Many applications provide a combination of the two and also incorporate software techniques. A number of manufacturers have implemented the MPEG algorithm. They are listed in Appendix A. There are several ways of implementation: chip set for video encoder or decoder, chip set for audio encoder or decoder, software based, and systematic implementation including video/audio by using the chip sets or DSP chip set on PC or dedicated platforms. Since many end users need decoders only, most of the manufacturers have produced audio decoder and/or video decoder. Recently, encoder chip sets have come into the market and real-time implementation of the combined system has become possible. Also, single chip implementation has become feasible. We will proceed to introduce some trends in video and audio implementations.

Video codec

At first, research was under way to implement a video decoder chip set rather than an encoder, which is much more complicated. To have a cost-effective implementation of such a decoder, state-of-the-art VLSI technology and conventional algorithms that can be easily modified for MPEG can be used. DCT and intraframe processing techniques can be implemented by using the chip sets used for JPEG or H.261 [474]. The DCT chip, in particular, has been implemented by a number of manufacturers [384]. The remaining computations, except DCT, can be realized by designing custom ICs such as digital signal processor chip sets [500].

For efficient hardware design, the encoding and decoding algorithms can be partitioned into several parts subject to their properties. For example, DCT and quantization used in the encoder, and IDCT and inverse quantization used in the decoder can be handled with parallel arithmetic operations for (8×8) block data. On the other hand, variable-length coding and decoding are essentially sequential processes, which means that different approaches should be employed. In accordance with the basis for sharing the burden among several programmable chip sets, the MPEG encoder and decoder could be implemented with real-time operation for the SIF image at 30 frames/s [489].

There are not many manufacturers who make a single-chip encoder or decoder. This trend, of course, is changing rapidly. At present many application products utilize the chip set CL450 made by C–Cube Microsystems. The CL450 is a fairly generic video decoder with a transistor count of about 400,000. The CL450 performs real-time MPEG bit streams at resolutions of 352H$\times$240V pels or 352H$\times$288V pels. Compressed data rates of 1.2 to 3 Mbps are typically implemented. The chip set has three external interfaces: host, DRAM, and video interfaces as illustrated in Fig. 10.41. The host interface is used to send compressed data and high-level commands into the CL450. The DRAM interface connects directly to an 80 ns fast page mode DRAM. The video interface outputs decompressed video in 24-bit RGB or 16-bit YC_bC_r. The MPEG-1 video encoder CLM4500 entered the market towards the end of 1993. The chip set runs microcode on two video-compression processors.

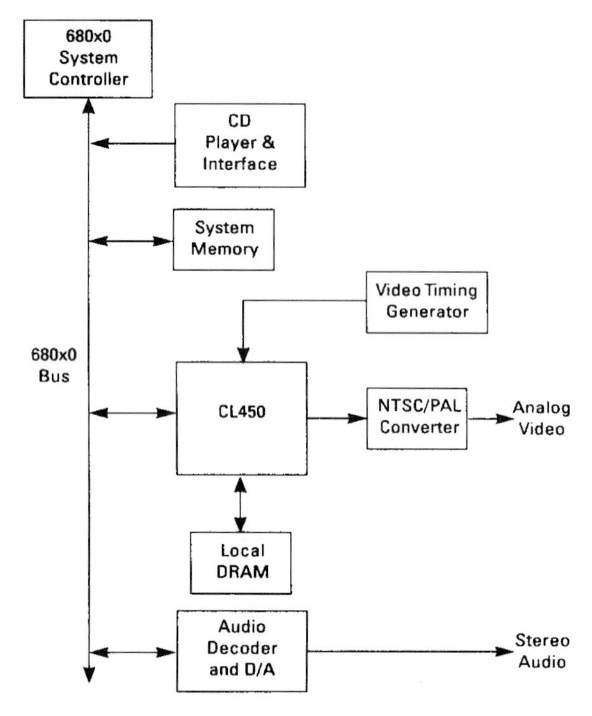

Figure 10.41 A typical system implementation of the MPEG-1 decoder.

Some manufacturers have developed chips that implement specific DSP algorithms besides programmability over a wide range of parameters. The video-flow architecture developed by Array Microsystems and Samsung Electronics consists of two processors: an image-compression coprocessor (ICC–a77100) and the companion motion-estimation coprocessor (MEC–a77300). The two chips readily tie into a host processor and local controller as shown in Fig. 10.42 [393].

The chip set is based on a high-performance parallel processing architecture implemented using a unique vector dataflow architecture, VIDEOFLOW™. This parallel processing architecture embodied in the chip set permits it to deliver exceptional performance for real-time implementation of the MPEG, H.261, and JPEG standards. The image-compression coprocessor (ICC) is a programmable IC optimized for the execution of these DCT-based video-compression standards. The ICC implements all algorithmic functions of the encoder such as DCT, quantization, zero-runlength count and all arithmetic operations like prediction, averaging, and so forth.

The motion-estimation coprocessor (MEC) is optimized to implement block-matching algorithms for motion estimation. The MEC has an embedded RISC microcontroller that controls the block-match processor, video memory interface, and communications with the ICC. The block-match processor is a 7.2 BOPS-pipelined processor that performs add, subtract, and mean absolute difference

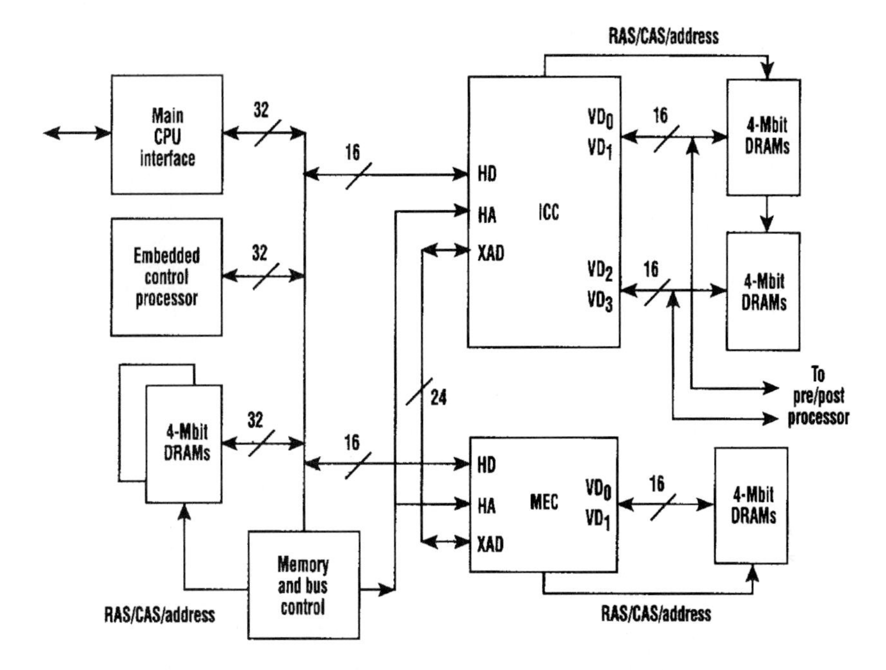

Figure 10.42 A block diagram of the two-chip solution developed by Array Microsystems and Samsung Electronics. RAS/CAS: Row address strobe/Column address strobe, HD: Host data, HA: Host address, XAD: X-bus address/data, VD: Video-memory data bus, ICC: Image compression coprocessor, MEC: Motion estimation coprocessor.

(MAD) calculations associated with motion estimation. It is capable of performing three (8×8) block comparisons in 100 nsec, three (16×16) block comparisons in 320 nsec, or three interpolated block comparisons in 480 nsec. The MEC can be programmed to implement a variety of block-matching algorithms like exhaustive search and two-level hierarchical search over a user-programmed search range at both full- and half-pel search resolutions.

Microcode programmable processors can be integrated into a system of image compression/decompression. The processors can be designed for coding the images. For example, the Vision Controller [390] performs Hufman encoder/decoder, video pre- and postprocessing, and host interfacing, whereas the vision processor performs the coding techniques such as DCT, quantization, motion estimation, and so forth. Since all functions can be programmed, the chip sets can be used for multistandards. Real time implementation of encoder, however, can be designed by dedicated chip set that includes a group of general-purpose programmable processors and a set of special-purpose processors [509]. For desktop multimedia computing on PCs/workstations, a programmable system architec-

Table 10.26 Main features of typical board-level applications

Maximum video window size	704H×480V
GOP structure	$N = 15$, $M = 3$, programmable
Time to get 1 sec video	40 sec using logarithmic search 100 sec using full search BMA
MV range	Encoder: up to ±30 pels decoder: ±7 pels

ture is more desirable. Texas Instruments TMS320C80, known as the Multimedia Video Processor (MVP), is a single-chip multiprocessing device with highly parallel internal architecture [988]. It consists of a RISC CPU, four DSP processors, an intelligent DMA controller, video controllers, and 50 Kbytes of SRAM on a single chip.

Some vendors produce the compression/decompression boards using chip sets such as DSP56001 or CL450. The board-level solution on PCs can capture, digitize, compress, and playback (decompress) full-motion video. Performance, however, depends on the speed of the workstations. This means that real-time implementation is not easy with general-purpose PCs. Full-search block-matching motion estimation requires more time as shown in Table 10.26.

It is fairly hard to get real-time implementation of such a complex algorithm using software only. Some companies have demonstrated reasonably good quality MPEG decoding software (sound is also synchronized). Since the encoder is more complicated than the decoder, the decoder has been extensively implemented. Some of the software is available from the public domain networks, such as the Internet.

Audio codec

MPEG audio is a subband-based algorithm with adaptive quantization. A psychoacoustic model estimates the ear's sensitivity to quantization noise in each subband. Therefore, the encoder is a relatively complex algorithm. Furthermore, the decoder is of primary interest to the end users, and MPEG audio specifies the decoding algorithm in detail.

Most of the encoders have been implemented by using a general-purpose DSP chip set such as the TMS320 series. The chip sets operate at 30 to 40 MHz for a peak performance of 40 MFLOPS (million floating-point operations per second). It can as be used as a stand-alone with additional memory. It also allows operation with any host platform. Manufacturers provide an audio processing board using the DSP chip sets and users can utilize it to encode or decode the

MPEG algorithm. For providing a low-cost decoder, the DSP chip sets can be application specific. Manufacturers implement audio decoder chips using CMOS submicron technology. The receiver decodes the compressed audio bit stream in real time. It can also demultiplex the audio portion from the MPEG bit stream prior to decoding. General specifications of audio decoder chip sets are

- Compatibility with layers I and II (layer III is not mandatory)
- Preferred bit rates, for example, 64, 94, 128, and 192 Kbps for a mono channel (for stereo, twice these bit rates) [522]
- Supports up to 1-second audio delay for audio/video synchronization
- Supports three sampling rates (32, 44.1, and 48 KHz)
- Joint stereo coding of intensity stereo signal

Figure 10.43 shows the typical audio decoder block diagram, which is composed of decoding blocks and peripherals. The decoder receives data from the channel buffer and performs all these functions: degrouping, dequantization, denormalization, and subband synthesis. Fast algorithms for the subband filtering are required, since the subband filter synthesis operation takes 40 % of the overall

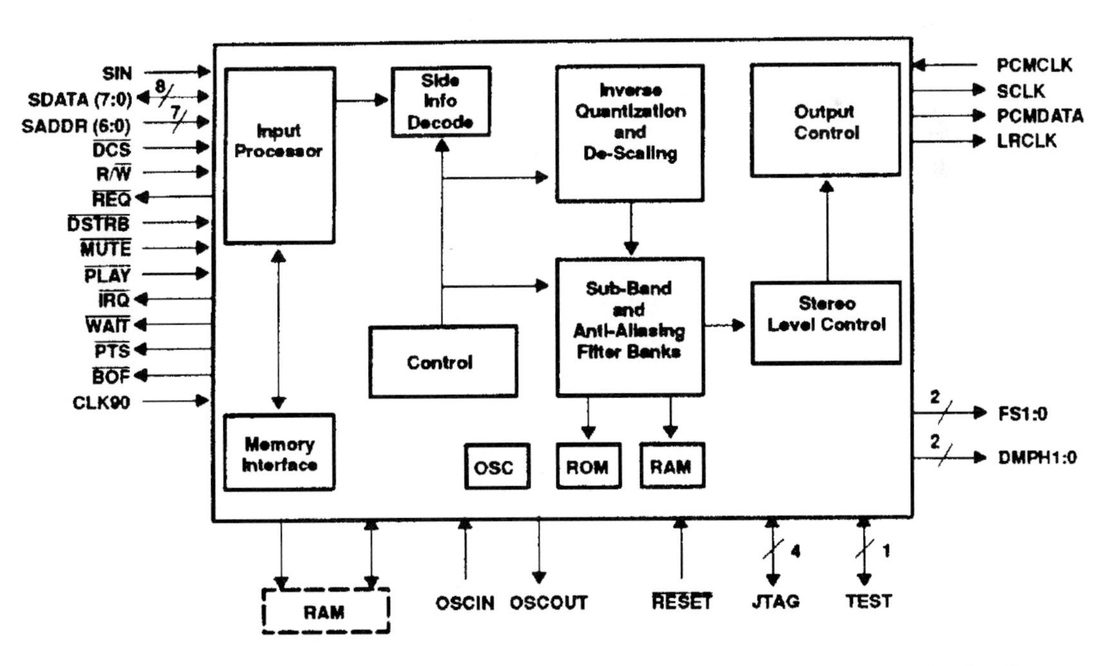

Figure 10.43 An example for MPEG audio decoder implementation [523]. © 1993 TI.

decoding time [548]. Usually, ROM provides lookup tables for scale factors, quantization values, and window coefficients. RAMs can be used for storing dequantized coefficients or vectors generated in the subband filter bank [539]. The decoder can be used in two modes. In a stand-alone audio system, the decoder functions as a single-chip decoder with no support circuitry except crystal clock input [523]. If the application also includes video decoding, the external DRAM may be used to delay the audio stream for synchronization purposes. A detailed list of manufacturers and products for the MPEG audio and/or video is included in Appendix A.

Chapter 11
MPEG-2 Generic Coding Algorithms

Summary

As a second-phase project by the ISO/IEC MPEG committee, MPEG-2 is targeted at higher bit rates and broader generic applications, including consumer electronics, telecommunications, and next-generation broadcasting. The syntax and technical contents were frozen in November 1993 as ISO/IEC Committee Draft (CD) 13818, "Generic coding of moving pictures and associated audio" (591), which has four parts: Part 1, systems; Part 2, video; Part 3, audio; and Part 4, conformance testing. Part 2, video of MPEG-2, was finalized jointly by MPEG and the Experts Group for ATM Video Coding in the ITU-T SG 15. Thus it is also called ITU-T Recommendation H.262. Parts 1 through 3 attained international standard (IS) status in November 1994.

In this chapter we will discuss the generic coding algorithms based primarily on MPEG-1 (see Chapter 10), including the extension up to HDTV. We have tried to avoid duplicate explanations of MPEG-1 algorithms and have focused mainly on the extended functions of MPEG-2. Video coding is treated in Section 11.2 and audio coding is described in Section 11.3. At the end, implementation trends and techniques will also be covered.

11.1 Introduction

The compression schemes discussed in previous chapters provide specific applications at corresponding coding efficiencies. ITU-T H.261 has been designated for video coding and transmission of slow-moving videophone and videoconferencing signals. MPEG-1 is aimed at systems working with 1.5 Mbps DSM and low-resolution SIF displays. Consequently, a full-motion video and audio coding standard has been in development to meet a number of requirements. It is needed to support digital video transmission in the range of 2 to 15 Mbps, including applications for DSM, existing TV (PAL, SECAM, and NTSC), cable, electronic news gathering, electronic cinema, direct broadcast satellite, EDTV, and HDTV. Computer graphics, multimedia, and video games are also included as the new application areas. Hence, it provides a "generic solution" for video/audio coding (storage and/or transmission) worldwide. The standard is flexible enough to allow both high-performance–high-complexity and low-performance–low-complexity codec systems. The generic coding standard, thus, is designed to meet a wide spectrum of bit rates, resolutions (both spatial and temporal), quality levels, and services.

A number of applications may use the high-quality video/audio coding standard, not only existing ones but also those emerging in the future. These can be classified into three major categories. First, application to DSM can be considered. This includes the H.261 and MPEG-1 functionalities. Manipulation in a multimedia computer may fall into this category. Scalability plays an important role in this field, by which only needed pictures can be retrieved with certain qualities and resolutions. The D pictures specified in MPEG-1 that provide fast decoding and searching modes are not part of MPEG-2, as the latter can have extended scalabilities. The second category is distribution of coded audiovisual signals from a single source to multiple receivers through the transmission media. The typical media can be cable, terrestrial, or satellite channel. The third category is communication that includes videophone and videoconferencing, i.e., typical functions in H.261 (Chapter 9). Audiovisual signals are transmitted both ways between two terminals through communication networks such as B-ISDN, for which ATM is preferred.

MPEG-3, which was originally intended for EDTV and HDTV at higher bit rates, was merged with MPEG-2. Hence, there is no MPEG-3. MPEG-4, originally designed for very low–bit-rate audio/video coding (tens of Kbps), is presented in Chapter 12.

Brief history of MPEG-2

MPEG-2 activities have been set forth with the completion of MPEG-1, designed primarily for DSM. Meanwhile, the Experts Group for ATM Video Coding (AVC) was formed in 1990 to develop a video coding standard appropriate for B-ISDN using ATM transport, in the ITU-T (the ITU Telecommunication Standardization Sector, formerly CCITT), just after formulating Recommendation H.261. Because there was an overlap in technology and applications, liaison statements between the MPEG-2 and the AVC groups were exchanged and the CCITT delegates began to participate in joint sessions. Finally, they agreed to establish the same standards, i.e., ITU-T Recommendation H.262 and ISO/IEC 13818. Therefore, we also include the AVC group activities in this chapter.

Parts 1 through 3 of MPEG-2 attained international standard status in November 1994. The brief historical developments are as follows:

- July 1990: AVC (ATM Video Coding Experts Group) established in CCITT SG XV.
- Nov. 1991: Subjective tests (32 proposals) submitted [607, 937].
- May 1992: Test Model 1 published.
- Jan. 1993: Test Model 4 published [579].
- Mar. 1993: The first working draft (WD) specification issued. Technical contents of the Main Profile and Main Level were frozen.
- Apr. 1993: Test Model 5 published [583].
- July 1993: The third WD issued.
- Nov. 1993: CD 13818 adopted [591].
- Mar. 1994: DIS for Parts 2 (video) and 3 (audio) adopted.
- July 1994: DIS for Part 1 (systems) adopted.
- Nov. 1994: IS for Parts 1–3 (systems, video, and audio), Part 4 (CD for conformance testing), Part 5 (simulation software, proposed draft technical report), WD for Part 6 (digital storage media control and command), and WD for Part 9 (real-time interface) adopted.
- July 1995: IS for Part 4 (conformance testing) and Part 5 (simulation software) adopted.
- Sept. 1995: IS for Part 6 (digital storage media command and control, protocol for set-top to server and set-top to network) adopted.
- Nov. 1995: IS for Part 9 (real-time interface) adopted.
- July 1996: IS for Part 8 (10-bit audio) adopted.
- Mar. 1997: IS for Part 7 (nonbackward compatible audio) adopted.

Video coding requirements

The MPEG committee suggested the extended video/audio coding requirements for MPEG-2. A number of proposals that essentially satisfied these requirements were subjectively tested in November 1991 [494, 608]. The requirements, however, have been elaborated and enhanced subsequently. The final MPEG-2 CD is basically a fully generic system for audiovisual interactive services.

Multiple video formats can be used in MPEG-2 video coding, including 4:2:0, 4:2:2, and 4:4:4 formats (Fig. 2.3). SIF specified in MPEG-1 is also a subset of MPEG-2 as a low-level coding. Originally, MPEG-2 was intended for coding ITU-R (formerly CCIR) 601 level of pictures at less than 10 Mbps with the following objectives: 3 to 5 Mbps for standard TV quality, and 8 to 10 Mbps for performance close to ITU-R 601 [593]. The higher bit rates that have been associated with MPEG-3 have been merged in MPEG-2 since July 1992. As a consequence, MPEG-2 covers a wide range of picture formats and bit rates. It supports field or frame mode processing and an alternate scan (see Fig. 11.20 later in this chapter).

The MPEG-2 algorithm can be applied for a wide range of application fields, from low bit rate to high bit rate, from low-resolution to high-resolution, and from low picture quality to high quality. Although the algorithm can meet with these various purposes, it is too complex to integrate all the applications into a single syntax. Considering the practicality of implementing the full syntax of the specification, a limited number of subsets of the syntax are represented by means of "profile" and "level." Five profiles are defined in IS 13818-2 [591]. Furthermore, several levels are defined in a profile. Thus a number of typical applications can be defined.

Some of the prominent features in MPEG-2 are compatibility and scalability, as compared in Table 11.1 to the function of MPEG-1. The four compatibilities can be defined as shown in Fig. 11.1.

- The system is called forward compatible if a new generation decoder, such as MPEG-2, is able to decode the signal or part of the signal of an existing standard encoder, such as MPEG1 or H.261.
- The system is backward compatible if an existing standard decoder is able to decode the signal or part of the signal of the new standard encoder.
- The system is upward compatible if a higher-resolution decoder, such as MPEG-2, is able to decode the bit stream produced by a lower-resolution encoder, such as MPEG-1 or H.261.
- The system is downward compatible if a lower-resolution decoder is able to decode the bit stream or part of the bit stream produced by a higher-resolution encoder.

Table 11.1 Functional comparison between MPEG-1 and MPEG-2 video

	MPEG-1	MPEG-2
Video format	SIF progressive	SIF, 4:2:0, 4:2:2, 4:4:4, progressive/interlaced
Picture quality	VHS	Distribution/contribution
Bit rate	Variable ($\leq$ 1.856 Mbps)	Variable up to 100 Mbps
Low delay mode	$<$ 150 ms	$<$ 150 ms (no B pictures)
Accessibility	Random access	Random access/channel hopping
Scalability		SNR, spatial, temporal, simulcast, data partitioning
Compatibility		Forward, backward, upward, and downward
Transmission error	Error protection	Error resilience
Editing bit stream	Yes	Yes
DCT	Noninterlaced	Field (progressive) or frame (interlaced)
Motion estimation	Noninterlaced	Field, frame, and dual-prime based. Top (16 $\times$ 8) block and bottom (16 $\times$ 8) block.
Motion vectors	Motion vectors for P, B pictures only	Concealment motion vectors for I pictures besides MV for P & B
Scanning of DCT coefficients	Zigzag scan	Zigzag scan, alternate scan for interlaced video

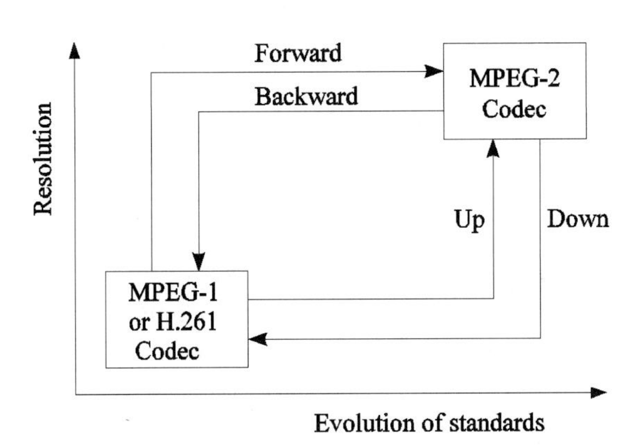

Figure 11.1 Compatibilities (forward, backward, upward, and downward).

The benefits of compatibility are cost reduction, not forcing customers to invest in a new standard decoder, and providing various quality or bit-rate services.

A scalable bit stream is defined as one having the property that part of it can be retrieved and decoded independent of the remaining bit stream. One reason

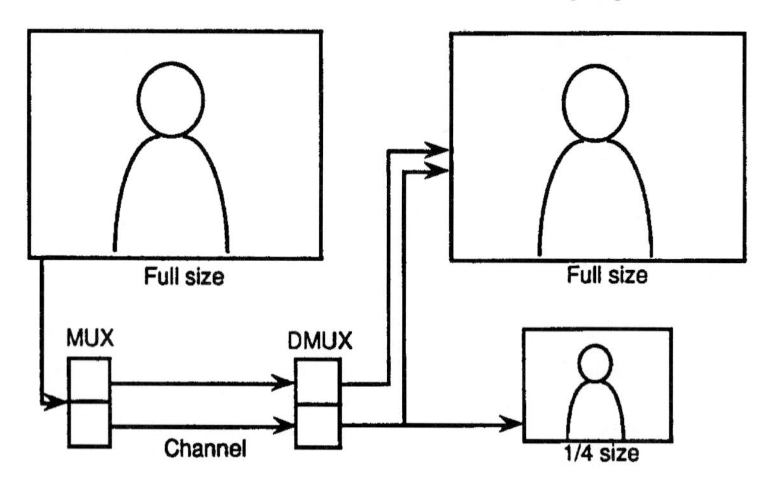

Figure 11.2 Spatial scalability with two different resolutions.

for the scalability is to permit decoders with less processing power (cheaper decoders) to display video at lower resolution and/or at lower quality. In view of this, a compatible scheme is likely to be scalable. MPEG-2 provides several different forms of scalabilities: spatial scalability, SNR scalability, temporal scalability, data partitioning, and hybrid scalability, which is a mix of these techniques. The scalability properties are applicable to various fields involving video telecommunications, database browsing, interworking of TV and HDTV, and so forth. Data partitioning is a tool intended for use in ATM networks, when two channels are available for transmission and storage. These properties give users much flexibility in the choice of video formats and the ability to provide resilience to transmission errors.

Spatial scalability includes generating two or more spatial resolution layers from a single video source, as depicted in Fig. 11.2, so that the lower layer is coded to support the basic resolution, and the enhancement layers provide the higher resolutions. SNR scalability occurs when the bit stream has different-quality layers at constant spatial resolutions. Temporal scalability involves partitioning video frames into layers, so that the lower layer is coded to provide the basic temporal rate and the enhancement layer provides full temporal resolution. The lower temporal resolution systems can generate the enhancement layer by using temporal prediction at the decoder. For data partitioning scalability, more critical parts of the bit stream (such as headers and DC and low-frequency coefficients) are transmitted in the channel with better error performance, and less critical data (such as high-frequency coefficients) are transmitted in the channel with poor error performance.

Some techniques for frequency scalability have been simulated in test models [579, 583]. Frequency scalability, however, has some drawbacks: less efficiency in field-based motion compensation, and drift effect when using multiple prediction loops [590]. Furthermore, spatial scalability can be used for implementation of different-quality services. A comparison of spatial scalability and frequency scalability is shown in [609]. In CD 13818 [591], frequency scalability has been partially adopted as a data partitioning method that generally divides DCT coefficients into different layers (channels), subject to their transmission error sensitivities.

The basic video coding algorithm for MPEG-2 does not significantly differ from that in the MPEG-1 coder as shown in Fig. 10.13, which consists of motion-compensated interframe prediction that removes the temporal redundancy. Motion-compensated interframe prediction also includes both forward and backward predictions. This algorithm has been proven effective for a wide range of picture formats and bit rates.

To improve coding efficiency the following techniques have been adopted for interlaced video coding:

- Field/frame adaptive DCT
- Field/frame adaptive motion compensation prediction
- Special prediction (dual prime (16×8) block motion compensation)

In the quantization process, a new nonuniform quantizing step size assignment for luminance or chrominance has been defined to allow finer quantization control for high-quality coding. A new VLC table has been adopted for intra pictures as well as for nonintra pictures. In addition to zigzag scanning of DCT coefficients, a new scanning method has been adopted to cope with field pictures (see Fig. 11.20 later in this chapter).

Audio coding requirements

MPEG-2 audio, ISO/IEC 13818-3, aims to support all the normative features listed in MPEG-1 audio, ISO/IEC 11172-3. It also provides extension capabilities—multichannel configurations, up to 5 audio channels (left, center, right, left surround, and right surround) plus an optional low-frequency enhancement (LFE) channel and up to seven optional multilingual channels. Therefore, MPEG-1 forms the "maximum core," and MPEG-2 audio is structured to utilize the ancillary data field of MPEG-1 to meet the additional requirements. The syntax supports PCM resolution of 16 bits and higher resolutions up to 24 bits, and sampling rates of 48, 44.1, and 32 KHz. The additional sampling rates are 24, 22.05, and 16 KHz. For each layer, 14 bit rates are defined ranging from 8 to 256 Kbps, with one free bit rate that need not be a fixed bit rate.

MPEG-2 audio provides complete backward and forward compatibilities with MPEG-1 audio. An MPEG-2 multichannel decoder correctly decodes an

Table 11.2 Functional comparison between MPEG-1 and MPEG-2 audio (LFE: low-frequency enhancement channel)

	MPEG-1	MPEG-2
Input resolution	16 bit uniform	16 bits and up to 24 bits
Sampling rate	48, 44.1, 32 KHz	48, 44.1, 32, and 24, 22.05, 16 KHz
Bit rate	Free and up to 448 Kbps	Free and up to 256 Kbps
No. of channels	Mono, stereo, dual, joint stereo	Channels 1, 2, 3, 4, 5, and LFE, up to 7 multilingual channels
Compatibility		Forward and backward
Scalability		Channels 1, 2 decodable independently

MPEG-1 mono or stereo bit stream, and an MPEG-1 stereo decoder correctly decodes an MPEG-2 multichannel bit stream. In the latter case, the MPEG-1 stereo decoder reproduces a meaningful down-mix of the original five channels. MPEG-2 audio also provides scalability in that channels 1 and 2 of the MPEG-1–compatible signal can be decoded without decoding any additional channels, as compared in Table 11.2.

For the purpose of providing signaling and controlling of multichannel audio configurations, MPEG-1 ancillary data are supported in the bit stream. It also uses the same error check as provided by MPEG-1 audio. The details are described in Section 11.3.

11.2 Video Coding

This section describes video coding topics related to MPEG-2. Valuable algorithms and techniques proposed for the subjective testing that are not necessarily part of MPEG 2 are mentioned for further study. We discuss extended video formats and coding algorithms for different picture modes. Mostly, enhanced properties from MPEG-1 are emphasized (for example, compatibility and scalability).

11.2.1 Proposed algorithms

A number of proposals were presented for the possible MPEG-2 video coding scheme, and they were required not only to provide good picture quality and to satisfy all the features of MPEG-1, but also to support additional features such as compatibility with MPEG-1, ability to generate multiresolution scales, robustness to cell loss for transmission on ATM networks, and so forth. The proposals have been registered with document nos. MPEG91/202 through MPEG91/231, or AVC-142 through AVC-168. Some of the proposals can be found in [580].

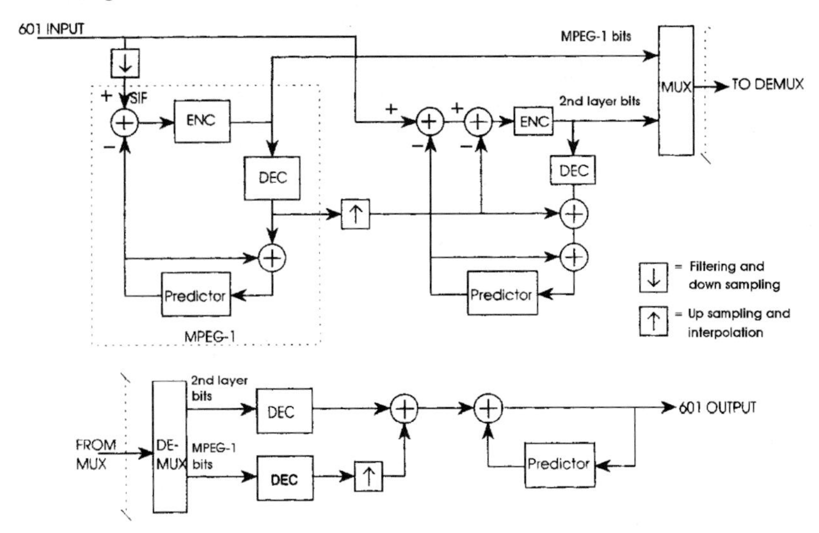

Figure 11.3 Compatible coding with embedded prediction (DEC: Decoder, ENC: Encoder) [598]. © 1993 Signal Processing.

There are several methods to achieve compatibility between the existing standard coder and the new standard coder. First, the simulcast method achieves compatibility at the service level. The required quality levels are obtained from parallel encoders operating independently and producing separate and simultaneous bit streams. This is a simple way to achieve compatibility. The total number of bits to be transmitted is the sum of the bits for each channel.

A syntactic extension method can be used for compatibility. The encoder produces only one data stream and its syntax is an extension of the existing standard. This permits upward and forward compatibilities. Downward and backward compatibilities, however, are not guaranteed if the existing standard decoder is not equipped to deal with the extended syntax of the new standard.

The embedded bit stream method is a kind of layered coding. Two different layers are treated separately but are coded in a bit stream. Syntax structures of lower and higher layers are partially the same. Downward and backward compatibilities are able to ignore the second component of the bit stream. In principle, there is no waste of bit rates since the two components contain complementary information. Therefore, several proposals presented this kind of compatibility [596, 597, 598, 602] and a block diagram is shown in Fig. 11.3, in which the prediction errors are embedded.

One of the goals of MPEG-2/H.262 is scalable extension at different quality levels. The scalability can be achieved by using the following parameters: spatial resolution, temporal resolution, and bandwidth. Spatial resolution can be used for both compatibility and scalability. The output bit stream contains

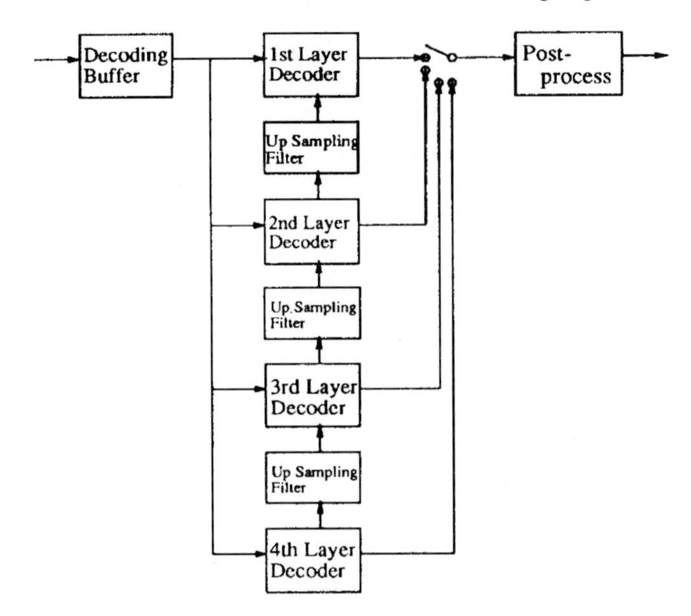

Figure 11.4 A decoder example with four-layer spatial resolution scalability [602]. © 1993 Signal Processing.

multiresolution video information. Most of the proposals adopted this spatial scalability with decimation and interpolation technique [596, 598, 602, 603] or DCT coefficient selection [594]. In the former, the encoder has as many layers as there are layers having different spatial resolutions. The decoder can select one of the layers, as shown in Fig. 11.4. This can decode a four-layered bit stream.

In the latter, (8×8) DCT coefficients are subdivided for the lower layers as depicted in Fig. 11.5. The (4×4) low-frequency coefficients are extracted for 1/4 resolution and so on. The technique is a kind of spectral selection for progressive image transmission (See Section 8.5). The same quantization matrix can be used for lower layers for simplicity.

Temporal scalability can be achieved by reducing the number of fields or frames per second with an appropriate motion compensation. For the frequency scalability, a nonrectangular block size of (16×8) is suggested in [597]. Low-frequency (8×8) and high-frequency (8×8) coefficients are transmitted on two different channels. Prediction of the odd fields is based on previous odd fields only and the SIF pictures are obtained by an (8×8) IDCT at the decoder.

Various field/frame mode decisions are reported in [595, 596]. The decision to select field versus frame coding for a macroblock can be made independent of the mode chosen for motion compensation. One technique is that line-to-line correlation is used for the field/frame mode decision. If the correlation coefficient

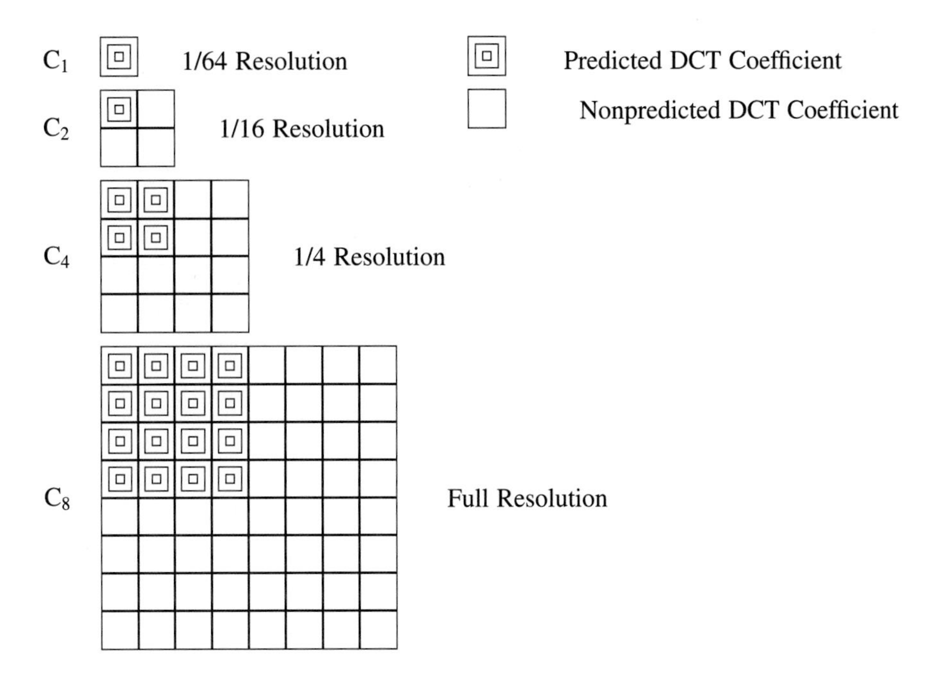

Figure 11.5 Hierarchical scales of the (8×8) DCT coefficients for decoding a four-layered bit stream [594]. © 1993 Signal Processing.

of a frame is greater than that of the fields, then frame coding mode is selected. Otherwise, field mode is chosen. Different quantization matrices are applied for field and frame modes.

Block sizes for estimating the motion vectors have been simulated using (16×16) for frame mode, (16×8) for field mode [596], and for super macroblocks of size (16×32) [595]. All the motion vectors are predicted with half-pixel resolution. The motion vectors for even fields are predicted from the previous even fields and the MVs for odd fields are predicted from the previous odd fields. Full search is generally adopted, but the hierarchical motion estimation technique is adopted in [602] (Fig. 11.6). In each layer the motion is detected using the block-matching method, with initial motion vectors at a given layer chosen as twice the result of motion estimation at the previous lower layer.

The efficiency of DC coefficient prediction for intra MBs can be increased by an adaptive process described in [596]. Two DC predictors, one for the top row of (8×8) blocks and second for the bottom row of (8×8) blocks can be employed. All DC predictors are reset at the beginning of a slice or whenever nonintra MB is encountered. In Fig. 11.7, DC coefficients are predicted between

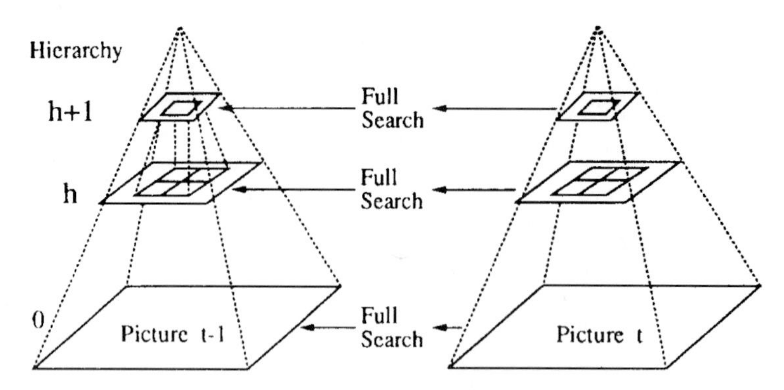

Figure 11.6 Hierarchical motion estimation [602]. © 1993 Signal Processing.

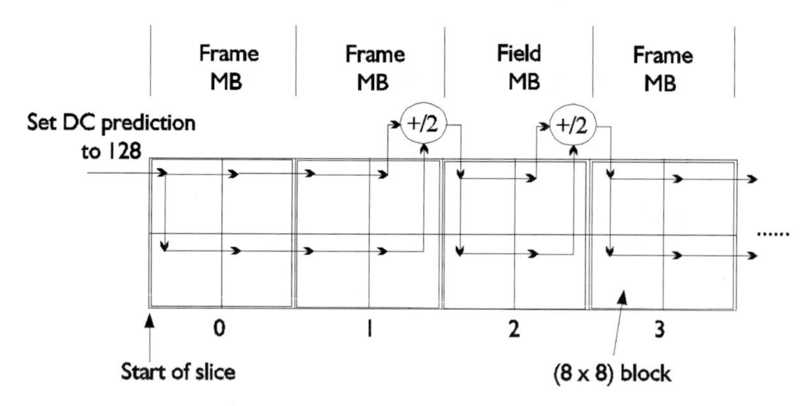

Figure 11.7 An example of DC prediction for intraframe/field luminance macroblocks [596]. © 1993 Signal Processing.

frames or frames/fields. When there is a change in the mode, an average of the DC coefficients of the top right and bottom right (8×8) blocks of the previous MB is used to predict the DC coefficient of the first block of the following MB.

In most of the video coding standards, 2D (8×8) DCT has been adopted for the correlation reduction in the spatial domain. Other techniques to achieve this have been also proposed, however. Subband coding [599] can be used, as shown in Fig. 11.8. The subband filter (analysis section) is located before prediction and quantization. The input signal is predicted in the subband domain. The field merged frames are coded in 16 subbands and the macroblock structure is not needed. The prediction errors of the subband signals are quantized and coded only. An efficient bit rate can be obtained by using different quantizers for the subbands that have different activities.

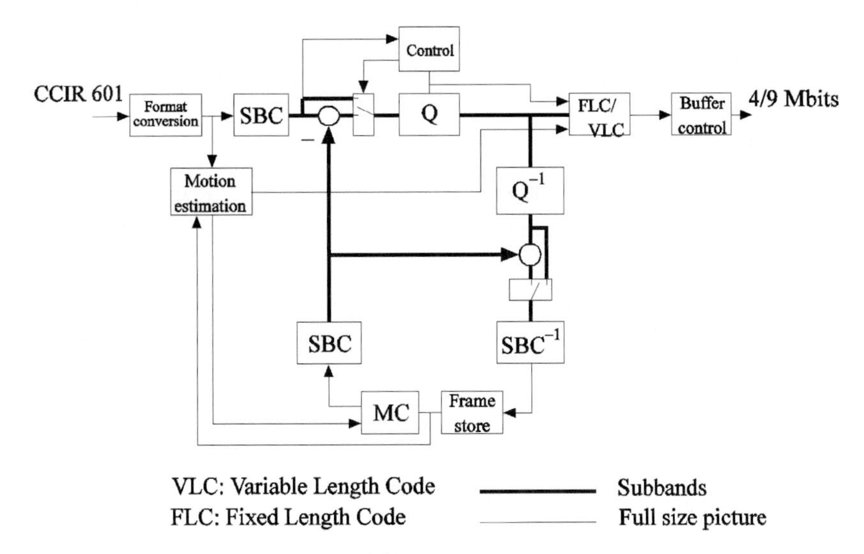

Figure 11.8 A subband-based video encoder (SBC: subband coding) [599]. © 1993 Signal Processing.

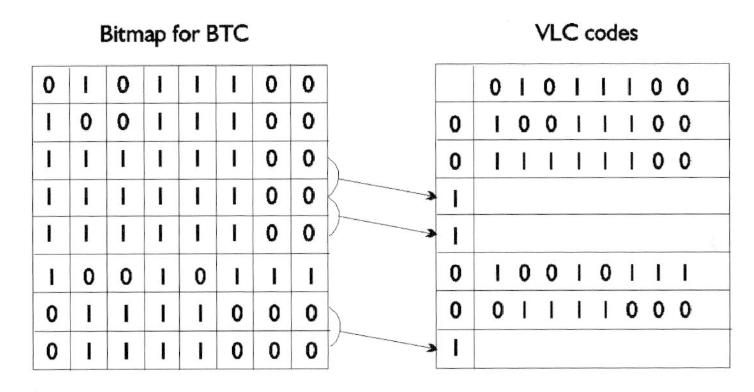

Figure 11.9 A modified BTC bi-level coding example [603]. © 1993 Signal Processing.

As a non-DCT coding technique, a modified block truncation coding (BTC) can be used for intraframes [603]. Each pixel in the MB is first truncated into one of two levels, 0 or 1 as illustrated in Fig. 11.9. Each line (row) in the MB is compared with the previous one. If the corresponding (truncated) pixel values are identical, only a 1-bit code is transmitted, resulting in bit-rate reduction. This does not improve the SNR significantly, but it can eliminate the block artifacts in regions of the video containing sharp edges or bi-level graphics. If the edge phenomenon is significant in a block, then only DPCM with a noise-shaping technique [611] can be effective, as proposed in [601]. The coding performance would be increased by adaptive quantization and adaptive VLC [600].

11.2.2 MPEG-2 video structure

This section gives a description of the various source formats and examples for their conversion from and to ITU-R 601. An extended version of the hierarchical video structure in the MPEG-1 system is discussed. Common items of the fundamental formats, CIF or SIF, and bit-stream structure are described in Sections 9.2.1 and 10.2.2, respectively.

Source formats

The original or reconstructed picture consists of three components: a luminance matrix (Y) and two chrominance matrices (C_b and C_r). The ITU-R 601 video format has been the most popular, and many applications originate in this format. The actual ITU-R 601 format consists of a luminance component that is sampled at 13.5 MHz and chrominance components that are sampled at half that of luminance. This is called a 4:2:2 format (Fig. 2.3). A number of variants from this format are possible, and they are briefly outlined as follows:

- HDTV 4:2:0, the actual HDTV signal at 50/60 Hz
- EDTV 4:2:0, HDTV with half-chrominance resolution at 50/60 Hz
- 4:2:2, the actual ITU-R 601 at 50/60 Hz
- 4:2:0, ITU-R 601 with half-chrominance resolution at 50/60 Hz
- CIF/SIF, formats in H.261/MPEG-1
- HHR, interlaced format with half-horizontal resolution of the 4:2:0
- SIF–I, interlaced format with 1/8 of the 4:2:0 resolution

We have discussed CIF and SIF, which can be derived from ITU-R 601. The number of C_b and C_r matrices in 4:2:0 format shall be one-fourth the number of the Y matrix, respectively. The Y matrix has an even number of rows and columns. The luminance and chrominance samples are positioned as shown in Fig. 11.10. In field-based processing, the chrominance samples are located in an alternative way, equally decomposed from the frame mode positioning as shown in Fig. 11.10 (compare with Fig. 9.4).

In the 4:2:2 format, the number of C_b and C_r matrices is one-half the number of Y matrices in the horizontal direction and the same as the number of Y matrices in the vertical direction. The number of luminance matrices and chrominance matrices is same in the 4:4:4 format. There is no reduction in the number of chrominance matrices. The 4:2:0 or 4:2:2 format is preferred in most of

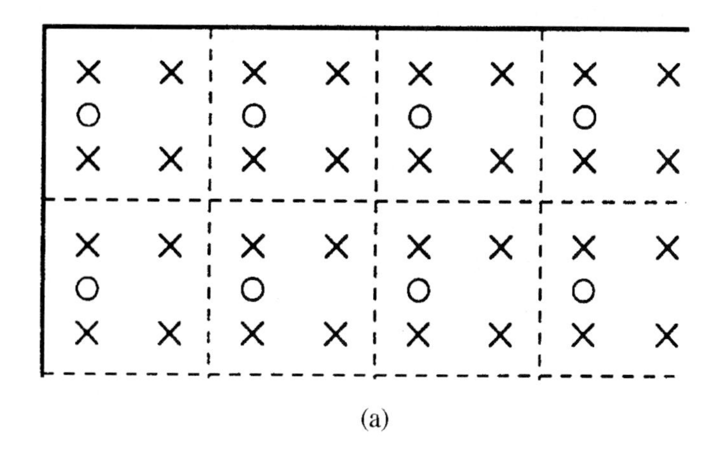

(a)

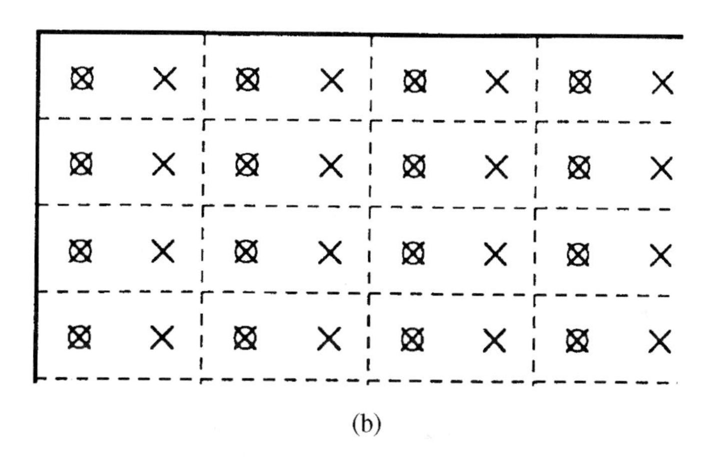

(b)

Figure 11.10 The positions of luminance and chrominance samples in (a) a 4:2:0 and (b) a 4:2:2 format (x = luminance pels, o = chrominance pels).

the applications, however. Table 11.3 shows the active parameters in 4:2:2 ITU-R 601 format and 4:2:0 format. Since human vision is less sensitive to chrominance components than to luminance components, no MPEG-2 profiles are defined for the 4:4:4 format. This may be added later on as the need arises.

A given source format has to be converted to the required format at the preprocessing stage. High-resolution formats are down-sampled with low pass filtering if they are input to the low-resolution coders. Low-resolution formats are up-sampled with low pass filtering again in order to reduce aliasing effect

Table 11.3 Adaptive 4:2:0 and 4:2:2 ITU-R 601 formats

	4:2:2 /625	4:2:0 /625	4:2:2 /525	4:2:0 /525
Number of active lines				
Luminance (Y)	576	576	480	480
Chrominance (C_b, C_r)	576	288	480	240
Number of active pels/line				
Luminance (Y)	720	720	720	720
Chrominance (C_b, C_r)	360	360	360	360
Frame rate (Hz)	25	25	30	30
Frame aspect ratio (hor:ver)	4:3	4:3	4:3	4:3

if they are to be interpolated from the low-resolution decoders. An example for NTSC/CIF conversion can be found in [413] and in Section 9.2.1. The conversion algorithms between ITU-R 601 and SIF are discussed in Section 10.2.2. Also, the ITU-R 601 format can be converted to HDTV format [624], and several HDTV formats can be converted from one another [791].

Since the MPEG-2 standard provides both progressive and interlaced formats, conversion techniques between the two formats are useful for practical applications. Progressive-to-interlaced conversion implies down-sampling in the vertical direction only. This is followed by vertical filtering. The interlaced image should be padded with zeros as shown in Fig. 11.11 to form a progressive one and then filtered using the appropriate interpolation filter. This is an example for down-converting the interlaced 60 Hz HDTV-1050 format to the progressive 30 Hz TV-525 format and up-converting it through the zero padding and interpolation techniques. Due to the filtering procedures, the reconstructed signal may not be equivalent to the input signal even though there is no coder/decoder. Interlaced-to-interlaced conversion (e.g., SIF-I to ITU-R 601, ITU-R 601 to interlaced HDTV) can be implemented similarly.

Extended bit-stream hierarchy

The MPEG-2 video bit-stream structure is a syntactic superset (extension) of the MPEG-1 structure [591, 622]. It is built on the general concepts of layered structure of GOP, motion compensation, and 2D (8×8) DCT with adaptive quantization and VLC. The MPEG-2 bit stream consists of extended layers from the basic system of MPEG-1, so that it accommodates more features than MPEG-1. First of all, it should include functions of MPEG-1, i.e., compatibility. Figure 11.12 shows the two possible paths, one is exactly following MPEG-1 procedures and the other is an extended version. Designers may select one of two ways, but the extended one represents every feature of MPEG-2 at the expense of high complexity. Therefore, for the low-resolution scheme using SIF, it is simply

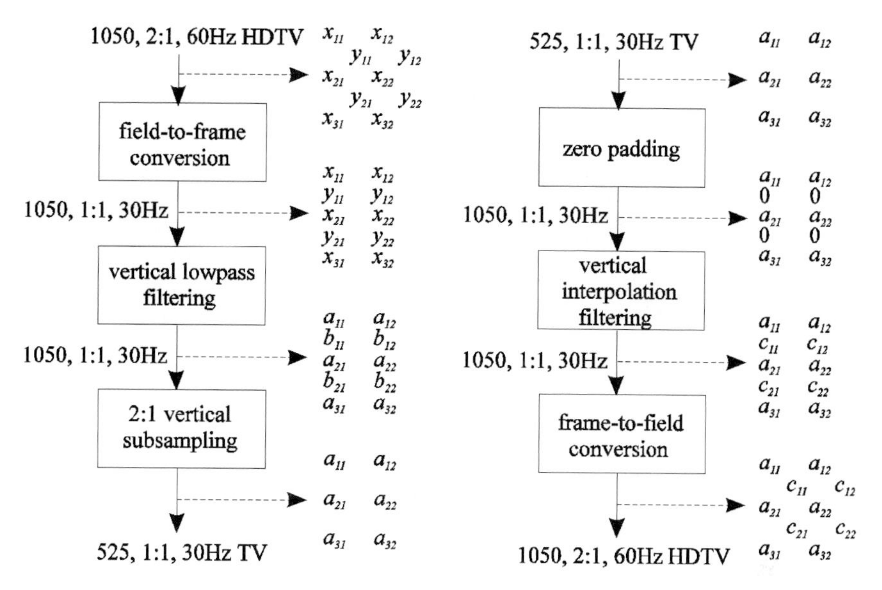

Figure 11.11 A down- and up-conversion example for interlaced HDTV and progressive TV (interlaced 2:1, progressive 1:1) [624]. © 1993 IEEE.

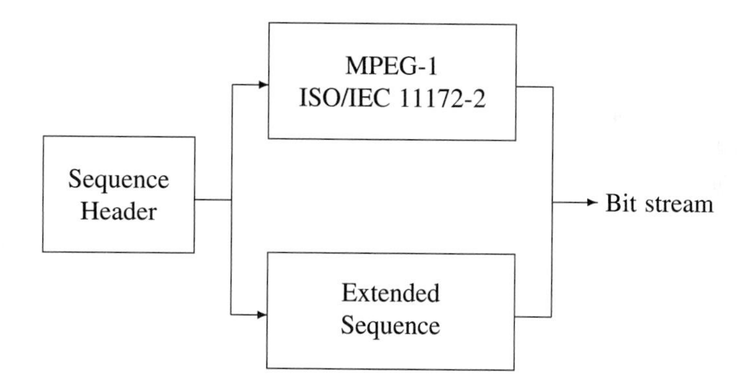

Figure 11.12 Possible routes in MPEG-2 video bit stream syntax.

implemented by the MPEG-1 algorithm. Nevertheless, MPEG-2 is able to encode and decode low-resolution video, showing better quality for fast-moving video and subjectively better quality for slow-moving video due to the frame/field adaptive coding [623].

One of the major differences between the two standards is that the MPEG-2 is capable of handling interlaced video sequences such as the ITU-R 601 format. The coding scheme can be adaptive with frame or field mode select, though

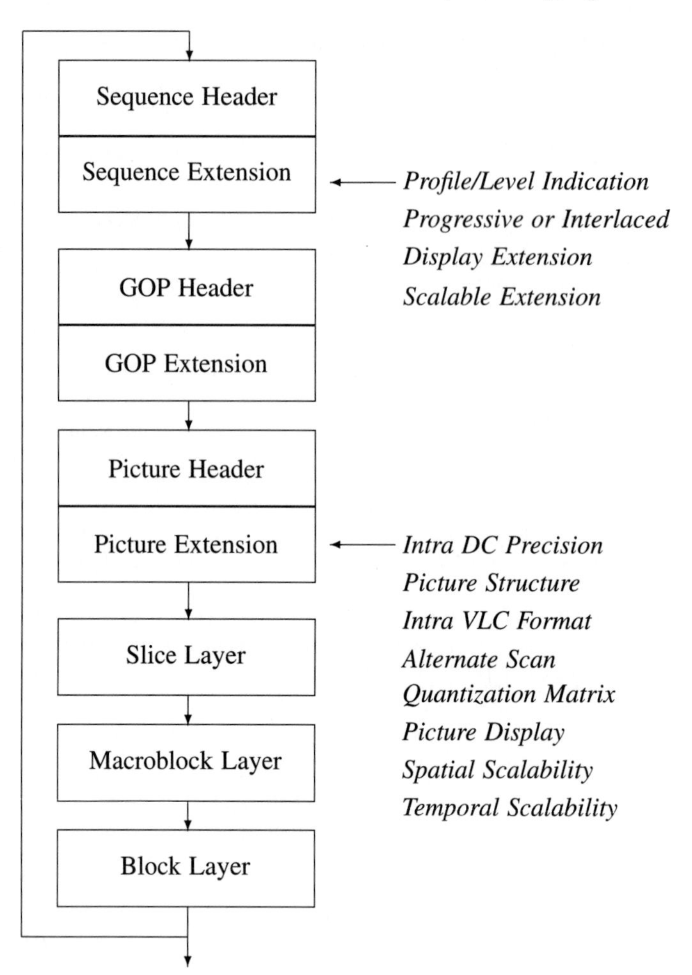

Figure 11.13 The MPEG-2 video hierarchy and extended functions.

MPEG-1 deals only with one fixed mode. The other features include scalability, compatibility, error resilience, and very high-resolution video coding.

A coded video sequence commences with a sequence header, which may optionally be followed by a sequence extension and groups of pictures. Figure 11.13 shows a simplified block diagram of the MPEG-2 video coding hierarchy. If the sequence extension is not defined (no extension start code), the following layers execute the same procedure as in MPEG-1. This is forward compatibility. If the sequence extension is defined, extended features are available for more efficient coding. The sequence extension header mainly includes the following: profile and level indication among the five profiles and four levels, progressive or interlaced indicator, chroma format indicating 4:2:0 or 4:2:2 (4:4:4 format is not defined in any of the profiles), extension of horizontal or vertical display size, and scalable extension representing the scalable mode among the 4 possible modes and the subsequent parameters.

The group-of-pictures (GOP) header has functions similar to the MPEG-1 header and no special functions are needed in the GOP extension header. The important parameters for coding extended pictures are defined in the picture coding extension header. Precision of the intra DC coefficient may be extended up to 11 bits. Since both interlaced and progressive pictures are available, picture structure indicates top or bottom fields, or frame picture. For intra macroblock coding, an alternate VLC table is available (selecting intra VLC format), whereas in MPEG-1, the same VLC table is used for intra and non-intra modes. Also, alternate scanning (see Fig. 11.20b) of DCT coefficients is available. It is especially efficient for field picture coding that is less correlated in the vertical direction.

Quantization matrices can be updated even though each quantization matrix has a default matrix. For 4:2:0 data, only two matrices are used, one for intra and the other for nonintra blocks. For 4:2:2 data, four matrices can be used, not only for luminance but also for chrominance coefficients. The display size on the screen can be varied by setting horizontal and vertical offsets. For spatially scalable coding, the horizontal and vertical sizes of the lower layer are informed in units of the enhancement layer. Also, field or frame, interlaced or progressive information is indicated. In temporal scalable extension, a forward/backward temporal reference picture in the lower layer is indicated.

The three layers below the slice layer are not significantly affected by picture coding extension. Slices cover all macroblocks in a picture, the so-called restricted slice structure as shown in Fig. 11.14. A general slice structure that does not need to encode certain areas in a picture (no slices exist in these areas) is also a part of MPEG-2.

The main difference in coding the macroblock from MPEG-1 is frame/field-based motion estimation. Assuming the significant pel area of ITU-R 601, the number of (16×16) macroblocks is (44×30) in the 525/60 system and (44×36) in the 625/50 system (see Table 11.3). A smaller size macroblock (16×8) in the field mode is available. In addition, dual-prime motion estimation is included. We will discuss these further in the subsequent sections.

Although MPEG-2 is intended to become a generic standard that is able to meet the requirements of every possible application, the ISO/IEC JTC has decided to introduce the *profiles* and *levels* approach to meet a wide spectrum of applications, such as videophone, videoconferencing, TV, and HDTV. Profiles and levels are used to define a limited number of subsets to promote interworking among different applications. Several functional requirements for these applications are addressed by the concept of profiles. In CD 13818-2 [591], MPEG has specified five profiles: the simple profile (SP), the main profile (MP), the SNR pro-

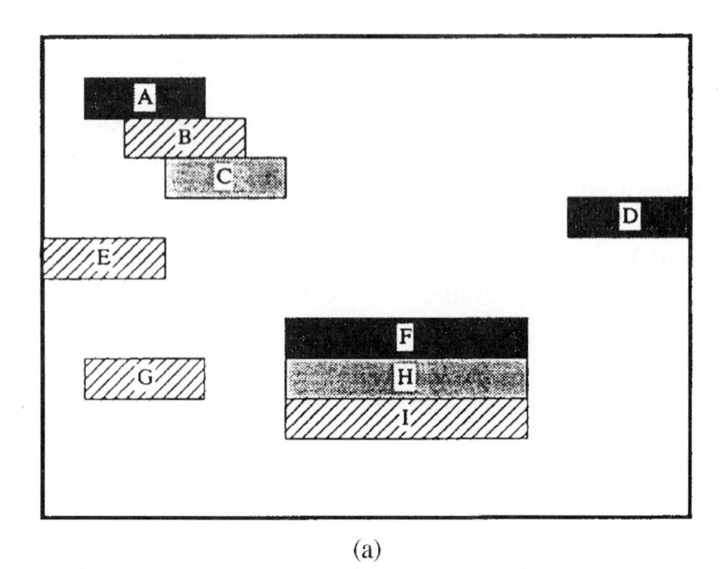

(a)

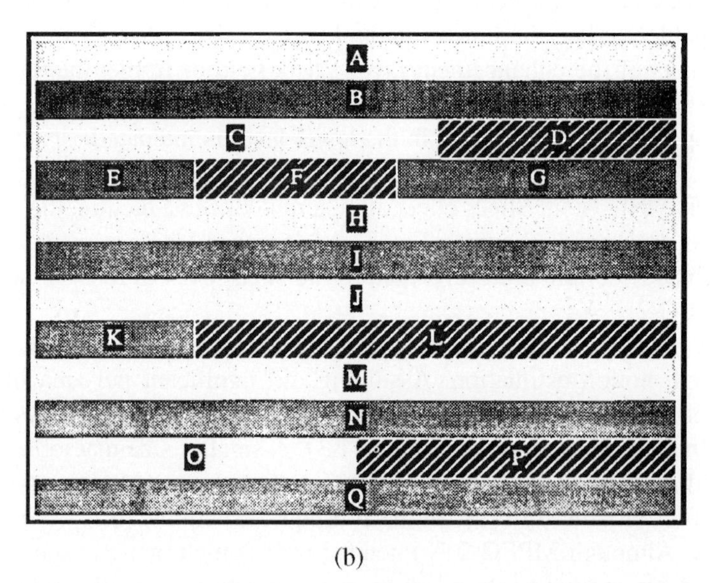

(b)

Figure 11.14 Possible slice structures. The entire picture need not be enclosed by slices in (a) the general structure, but it needs to be enclosed in (b) the restricted structure [592]. © 1995 ITU-T.

Table 11.4 Upperbound parameters in profiles and levels (VBV: video buffering verifier)

Profile	Level	H.size (pels)	V.size (pels)	Frame rate (Hz)	Bit rate (Mbps)	VBV size (Mbits)	MV range (pels)
Simple	Main	720	576	30	15	1.835	−128~ 127.5
Main	Low	352	288	30	4	0.489	−64~ 63.5
	Main	720	576	30	15	1.835	−128~ 127.5
	High 1440	1440	1152	60	60	7.340	−128~ 127.5
	High	1920	1152	60	80	9.787	−128~ 127.5
SNR scalable	Low	352	288	30	3 (4)	0.367 (0.489)	−64~ 63.5
	Main	720	576	30	10 (15)	1.223 (1.835)	−128~ 127.5
Spatially scalable	High 1440	720 (1440)	576 (1152)	30 (60)	15 (40) (60)	1.835 (4.893) (7.340)	−128~ 127.5
High	Main	352 (720)	288 (576)	30 (30)	4 (15) (20)	0.489 (1.835) (2.447)	−128~ 127.5
	High 1440	720 (1440)	576 (1152)	30 (60)	20 (60) (80)	2.447 (7.340) (9.786)	−128~ 127.5
	High	960 (1920)	576 (1152)	30 (60)	25 (80) (100)	3.036 (9.787) (12.233)	−128~ 127.5

Note: Numbers in parentheses refer to the enhanced layers.

file, the spatially scalable profile, and the high profile (HP). Each profile can be then classified into four levels: the low level (LL), the main level (ML), the high 1440 level (H-14), and the high level (HL). The principal parameters in these profiles and levels are summarized in Table 11.4.

The profiles and levels are forward compatible. Since the main profile is a syntactic superset of the simple profile, a main profile decoder should be able to decode bit streams generated by a simple profile encoder. In Table 11.5, the high

Table 11.5 Forward compatibility between different profiles and levels [592]

Profile & Level indication in bitstream	Decoder										
	HP @ HL	HP @ H-14	HP @ ML	Spatial @ H-14	SNR @ ML	SNR @ LL	MP @ HL	MP @ H-14	MP @ ML	MP @ LL	SP @ ML
HP@HL	o										
HP@H-14	o	o									
HP@ML	o	o	o								
Spatial@H-14	o	o		o							
SNR@ML	o	o	o	o	o						
SNR@LL	o	o	o	o	o	o					
MP@HL	o						o				
MP@H-14	o	o		o			o	o			
MP@ML	o	o	o	o	o		o	o	o		
MP@LL	o	o	o	o	o	o	o	o	o	o	o*
SP@ML	o	o	o	o	o		o	o	o		o
ISO/IEC 11172	o	o	o	o	o	o	o	o	o	o	o

o indicates the decoder shall be able to decode the bitstream including all relevant lower layers.
* Note that SP@ML decoders are required to decode MP@LL bitstreams.

Source:© 1995 ITU-T.

profile–high level decoder can decode all profiles and levels. The simple profile aims at ITU-R 601 video directly, i.e., there are no B pictures. The main profile deals with four different application fields, from SIF in MPEG-1 to HDTV format. All features in MPEG-2 are applicable in this profile with the exception of scalability.

The other three profiles are available for designing scalable coders. MPEG-2 provides basically four scalability modes. Only spatial scalability and SNR scalability are currently included in the CD [591]. Data partitioning and temporal scalability are also standardized coding tools (not yet included in any of the defined profiles). The ITU-T SG-15 ATM Video Coding Experts Group has more interest in data partitioning scalability as a candidate to cope with ATM cell losses.

Obviously other profiles/levels and scalabilities may be needed in the future. The MPEG-2 standard will include them in future amendments, keeping up with the developments in technology and consumer electronics. In the market, products interworking between different formats or functions may appear in the near future.

11.2.3 Coding I, P, and B pictures

In this section, the coding techniques for various picture modes are discussed. Compared with the MPEG-1 algorithm, in MPEG-2, the encoder and decoder perform more complicated operations, including frame and field processing and motion estimation/compensation.

Motion estimation and compensation

In the motion estimation process, motion vectors for predicted and interpolated pictures (Fig. 10.8) are coded differentially between macroblocks. Basically, the two motion vector components, the horizontal one first, followed by the vertical one, are coded independently. In B pictures, four motion vector components corresponding to forward and backward MV are transmitted. The prediction MV is set to zero in the macroblocks at the start of a slice or if the last MB was coded in intra mode.

The motion compensation process forms predictions from previously decoded pictures, using the motion vectors that are of integer or half-pel resolution. For the half-pel resolution, decoded motion vectors are linearly interpolated in the previous or future picture.

In the case of I picture or intra coded macroblocks, no prediction (no MV) is formed. They may, however, carry motion vectors known as *concealment vectors*, which are intended for use when the bit stream errors prevent the decoding of coefficient information and motion vectors. If concealment is used in an I picture, then the decoder performs prediction similar to a P picture.

Predictions for motion compensation are formed by predicting pels in the current MB from the reference fields or frames. A positive value of horizontal or vertical component of a MV indicates that the prediction is made from pels to the right or below in the reference field/frame, respectively.

Figure 11.15 depicts the motion compensation process. The input bit stream carries differential motion vectors coded by VLC. Dual-prime arithmetic yields motion vector refinements. Motion vectors for color components are obtained from the MV for the luminance as follows:

$$\begin{aligned}
\text{For 4:2:0,}\quad & MV(hor)_{chro} = MV(hor)_{luma}/2 \\
& MV(ver)_{chro} = MV(ver)_{luma}/2 \\
\text{For 4:2:2,}\quad & MV(hor)_{chro} = MV(hor)_{luma}/2 \\
& MV(ver)_{chro} = MV(ver)_{luma}
\end{aligned} \tag{11.1}$$

Reference pictures, (forward or backward, field or frame based) are referred for the motion compensation with integer or half-pel resolution. Motion compensation is carried out both in the encoder and in the decoder. In both cases, the IDCT output signal $f[y][x]$ and combined predictions $p[y][x]$ from the reference pictures are added to form the decoded compensated picture $d[y][x]$ whose pel intensities are clipped to 0 to 255. If a block is not coded, then $f[y][x] = 0$, and $d[y][x] = p[y][x]$. If a MB is intra coded, then $p[y][x] = 0$, and $d[y][x] = f[y][x]$.

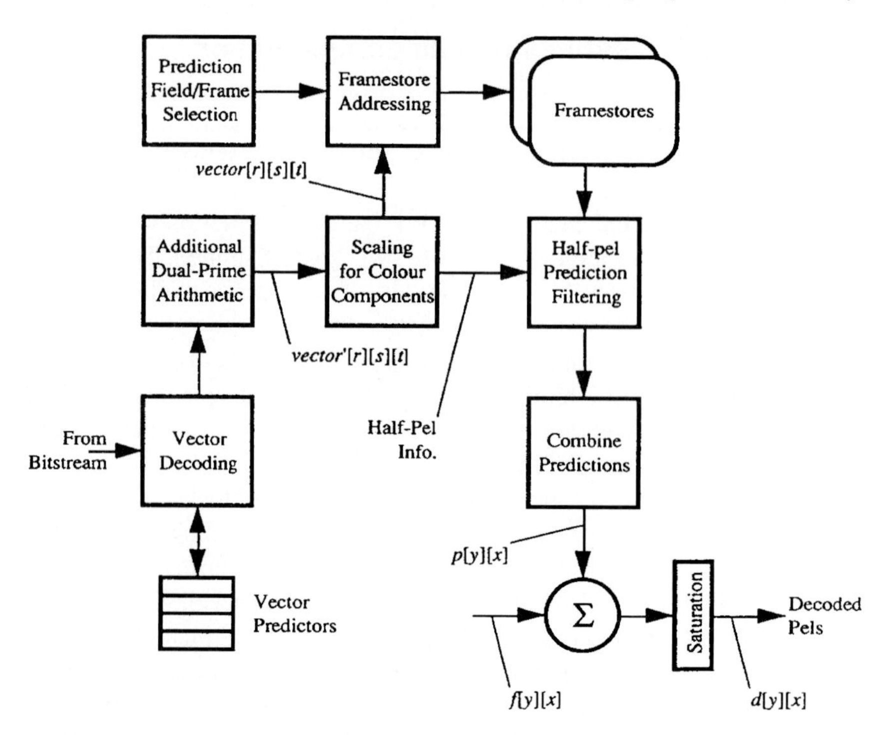

Figure 11.15 The motion compensation process in MPEG-2 [592]. © 1995 ITU-T.

Motion vectors are coded differentially with respect to previously decoded motion vectors in order to reduce the number of bits required to represent them. Each differential MV component is coded by using a VLC for the allowable vector range and residual value, as discussed in Example 10.5. The MV components are decoded as follows:

$$delta = ((\text{Abs(motion code)} - 1) \times f) + \text{motion residual} + 1 \qquad (11.2)$$

where f is derived from the MV range (Table 10.13) divided by 32. For example, the value of f is 8 if the MV range is 256 with the *f-code* of 4. This value, *delta*, is added to the predicted MV (previously decoded) to get the final reconstructed MV. When predicting the vertical component of MV with field format in frame pictures, scaling is necessary.

There are four prediction modes in MPEG-2 since it also provides field-based processing, while MPEG-1 provides only progressive-based (noninterlaced) prediction. The four modes are given by

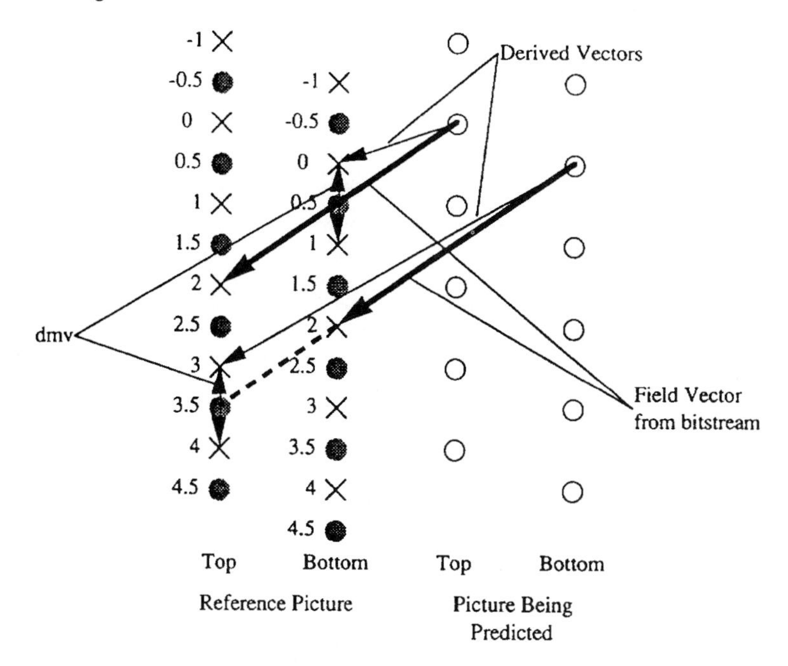

Figure 11.16 Vertical motion vector scaling for dual-prime prediction [592]. © 1995 ITU-T.

- *Field prediction*: Predictions are made independently for each field by using data from one or more previously decoded fields. A frame is composed of two fields (top and bottom) in typical applications. The top field has parity zero and the bottom field has parity one. Prediction can be made between the same-parity fields or the opposite-parity fields.
- *Frame prediction*: Predictions are made from one or more previously decoded frames. Even in a frame picture, field predictions can be selected on an MB-by-MB basis.
- (16×8) *motion compensation*: Two motion vectors are used for each MB. The first motion vector is applied to the upper 16×8 region and the second to the lower (16×8) region. This method can be used with field pictures only.
- *Dual-prime prediction*: This is a motion vector refinement process in P pictures only. Predictions are made from two reference fields and are averaged to form the final prediction.

Dual-prime prediction (Fig. 11.16) reduces the coding overhead associated with field motion type. It provides higher prediction efficiency, assuming a linear motion trajectory like panning. The four motion vector components (horizontal and vertical forward motion vector components for the two fields) are decoded as described earlier.

Predictions between the same-parity fields are normally implemented. It is not necessary to scale the motion vectors. To form a MV for the opposite parity (top-to-bottom, bottom-to-top), the decoded same-parity–based motion vectors are scaled to reflect the different temporal distance between the fields. A correction is made to the vertical component, reflecting the vertical shift between the opposite-parity fields. Then the differential motion vector decoded from the bit stream is added to correct the error due to scaling.

The vertical correction, c_v, has the value -1 for the top-to-bottom prediction, and the value $+1$ for the bottom-to-top prediction. A small differential motion vector *dmv* takes one of the values -1, 0, or $+1$. The field distance d_f is 1 for the top-to-bottom and 3 for the bottom-to-top, assuming the top field is first. The MV multiplied by d_f is divided by 2 with nearest-integer arithmetic as follows:

$$\begin{aligned} MV_{hor} &= ((mv_{hor} \times d_f)//2) + dmv_{hor} \\ MV_{ver} &= ((mv_{ver} \times d_f)//2) + c_v + dmv_{ver} \end{aligned} \tag{11.3}$$

Example 11.1 *For the dual-prime prediction shown in Fig. 11.16, given that the vertical top field motion vector, $MV_{ver,t}$ is 4 in half-pel units (2 in integer pel units in the figure) and the same for the bottom field, and dmv_{ver} is -1, we obtain the top-to-bottom prediction, $MV_{ver,tb}$ and the bottom-to-top prediction, $MV_{ver,bt}$ as follows:*

$$\begin{aligned} MV_{ver,tb} &= (4//2 - 1) - 1 = 0 \\ MV_{ver,bt} &= ((4 \times 3)//2 + 1) - 1 = 6 \end{aligned}$$

Similarly, the horizontal component of the MV can be predicted by using Eq. 11.3. Final prediction is formed by the nearest integer average of the same-parity prediction and the opposite parity-prediction.

Mode selection

In MPEG-2, there are basically three picture coding modes, I, P, and B pictures, as in MPEG-1. Although in principle, freedom could be allowed for choosing one of the three modes for any specific picture, predetermined periodic structure is frequently used [579, 583]. Typical formats of the group of pictures are $N = 12$, $M = 3$ for the 25 Hz, and $N = 15$, $M = 3$ for the 30 Hz system, as discussed in Example 10.2.

Macroblock types that will be VLC coded are determined for all picture modes. In I pictures, intra mode and intra with modified quantizer mode are used. In P pictures, there are some modes to be selected: MC/no MC, coded/not coded, intra/inter, and quantizer modification or not. The way to determine the modes is similar to that of MPEG-1 (Fig. 10.21). As before, the MPEG-2 does not specify how to make these decisions. In B pictures, interpolative/forward/backward decisions are added. For the scalability in the three picture modes, the decision compatible or not is included.

The decision as to field or frame processing does not directly appear in the MB type table. It can, however, be analyzed with the MC/no MC decision. DCT coding mode is changed by the selection. For P pictures, the decision of selecting the frame-based MV or field-based MV is made by the following error signal (ES) comparison:

$$ES_{frame} \leq (ES_{field1} + ES_{field2})$$
$$ES_{MC} < ES_{No\ MC} \tag{11.4}$$

where the error signal arises from the motion compensation. In Eq. (11.4), if the conditions are true, then the picture is frame-based and MC mode.

Field-based or frame-based DCT coding is used based on the following algorithm. If the field-based variance is smaller than the frame-based one, then the MB is coded by using the field-based DCT.

```
  var1 = var2 = 0.;
  for (pix=0; pix<16; pix++){
  for (line=0; line<16-2; line+=2){
    diff1 = x(pix,line) - x(pix,line+1);
    diff2 = x(pix,line+1) - x(pix,line+2);
    var1+ = (diff1*diff1) + (diff2*diff2);
/* frame-based variance */
    diff1 = x(pix,line) - x(pix,line+2);
    diff2 = x(pix,line+1) - x(pix,line+3);
    var2+ = (diff1*diff1) + (diff2*diff2);
/* field-based variance */
      }
  }
```

The intra/inter decision is based on the comparison of signal variance and difference signal variance as described in Fig. 9.9. The MC/no MC decision can also be found in Fig. 9.12. The compatible or not compatible MB decision is based on the lowest MSE criterion. Quantizer modification is based on the output buffer status as discussed in MPEG-1.

Encoding and decoding process

The basic coding scheme of MPEG-2 is similar to that of MPEG-1. It uses DCT and motion-compensated interframe coding and supports I, P, and B picture coding types (D pictures are not supported). Therefore, the block diagram shown in Fig. 10.13 is sufficient to understand MPEG-2 data flow. Minor changes of parameters and modes are required to guarantee the extended coding. There are also several new blocks, such as the frame/field coding analyzer, the frame/field formatter and deformatter, and the motion type analyzer.

The input frames are reordered at first in order to form the bit stream (coding) order. Motion-compensated prediction is differenced from input frames, and the resulting error signal is formatted as field or frame units determined in the frame/field coding analyzer. These units go through the DCT, quantization and VLC coding. In the decoder, the reverse process is the inverse frame/field formatter as shown in Fig. 11.17. The inverse transformed signal in the encoder is also inverse frame/field formatted. The extended quantizer table and quantizer scale can be applied based on intra/inter mode. Alternate scanning order is available for field mode coefficients (see Fig. 11.20). For VLC coding of the scanned coefficients, a new VLC table for intra coefficients is defined. The motion analyzer determines the best prediction modes among the frame, field, (16×8), and dual-prime predictions. In the decoder, motion compensation is performed with MV and motion type, determined by the motion analyzer in the encoder.

Figure 11.17 is an example for the main profile decoder that supports not only frame pictures but also field pictures. The operation of the decoder is similar to that of the local decoding loop in the encoder. The signal from the variable- and fixed-length decoder is inverse scanned and dequantized. The macro-block declassifier decodes MB type code and gives CBP, MV, inter/intra, and quantizer scale. The decoded frames are reordered to normal display order before being displayed on the monitor.

Transform and quantization

The (8×8) blocks are transformed with a 2D-DCT as in the MPEG-1 coder. Each block of (8×8) pels thus results in a block of (8×8) transform coefficients. Then the coefficients are quantized by using a default or modified matrix. Since DCT is more efficient for highly correlated sources, intra mode picture can be more efficiently compressed than inter mode, with or without motion compensation. If an MB is not skipped and the coded block pattern is 1 for a given block, FDCT and IDCT are performed in the encoder and decoder, respectively. Reconstructed data are clipped to the range from -255 to $+255$.

In frame pictures, both frame and field DCT coding may be used, subject to the mode decision, before actually performing the DCT. This decision coincides

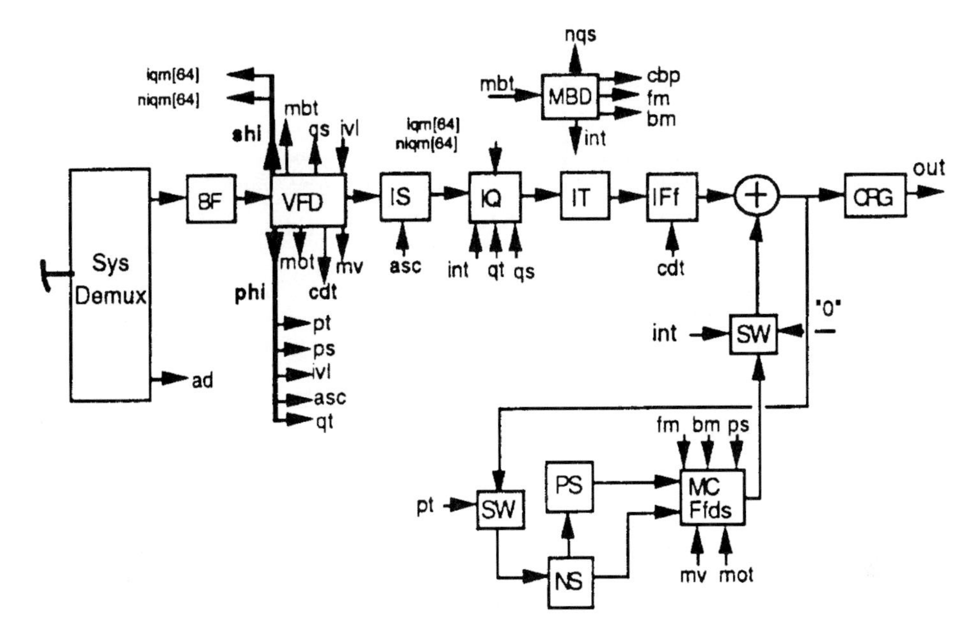

BF: buffer; **VFD**: variable- and fixed-length decoder; **IS**: inverse scanner; **IQ**: inverse quantizer; **IT**: IDCT; **IFf**: inverse frame/field formatter; **SW**: intra/inter or picture-type switching; **PS**: previous picture store; **NS**: next picture store; **MC**: motion compensation; **MBD**: macroblock declassifier; **ORG**: frame reorganizer; **ad**: bit stream to audio decoder; **iqm**: intra quantizer matrix; **niqm**: nonintra quantizer matrix; **shi**: sequence header information; **mbt**: macroblock type; **qs**: quantizer scale; **ivl**: intra vlc type; **mot**: motion type (frame, field, (16×8), dual); **cdt**: coding type (frame, field); **phi**: picture header information; **pt**: picture type (I, P, B); **ps**: picture structure (frame, field); **asc**: alternate scan; **qt**: quantizer table; **int**: intra mode; **nqs**: new quantizer scale; **fm**: forward motion; **bm**: backward motion; **mv**: motion vector.

Figure 11.17 Video decoder (main profile) block diagram [622]. © 1993 SPIE.

with the frame/field motion compensation. Figure 11.18 shows frame/field block organization for DCT in frame pictures. Field-based blocks consist of pels of either the first field only (white area) or the second field only (dotted area). If correlation is higher between the two fields of a frame, then the field-based blocks are used.

Each quantization matrix has a default matrix. The default matrix for intra blocks (both luminance and chrominance) is given in Table 10.6 and the default matrix for nonintra blocks (both luminance and chrominance) has all its components = 16. User-defined matrices may be downloaded and can occur in the sequence header or in the quant matrix extension header. For the 4:2:0 format, two matrices, one for intra and one for nonintra, are used. For 4:2:2 or 4:4:4

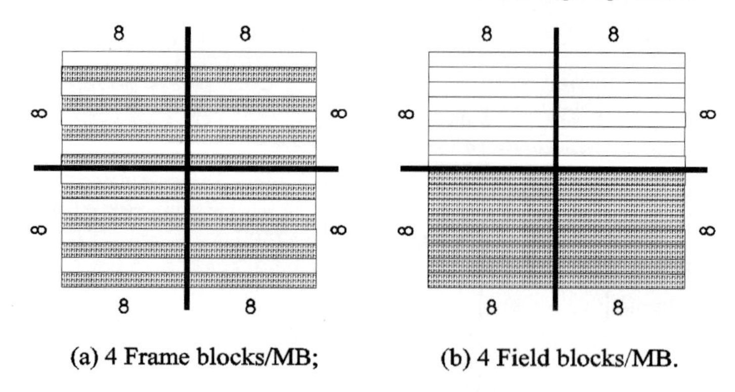

Figure 11.18 Frame/field block organization for (8×8) DCT in frame pictures.

formats, four matrices (intra or nonintra for luminance or chrominance blocks) may be used. Thus, different quantization matrices can be used for luminance and chrominance data.

Quantization of DC coefficients: The quantizer step size for the DC coefficient of the luminance and chrominance components is 8, 4, 2, and 1, according to the intra DC precision of 8, 9, 10, and 11 bits. (The dynamic range of DCT coefficients for 8-bit precision input signals goes to 11 bits.) A typical source (video) precision is 8 bits and the step size is 8 as in MPEG-1. The DC coefficients are divided by the given step size for quantization and multiplied by the same step size for dequantization

$$\hat{S}_{dc} = step\ size \times S_{qdc} \tag{11.5}$$

Quantization of AC coefficients: The corresponding header information to the quantizer scale factors are the quantizer scale type in the picture coding extension header and the quantizer scale code in the slice and MB headers. If the scale type = 0, then the quantizer scale factors are same as those in MPEG-1. The MPEG-1 quantizer scale is linear in range and can take values from 1 to 31. While this dynamic range is sufficient for lower-resolution video, it may be a limitation for higher-resolution video.

If the scale type = 1, then the quantizer scale is designed to be nonlinear and covers a larger range, between 0.5 and 56. It is more precise for small data values as it increases in units of 0.5, and is much coarser for larger data values as illustrated in Fig. 11.19. To minimize problems for implementation due to 0.5 precision, the quantization scales are multiplied by 2 to generate the step sizes (y-axis in the figure) and divided by 2 as given in Eq. (11.7).

One way to quantize the DCT coefficients in the encoder is given as follows:

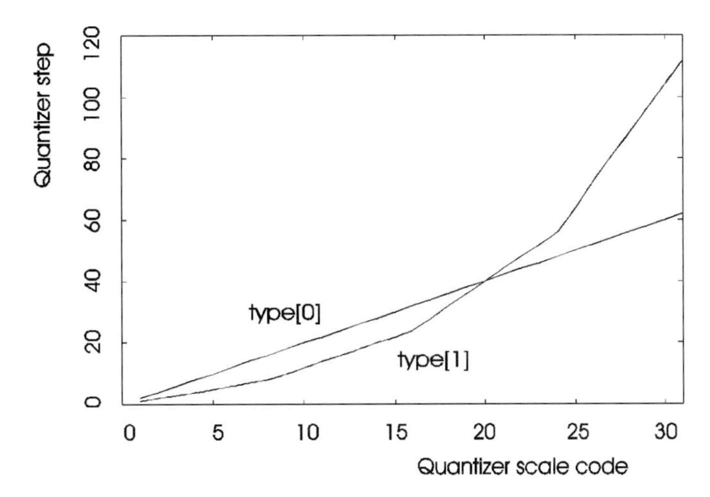

Figure 11.19 Linear and nonlinear quantizer step sizes (type[0]: linear, type[1]: nonlinear).

$$S_{quv} = ((32 \times S_{uv})//Q_{uv} + k)/(2 \times qs) \tag{11.6}$$

where $k = 0$ for intra blocks $Sign(S_{quv})$ for nonintra blocks, and qs denotes the quantizer scale code for the linear or nonlinear quantizer step size as shown in Fig. 11.19 and Table 8.8. Readers may compare this with Eq. (10.3) for MPEG-1 quantization. The operator / means an integer division with truncation of the result toward zero and the operator // means an integer division with rounding to the nearest integer.

The following equation specifies the arithmetic to reconstruct $\hat{S}_{uv}$ from S_{quv} for all coefficients (intra AC and inter DC/AC) except the intra DC coefficient:

$$\hat{S}_{uv} = ((2S_{quv} + k) \times Q_{uv} \times qs)/32 \tag{11.7}$$

DCT coefficients are inverse transformed in the encoder for the motion-compensated prediction and in the decoder for data reconstruction. We discussed the IDCT mismatch problem in Section 9.3.1, where the mismatch error criteria are described. In the MPEG-2 decoder, another solution to this problem is suggested: the sum of all DCT coefficients in a block should be odd in order to avoid all-zero outputs of IDCT. The largest frequency coefficient $\hat{S}_q[7][7]$ is manipulated by adding or subtracting as follows:

$$\hat{S}_q[7][7] = \begin{cases} \hat{S}_q[7][7], & \text{if } \hat{S}_q[7][7] \text{ is odd} \\ \hat{S}_q[7][7] - \text{Sign}(\hat{S}_q[7][7]), & \text{if } \hat{S}_q[7][7] \text{ is even} \end{cases}$$

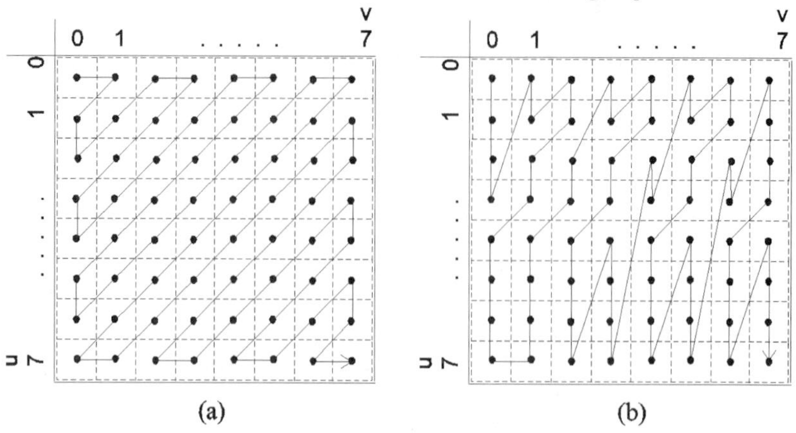

Figure 11.20 Two scanning methods of the DCT coefficients in MPEG-2. (a) Zigzag scan (scan[0][u][v]). (b) Alternate scan (scan[1][u][v]). © 1995 ITU-T.

Scanning and VLC coding

Quantized coefficients are scanned and converted to a one-dimensional array. Two scanning methods are given: zigzag scan, which is typical for progressive (noninterlaced) mode processing, and alternate scan, which is more efficient for interlaced format video. Figure 11.20 shows the two methods. The structure of alternate scan seems like vertical scan, since the correlation along the horizontal direction is higher than along the vertical direction.

Intra DC coding: After the DC coefficient of a block has been quantized as stated earlier, it is coded losslessly by a DPCM technique. Coding of the blocks within a macroblock follows the normal scan as discussed in Fig. 10.20, i.e., the DC value of block 3 becomes the DC predictor for block 0 of the following macroblock. Three independent predictors are used for the luminance and two chrominance components. For 4:2:2 and 4:4:4 formats, the number of chrominance blocks are different, as shown in Fig. 11.21 (compare with Fig. 10.20). The order of numbered blocks indicates the corresponding spatial position and does not imply the order of DC prediction.

There are several cases for which the DC predictors are reset to 128, 256, 512, or 1024 according to the intra DC precision 8, 9, 10, or 11 bits, respectively: at the start of a slice, whenever a nonintra MB is decoded, and whenever a MB is skipped (the MBA increment is greater than 1). For the remaining MBs, the DC prediction is simply the previously coded DC value of the same type of components.

The differential DC values are categorized and VLC coded according to their sizes (using the same table as shown in Table 10.7 but extended up to size 11).

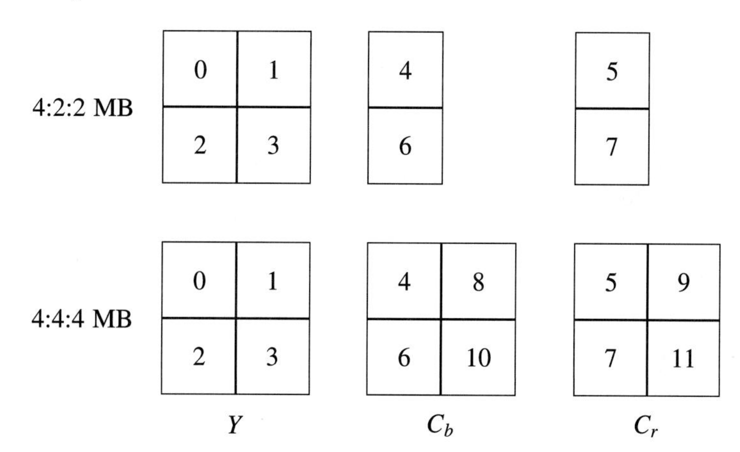

Figure 11.21 Macroblock structure in 4:2:2 and 4:4:4 formats.

Next, for each category, enough additional bits are appended to the size code to uniquely identify which difference in that category actually occurred. (Table 10.8 also needs to be extended as was Table 10.7.) If we design an 8-bit precision coder, however, the intra DC coding procedure is the same as in MPEG-1.

At the decoder, the differential value is added to the predicted value. Once the size category is decoded by the VLC table, the half dynamic range within the size is determined. The differential value is obtained as follows:

```
if (DC differential = half range)
  differential = DC differential;
else
  differential = (DC differential + 1) - (2 * half range);
```

This is because the additional code for the negative value in Table 10.8 is read as positive but less than half the range.

Example 11.2 *Given the quantized DC value* 3 *and the prediction* 143, *then the differential value is obtained as* $3 - 143 = -140$, *and it is encoded using Tables 10.7 and 10.8 as*

1111 110 / 0111 0011

assuming the 8*-bit precision luminance. The decoded differential value is calculated as*

$$\textit{differential} = (115 + 1) - (2 \times 128) = -140 \qquad (11.8)$$

The quantized DC value is reconstructed as $143 + (-140) = 3$.

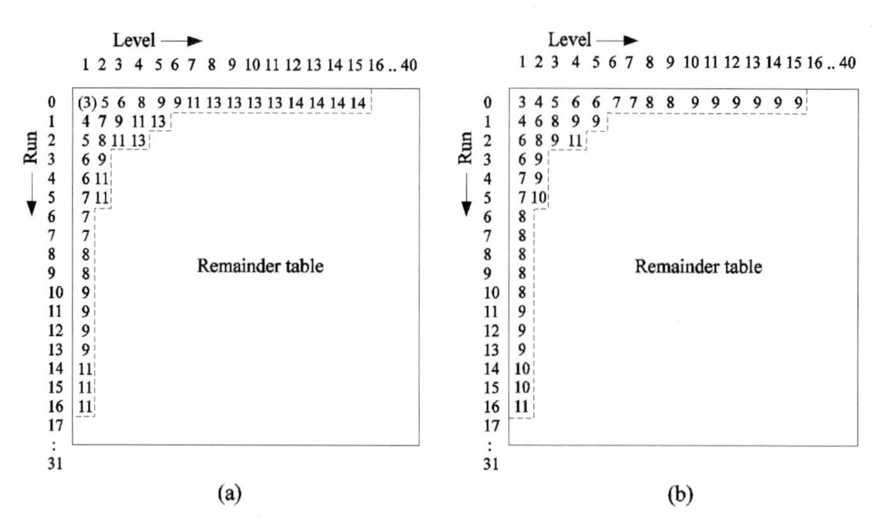

Figure 11.22 VLC code length comparison between (a) table zero and (b) table one for coding DCT coefficients.

Intra AC coding: Coding of AC coefficients is based on the run/level coding technique, as discussed in previous chapters. The technique is employed not only for intra coefficients but also for nonintra coefficients. In MPEG-1, the VLC table for intra AC coefficients is same as for nonintra coefficient coding (Table 10.10). An alternate VLC table is, however, available for intra AC coefficients in MPEG-2. To get the minimum difference from the MPEG-1 table, only the 42 most frequently occurring combinations are modified to relatively short code words, since in general the correlation of intra pictures is higher than that of nonintra pictures. Figure 11.22 shows the code word length of the two VLC tables: Table zero is the same as in MPEG-1. In case of run/level $(0, \pm 1)$, the length is 2 if it is for the DC coefficient. Table one is newly designed for intra AC coefficients. The complete tables are shown in Appendix C.

One of the major differences is in the length of the code word allocated to the end-of-block (EOB) signal, which for intra macroblocks can be assigned a longer codeword (four bits instead of two bits). It is known that the alternate intra VLC table improves the coding efficiency of intra macroblocks by about 0.3 dB in SNR at 12 Mbps as compared with the MPEG-1 VLC table [622].

In the case of escape code, the values of run and signed level are fixed-length coded. A code word (24 bits) longer than in MPEG-1 is needed to encode an escape code as follows:

$$Escape\ code(6) + run(6) + level(12) = 24\ \ bits \tag{11.9}$$

when the signed level is between −2047 to +2047 (12 bits) and the length of run is limited to 63 (6 bits).

Nonintra coefficient coding: Nonintra macroblocks can occur in P and B pictures as shown in Table 10.10. All DC and AC coefficients are coded by using the VLC table used in MPEG-1 (table zero in Fig. 11.22 and in Appendix C).

11.2.4 Scalable coding techniques

Scalable video coding is useful for a number of applications in which video needs to be decoded and displayed at a variety of resolutions (temporal or spatial) and quality levels, for example, multipoint video conferencing, windowed display on workstations, video communications on asynchronous transfer mode (ATM) networks, and HDTV with embedded standard TV.

The easiest way to cope with these demands is the simulcast technique, which is based on transmitting a set of various contents of an encoded video sequence simultaneously [604]. In this case, channel bandwidth is not used efficiently, since the bandwidth should be shared among various resolutions.

An efficient alternative to simulcast is scalable video coding, in which the bandwidth allocated to a given scale can be reused (though partially) for coding other scales. Multiuse of the bandwidth is a condition for scalable video coders. Actually, the enhancement layer is coded by utilizing information from the base layer encoder. In MPEG-2, if there are several scales (layers), the lowest layer is called the base layer and the others are called enhancement layers.

Scalable video coding can be achieved in the spatial, temporal, and frequency domains. Spatial scalability deals with different resolutions, which can be varied by the decimation and interpolation techniques. Temporal scalability deals with different frame rates (temporal resolution). These techniques have been proposed (see Section 11.2.1) in the first test model [585].

The fundamental idea of frequency scalability is that the DCT coefficients can be utilized for filtering in the frequency domain [581]. Layered transmission results in different quality levels and resolutions. Although frequency decomposition is possible in the subband domain [606], encoding of selective coefficients in the DCT domain can be used for HDTV/TV conversion [605]. A main problem that occurs in such frequency-scaling schemes is called the drift problem; the cause of this is that lower layers may not have interframe feedback loops and may accumulate errors due to inaccuracies of motion vectors.

Some modifications were suggested to minimize the drift problem by using block-based filtering [589], and multiple prediction loops [590, 586] in the DCT or subband domain [612]. Furthermore, the drift problem has been extensively simulated in the test models [583].

Such a problem, however, has not been completely solved yet, and the CD 13818 includes basically four scalability modes:

- Spatial scalability
- SNR scalability
- Temporal scalability
- Data partitioning

It is possible, as well, to hybridize some of these scalabilities. Only SNR and spatial scalability are defined in the profiles (see Table 11.4).

Spatial scalability

Figure 11.23 shows a two-layer coder structure for spatial scalability. Input video with high resolution is first spatially decimated into lower-resolution video. The video at each resolution is encoded by the corresponding encoder. The locally decoded base-layer video is spatially interpolated and used for prediction in the enhancement layer. The two bit streams are multiplexed into a single bit stream for storage or transmission. The output of the lower-layer decoder (Fig. 11.24) is interpolated again and used for prediction in the enhancement layer. Therefore,

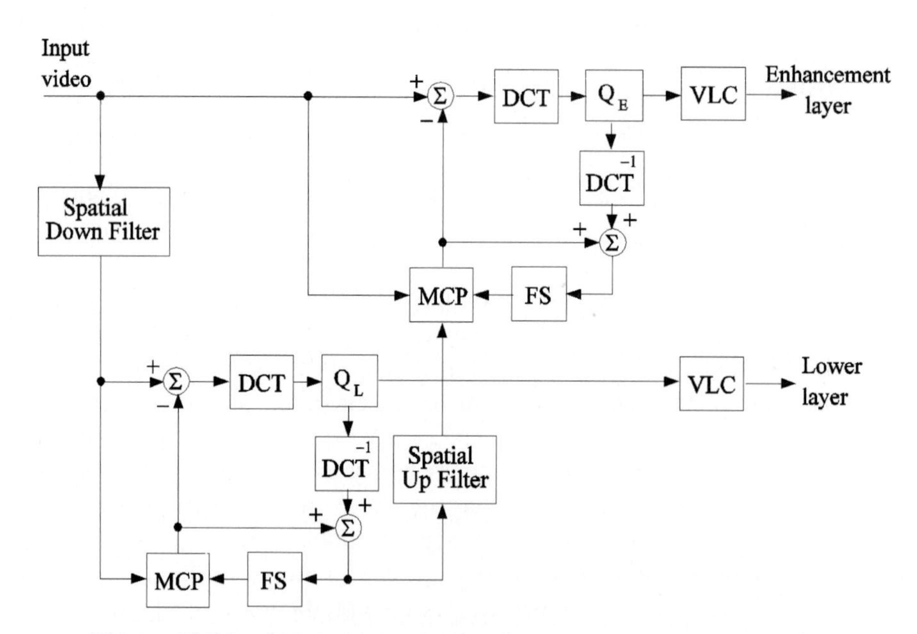

Figure 11.23 An example of a two-layer coder structure for spatial scalability (MCP: motion compensated prediction; FS: frame store).

from one input video bit stream at the encoder, two outputs (base and enhancement) are decodable at the decoder.

Some important steps for the implementation of the spatial scalable extension are:

- Generate two spatial-resolution layers from a single video source.
- Code the base layer by itself to provide basic spatial resolution.
- Code the enhancement layer using the spatially interpolated base layer, leading to full spatial resolution.
- Multiplex the two layer bit streams.
- Use the decoded base layer for decoding the enhancement layer.

The enhancement layer (Fig. 11.24) is predicted not only from previously decoded pictures (temporal prediction), but also from decoded and up-sampled pictures of the lower layer. The two predictions can be selected individually or together, by the relative macroblock type information (spatial/temporal weight class). The decision should be made in the encoder and notified in the macroblock header.

The spatial prediction from the lower layer should be coincident (in display time) or very close in time to work efficiently. The up-sampled (interpolated) prediction of the lower layer should be entirely within the display window of the enhancement layer.

For interlaced-to-interlaced scalability (for example, when the base layer employs ITU-R 4:2:0 interlaced frames and the enhancement layer employs interlaced HDTV frames), somewhat sophisticated decimation and interpolation operations are necessary for good performance (Fig. 11.25). The two base fields are first deinterlaced to produce two progressive fields. Here a field in the enhancement layer is regarded as a progressive version of a field in the base layer. Then the two fields in the base layer are resampled vertically and horizontally to fit the size of the enhancement layer. The deinterlacing procedure involves filling the same number of zeros as the number of lines (zero padding) and filtering vertically and temporally.

For interlaced-to-progressive scalability, the deinterlacing and resampling procedure of base layer fields is similar to interlaced-to-interlaced scalability. Since the two fields belong to one frame in the enhancement layer, only one field is selected by the designer.

The requirements for the spatially scalable profile are defined in Table 11.4, where only one enhancement layer is used for High-1440 level, and the third layer is possible using the hybrid mode with SNR scalability.

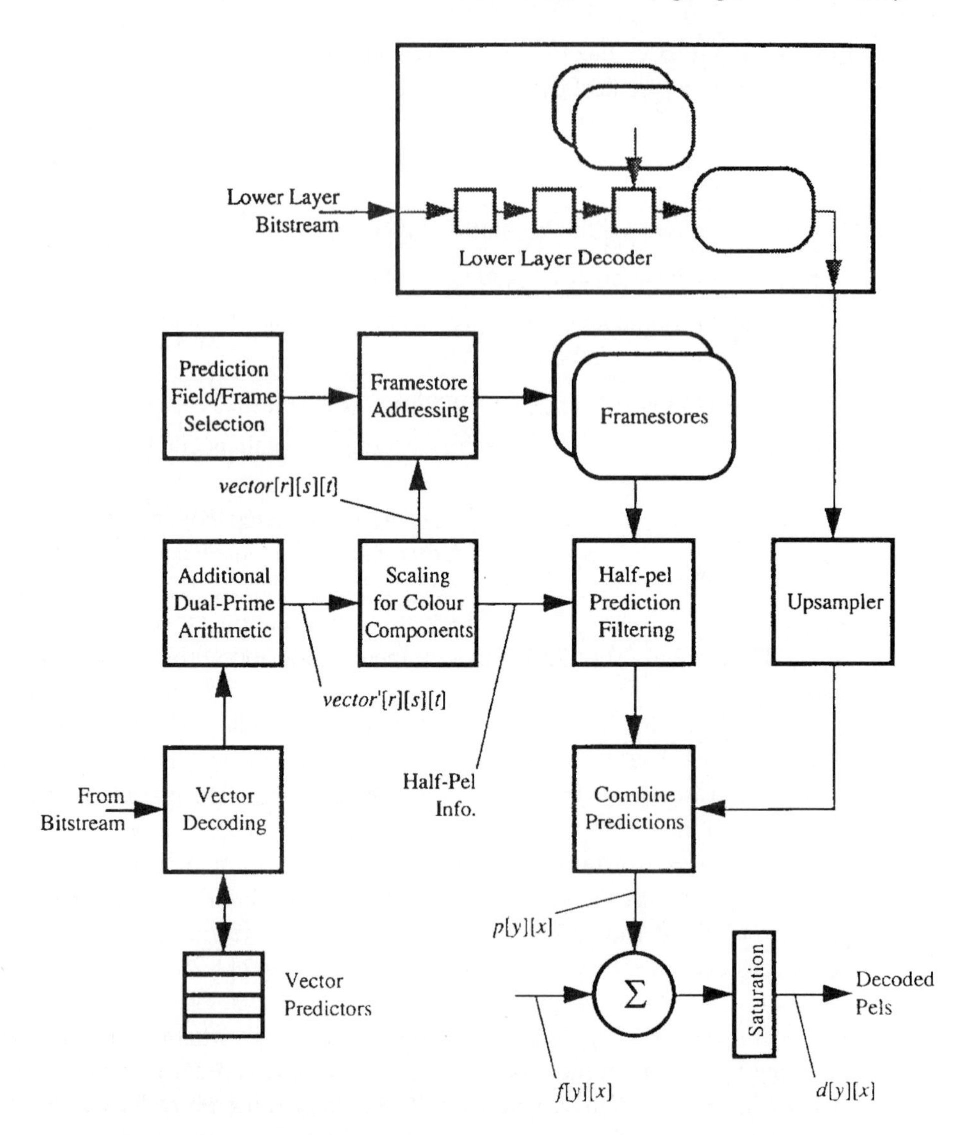

Figure 11.24 The motion compensation process for spatial scalability [592]. © 1995 ITU-T.

SNR scalability

SNR scalability provides two or more video layers of the same spatial resolution but at different qualities. The enhancement layers contain only coded refinement data for the DCT coefficients of the base layer. As illustrated in Fig. 11.26, data from the two layers are combined after the inverse quantization

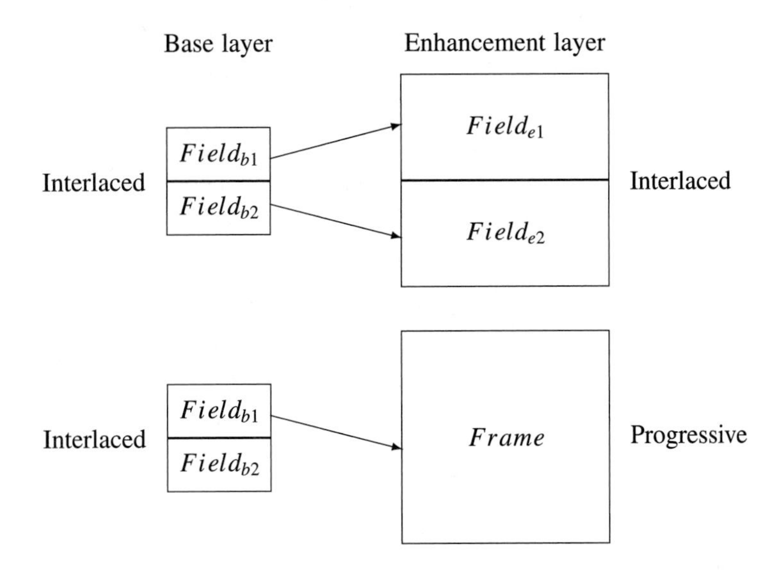

Figure 11.25 Interlaced and progressive prediction from the base layer in a spatially scalable coder.

process by adding the DCT coefficients as

$$F''[v][u] = F''_{lower}[v][u] + F''_{enhance}[v][u], \quad \text{for all } u, v. \tag{11.10}$$

The lower layer derived from the first bit stream can itself be either nonscalable or may require the spatial or temporal scalability (hybrid mode) decoding process. The enhancement layer derived from the second bit stream (Fig. 11.26) mainly contains coded DCT coefficients and a small overhead. Therefore, the increase in the total bit rate is not large. It is not possible to reconstruct the enhancement layer without decoding the lower layer in parallel. After summation of the DCT coefficients, the decoding process is the same as in the nonscalable decoder.

The chroma simulcast process is analogous to that of SNR scalability. It can, therefore, be explained as a kind of SNR scalability. Chroma simulcast is used, for example, to enhance a 4:2:0 signal after processing the enhancement layer data. If the chroma simulcast mode is set, the luminance blocks are processed as described earlier but the chrominance blocks are differently processed. The DC chrominance coefficient of the lower layer is used as a prediction of the DC coefficient in the enhancement layer, but the AC chrominance coefficients of the lower layer are discarded and replaced by the AC coefficients of the enhancement layer, which have a higher chrominance resolution.

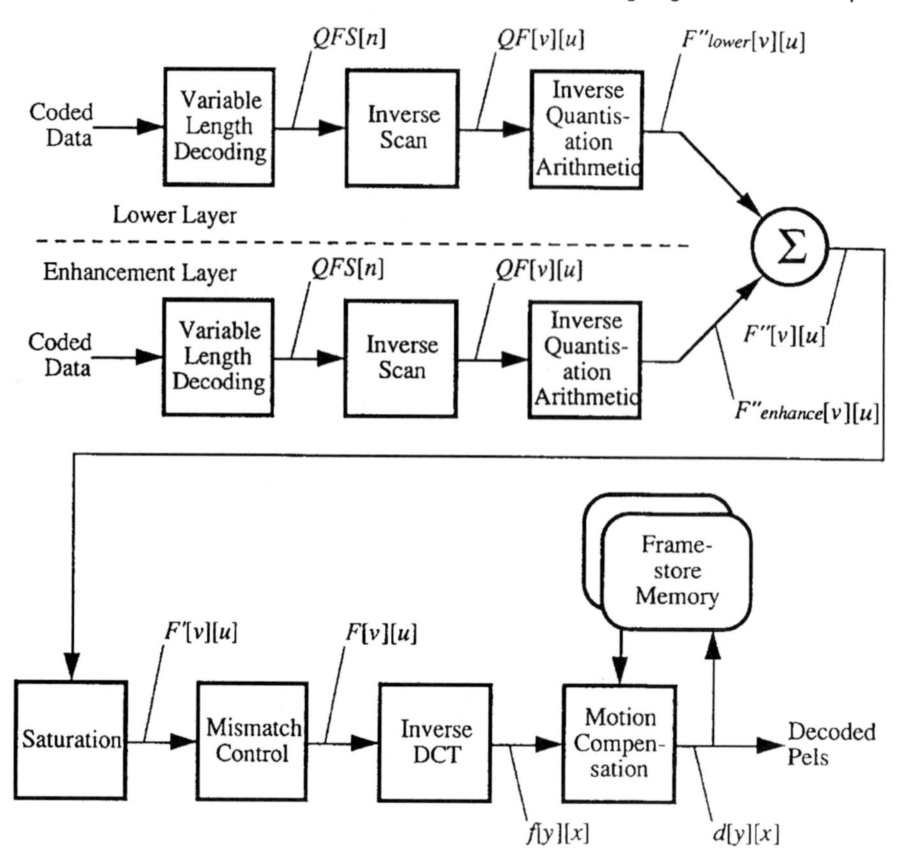

Figure 11.26 The decoding process for SNR scalability [592]. © 1995 ITU-T.

The SNR scalable encoder can be implemented by adding a quantizer refinement with or without feedback loop. Figure 11.27 shows the encoder with the quantizer refinement. The encoder outputs two-layer pictures, the lower layer generating a basic MPEG-2 bit stream (lower quality) and the enhancement layer (higher quality) generating a quantized version of the quantization error in the lower layer. The quantization error is fed back into the motion-compensated loop.

Since the motion compensation for the lower layer is performed from the decoded picture of the enhancement layer, a drift problem occurs between the motion-compensated images at the encoder (refined) and at the decoder (non-refined lower layer). This drift is limited to acceptable levels for N up to 15 and $M = 3$ [583]. The encoder, which has no feedback loop, does not produce the drift. In this case, the enhancement-layer encoder simply encodes the residual prediction error of the lower-layer encoder. At the decoder, the lower-layer decoder is a basic

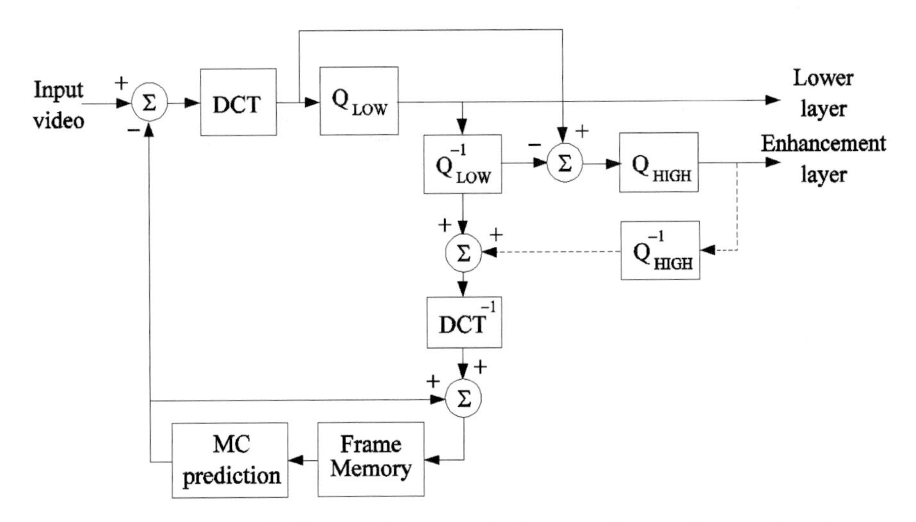

Figure 11.27 SNR scalable encoder by quantizer refinement and feedback in the prediction loop.

MPEG-2 decoder. The enhancement-layer decoder has refinement information of the lower-layer pictures.

In the SNR scalable profile, the spatial or temporal scalable extension mode does not exist, since the layers have the same resolutions. Therefore, the spatial enhancement layers are not included in the SNR scalable mode and the number of layers defined in the SNR scalable profile is low. Furthermore, the upper bound for the spatial resolution is limited to (720×576) in the main level and (352×288) in the lower level.

Temporal scalability

Temporal scalability involves at least two layers. Both the lower and the enhancement layers process the same spatial resolution pictures but at different temporal resolutions. The enhancement layer has a full temporal picture rate that is enhanced from the temporal resolution of the lower layer.

The decoding process for both layers shown in Fig. 11.28 is quite similar between both layers. The enhancement layer is decoded with the prediction of the lower layer. The reference pictures for prediction are selected by the reference select code (transmitted to the decoder). For example, in P pictures, the reference can be selected from (a) the most recent decoded enhancement picture, (b) the most recent lower-layer picture, or (c) the next lower-layer picture as shown in Fig. 11.29. The order of the lower-layer pictures is in the display order.

When the most recent frame in the lower layer is used as the reference, the frame is temporally coincident with the frame or the first field in the enhancement layer. The predicted pictures are interleaved with the relevant filtering in order to

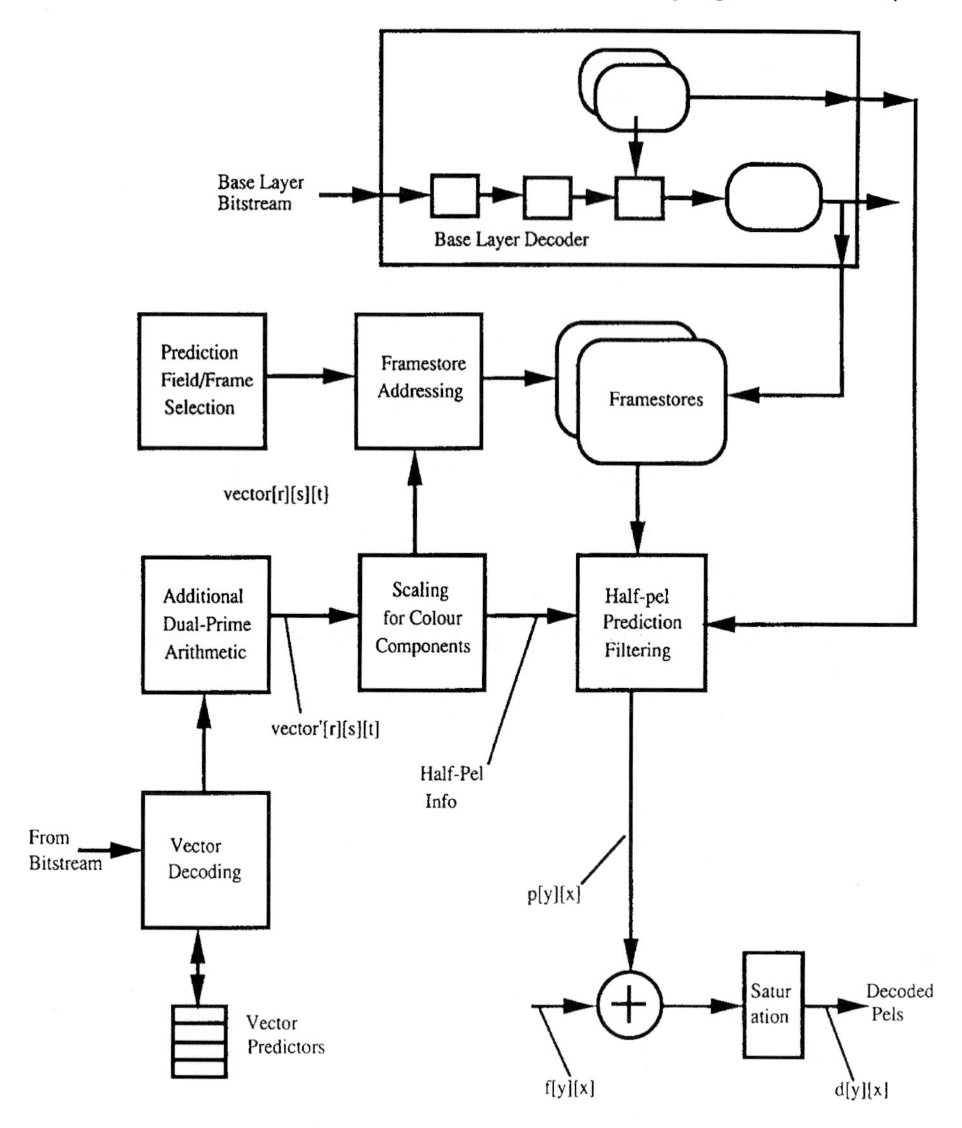

Figure 11.28 The decoding process for temporal scalability [592]. © 1995 ITU-T.

generate higher–temporal resolution video in the enhancement layer output, where the lower layer is decoded independently.

Data partitioning

Data partitioning is a technique that splits a video bit stream into two or more layers, called partitions. A priority breakpoint in the slice layer header indicates

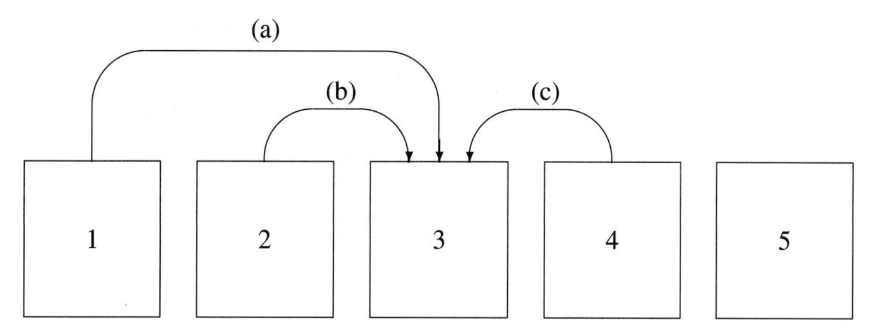

Figure 11.29 Prediction selection for the temporally scalable enhancement layer in P pictures.

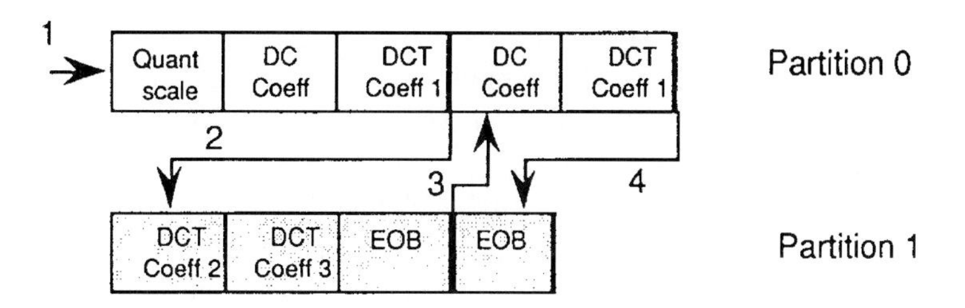

Figure 11.30 A segment from a bit stream with two partitions with priority breakpoint set to 64 (the DC and the first AC coefficient belong to partition 0) [591]. © 1995 ITU-T.

which syntax elements are placed in partition 0 or partition 1. Partition 0 is called the base partition or high-priority partition. Partition 1 is called the low-priority partition. For the DCT coefficients, the lower the frequency of the coefficients, the more important is the picture quality and the higher the priority.

Figure 11.30 shows an example for data partitioning with priority breakpoint 64, which means that all headers, the DC and the first AC coefficient, belong to partition 0. In the second block, which has only two nonzero coefficients, only the EOB signal belongs to partition 1. Therefore, the base layer can be reconstructed with lower quality, even when the data in partition 1 cannot be recovered due to channel errors.

This type of scalability is useful for data transmission in channels that have different channel error probabilities, particularly in ATM networks [584]. Since the DCT coefficients are partitioned, the data partitioning is a kind of frequency scalability that has not been adopted in any of the profiles.

11.3 Audio Coding

This section describes the audio coding algorithm in MPEG-2. Essentially, MPEG-2 audio includes MPEG-1 audio (see Section 10.3) and extends to multichannel configurations and lower sampling frequencies. We emphasize the extended properties in this section.

Extension to multichannel audio

Regarding efficient stereo representation, MPEG-2 recommends multichannel configurations, such as *3/2 stereo* (three front and two rear surround loudspeaker channels). In addition to the two-channel (L, R) stereo, an additional center channel (C) and two surround channels (LS, RS) are used for five-channel audio signals. The 3/2 stereo format allows improved realism of auditory ambiance and outperforms the two-channel stereo representation. The multichannels are named as T_1 to T_5, where the T_1 and T_2 channels are compatible with MPEG-1 audio and the T_3 to T_5 channels are extension channels. Any three out of the five signals (L, C, R, LS, RS) can be selected for the extension channels.

The different combinations of audio channels are shown in Fig. 11.31. Up to five channels can be used for one program as shown in Fig. 11.31a through g or for two programs as shown in Fig. 11.31h, i, and j. The latter configurations can provide bilingual programs or multilingual commentaries. Special services for the hearing impaired can be provided. For example, a bilingual 2/0 stereo program or one 2/0, 3/0 stereo sound plus accompanying services, i.e., clean dialog for the hard of hearing and commentary for the visually impaired can be easily configured. The lower number of channel configurations including mono presentation can also be applied. After multichannel decoding, up to five audio channels are recovered and can then be represented in any convenient format.

The low-frequency enhancement (LFE) channel is capable of handling signals in the range from 15 Hz to 120 Hz. This channel is optionally used to enable listeners to enhance the low-frequency content of the program in terms of both frequency and level. The LFE channel is an option at the receiver and is not included in any dematrixing operation in the decoder.

Downward compatibility is accomplished by using downward mixing equations, Eq. (11.11), in which two channel signals can be generated from the 3/2 stereo:

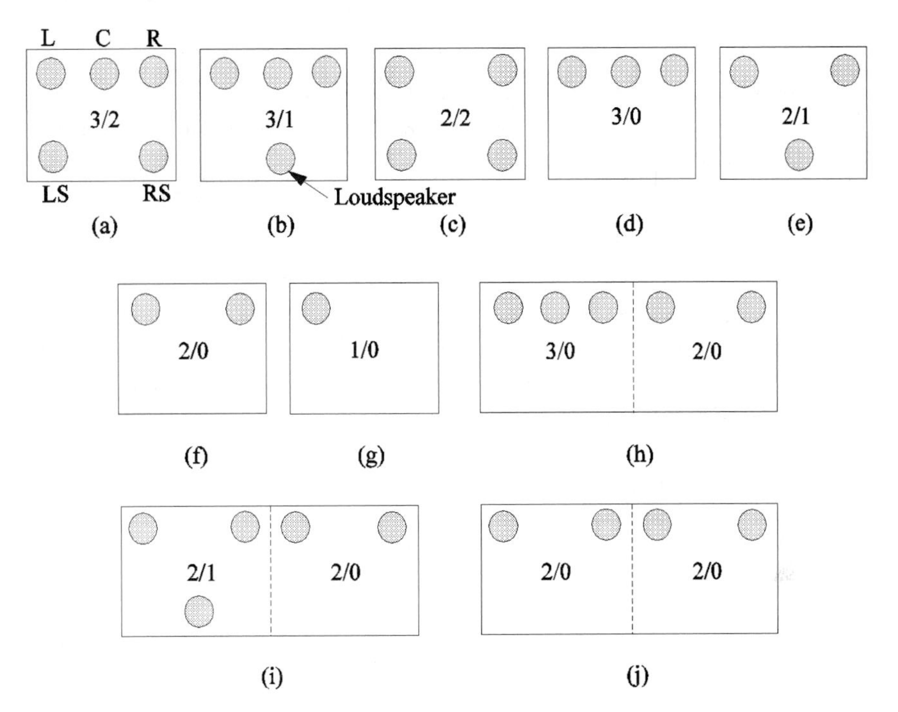

Figure 11.31 The possible configurations of multichannel audio for one program or two programs. (L: left; R: right; C: center; LS: left surround; RS: right surround).

$$L_0 = L + x \times C + y \times LS$$
$$R_0 = R + x \times C + y \times RS \qquad (11.11)$$

where the possible values of the parameters x and y are $1/\sqrt{2}$.

Backward compatibility means that an MPEG-1 decoder is able to decode properly an MPEG-2 audio bit stream. It is possible to reproduce the two conventional stereo signals using Eq. (11.11). Forward compatibility is established when the multichannel decoder produces the complete 3/2 stereo presentation from the basic stereo signal. Upward and downward compatibilities are also provided. These compatibilities are realized by coding the basic stereo information in conformance with MPEG-1 audio and exploiting the ancillary data of the MPEG-1 audio frame (see Fig. 10.24) for the multichannel extension. Figure 11.32 shows the MPEG-2 audio encoder/decoder compatible with the basic MPEG-1 system. Channels T_1 and T_2 can be the basic two channel stereo (L_0, R_0). The backward compatibility may compromise the audio quality of the multichannel encoder. Hence MPEG is evaluating proposals for nonbackward

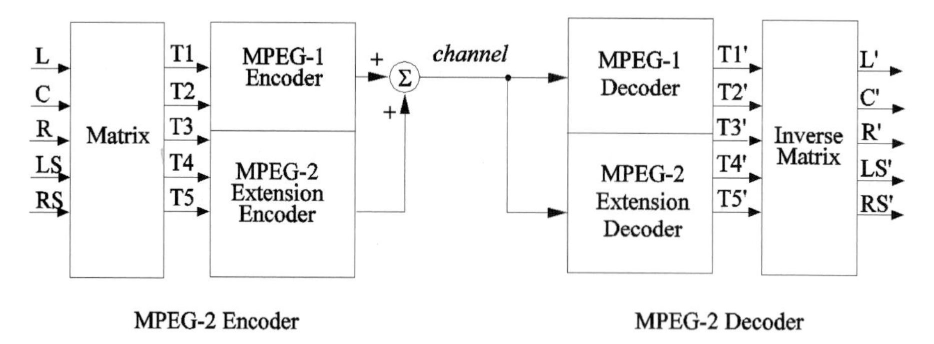

Figure 11.32 The MPEG-2 audio codec compatible with MPEG-1 audio.

compatible (NBC) multichannel audio coding, with a view to improve the overall audio quality at lower bit rates.

The MPEG-1 audio frame incorporates different types of information: header information (32 bits), cyclic redundancy check (16 bits), coded audio data including bit allocation, scale factor select information, scale factors, and the subband samples, and finally ancillary data. The variable length of the ancillary data field allows the large number of different applications and enables packing the extension information of the channels 3, 4, and 5 into the first part of the ancillary data field. The remaining part of the field can be used for other ancillary data, if needed.

Extension to lower sampling frequencies

MPEG-2 audio is capable of using lower sampling frequencies (below 32 KHz) identified in the header information. Its differences from the MPEG-1 audio encoder are described here.

Layer I: A slot consists of 32 bits, though it equals 8 bits in layers II and III, as in MPEG-1 coding. The number of slots in a frame depends on the sampling frequency and the bit rate. Each frame contains information on 384 samples of the original input signal. Frame size at the sampling frequency 24 KHz is calculated by 16 ms = 384 × (1/24 KHz), and the number of slots in a frame is calculated by $N = \mathit{bit\ rate} \times (384/32)/f_s$. Assuming the bit rate of 64 Kbps as in Example 10.7, frame size and the number of slots for lower sampling frequencies are given by

f_s (KHz)	Frame rate (frames/sec)	Frame size (ms)	No. of slots
24	62.5	16	32
22.05	57.42	17.415...	34.8 ⇒ 35
16	41.67 ...	24	48

Layer II: The coding algorithm follows the MPEG-1 layer II. The differences in MPEG-2 layer II are the formatting, the possible quantizations, and the psychoacoustic modeling. A slot consists of 8 bits, and the number of slots in a frame depends on the sampling frequency and the bit rate. Each frame contains information on 1152 samples of the original input signal. The frame size and the number of slots for lower sampling frequencies are as follows:

f_s (KHz)	Frame rate (frames/sec)	Frame size (ms)	No. of slots
24	20.83	48	384
22.05	19.14	52.5...	418
16	13.89 ...	72	576

Layer III: The differences from MPEG-1 layer III coding are the changed scale factor band tables, the omission of some side information due to the changed frame layout, and some changed tables in the psychoacoustic modeling. There are 21 scale factor bands at each sampling frequency for long windows and 12 bands each for short windows. The scale factor band is a set of frequency lines in layer III that are scaled by one scale factor. Long windows contain 576 frequency lines from the long block (18 frequency lines output) MDCT of 32 subbands, while short windows contain 192 frequency lines from the short block (6 frequency lines output) MDCT operation.

11.4 MPEG-2 Implementations

MPEG-2, which has been an international standard since November 1994, is in fact at the commercially applicable stage and has already been adopted for a number of application areas. Linking with ATM standardization and HDTV broadcasting are the topics of interest. Since July 1990, the ITU-T ATM group has had joint work with MPEG-2. In the beginning of 1994, MPEG-2 was adopted by the GA HDTV system in the United States (Chapter 13) and DVB in Europe. U.S. broadcasting and cable TVs (video-on-demand) are the first commercial use of MPEG-2 technology. Hence, MPEG-2 has become the common technology for coding digital video in the spectrum of applications ranging from communication/broadcasting to computers/electronic games.

The MPEG-2 algorithm, which supports many spatial and temporal resolutions including ITU-R 601 video, requires more than four times the complexity of the MPEG-1 algorithm. Moreover, as MPEG-2 can process interlaced video, motion estimation/compensation needs more than six times the complexity of that in MPEG-1 [627]. Hence, hardware design must be made with faster and higher performance. RISC processing and array processing [633] are the common techniques to accelerate the processing speed.

Implementation of MPEG-2 is possible in several ways: software-only, hardware-assisted, decoder-only, and DSP-based encoder/decoder implementation. A number of the MPEG-2 decoder LSIs became commercially available toward the end of 1994. Since then, new application fields have been introduced, such as video-on-demand, news-on-demand, pay-per-view, set-top box for cable TVs, and computer games. Software-only video decoders can decode and display a compressed video stream containing frames of size equal to or greater than (320×240), at reduced frame rates, on desktop personal computers. Because MPEG-2 video provides higher quality but is too complex for software-only playback, hardware-assisted codecs provide substantially better image quality and better coding and playback performance than software-only video codecs.

Hardware-assisted codecs often run on programmable DSPs, wherein algorithms for various purposes can be micro-coded. TI's TMS320 series and Array Microsystem's ICC/MEC are typical DSPs for the programmable codecs. These chip sets can be implemented on a PC plug-in card [631] or a dedicated hardware system.

Most of the chip set manufacturers produce decoders that are capable of decoding MPEG-1 and MPEG-2 bit streams. The complete decoding function is realized with the addition of a microcontroller and a bank of DRAM memory. SGS-Thomson's STi3500 needs the memory configuration of four 256K$\times$16 DRAMs [628]. The system has two global buses, a high-speed, 64-bit memory data bus and a 16-bit control bus to transfer data to and from the external DRAM, where the microcontroller interface has an 8-bit data bus and a 6-bit address bus. Matsushita's VDSP2 is an image compression digital signal processor that is capable of both encoding and decoding the MPEG-2 algorithm by changing programs on the same chip [627]. It consists of four types of processors: a DSP core processor, a data I/O processor, a DCT processor, and a VLC/VLD processor. The DSP core processor runs by RISC operation and parallel processing using pipeline and SIMD operation. Motion estimation is performed in steps: first, two-pixel resolution ME using dedicated LSIs, and then full/half-pixel resolution by using programmable DSPs. AT & T Microelectronics has launched a single-chip decoder that integrates all the functions necessary to decode an MPEG-2 bit stream into 4:2:2 raster output. It supports MP@ML (main profile at main level, see Table 11.5) video, but no B pictures are used. This means that the capacity required for the external memory is only 8 Mbits (half that needed for MP@ML with B pictures).

Some MPEG-2 decoders are designed for offering auxiliary functions, such as on-screen display function, which is used for the superimposed display of messages. These functions are expected to be invaluable for multichannel cable TV and satellite broadcasting uses. The size of external memory to store the auxiliary

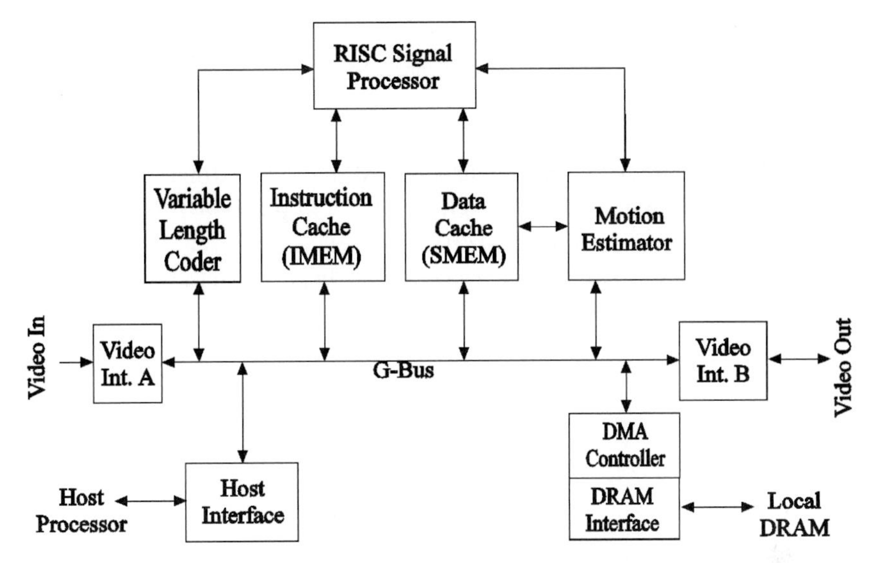

Figure 11.33 VideoRISC processor block diagram of C-Cube's CLM4600 chip set.

screens becomes quite large. When receiving movies, a 3:2 pull-down function (Fig. 13.6) is necessary to convert the 24 frames/sec temporal rate of movies into the 30 frames/sec rate. Decoders by AT & T Microelectronics, Matsushita Electric Industrial, Mitsubishi Electric, Pioneer Video Corp. of Japan, and others support the 3:2 pull-down function. The function to convert aspect ratio (from 4:3 to 16:9) can be added to meet wide-aspect TVs. Most of the decoders (from C-Cube, AT & T, LSI Logic, SGS-Thomson, etc.) support this function.

Furthermore, a number of manufacturers are accelerating the development to introduce codecs to the market ahead of competitors: AT & T Microelectronics, C-Cube Microsystems, Hitachi, Mitsubishi Electric, NEC, Pioneer Video, Sony, SGS-Thomson Microelectronics, Toshiba, IBM Microelectronics Division, LSI Logic, IIT, and so forth. In Appendix A, the reader can acquire valuable information on the manufacturers and their chip sets.

Hardware implementation of encoders is not crucial for general users, but is necessary for program providers. Therefore, the manufacturers are eager to implement decoding chip sets instead of encoding chip sets. Several chip sets are designed to implement the MPEG-2 encoder. It is possible to use the programmable DSPs, of course. C-Cube CLM 4600 compresses digital video into MPEG-2 syntax at broadcast resolutions for NTSC and PAL. It consists of 10 C-Cube Video RISC™ processors (VRP) with 20 Mbytes DRAM. The VRPs work in parallel. Each VRP consists of a 32-bit RISC processor, half-pixel motion estimator, variable-length coder, and so forth, as shown in Fig. 11.33. The RISC processor includes both general-purpose computing instructions for managing the MPEG-2 encoding process and DSP instructions for performing DCT, pixel averaging, and so on.

A two-chip solution for MP@ML encoding is available in Matsushita's VDSP2. It does not have an internal motion estimator, so this circuit must be provided externally. IIT, TI, and LSI Logic [990] are also developing chip sets capable of encoding MPEG-2. Even though real-time MPEG-2 encoding, especially by a single-chip system, has been implemented by some companies such as IBM, it is a technological challenge.

Chapter 12
MPEG-4 and H.263 Very Low Bit-Rate Coding

Summary

ISO/IEC MPEG-4 and the ITU-T Experts Group on Very Low Bit-Rate Visual Telephony (LBC) have been working on the generic coder for very low bit-rate video/audio coding applications. MPEG-4 has subsequently expanded its role in terms of tools, profiles, algorithms, functionalities, and bit rates. This chapter gives a survey of the status of MPEG-4 and the LBC group, including its planned schedule, and initial ideas about new coding concepts: analysis/synthesis, model-based, fractal, and morphological-based approaches. Numerous trials using the conventional waveform-based coding techniques are also evaluated. ITU-T Draft Recommendation H.263 is also discussed as the near-term solution for very low bit-rate coding. The existing and proposed audio coders are also discussed in this chapter.

12.1 Introduction

MPEG, a working group in ISO/IEC, has already produced two important standards, MPEG-1 (Chapter 10) and MPEG-2 (Chapter 11), and started work on the MPEG-4 standard at the MPEG meeting in Seoul (November 1993), with completion scheduled for 1998. Additional meetings have been held in Paris [701]

Table 12.1 Time schedule for ITU-T SG15 LBC and ISO/IEC MPEG-4 activities

ITU-T LBC		ISO/IEC MPEG-4	
Date	Milestones	Date	Milestones
09/93	New work item	11/92	New work item
11/94	Call for proposals Draft Rec. H.263	11/94	Call for proposals Proposal package description
11/95	Subjective testing	11/95	Subjective testing Second call for proposals
		01/96	Verification models (VM 1)
		03/96	Collaboration on VMs resulting in version 2
11/96	Outline recommendation	11/96	First working draft (WD)
		07/97	Final version of WD
11/97	First draft recommendation	11/97	Committee draft (CD)
03/98	Second draft recommendation	03/98	Draft international standard (DIS)
11/98	Recommendation	11/98	International standard (IS)

(Mar. 1994), Grimstad [702] (July 1994), Singapore [703] (November 1994), where the call for proposals [707] was issued, and Lausanne. Additional meetings were held in Tokyo (July 1995), Dallas (November 1995), and Munich (January 1996).

As it happened with JPEG, MPEG-1, and MPEG-2, MPEG-4 will evolve in close collaboration with the ITU-T, which specializes in telecommunications standards, and particularly with its Study Group 15, Working Party 15/1, the Experts Group on Very Low Bit-Rate Visual Telephony (LBC). This group is finalizing a set of near-term recommendations for PSTN video telephony at the end of 1995, and has already released the first Draft Recommendation H.26P [708] (ITU-T SG15 allocated the number H.263 in Feb. 1995). Test models for the near-term solution (TMN1 through TMN5 [709]) have also been created, providing encoder reference. The long-term solutions of the ITU-T LBC group, however, are timed to be concurrent with those of MPEG-4. In particular, the video coding algorithm of the LBC group will be developed in collaboration with MPEG-4, while network related matters (error protection, error concealment) and specific requirements for the videophone application are primarily the responsibility of ITU-T SG15. Therefore, the MPEG-4 meetings and the LBC meetings are held to overlap both in location and time, as shown in Tables 12.1 and 12.2.

According to the current time schedule, MPEG-4 will be presented in a two-round process as will the ITU-T LBC group. First, a call for proposals

Table 12.2 Low bit-rate videotelephony standards (emerging/established) in ITU-T and ISO/IEC

ITU-T SG15		ISO/IEC
H.324 (< 64 Kbps)	H.321 (> 64 Kbps)	MPEG-4 (up to 1024 Kbps)
H.263 (video compression)	H.261 (video compression)	Video compression (to replace H.263)
G.723 (speech compression)	H.251 (audio compression)	Audio compression (separate work at 4 Kbps)
H.223 (multiplexing)	H.221 (multiplexing)	System multiplexing
V.34 Modem (up to 28.8 Kbps)	N/A	N/A

was issued in Nov. 1994 with a detailed description of requirements and test procedures. Subjective evaluation and formal testing of all proposals (both video and audio) were completed in November 1995. The first verification model (VM) was released in January 1996. The second call for proposals was scheduled in November 1995 and a second formal test in November 1996. At that time, the competitive phase will end, followed by the start of the collaborative phase.

MPEG-4 was originally intended to provide generic very high compression ratio coders, resulting in bit rates less than 64 Kbps, aiming at usable signals at rates as low as 10 Kbps. For such an ambitious target of a new standardization effort to be justified, the performance should be significantly better than that of existing schemes such as H.261 or MPEG-1. For markedly improved results to be achieved over traditional techniques, new video/audio coding techniques need to be introduced. The new disciplines can be derived from computer vision, computer graphics including image and video analysis/synthesis, modeling, fractaling, morphing, and so forth. It is possible that some sort of modeling techniques will be used, when the assumptions of a particular model are satisfied in the image content that generally appears in the videophone application, such as a head-and-shoulders scene.

Whereas MPEG-4 was initially aimed at very low bit-rate coding (< 64 Kbps), its activities during 1995 have resulted in extended applications, tools, algorithms, profiles, and bit rates. The objective is to develop a flexible and extensible coding standard. It will also facilitate the user's ability to achieve various forms of interactivity with the audiovisual content of a scene and to mix synthetic and natural audio information in a seamless way. At the July 1995 meeting held in Tokyo, Japan, the following relevant documents were finalized:

1. Requirements for the MPEG-4 Syntactic Description Language (MSDL), Draft Revision 2
2. MPEG-4 Proposal Package Description (PPD), Revision 3a
3. MPEG-4 Testing and Evaluation Procedures Document, Final Revision
4. MPEG-4 call for proposals

MSDL will allow not only for description of the bit stream structure but also for configuration and programming of the decoder. MSDL is a flexible and extensible description language that allows selection, description, and downloading of tools, algorithms, and profiles. The functionalities described in PPD can be summarized as follows:

- Content-based interactivity
  - Content-based multimedia data access tools
  - Content-based manipulation and bit stream editing
  - Hybrid natural and synthetic data coding
  - Improved temporal random access
- Compression
  - Improved coding efficiency
  - Coding of multiple concurrent data streams
- Universal Access
  - Robustness in error-prone environments
  - Content-based scalability

In addition to the new or improved functionalities, there are several other important functionalities that are needed to support the envisioned audiovisual applications. The functionalities listed below may be provided already by existing or other emerging standards.

- Synchronization
- Auxiliary data capability
- Virtual channel allocation flexibility
- Low (end-to-end and/or decoder) delay mode
- User controls
- Transmission media interworking
- Interworking with other audiovisual systems
- Security
- Multipoint capability

- Content
- Format
- Quality

The PPD also describes three examples of application classes that will significantly benefit from the new functionalities provided by MPEG-4, to give an idea of how the functionalities can be used in future products. The following three examples of application classes can be classified as essential or optional.

(a) Audiovisual database access An audiovisual terminal, e.g., personal digital assistant (PDA), in low-capacity networks (e.g., future cellular, PSTN) supporting services like electronic news, electronic yellow pages, and information for travelers. Video storage on solid-state memory for playback/recording on a portable device. Teleshopping from an interactive video catalog.

(b) Audiovisual communications and messaging Video telephone, multipoint videoconferencing systems, video answering machines, and video email are some of the applications.

(c) Remote monitoring and control Security monitoring of buildings and construction sites, and vehicular traffic monitoring are typical applications.

The third document addresses the testing and evaluation procedures for tools and algorithms as defined in the PPD. Some guidelines about the assessment of MSDL are also presented. Details on the audio and video test sequences, subjective tests for quality assessment of audio/video, testing procedures for the algorithms, a test methods submission checklist for foward testing, and so forth, are provided in this document.

The call for proposals describes how all the MPEG-4 documents can be obtained in electronic form. It also has the registration forms for the MPEG-4 first round of testing and evaluation.

The MPEG-4 workplan (Table 12.1) calls for two competitive stages and a phased collaborative stage culminating in the International Standard (IS) by November 1998. The workplan includes developing verification models (VM), a second call for proposals, revised testing and evaluation procedures, a collaborative effort on MSDL, and so forth, which will be part of the IS.

When successfully completed, the MPEG-4 will provide generic tools in video coding, audio coding, and system configurations. No algorithmic approach in video/audio coding has yet been endorsed by MPEG-4, but only a set of requirements has been specified in [694]. There are 12 identified video requirement areas: content, format, quality, bit rate, error resilience, delay, codec complexity, extensibility, user controls, transmission media interworking, terminal interworking, and scalability. Similar items for audio requirements are addressed.

A detailed description of the requirements can be found in [694]. In this chapter, we only describe the new coding approaches for video and audio signals. Since the Draft Recommendation H.263 of ITU-T LBC has been issued, we compare this with the existing Recommendation H.261 (Chapter 9).

12.2 Video Coding

12.2.1 Waveform-based coder

The waveform-based coding approach may offer the near-term solution for MPEG-4 video coding. The near-term recommendation of ITU-T H.263 is based on this approach for very low bit-rate visual telephony. The picture format of low resolution (QCIF, Fig. 9.3) is considered and the coding algorithm is an extended version of the existing standards: the H.261/MPEG or Cost 211 reference model. A lot of research still needs to be done on topics like alternative transforms, advanced motion compensation techniques and joint source/channel coding. These approaches are not limited to the type of scene, though the other algorithms, such as model-based coding, are limited to a particular model [740].

For choosing the enhanced transforms, the designer needs to consider the computational complexity and performance. Block transforms, such as the DCT, suffer an undesirable effect commonly present in low bit-rate coding, which is called the blocking effect [384]. This is caused by artificial edges at the block boundaries, having the appearance of bricks in the image. Lapped orthogonal transform (LOT) presents higher coding gain [743], and the blocking effect is largely reduced [742]. When motion compensation is performed in the transform domain [741], the prediction errors are comparable to or smaller than those of the standard method using exhaustive search block matching in the spatial domain. On the other hand, LOT demands larger implementation complexity, establishing a trade-off that has to be considered before choosing the appropriate transform for a specific application.

Wavelet transform (WT) techniques have been investigated for low bit-rate coding. Much lower bit-rate and reasonable performance are reported in [725, 724, 722], based on the application of these techniques to still images. A combination of WT and VQ [746, 727] gives better performance. Wavelet transform decomposes the image into a multifrequency channel representation, each component of which has its own frequency characteristics and spatial orientation features that can be efficiently used for coding [726]. Vector quantization makes use of the correlation and redundancy between nearby pixels [727] or between frequency bands [746]. Figure 12.1 shows a block diagram for the WT and classified VQ [746]. This scheme exploits the residual correlation among different layers of the WT domain using block rearrangement to improve the coding efficiency. Further improvements can also be made by developing the adaptive threshold techniques

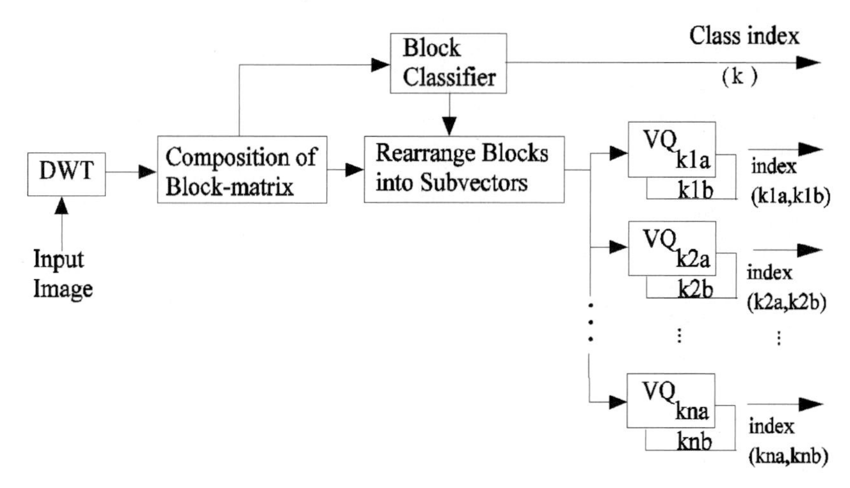

Figure 12.1 A block diagram for the classified WT with VQ of images [746]. © 1995 IEEE.

for classification based on the contrast sensitivity characteristics of the human visual system. Joint coding of the WT with *trellis coded quantization* (TCQ) [722] as a joint source/channel coding is another area to be considered.

Additional video coding research applying the WT on a very low bit-rate communication channel is performed in [723, 749, 729]. The efficiency of motion-compensated prediction can be improved by overlapped motion compensation [729] in which the candidate regions (two or more blocks) from the previous frame are windowed to obtain a pixel value in the predicted frame. Since the WT generates multiple frequency bands, multifrequency motion estimation [749, 721] is available for the transformed frame. It also provides a representation of the global motion structure. Also, the motion vectors in lower-frequency bands are predicted with the more specific details of higher-frequency bands. This hierarchical motion estimation can also be implemented with the segmentation technique that utilizes edge boundaries from the zero-crossing points in the WT domain [723]. Each frequency band can be classified as temporal-activity macroblocks or no-temporal-activity macroblocks [738]. The lowest band may be coded using an H.261-like coder, which uses DCT, and the other bands may be coded using VQ or TCQ.

The block size for motion estimation is another variable. The smaller block size is more locally adaptive, representing a few transform coefficients. It needs, however, more overhead bits for side information [739]. The larger block size can be chosen with respect to the improvement of the coding efficiency and visual appearance of the reconstructed pictures. Simulation results are reported at less than 64 Kbps with 5 to 10 frames/sec. for a typical videoconferencing sequence using these waveform-based coders.

Very low bit-rate coders are applicable to mobile wireless networks. Wireless video transmission over a dynamic network requires adaptation to changes in bandwidth, network traffic, and channel characteristics. For these purposes, new computing techniques are needed that enable low-power, flexible, adaptive, and robust video communication in hostile environments with no access to an established communications infrastructure [744]. The communications channel in the wireless context is likely to be poor, necessitating that careful attention be made to error correction and tolerance to fading effects. It is worthwhile to use integer-coefficient filters in the subband decomposition techniques and adaptive delivery of the compressed video at a tens of Kbps transmission rate.

12.2.2 Object-oriented coder

Object-based coding utilizes general models as planes or smooth surfaces for the object model. Information, such as surface orientation and motion, is estimated from image sequences and is utilized for motion compensation or motion interpolation. It does not require a recognition algorithm. Object- oriented analysis-synthesis techniques have been developed by Musmann [753], Hötter [755], and Ostermann [759] at the University of Hannover. Hötter et al. [757, 755, 753] have proposed a method utilizing a segmented surface model, in which changed regions caused by object motion are analyzed and modeled by planar patches. Ostermann [759] and Morikawa [772] have proposed a method utilizing global surface models, in which a smooth surface model of the object is estimated from an image sequence. Object-oriented analysis-synthesis coding subdivides an image into moving objects and defines the three sets of parameters: *motion*, *shape*, and *color* information of the object [753]. This task is performed in the image analysis block of Fig. 12.2. The object parameters depend on the type of source model. The synthesized image is used to estimate the motion parameter. The three parameters are coded by parameter coding (Fig. 12.3). Instead of the frame memory of block-oriented coding, object-oriented coding requires a memory to store the parameters of the objects. Decoded parameters are stored and used for the parameter coding/decoding of the following image frame and for image synthesis. The parameter coding depends on a receiver model that includes assumptions about the visibility of coding errors. The reconstructed (synthesized) image is used for analysis of the following image, generating motion, shape, and color parameters.

The object-oriented coder can be viewed as a generalization of block-oriented hybrid coding as shown in Fig. 12.4. The motion estimation of a block-oriented coder represents the motion vector for luminance and chrominance components (motion and color parameters in an object-oriented coder). The object-oriented coder additionally transmits the shape of each object. For maintaining the same bit rate, the bit rates for motion and color information have to be relatively reduced in the object-oriented coder. This requirement can be justified by some techniques: the synthesis of the color information is improved at object boundaries

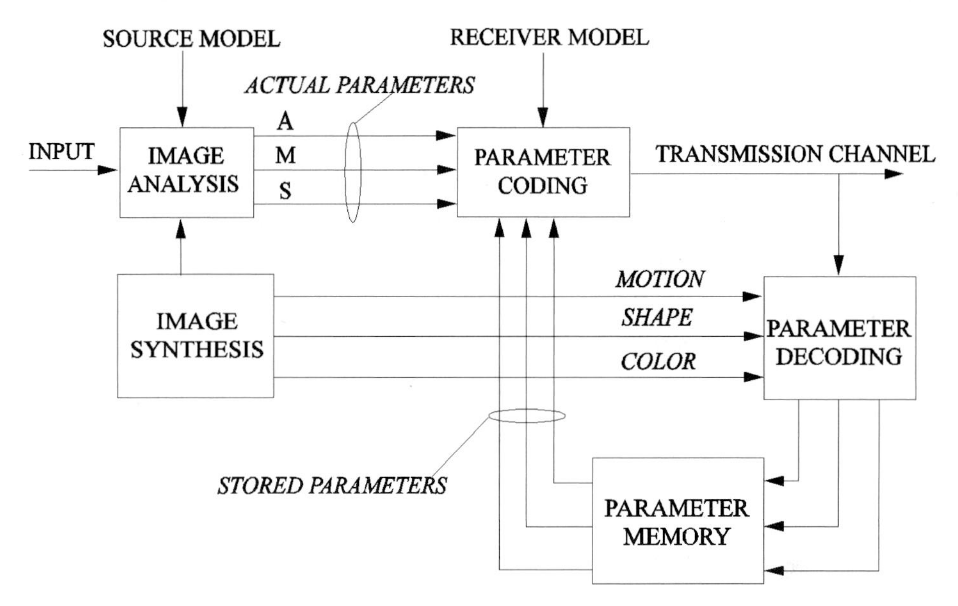

Figure 12.2 A block diagram of an object-oriented analysis-synthesis coder (A: motion; M: shape; S: color) [753]. © 1989 Signal Processing.

using shape information, and the definition of *model failure objects* (MF objects) and *model compliance objects* (MC-objects) improve's the coding efficiency [754]. The decoded image is stored in image memory and used for motion estimation and parameter coding of the following image frame. The function of parameter coding is similar to that of block-oriented coding. Simulation results [754, 760] show better performance than H.261 scheme at very low bit-rates of less than 10 Kbps.

Image analysis/synthesis

The goal of image analysis is to estimate temporal changes between a synthesized previous image and a new input image. Object motion and object shape estimation are the main issues of interest. The hierarchical block-matching technique can be used for the object motion [755]. In the first step of the hierarchy, temporally unchanged image regions, i.e., the static background, are detected. In further steps of the hierarchy, the changed regions are analyzed and the temporal changes are described by three parameter sets: motion, shape, and color [754]. As the source model of moving objects, a 2D object model [755] or a 3D object model [759] can be used. In the 2D modeling, the object shapes are approximated by a polygon/spline representation. The vertices of the representation are predictively

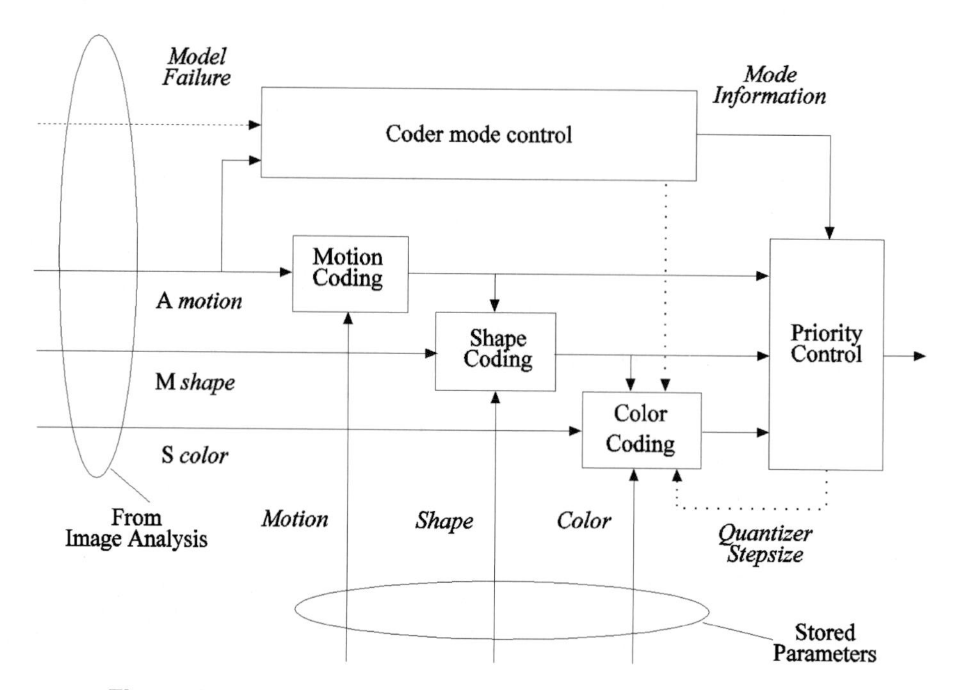

Figure 12.3 Parameter coding for the motion, shape, and color parameters [754]. © 1994 IEEE.

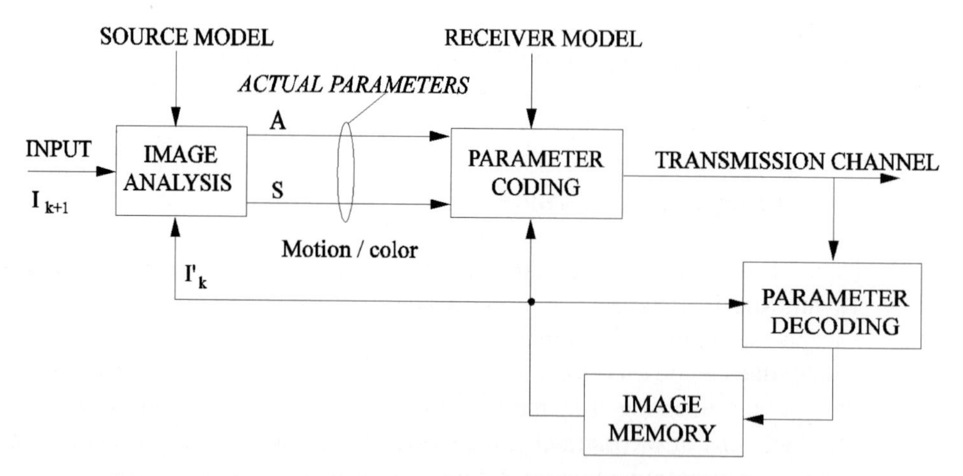

Figure 12.4 A generalized block-oriented hybrid coder in view of object-oriented analysis-synthesis coder [754]. © 1994 IEEE.

coded using the actual motion compensation and the stored shape information from previous image frames [756].

The object-oriented analysis-synthesis coding is basically efficient for some images: only a few objects are moving moderately, the moving objects cover a part (roughly half [754]) of the image, and there is no camera motion, zoom or pan. Moreover there are *covered* regions and *uncovered* regions, such as an opening mouth or closing eyes, in an object. Hence, the motion description is only valid in a part of the object and these small objects may not be represented with sufficient accuracy by displacements. These objects are defined as MF objects (model failure). In contrast, the objects that can be represented with sufficient accuracy are defined as MC objects (model compliance). The initial 3D object model can be derived from the image sequence by employing the object silhouette [759]. The object silhouette is obtained by image segmentation and the border of an object silhouette is interpreted as the outermost contour of an object. The 3D model is obtained in the transmitter and is used to synthesize 2D image sequences in the receiver. One way to arrive at a 3D model is by rotating the 2D silhouette [763]. For an accurate motion detection, the object boundaries have to be known, while for a correct object boundary detection, an accurate motion estimation is required. Hence, motion estimation and shape detection are to be treated jointly, because they influence each other [753].

Image synthesis is performed both at the encoder and at the decoder. Principally, two techniques can be employed for image synthesis. First, the filter concatenation technique [750] synthesizes the following image by means of the previously synthesized image and the actual motion parameters. If displacement vectors with half-pixel resolution are employed, the image has to be interpolated by means of spatial filtering. This technique is usually applied in block-oriented hybrid coding techniques (Section 9.3). The main disadvantage of this technique is that synthesis errors due to the interpolation filter influence the next filtering. In contrast, the parameter concatenation technique [756] synthesizes the following image by means of the stored signal of the object memory and an appropriate addition of all motion parameters that have been transmitted. In this case, the signal of the first image of the sequence is stored in an object memory. The main advantage of this technique is that the interpolation filter is applied from the stored object memory for each image to be synthesized.

Parameter coding

The motion, shape, and color parameters are properly coded in parameter coding block (Fig. 12.2). The coder modes and the priority control of the parameters are included in Fig. 12.3. The coder mode control selects the parameter sets to be transmitted (Table 12.3). In the case of motion-failure objects, the motion information is not coded and only shape and color information is transmitted.

Table 12.3 Coder mode control for the motion, shape, and color parameters [754]. (MC: model compliance; MF: model failure)

Parameters	MC objects	MF objects	Uncovered image areas
Motion	Coded	Uncoded	Uncoded
Shape	Coded	Coded	Uncoded
Color	Uncoded	Coded	Coded
Distortions	Geometric error	Quantization error	

Source: © 1994 IEEE.

In the case of motion-compliance objects, color parameters need not be coded. For uncovered image areas, the shape information is determined by the shape and motion parameters of the MC objects. Geometrical errors are more important for MC objects and quantization errors are important for the other two modes. Essentially, the motion parameters are transmitted first, then the shape update parameters, and finally the color parameters of the MF objects. To guarantee the predictive coding of all parameter sets, uncovered image areas are finally transmitted.

12.2.3 Model-based coder

Model-based coders update a head-and-shoulders model of the speaker at the receiver on the basis of parameters extracted and quantized at the transmitter. These coders offer the promise of fairly good quality at very low–bit-rates, but the complete systems are still under study and the coders will require considerable computational resources. Furthermore, the reported systems are limited to a particular model, such as a head and shoulders and a stationary background. Hence, special knowledge obtained from a known a priori model of the object is used in the coding system [767, 761]. The analysis task is simplified by using the knowledge of the objects. Modeling of the scene will still be complicated if the object is not rigid and contains a lot of details that are not known beforehand.

The model-based coder analyzes the input images at the encoder and synthesizes the output images at the decoder on the basis of the general knowledge about the subjects to be coded and the three-dimensional shape model [777]. The coding scheme treats an image as a two-dimensional projection of the 3D world and not as a statistical signal. The encoder transmits the first image frame and estimates the 3D structure of the subsequent subjects. For the following image frame, the system transmits the analyzed parameters (such as the location of feature points) instead of the image itself. Output images are synthesized by deforming the shape model from the transmitted parameters. That is, most of the

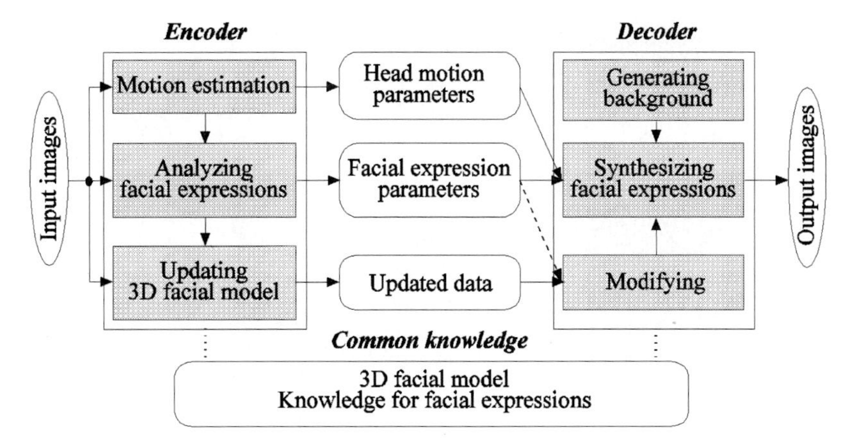

Figure 12.5 A model-based coding system for facial image sequences [777]. © 1994 IEEE.

information is contained in the first frame, resulting in a very low bit-rate coding scheme. In this method, both the transmitter and the receiver have the facial model as shown in Fig. 12.5.

The following issues need to be mastered so as to achieve the implementation and adaptation of a model-based coder:

- 3D solid modeling of input images
- Modeling of facial action and 3D motion estimation
- Fitting to the contours of facial features
- Real-time image synthesis

The automatic modeling and analysis of general scenes are major problems. One approach is that, when the model and the initial position of the face is given, the facial motion can be tracked by using the facial feature points. Choi et al. [766, 777] and Li et al. [770] proposed some automatic analysis methods for facial movements. For the modeling of the image, a triangle patching technique is frequently used, as shown in Fig. 12.6, structuring the fundamental 3D wireframe model. A lot of triangle patches, generally four to seven hundred [768], are produced. The patches follow the contour lines of the facial area, such as eyebrows, eyes, nose, mouth, and jaw. Thirty-four feature points can then be allocated for motion areas [771]. The displacement parameters of the feature points are transmitted, and the moved position is calculated on the wireframe model at the decoder.

The model-based coder is applicable not only in the low bit-rate coding systems, but also in the speech- and text-driven synthesis system [776]. The speech-

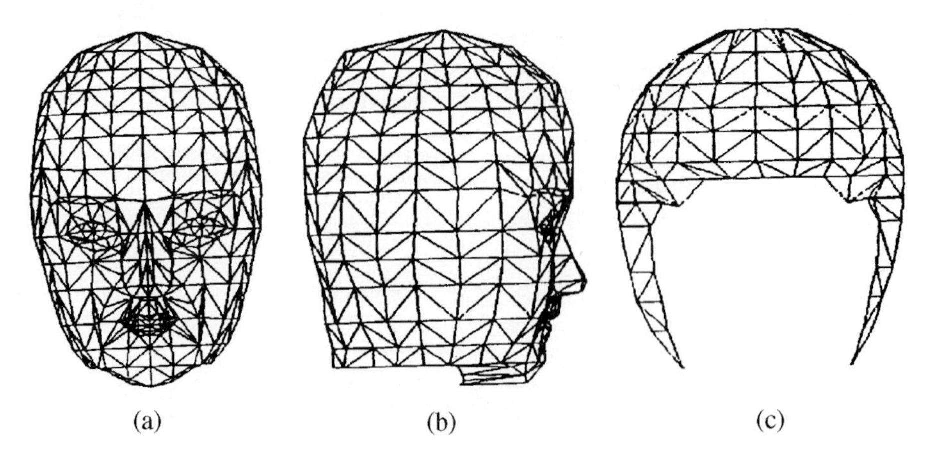

Figure 12.6 An example of 3D surface model for a person's face. (a) Front view. (b) Lateral view. (c) Hair [777]. © 1994 IEEE.

driven system needs to convert the phoneme features into image movement. Each pronunciation class has wide opening, narrow opening, or both types of mouth shape code. These narrow or wide codes can describe local facial movements. This technique, however, is likely to synthesize an artificial, unpleasant image if the coder uses too few feature points and neglects deeper muscle structure.

Since the 3D model-based coding uses parameterized models and is more graphics oriented, the reconstructed image quality is generally considered to be good. There seem to be many applications that can not be achieved by waveform-based coding. The application areas include videophone communication, speech driven image synthesis, virtual space image conference, and detection of facial change at age. The problems to be solved for the successful systems are how to obtain generality with unmodeled objects and how to reduce serious effects caused by analysis errors.

12.2.4 Fractals and morphological coders

Fractal image coding as a completely new concept has received considerable attention for very low bit-rate image coding. The basic idea behind fractal image coding is to represent an image sequence by a transformation using self-similarity. A fractal image can be generated from an arbitrary (start) image by applying the set of mappings on the image iteratively until the image is stable. Fractal images are also redundant images in the sense that they are made up of transformed copies of either themselves or parts of themselves. In the context of image compression, it is assumed that image redundancy can be exploited by modeling it as the redundancy present in fractal images. Therefore, if the transformation describes the image well, the image context can be compactly represented by the transformation. Thus, the problem of fractal image coding is to find a necessary transformation.

Barnsley and Sloan [781] first presented the idea of fractal image coding and a number of researchers have followed this for finding an efficient image/video coder. This research is mainly based on the Barnsley's iterated function system (IFS), which is a typical example of the self-transformation system (STS). When applying the IFS to compress natural images, however, compression performance may not be as good as in graphic images. The self-similarity can be found in a small area or in a large area. Monro and Dudbridge [784] proposed a blockwise mapping method. The smaller block mapping rather than the entire image as the whole function results in a good approximation. Jacquin developed the first automatic fractal image coder based on the piecewise transformation system (PTS) [782, 793]. The whole function in the PTS is composed of transformations of pieces of itself, while the STS maps the entire function to each part of the function. The PTS is a block-coding method that relies on the assumption that image redundancy can be efficiently utilized through blockwise self-transformation, called fractal block coding. Image partition methods not only affect the visual artifacts but also affect the coder performance. Novak et al. developed a triangular partition method in [785]. Fisher [783] tried several different kinds of image partitioning. The reader may refer to numerous references cited in this book.

In comparison to still-image coding, little work has been done on fractal-based video coding. Beaumont [786] directly extended Jacquin's 2D coder to 3D image sequence coding. The huge computational burden in the 3D coder is a problem, however. Instead of the PTS model, the 3D STS model with no search processing [789] reduces the computational complexity. A 3D image multiviewed by multiple cameras can be modeled efficiently [805]. To synthesize intermediate images from the multiple images, interpolation of viewing angle images contributes to the complexity reduction. Paul [807] proposed a fractal-based video coder that estimates motion by predicting the IFS map of a given range block using that of a parent range block along the motion trajectory. The IFS mapping can be modeled for hierarchical video coding [806] that combines the pyramid and the interframe video coding, obtaining the functional equations in both the spatial and time domains.

The morphological coder has been known to work well in the graphics image. Morphological segmentation, however, is a new technique that can be applied to segmentation of natural still images and video sequences. It can be a good candidate for implementing the very low bit-rate video coder. It is mainly based on segmentation and the corresponding representation of coding parameters. The motion parameter can be adjusted with the segmented images [810] and the image is split and merged. Subband decomposition is also possible using the morphological filters [809, 728].

12.3 ITU-T Draft Recommendation H.263

The ITU-T (formerly CCITT), particularly its Experts Group for Very Low Bit-Rate Video Telephony (LBC), has produced a set of near-term recommendations for PSTN (public switched telephone network) video telephony. The near-term solution has been finalized in December 1995 and its draft recommendation, named H.263, was developed in April 1995 [708]. In the meantime, it was named H.26P and has been extensively simulated by ITU-T SG 15. Test models for the near-term (TMN1 through TMN5 [709]) solution have also been created, providing an encoder reference. TMN5 has been available since March 1995, including unrestricted motion vectors, syntax-based arithmetic coding, advanced prediction mode, and P/B frames. Source code and documents are available in the public domain (see Appendix B). In this section, we describe the H.263 designed for very low bit-rate coding applications.

The video coding algorithm of Draft Recommendation H.263 is an extended version of ITU-T H.261 (Chapter 9). It is based on motion-compensated hybrid DPCM/DCT coding with considerable improvements to fit bit rates less than 64 Kbps, as shown in Fig. 12.7. Reference Model 8 [451] was the last reference model used for simulating H.261, and the TMN series is the test model for H.263. We describe and compare these two models in this section. The main structure of TMN5 is the same as that of RM8. Some important deviations from RM8 are

- Include various video formats such as sub-QCIF, 4CIF, 16CIF
- Advanced prediction mode: half-pixel motion estimation, median-based MV prediction, 4 MVs per macroblock, and overlapped block MC
- Unrestricted MV mode: when MV points outside the picture area, use edge pixels
- A syntax-based arithmetic coding (SAC) mode is possible to change the given VLC tables
- PB-frames mode (forward and bidirectional prediction): similar to those in MPEG
- Weighted quantizer matrix for B-blocks
- No loop filter; no macroblock addressing.
- 1-bit coded or not-coded macroblock information in MB layer (separate coded block patterns for luminance (CBPY) and chrominance (MCBPC) components and for intra/inter mode)
- 2-bit differential quantizer information in MB layer and 5-bit quantizer information in picture layer and in GOB layer

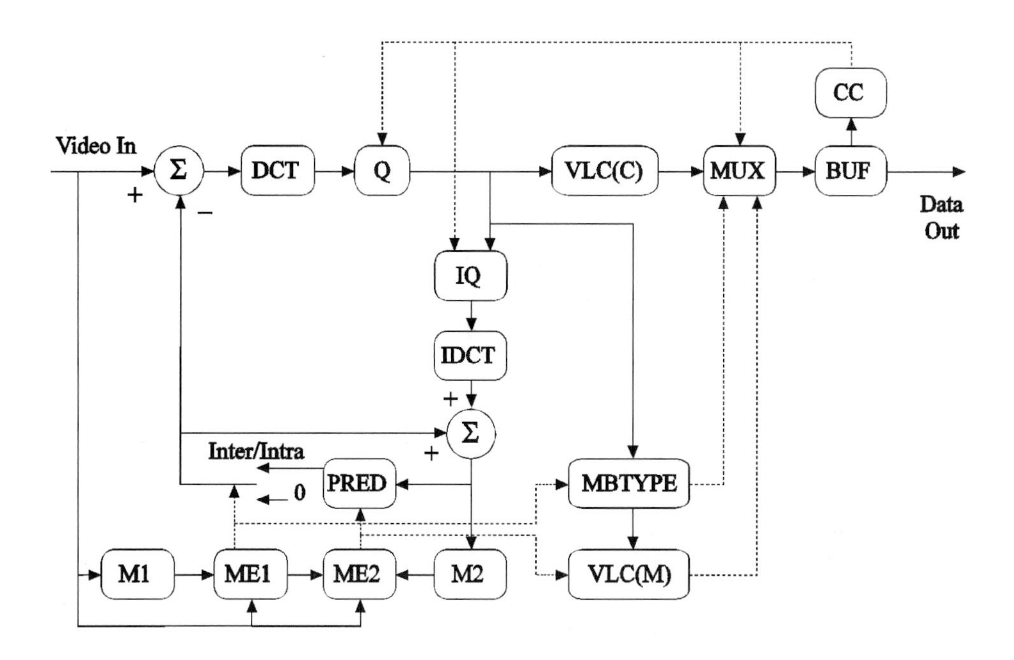

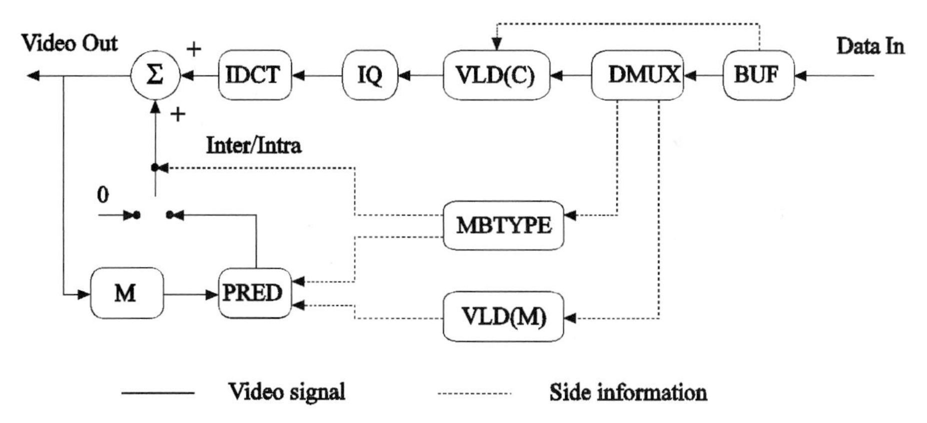

Figure 12.7 A block diagram of the H.263 encoder and decoder. (ME1: Integer pixel motion estimation and intra/inter decision; ME2: half-pixel motion estimation; M1: input frame store; M2: decoded frame store; PRED: make prediction block; MBTYPE: decide block type and block pattern; VLC(C): variable-length coder for transform coefficients; VLC(M): VLC for motion vectors; CC: coding control; DCT: discrete cosine transform; Q: quantizer; VLD(C): variable-length decoder for transform coefficients; VLD(M): VLD for motion vectors; M: decoded frame store) [708]. © 1995 ITU-T.

- 3D VLC (Last-Run-Level) for coding the transform coefficients
- VLC for B-blocks

Analog input pictures are sampled at an integer multiple of the video line rate, and the input video format in the source coder is a common intermediate format (CIF), as that of H.261, to permit a single recommendation for both 625- and 525-line television standards. The CIF video is converted into a QCIF picture in the source coder, where the luminance sampling structure is 176 pels/line and 144 lines/picture in an orthogonal arrangement as shown in Fig. 9.4. The QCIF is the first picture format used in H.263. But other picture formats can be used: sub-QCIF (128 pels/line and 96 lines/picture), CIF (352 pels/line and 288 lines/picture), 4CIF (704 pels/line and 576 lines/picture), and 16CIF (1408 pels/line and 1152 lines/picture). The pixel aspect ratio maintains the 4:3 ratio, i.e., $(4/3) \times (144/176)$. All decoders must be able to decode QCIF and sub-QCIF pictures (CIF is optional). Encoders must be able to operate with at least one of the formats QCIF or sub-QCIF (the others are optional). Color format and positioning of luminance and chrominance samples are same as in H.261 (Chapter 9).

Application areas of H.263 are aimed at bit rates up to 64 Kbps. The coder may be used for bidirectional or unidirectional visual communications, such as videophone, videoconferencing, and mobile telephone. Due to these very low transmission rates, no error handling is included in draft Recommendation H.263.

In the TMN4, a 7-tap filter is used for conversion from CIF to QCIF, where the filter taps are $(-1, 0, 9, 16, 9, 0, -1)/32$. The operator / means integer division with truncation toward zero. The conversion from CIF to QCIF (the down-sampling process) is performed by using this filter, in both horizontal and vertical directions, for the luminance and chrominance components. The conversion from QCIF to CIF (the up-sampling process) is performed by using the same filter weighted by 1/16. At the edges, mirroring around the end pixels is performed. As with all the other standards, the actual conversion process CIF to/from QCIF is not part of H.263.

A four-layered structure is used in the coding of H.263 video: picture layer, group-of-blocks layer (GOB), macroblock layer, and block layer as shown in Fig. 12.8. Most of the abbreviations are defined in Section 9.2.2. A picture layer consists of a picture header followed by data for a GOB layer. At the end of a sequence, a GOB layer is followed by an EOS (end of sequence, 0000 0000 0000 0000 1111 11) code and a number of PSTUF (stuffing) bits. A QCIF picture contains 99 macroblocks as shown in Table 12.4. A GOB layer consists of a header and one or more macroblock rows. A macroblock consists of six blocks (four luminance blocks and two chrominance blocks) in the same order as shown in Figs. 9.5 and 10.20. A block consists of (8×8) pels.

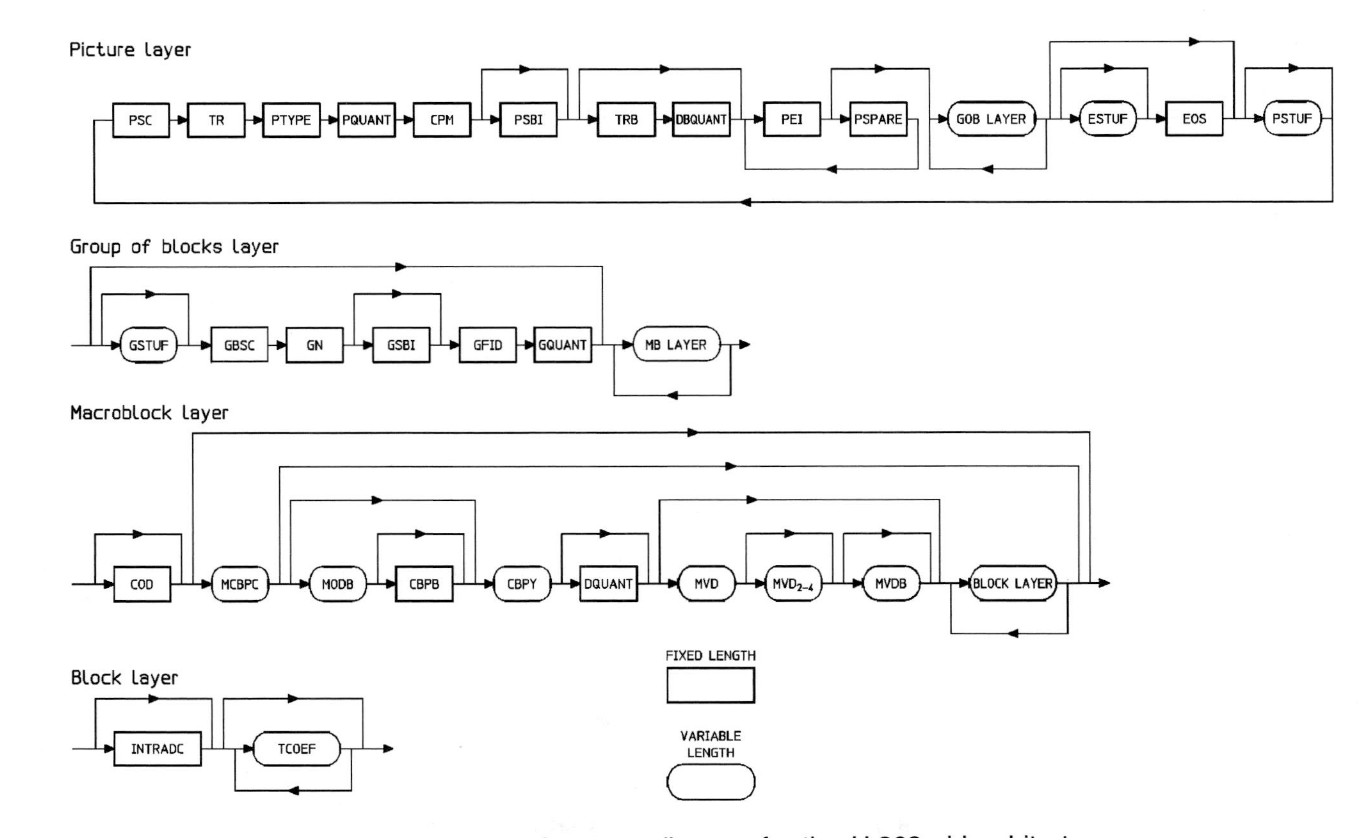

Figure 12.8 A syntax diagram for the H.263 video bit stream [708]. © 1995 ITU-T.

Table 12.4 Arrangement of 99 macroblocks in a QCIF picture [708]

Row number											
0	1	2	3	4	5	6	7	8	9	10	11
1	12	13	14	15	16	17	18	19	20	21	22
2	23	24	25	26	27	28	29	30	31	32	33
3	34	35	36	37	38	39	40	41	42	43	44
4	45	46	47	48	49	50	51	52	53	54	55
5	56	57	58	59	60	61	62	63	64	65	66
6	67	68	69	70	71	72	73	74	75	76	77
7	78	79	80	81	82	83	84	85	86	87	88
8	89	90	91	92	93	94	95	96	97	98	99

Source: © 1995 ITU-T.

Table 12.5 Quantization information for B-block coding (the operator / means division with truncation)

DBQUANT	BQUANT
00	$(5 \times \text{QUANT})/4$
01	$(6 \times \text{QUANT})/4$
10	$(7 \times \text{QUANT})/4$
11	$(8 \times \text{QUANT})/4$

Source: [708]. © 1995 ITU-T.

The picture start code (PSC, 22 bits) is given by 0000 0000 0000 0000 1000 00. The TRB (temporal reference for B-frames, 3-bit FLC) indicates the number of nontransmitted pictures since the last P-picture and before the B-picture when PB-frames mode is used.

The picture type information (PTYPE) (13 bits) in the picture layer indicates:

- Bit 1 Always 1, in order to avoid start code emulation
- Bit 2 Always 0, for distinction with H.261
- Bit 3 Split screen indicator, 0 off, 1 on
- Bit 4 Document camera indicator, 0 off, 1 on
- Bit 5 Freeze picture release, 0 off, 1 on
- Bit 6–8 Source format, 001 sub-QCIF, 010 QCIF, 011 CIF, 100 4CIF, 101 16CIF
- Bit 9 Picture coding type, 0 intra, 1 inter
- Bit 10 Optional unrestricted motion vector mode, 0 off, 1 on
- Bit 11 Optional syntax-based arithmetic coding mode, 0 off, 1 on

Table 12.6 Variable-length code words for CBPY (coded block pattern for luminance) [708]

Index	CBPY (**intra**) (Y_0Y_1 Y_2Y_3)	CBPY (**inter**) (Y_0Y_1 Y_2Y_3)	Number of bits	Code words
0	00 00	11 11	4	0011
1	00 01	11 10	5	0010 1
2	00 10	11 01	5	0010 0
3	00 11	11 00	4	1001
4	01 00	10 11	5	0001 1
5	01 01	10 10	4	0111
6	01 10	10 01	6	0000 10
7	01 11	10 00	4	1011
8	10 00	01 11	5	0001 0
9	10 01	01 10	6	0000 11
10	10 10	01 01	4	0101
11	10 11	01 00	4	1010
12	11 00	00 11	4	0100
13	11 01	00 10	4	1000
14	11 10	00 01	4	0110
15	11 11	00 00	2	11

Source: © 1995 ITU-T.

- Bit 12 Optional advanced prediction mode, 0 off, 1 on
- Bit 13 Optional PB-frames mode, 0 normal I- or P-picture, 1 PB-frames mode

Bit 5 is also used for packet retransmission or fast update request causing intra mode and allows a decoder to display decoded pictures in the normal manner. Bit 3 indicates that the upper and lower half of the decoded picture could be displayed side by side. If source format (bit 6-8) is different from the previous picture, the current picture shall be an I-picture. If bit 9 is set to 0, bit 13 shall be set to 0 as well. Bit 10-13 refer to optional modes that are only used after negotiation between encoder and decoder.

Quantizer information is indicated by PQUANT, DBQUANT, GQUANT, or DQUANT. PQUANT (5-bit FLC) indicates the quantizer number to be used for the picture until updated by any subsequent GQUANT in a GOB layer or DQUANT (differential QUANT) in a macroblock layer. The code words are the natural binary representations of the values of QUANT, which, being half the step sizes, range from 1 to 31. Since B-pictures can be coarsely quantized, BQUANT (quantizer for B-pictures) is designed from a given quantizer as shown in Table 12.5, and DBQUANT represents the quantization information for the B-block. GQUANT (5 bits) is defined in same manner as PQUANT. This can be updated by any subsequent DQUANT. The 2-bit DQUANT indicates the differential values of quantizer number by -1, -2, $+1$, or $+2$.

Each GOB is composed of one row of macroblocks. For the first GOB, starting at the begining of MB row number 0 (Table 12.4), the GOB header is not transmitted (only MB data are present). The group-of-block start code (GBSC, 17 bits) is defined as 0000 0000 0000 0000 1.

The group number (GN) is a fixed-length code word of 5 bits, indicating the number of the first MB row of the group of blocks. The GFID (GOB frame ID, 2-bit FLC) has the same value in every GOB header of a given picture. It has also the same value as that in the previous picture that has the same PTYPE information. GN and GQUANT are only present when GBSC is present. Decoders must be designed to discard GOB headers but may use them for error resilience.

The coded macroblock indication (COD) (1-bit) is set to zero if at least one block of the macroblock is coded and set to one if the macroblock is not coded. The contents of the noncoded macroblock are copied from the same position of the last picture. Macroblock types, coded block pattern for luminance (CBPY) (Table 12.6), and coded block pattern for chrominance (MCBPC) (Table 12.7) are represented by variable-length codes. If the macroblock type is 0 or 3 (Tables 12.7 and 12.8), the MB is coded in intra mode (all the blocks are coded). CBPC indicates that the chrominance block (C_b or C_r) is coded or not. If any non-INTRADC coefficient is present (coded) for a chrominance block, the CBPC is notified as 1, otherwise 0. The 6-bit CBPB (coded block pattern for B-blocks) indicates which of the six blocks in a macroblock is coded or not coded. Stuffing bits (indicated by MCBPC = stuffing) are used to adjust the bit rate. Separate code words for intra mode pictures and inter mode pictures are used for the coded chrominance block patterns as shown in Table 12.7.

If the PB-frames mode is indicated by PTYPE information, then TRB, DBQUANT, MODB, CBPB, and MVDB may be activated. MODB represents whether CBPB and/or MVDB (motion vector data for B-blocks) are transmitted for the macroblock. This means MVDB may not be transmitted while CBPB is present. TRB is the temporal reference for B-frames.

The variable-length code words for CBPY are given in Table 12.6. It gives information on which luminance blocks among the four blocks contain coded transform coefficients. $\text{CBPY}_n = 1$, if any non-INTRADC coefficient is present for block n, otherwise 0, for each bit CBPY_n in the coded block pattern. The shortest code word, 11, is assigned in the inter/no-blocks-coded and intra/all-blocks-coded case.

For intra block coding indicated by MCBPC, the DC coefficient is quantized with a step size of 8 and no dead zone. The resulting values are represented with 8 bits. TCOEF is present if indicated by MCBPC or CBPY. The EOB code used in H.261 is not included, since it can be replaced by 3D VLC coding for (8×8) coefficients. Table C.6 in Appendix C shows the 3D VLC code words for non-INTRADC coefficients. The last bit s denotes the sign of the level, 0 for positive and 1 for negative. An event is composed of three parameters: LAST,

Table 12.7 Variable-length code words for MCBPC (macroblock type and coded block pattern for chrominance (see Table 12.8 for MB type) [708]

MB type	CBPC (C_bC_r)	Code words: **Intra** pictures	Code words: **Inter** pictures
0	00		1
0	01		0011
0	10		0010
0	11		0001 01
1	00		011
1	01		0000 111
1	10		0000 110
1	11		0000 0010 1
2	00		010
2	01		0000 101
2	10		0000 100
2	11		0000 0101
3	00	1	0001 1
3	01	001	0000 0100
3	10	010	0000 0011
3	11	011	0000 011
4	00	0001	0001 00
4	01	0000 01	0000 0010 0
4	10	0000 10	0000 0001 1
4	11	0000 11	0000 0001 0
Stuffing		0000 0000 1	

Source: © 1995 ITU-T.

Table 12.8 Macroblock types and included elements (o means that the item is present for the MB coding) [708]

Picture type	MB type	Mode	COD	MCBPC	CBPY	DQUANT	MVD	MVD 2 – 4
Inter	0	Inter	o	o	o		o	
Inter	1	Inter+Q	o	o	o	o	o	
Inter	2	Inter+ 4MV	o	o	o		o	o
Inter	3	Intra	o	o	o			
Inter	4	Intra+Q	o	o	o	o		
Inter	Stuff-ing	–	o	o				
Intra	3	Intra		o	o			
Intra	4	Intra+Q		o	o	o		
Intra	Stuffing	–		o				

Source: © 1995 ITU-T.

Note: VLCs for MB type are shown in Tables 12.6 and 12.7.

Table 12.9 VLC table for differential motion vector data (MVD) [708]

Vector differences		Code words	Vector differences		Code words
−16	16	0000 0000 0010 1	0		1
−15.5	16.5	0000 0000 0011 1	0.5	−31.5	010
−15	17	0000 0000 0101	1	−31	0010
−14.5	17.5	0000 0000 0111	1.5	−30.5	0001 0
−14	18	0000 0000 1001	2	−30	0000 110
−13.5	18.5	0000 0000 1011	2.5	−29.5	0000 1010
−13	19	0000 0000 1101	3	−29	0000 1000
−12.5	19.5	0000 0000 1111	3.5	−28.5	0000 0110
−12	20	0000 0001 001	4	−28	0000 0101 10
−11.5	20.5	0000 0001 011	4.5	−27.5	0000 0101 00
−11	21	0000 0001 101	5	−27	0000 0100 10
−10.5	21.5	0000 0001 111	5.5	−26.5	0000 0100 010
−10	22	0000 0010 001	6	−26	0000 0100 000
−9.5	22.5	0000 0010 011	6.5	−25.5	0000 0011 110
−9	23	0000 0010 101	7	−25	0000 0011 100
−8.5	23.5	0000 0010 111	7.5	−24.5	0000 0011 010
−8	24	0000 0011 001	8	−24	0000 0011 000
−7.5	24.5	0000 0011 011	8.5	−23.5	0000 0010 110
−7	25	0000 0011 101	9	−23	0000 0010 100
−6.5	25.5	0000 0011 111	9.5	−22.5	0000 0010 010
−6	26	0000 0100 001	10	−22	0000 0010 000
−5.5	26.5	0000 0100 011	10.5	−21.5	0000 0001 110
−5	27	0000 0100 11	11	−21	0000 0001 100
−4.5	27.5	0000 0101 01	11.5	−20.5	0000 0001 010
−4	28	0000 0101 11	12	−20	0000 0001 000
−3.5	28.5	0000 0111	12.5	−19.5	0000 0000 1110
−3	29	0000 1001	13	−19	0000 0000 1100
−2.5	29.5	0000 1011	13.5	−18.5	0000 0000 1010
−2	30	0000 111	14	−18	0000 0000 1000
−1.5	30.5	0001 1	14.5	−17.5	0000 0000 0110
−1	31	0011	15	−17	0000 0000 0100
−0.5	31.5	011	15.5	−16.5	0000 0000 0011 0

Source: © 1995 ITU-T.

RUN, and LEVEL. If LAST is 0, there are more nonzero coefficients in the block. If it is 1, this is the last nonzero coefficient (i.e., EOB). Again the zigzag pattern (Fig. 8.4) is used for scanning the coefficients.

The 3D representation reduces the number of bits needed for an end-of-block code, assuming that a number of blocks are coded with a few nonzero coefficients. The number of successive zeros (RUN) and the nonzero value of the coded coefficient (LEVEL) are of the same concept as in H.261. The remaining combinations (those occurring infrequently) of LAST, RUN, and LEVEL are coded with a 22-bit FLC consisting of 7-bit ESCAPE (0000 011), 1-bit LAST (1 or 0), 6-bit RUN (0 to 63), and 8-bit LEVEL (-127 to 127, 0 is forbidden). This 22-bit FLC can also be used (optional) to code the events listed in Table C.6.

As previously mentioned, the H.263 decoder supports half-pixel motion estimation and a median-based MV predictor. Motion vectors are differentially coded using variable-length code words as shown in Table 12.9. The encoding algorithm of motion vectors depends on the encoder designer, but the motion vector is obtained by adding predictors to the vector differences (MVD) in the decoder. H.263 provides not only one MV per macroblock but also motion vectors for (8×8) blocks, i.e., four MVs per macroblock. In the case of four vectors per macroblock, called unrestricted motion vector mode, four MVDs are present in the macroblock layer. In the case of one vector per macroblock, the candidate predictors for the differential coding are taken from three surrounding macroblocks as shown in Fig. 12.9. At the borders of the current GOB or picture, the candidate motion vector(s) is/are set to zero (out of the left/right border) or to MV1 (out of the top border). The predictors are calculated separately for the horizontal and vertical components. For each component, the predictor is the *median* value of the three candidates, as described in Example 12.1. There is, of course, no motion vector in intra mode. When the corresponding macroblock is coded in intra mode, the candidate predictor is set to zero. MVD is included for all inter macroblocks.

Example 12.1 *From Fig. 12.9, suppose* $MV1 = (-1, 2)$, $MV2 = (0, 4)$, *and* $MV3 = (2, 1)$. *Then the prediction value of motion vector is given by*

$$Median(MV1x, \quad MV2x, \quad MV3x) = 0$$

$$Median(MV1y, \quad MV2y, \quad MV3y) = 2$$

If the motion vector for the current macroblock is $(1, 1)$, *then the differential motion vector is* $(1, -1)$ *and the VLC code words are represented as* $(0010, 0011)$, *from Table 12.9.*

The advantage of this median-based MV prediction is that the motion vector bits can be reduced (about 10 percent reduction was reported [709]). In the case of one MV per macroblock, the motion vector is used for all pixels in all four luminance blocks. Motion vectors for both chrominance blocks are derived by dividing the component values of the luminance motion vector by two, due to the lower chrominance resolution. Motion compensation with a half-pixel motion vector is performed using bilinear interpolation. Further discussion on how to decode the correct motion vector from VLC coded MVD is shown in Example 9.2.

Unrestricted motion vector mode: Motion vectors are allowed to point outside the picture. In this case, an edge pixel (the last full pixel) is used instead.

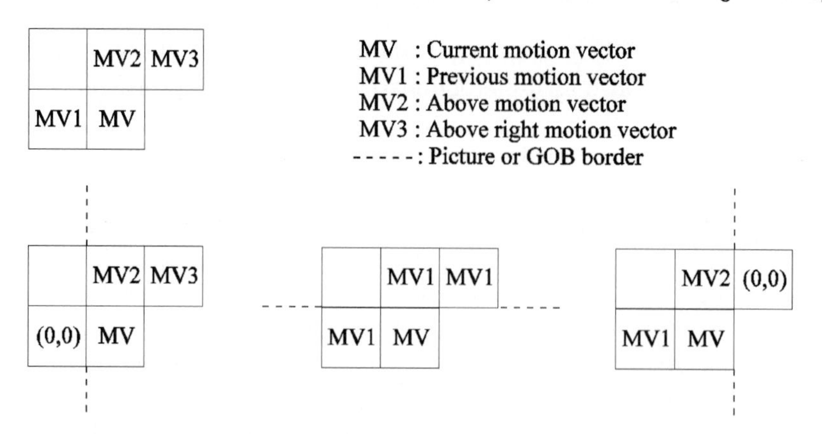

Figure 12.9 Motion vector prediction from surrounding macroblocks [708]. © 1995 ITU-T.

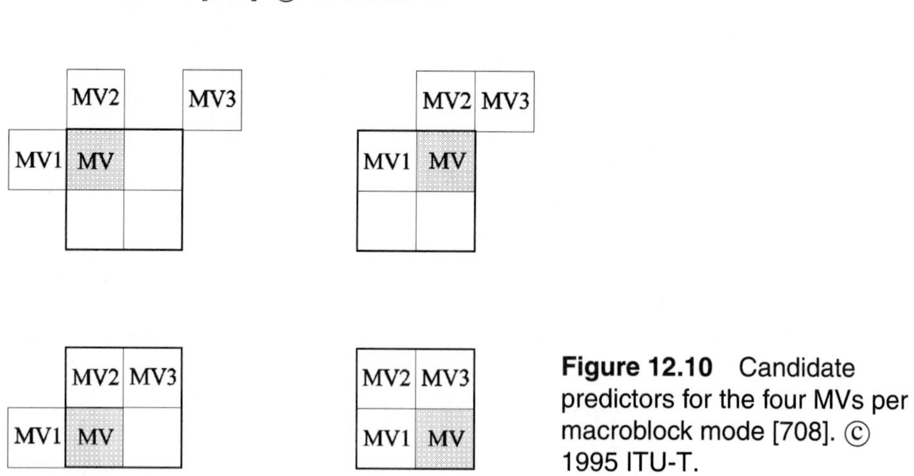

Figure 12.10 Candidate predictors for the four MVs per macroblock mode [708]. © 1995 ITU-T.

For instance, a motion vector using an overlapped block MC may exceed the picture area.

Syntax-based arithmetic coding (SAC) mode: Since further reduction of resulting bit rates can be achieved by arithmetic coding, all the corresponding VLC/VLD operations of H.263 are replaced with arithmetic coding/decoding operations in this mode.

Advanced prediction mode: This includes four MVs per macroblock and overlapped block motion compensation. The four MVs are MVD and MVD_{2-4}, indicated by PTYPE and MCBPC. In the case of one motion vector per macroblock, the MV is obtained by taking the median value from the three surrounding macroblocks as shown in Fig. 12.9. In the case of four vectors per macroblock, however, the candidate predictors MV1, MV2, and MV3 (from surrounding blocks) are redefined as indicated in Fig. 12.10. The motion vector for both chrominance

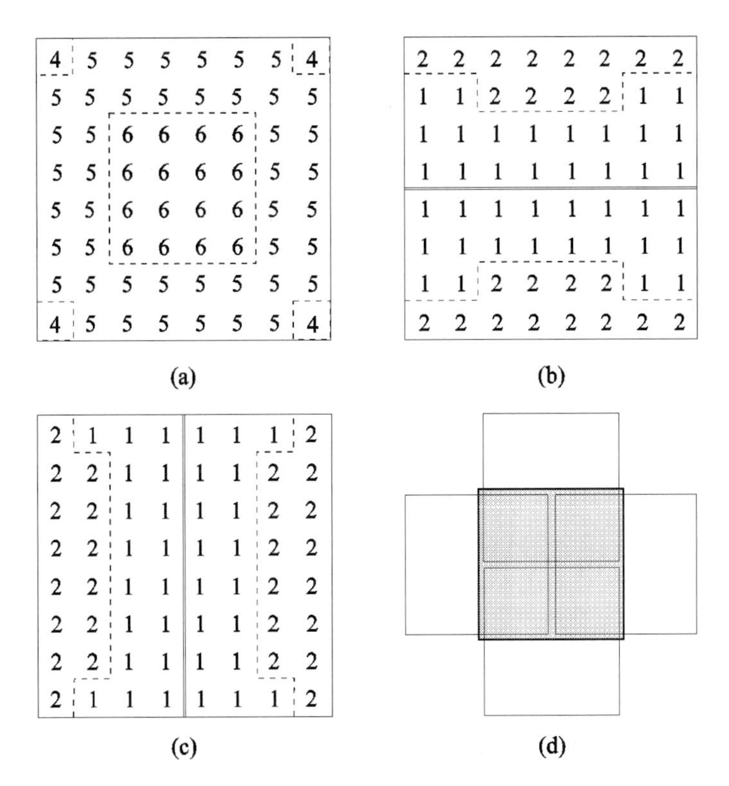

Figure 12.11 Weighting values for prediction with the motion vector of the (a) current, (b) top/bottom and (c) left/right luminance block. The prediction is overlapped (d) for the block motion compensation [708]. © 1995 ITU-T.

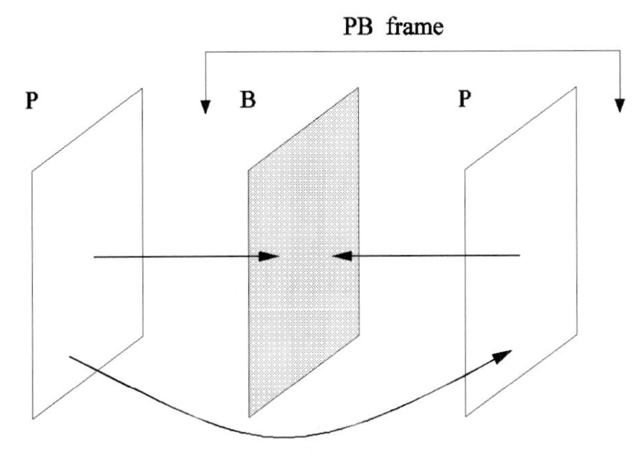

Figure 12.12 Prediction in PB-frames mode [708]. © 1995 ITU-T.

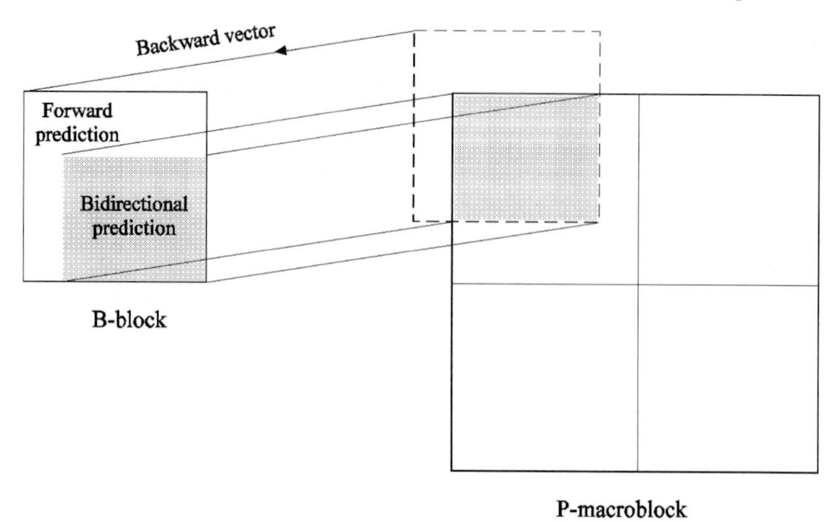

Figure 12.13 Forward and bidirectional prediction for a B-block [708]. © 1995 ITU-T.

blocks is obtained by calculating the sum of the four motion vectors divided by 8. The resulting sixteenth-pixel resolution vectors are modified to get the nearest half-pixel resolution motion vectors.

The motion vector prediction is obtained by overlapped block motion compensation. Each pixel in a luminance block is a weighted sum of three prediction values, divided by 8. To obtain the three prediction values, three motion vectors are used: the motion vector of the current luminance block and two overlapping motion vectors of the nearest left/right and top/bottom luminance block as described in Fig. 12.11. Note that the sum of the weighting coefficients is one. The motion vector is, of course, set to zero in the case of a not-coded or intra mode block. At the border of the picture, the corresponding motion vector is replaced by the current motion vector.

Optional PB-frames mode: A PB-frame (Fig. 12.12) consists of two pictures: a P-picture, which is predicted from the last decoded P-picture, and a B-picture, which is predicted from the last decoded P-picture and the P-picture currently being decoded. Parts of the B-picture may be bidirectionally predicted from the past and future P-pictures. The concept of PB frames comes from picture types in MPEG (Fig. 10.8). This optional mode is indicated by PTYPE (bit 11). For PB frames the coding mode intra implies the P-blocks are intra coded, and the B-blocks are inter coded with prediction as for an inter block.

Prediction of a B-block in a PB frame A block means an (8×8) block. The following procedure applies for luminance as well as chrominance blocks. First, the forward and backward motion vectors are calculated. It is assumed that the P-macroblock (luminance and chrominance) is first decoded and reconstructed. This macroblock is called P_{REC}. Based on P_{REC} and the prediction for P_{REC}, the prediction for the B-block is calculated. The prediction of the B-block has two modes (Fig. 12.13) that are used for different parts of the block: For pixels where the backward motion vector MV_B points inside P_{REC}, use bidirectional prediction. This is obtained as the average of the forward prediction using MV_F (forward motion vector) relative to the previously decoded P-picture and the backward prediction using MV_B relative to P_{REC}. The average is calculated by dividing the sum of the two predictions by two (division by rounding). For all other pels, forward prediction relative to the previously decoded P-picture is used.

The INTRADC coefficient is reconstructed as $\hat{S}_{uv} = 8 \times S_{quv}$. Reconstruction levels for the INTRADC coefficients are in the range of 8 to 2032 (quantized levels in the range of 1 to 254), as shown in Table 9.8. The codes 0000 0000 and 1000 0000 are not used, and the code 1111 1111 is used for the reconstruction level of 1024.

The reconstruction levels of all nonzero coefficients other than the INTRADC are obtained as follows:

$$\begin{aligned} \hat{S}_{uv} &= \text{Sign}(Q_n(2S_{quv} + 1)), \qquad \text{if} \quad Q_n \text{ is odd} \\ \hat{S}_{uv} &= \text{Sign}(Q_n(2S_{quv} + 1) - 1), \quad \text{if} \quad Q_n \text{ is even} \end{aligned} \tag{12.1}$$

where Q_n and S_{quv} denote the quantizer number (1 through 31) and the quantized level, respectively. Note that this process disallows even-valued numbers and prevents accumulation of IDCT mismatch errors. If $S_{quv} = 0$, then the reconstruction level $\hat{S}_{uv} = 0$. The reconstruction levels are clipped so that they fall within the range -2048 to 2047. The sign of the quantized version of the transform coefficient is notified at the end of the VLC code word (Appendix C, Table C.6).

12.4 Audio Coding

Speech coding technology at very low bit rates is already available. The bit rate is in the range of 4.8 to 8 Kbps, and research is being done for a lower bit rate of 2.4 Kbps. The bandwidth of speech is limited to 4 KHz for use in videophone or cellular-phone application. The bit rate of 128 Kbps is obtained for 16-bit PCM when sampled at 8 KHz, as shown in Table 12.10. The concept of a low bit-rate speech coder is defined in two groups, MPEG-4 and LBC, as the upper limit of 9.6 Kbps, to be reasonable for speech coding. For example, the ITU-T Recommendation G.728 16 Kbps LD-CELP coder and GSM RPE-LTP 13 Kbps coder have bit rates that are too high to be of use in a very low bit-rate videophone, but they have other characteristics that are useful benchmarks. Other coders that are available at present are the IS-54 8 Kbps VSELP coder, the Japanese digital cellular (JDC) 6.7 Kbps VSELP coder, the ITU-T 8 Kbps speech coder under standardization, the Inmarsat IMBE coder for satellite service, and the U.S. Federal Standard 1016 4.8 Kbps CELP for secure telephone service. Furthermore the AT & T 6.7 Kbps CELP+ coder, the Marconi 5 Kbps CELP coder, and the Comtech STC coders, Qualcomm CELP at variable bit rates (8, 4, 2, 1 Kbps), have also been developed.

The prevailing coding algorithms are therefore based on the Vector Excitation Coding (VXC) method. Two algorithms that have been adopted as standards in the United States are Code Excited Linear Prediction (CELP) [816] and Vector Sum Excited Linear Prediction (VSELP) [820]. The name *Code Excited Linear Prediction* has been used to reflect the fact that the decoder drives a prediction inverse filter with an input residual or excitation from a codebook. The name *Vector Excitation Coding* is a more generic definition for a large family of coding schemes in which the excitation is selected by searching a VQ codebook. Hence, it is an adaptive VQ coder, i.e., the codebook is dynamically updated by filtering a set of fixed codevectors through a time-varying, spectral-shaping filter to generate a new set of codevectors as shown in Fig. 12.14.

To solve the problem of using a large size codebook, an adaptive codebook selection subject to local signal characteristics is used. The reduced codebook size implies that a substantial drop in rate is needed for transmitting codevector indices, allowing a relatively large amount of side information. The VSELP coder reduces the weighting complexity by applying the weighting filter directly to original speech as shown in Fig. 12.15. It uses two codebooks, and each codebook has seven basis vectors. The effective codebook size is $2^7 + 2^7 = 256$ (8 bits). The weighted errors are minimized by adaptive techniques.

In closed-loop speech coding, the excitation is generated by a convolutional tree code. This was first developed in [815] and systemized in [816] as a CELP coder. The VXC coders require high complexity of the search process, beyond the capability of today's DSP chips. Subsequently, several complexity reduction methods have been proposed, for example, as reported in [817, 818, 820]. For more information, see [823].

For a number of end-user applications, a 4 Kbps speech coding algorithm is under development by the ITU-T. A draft standard will be finalized by 1997 [821]. The applications for the 4 Kbps speech coding algorithm include the following:

- Very low bit-rate PSTN visual telephony
- Personal communications
- Satellite-based personal communication systems
- Packet circuit multiplication systems
- Mobile visual telephony

Table 12.10 Trends in audio and speech coding

Band-width	Sampling freq. (KHz)	16-bit PCM (Kbps)	Coding method		Bit rate (Kbps)	Quality
Audio Coding (15–20 KHz)	48, 44.1, 32	768 705.6 512	MPEG-1, -2 AC-2, -3		up to 384/5CH	CD
Speech Coding (4 KHz)	8	128	Wave-form	Log-PCM (μ-law, A-law)	64	Toll
				ADPCM (G.721)	32	Toll
				ADPCM+SBC (G.722, 7KHz)	64	AM broadcast
			Hybrid	RPE	13	Commun.
				CELP	8, 4.8	Commun.
				VSELP	8, 6.7	Commun.
			Vocoder	LPC	2.4	Synthetic

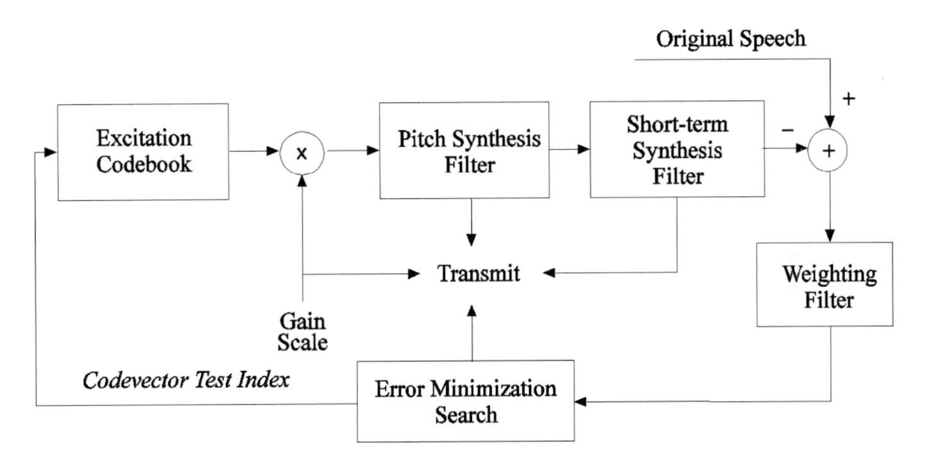

Figure 12.14 Vector excitation speech coding (encoder).

- Message retrieval systems
- Private networks

We describe here some of the existing audio/speech coding algorithms as follows: the *16 Kbps G.728 LD-CELP* coder is known to work well for videophone applications. Its very first application is in ISDN H.320 videophones. The coder has a low-delay (approximately 2 msec one-way delay) property if implemented as described in the specifications. It is, however, a high-complexity algorithm that takes 16 to 18 MIPS floating-point operations. Though the bit rate of this speech

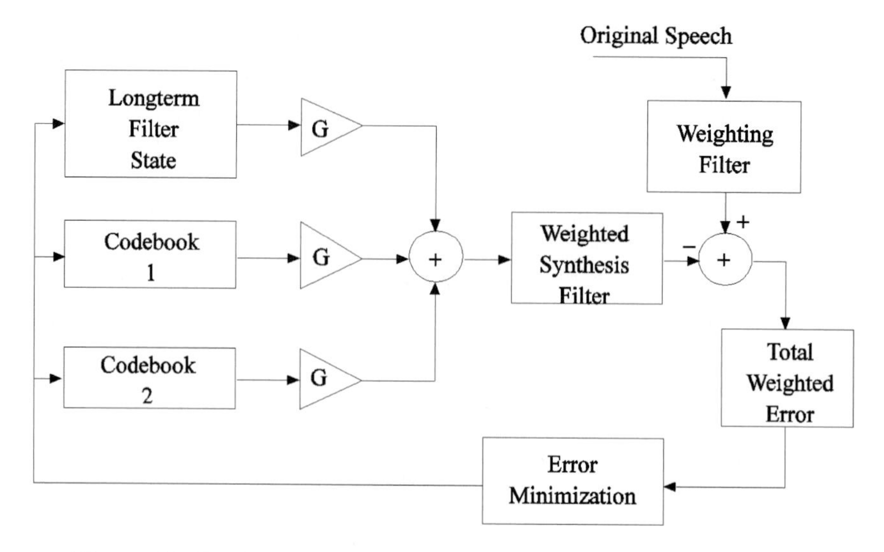

Figure 12.15 Vector sum excited linear prediction speech coder.

coder is 16 Kbps, many listeners have commented that it sounds better over a loudspeaker than the 32 Kbps ADPCM coder does [824]. The coder has been tested and no annoying artifacts were found with speech and nonspeech inputs.

The *GSM 13 Kbps RPE-LTP coder* is the first digital cellular speech coder. It is referred to as regular pulse excitation with long-term predictor (RPE-LTP). The source rate is 13 Kbps, and the transmission rate, including error detecting and correcting coders, is 22.8 Kbps [813]. Like all cellular coders, RPE-LTP is combined with channel error protection when it is used for the digital cellular application. The complexity of RPE-LTP is considered to be low. It can be implemented in 4 MIPS on a TMS320C25 DSP chip. A half-rate GSM coder [825] has been developed so as to use only the half-channel bit rate of 11.4 Kbps instead of 22.8 Kbps. Half-rate coders for existing audio coding standards are required to show full-rate quality and less than four times the complexity of full-rate coders.

The *IS-54 8 Kbps VSELP coder* runs at a bit rate of 7.95 Kbps, while the overall channel bit rate is 13 Kbps [826]. Its performance can be considered slightly better than that of RPE-LTP. The complexity of IS-54 is considered to be medium. It requires about 2 Kwords of RAM and less than 25 MIPS on 16-bit fixed-point DSP chips. IS-54 specifies functional description only so that the designer implements the system by himself.

The *JDC (Japanese Digital Cellular) VSELP coder* [827] is a modified version of the coder used for North American digital cellular transmission. Its source rate is 6.7 Kbps, while the transmission channel bit rate is 11.2 Kbps. The complexity of JDC VSELP is considered to be similar to that of the IS-54 coder and the quality is slightly inferior to the IS-54 coder. It has been also developed as a half-rate coder which has a channel bit rate of 5.6 Kbps.

The *Inmarsat IMBE coder* [828] is developed for Inmarsat maritime satellite communications. The IMBE (improved multiband excitation coder) is a vocoder, rather than a sophisticated waveform coder. This means that it is based on a parametric model of speech production. The IMBE coder has approximately half the complexity of the VSELP coder, making it a low-complexity coder. The speech coding bit rate is around 4 Kbps, while the total channel bit rate is 7.2 Kbps. The IMBE coder shows better performance than CELP-based coders in a radio channel environment.

The *Federal Standard 1016 CELP coder* [814] has a total bit rate of 4.8 Kbps, which includes a sync bit and a small amount of error detection bits in every 30 msec, 144-bit frame. The complexity of FS-1016 is moderate and the quality can be described as fair. It is a bit stream standard, but the designers need only to conform with those tables and bit stream specification. The U.S. government standard for speech coding based on the CELP technique uses a vector dimension of 60 samples when sampled at 8 KHz with a codebook size of 512 (9 bits). Thus, only 9/60=0.15 bit/sample or 1.2 Kbps is used for transmitting the code-vectors, while approximately 3.6 Kbps is used for the side information.

The *AT & T 6.8 Kbps CELP+ coder* is used in the AT & T videophones. The quality of this coder is considered to be equivalent to the IS-54 VSELP coder. Its complexity is roughly the same as VSELP. Hence, it appears to be very similar to VSELP coders, but it runs well at a lower bit rate of 6.8 Kbps.

The *Marconi 5 Kbps CELP coder* [829] is implemented on a 30 MIPS 16-bit fixed-point DSP chip. It requires about 29 MIPS for encoder and decoder operations. Since the total bit rate is 5 Kbps, there are still about 4 Kbps of bit rate available for the video coding on a 9.6 Kbps modem channel.

The *ITU-T G.723* [830] specifies a dual rate speech coder for multimedia telecommunication transmitting at 5.3 and 6.3 Kpbs. This recommendation describes a coded representation that can be used for compressing the speech or other audio signal component of multimedia services at a very low bit rate as part of the overall H.324 family of standards. The coder design was influenced by the very low bit-rate visual telephony as its major application. The higher bit rate has a greater quality. The lower bit rate gives good quality and provides system designers with additional flexibility. Both rates are a mandatory part of the encoder and decoder. It is possible to switch between the two rates at any 30 ms frame boundary.

Chapter 13
High-Definition Television Services

Summary

All-digital terrestrial broadcasting is now feasible. Rapid advances in video compression techniques and their implementation in VLSI have moved what seemed a cute but impractical idea toward a realistic solution. In the United States, the FCC Advisory Committee on Advanced Television Services (ACATS) has recommended a digital terrestrial HDTV system.

In this chapter, we describe video/audio coding in the Digital HDTV Grand Alliance (GA) system. The standardization efforts by MPEG described in Chapters 10 and 11 have a definite influence on this system. Several proponents of and the unified solution for the emerging U.S. HDTV terrestrial standard will be discussed.

13.1 Introduction

The three currently used color television standards of the world, NTSC (National Television System Committee), PAL (Phase Alternating Line), and SECAM (Sequentiel Couleur Avec Mèmoire), have been in use for more than 40 years. It is no surprise that new transmission standards are developed, as people prefer to watch improved systems. In the meantime, the success of direct broadcasting by satellites (DBS) has resulted in development of the MAC (Multiplexed Analog Components) system in Europe while the main three systems have operated in terrestrial broadcasting.

A number of activities aimed at setting new TV standards are taking place worldwide, especially in the Pacific Rim, Western Europe, and the United States. Common to the three HDTV standards are a widened aspect ratio (16:9 instead of 4:3), increased picture resolution, and audio of compact disc quality. All three regions, however, have decided to adopt their own techniques for coding video/audio data and transmitting data via satellite or terrestrial media.

The move to high-definition services was initiated in Japan because most parameters of the system to be used were already finalized there. The experimental HIVISION MUSE direct broadcast satellite system has been in operation since 1989. Due to early investments and focused attention on analog implementations, digital HDTV broadcasting will be considered as the second step when a next-generation satellite is launched. Meanwhile, studio equipment for digital HDTV is in operation for broadcast, affecting industries in other countries.

In Europe, HD-MAC transmission via DBS was developed. The signal processing of HD-MAC coding and decoding is digital. The transmission format, however, is analog and suffers from well-known drawbacks. The key demand is compatibility with conventional systems. The European Union has accepted the demise of HD-MAC and canceled a plan to provide $1 billion to HD-MAC programming. Instead, in September 1993, a group of 85 electronics companies, broadcasters, and satellite operators formed the Digital Video Broadcasting (DVB) project to develop digital technology for the European market [846].

The professional arena has moved steadily toward all-digital solutions due to their obvious advantages in processing and storage/distribution of video. North America has taken a new approach by formulating a fully digital HDTV standard rather than adopting analog or mixed analog/digital techniques already endorsed by others. While Europe and Japan have government support and cooperation in their endeavors, the United States depends primarily on private industry to fund the various projects and to develop new systems.

The U.S. monochrome television standard, i.e., the NTSC system, was adopted by the FCC (Federal Communications Commission) in 1941. July 1953 saw the introduction of color television, upon which all existing terrestrial transmitting systems are based. In both cases, an industry committee developed the technical standard that was then adopted by the FCC.

In 1987 the FCC chartered the advisory committee on advanced television service (ACATS) to recommend an advanced television (ATV) system [853] for the United States. The objective given to the ACATS by the FCC was:

> The Committee will advise the Federal Communications Commission on the facts and circumstances regarding advanced television system for Commis-

Table 13.1 Proposed digital HDTV terrestrial broadcasting systems

	DigiCipher	DSC-HDTV
Lines per frame	1050	787/788
Frames per second	29.97	59.94
Interlace	2:1 (Interlaced)	1:1 (progressive)
Horizontal scan rate	31.469 KHz	47.203 KHz
Aspect ratio	16:9	16:9
Active video pixels	1408 × 960 (luma) 352 × 480 (chroma)	1280 × 720 (luma) 640 × 360 (chroma)
Pixel aspect ratio	33:40	1:1
Bandwidth	21.5 MHz(luma) 5.4 MHz(chroma)	34 MHz(luma) 17 MHz(chroma)
Colorimetry	SMPTE 240M	SMPTE 240M
Video compression	MC DCT	MC DCT & VQ
Block size	8 × 8	8 × 8
Sampling frequency	53.65 MHz	75.5 MHz
Audio compression	Dolby AC-2	Dolby AC-2
Audio bandwidth	20 KHz	20 KHz
Audio sampling freq.	48 KHz	47.203 KHz
Dynamic range	85 dB	96 dB
No. of audio channels	4	4
Video data rate	17.47 Mbps(32QAM) 12.59 Mbps(16QAM)	8.45 to 16.92 Mbps
Audio data rate	0.503 Mbps	0.5 Mbps
Control data	126 Kbps	40 Kbps(spare)
Ancillary data	126 Kbps	413 Kbps
Sync	N/A	292 to 544 Kbps
Total data rate	24.39 Mbps(32QAM) 19.51 Mbps(16QAM)	11.14 to 21.0 Mbps
Error correction overhead	6.17 Mbps	1.3 to 2.4 Mbps
Modulation (terrestrial)	32 QAM or 16 QAM	2- and 4-level VSB
Bandwidth (terrestrial)	4.88 MHz	5.38 MHz
Modulation (satellite)	QPSK	QPSK
Bandwidth (satellite)	24 MHz/channel	24 MHz/channel
	AD-HDTV	**CCDC**
Lines per frame	1050	787/788
Frames per second	29.97	59.94
Interlace	2:1 (Interlaced)	1:1 (progressive)
Horizontal scan rate	31.469 KHz	47.203 KHz
Aspect ratio	16:9	16:9
Active video pixels	1440 × 960 (luma) 720 × 480 (chroma)	1280 × 720 (luma) 640 × 360 (chroma)
Pixel aspect ratio	27:32	1:1
Bandwidth	23.6 MHz(luma) 11.8 MHz(chroma)	34 MHz(luma) 17 MHz(chroma)
Colorimetry	SMPTE 240M	SMPTE 240M
Video compression	MC DCT	MC DCT
Block size	8 × 8	8 × 8
Sampling frequency	56.64 MHz	75.5 MHz

Table 13.1 *(cont.)*

	AD-HDTV	CCDC
Audio compression	MUSICAM	MIT-AC
Audio bandwidth	23 KHz	24 KHz
Audio sampling freq.	48 KHz	48 KHz
Dynamic range	96 dB	
No. of audio channels	4	4
Video data rate	17.73 Mbps	18.88 Mbps(32QAM) 13.60 Mbps(16QAM)
Audio data rate	0.512 Mbps	0.755 Mbps
Control data	40 Kbps	126 Kbps
Ancillary data	256 Kbps	126 Kbps
Sync	N/A	N/A
Total data rate	24.0 Mbps (4.8 HP, 19.2 SP)	26.43 Mbps(32QAM) 21.15 Mbps(16QAM)
Error correction overhead	5.66 Mbps	6.54 Mbps
Modulation (terrestrial)	SS QAM	32QAM or 16QAM
Bandwidth (terrestrial)	5.2 MHz	5.287 MHz
Modulation (satellite)	QPSK	QPSK
Bandwidth (satellite)	24 MHz/2 channels	24 MHz/channel

sion consideration of technical and public policy issues. In the event that the Commission broadcast television is in the public interest, the Committee would also recommend policies, standards and regulations that would facilitate the orderly and timely introduction of advanced television services in the United States.

Testing and data analysis recently were completed on the five high-definition television systems. The five systems are composed of the four digital systems (Table 13.1) and the narrow-MUSE analog system. The test procedure was focused on several viewpoints: spectrum utilization, cost to broadcasters and to consumers, and technology (audio/video quality, transmission error, extensibility, interoperability). The special panel of ACATS recommended that the analog-based systems be dropped and found no major differences in performance among the four digital proponents. Thus the ACATS encouraged the four proponents of the digital systems (Table 13.1) to form a grand alliance (GA) and submit a unified proposal to the FCC. Historical development of television systems in the United States is summarized as follows:

- Apr. 1941: The National Television System Committee (NTSC) standard is adopted (monochrome).
- July 1953: Color is added to the NTSC standard.

- Nov. 1987: The FCC charters the Advisory Committee on Advanced Television Service (ACATS).
- 1991–1992: Five proposed systems are tested at the Advanced Television Test Center (ATTC)
- Feb. 1993: The ACATS Special Panel eliminates Narrow-MUSE and directs retesting of all four digital systems with possible improvements.
- May 24, 1993: The Grand Alliance is formed by the companies representing the four remaining systems.
- Oct. 1993: Most of the key system elements (video compression, transport, scanning formats, and the audio subsystem) are approved.
- Feb. 1994: The final element, the modulation subsystem, is approved. The GA submits the technical description of the digital HDTV system.
- 1995: Extensive lab and field tests of the full system have been completed.
- 1996: Following completion of the laboratory tests and final field test verification of the system's performance, ACATS has recommended to the FCC adoption of the GA HDTV system as the terrestrial broadcasting standard for the United States. The FCC will act on this by late 1996.
- 1998: HDTV service and products are anticipated.

The ATV system proposed for use in broadcasting and cable networks in the United States contains several signal source formats, which can be categorized as interlaced and progressive, with two different image formats: 1920H $\times$ 1080V pixels or 1280H $\times$ 720V pixels. The temporal rates for the former are 23.976/24 Hz (1:1), 29.97/30 Hz (1:1), and 59.94/60 Hz (2:1). Corresponding rates for the latter are 23.976/24 Hz (1:1), 29.97/30 Hz (1:1), and 59.94/60 Hz (1:1), where (1:1), is progressive scan and (2:1) is interlaced scan. Any of these rates could be encountered by an ATV receiver on any channel in addition to the conventional NTSC signals. Six MHz terrestrial broadcast channels are allocated on NTSC-taboo channels that contain interference from distant NTSC transmitters in the same band or adjacent channel interference from nearby transmitters [852]. As a result, the FCC requires that the GA HDTV system provide acceptable viewer coverage without causing excessive interference to NTSC in the same and adjacent channels [857].

The GA system includes algorithms and components from each of the proposed digital systems (Table 13.1), e.g., perceptual modeling, bidirectional motion compensation, hierarchical motion estimation, and additive preprocessing. In addition, the Dolby AC-3 audio compression system was adopted. The design and construction of the prototype system is under way and a U.S. HDTV standard is anticipated by late 1996.

The four proponents are DigiCipher by GI Corp. and MIT [833], DSC-HDTV (Digital Spectrum Compatible) by Zenith and AT&T [834], AD-HDTV by ATRC (Advanced Television Research Consortium, consisting of Compression Labs Inc., David Sarnoff Research Center, Thomson Consumer Electronics, North American Philips, and NBC) [832], and CCDC (Channel Compatible DigiCipher) by MIT and GI Corp. [831]. We discuss these systems and the unified system proposed by the Grand Alliance in subsequent sections.

13.2 Proposed HDTV Systems

13.2.1 DigiCipher HDTV system

DigiCipher, proposed by the American Television Alliance (General Instrument Corporation and the Massachusetts Institute of Technology) is a digital simulcast system that requires a single 6 MHz television transmission channel. The DigiCipher system processes an analog *RGB* video source with 1050 lines, 2:1 interlace, and a 59.94 Hz field rate. The aspect ratio is 16:9, and the video sampling frequency is 53.65 MHz. The active luminance resolution is 960 lines by 1408 pixels. The four-channel, 16-bit audio data is processed using the Dolby AC-2 algorithm [856] and formatted into a single serial bit stream at 503 Kbps. Two transmission modes are supported: 32 QAM, the primary transmission mode, and 16 QAM. Both have a symbol rate of 4.88 Msymbols per second. The 32 QAM mode has a total transmission rate of 24.39 Mbps. Concatenated trellis coding, Reed-Solomon block coding, and adaptive equalization are used to protect against channel errors.

The DigiCipher system is an integrated system that can provide high-definition video, digital audio with CD quality, and data and text services over a single VHF or UHF channel. Figure 13.1 shows the overall system block diagram, which has encoder, multiplexer, forward error correction encoder, and QAM modulator. The encoder accepts *YUV* video, 4-channel audio, data/text, and control channel data. The *YUV* signals are obtained from analog *RGB* inputs by low pass filtering, A/D conversion, and an *RGB*-to-*YUV* matrix. The video encoder implements the compression algorithm and generates a video data stream. The digital audio encoder accepts four audio inputs and generates an audio data stream. The data/text processor accepts four data channels at 9600 baud and generates a data stream. The control processor interfaces with the control computer and generates a control data stream.

The multiplexer combines the four data streams into one data stream at 18.22 Mbps. The FEC encoder adds error correction overhead (6.17 Mbps) and provides 24.39 Mbps of data to the 16/32 QAM modulator.

At the decoder, an IF signal received from the VHF/UHF tuner is demodulated by the 32 QAM demodulator. The demodulator has an adaptive equalizer to

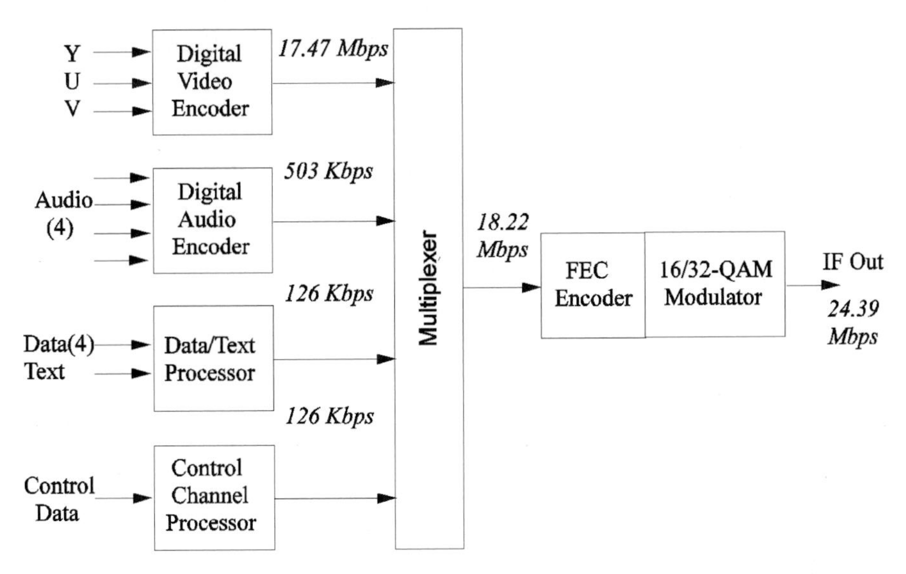

Figure 13.1 A DigiCipher HDTV system block diagram. Bit rates indicated here are for the 32 QAM case.

effectively meet multipath distortions that occur in terrestrial transmission channels. The FEC decoder provides error-free data to the Sync/Data selector, which detects synchronization signals and provides video and audio data and control data streams to the appropriate decoders.

The DigiCipher HDTV signal can be transmitted via various media as follows:

- Cable: DigiCipher transmission starting at the satellite uplink all the way to the cable subscribers.
- Satellite: Can be transmitted over C-band or Ku-band satellite channels using QPSK modulation.
- Other terrestrial distribution: Can be applied to other transmission media such as microwave distribution service (MDS), multichannel MDS, and fiber-optic cables.
- VCR and video disc recorders: All-digital recording and playback of the HDTV signal (data rate < 20 Mbps) is within the reach of current technology.

DigiCipher video coding

The overall video-compression algorithm is motion-compensated DCT as discussed in Sections 9.3 and 10.2. The compression process can be broken down into the following subprocesses:

- Video preprocessing (A/D conversion and *RGB*-to-*YUV* matrix)
- Decimation/interpolation
- Discrete cosine transform
- Motion estimation and compensation
- Integration of motion compensation with intraframe coding
- Adaptive field/frame processing
- Huffman coding and rate buffer control

Figure 13.2 shows the digital video encoder and Fig. 13.3 shows the decoder in the DigiCipher system. Input video signals (*RGB*) are converted into *YUV* signals before applying the compression algorithm. The luminance resolution is not changed. The resolution of chrominance information, however, can be reduced with only a slight effect on the perceived image quality. The U and V chrominance components are decimated by a factor of four horizontally and by a factor of two vertically, as illustrated in Fig. 13.4. Horizontal decimation is performed by applying a digital filter prior to subsampling (see Example 10.1). DigiCipher applies a 110-tap symmetrical filter for decimation/interpolation. Vertical decimation is performed by discarding one of every two fields. Thus, input chrominance resolution of (1408×960) pixels is decreased to (352×480) pixels. The decoder interpolates horizontally by zero-padding and applying the same filter with the interpolation increased by a factor of four. The decoder reconstructs the interlaced signal by repeating each chrominance field twice.

One of the main different properties in the DigiCipher system is the superblock concept. A superblock is an image of four (8×8) luminance blocks horizontally by two (8×8) luminance blocks vertically, coupled with one (8×8) chrominance block each for U and V derived from the same image area. A macroblock is an image area of 11 superblocks horizontally, as shown in Fig. 13.5. Motion estimation is based on the (32×16) Y block and is applied (appropriately scaled) to the U and V blocks.

Coefficient quantization is done by weighting the DCT coefficients. The weighting matrix is essentially based on that of the MPEG-1 coder (Table 10.6), except that the DC coefficient weighting is 16 instead of the 8 used in MPEG-1. Each DCT coefficient is initially represented as a 12-bit number that is then divided by the respective weighting factor. Thus, the DC coefficient is quantized with 8-bit resolution.

Additional scaling of the quantization matrix may still be necessary to achieve the desired data rate. The coefficients weighted by the quantization matrix are next divided by a quantization factor that is based on scene complexity and rate buffer status. The quantization factor ranges from 0 to 31. The factor 0 gives maximum precision and the factor 30 gives minimum precision. Factor 31

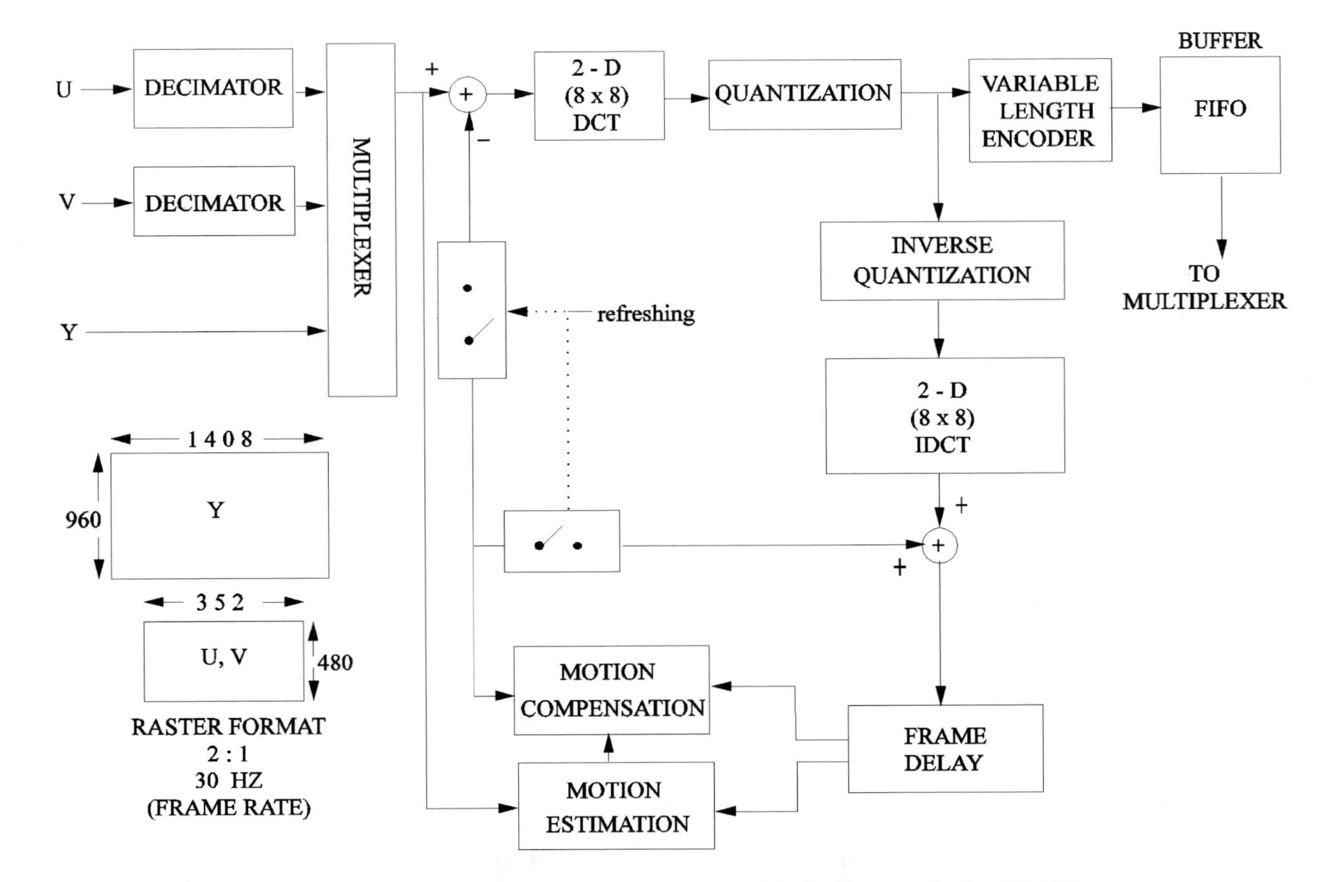

Figure 13.2 A digital video encoder block diagram in the DigiCipher system.

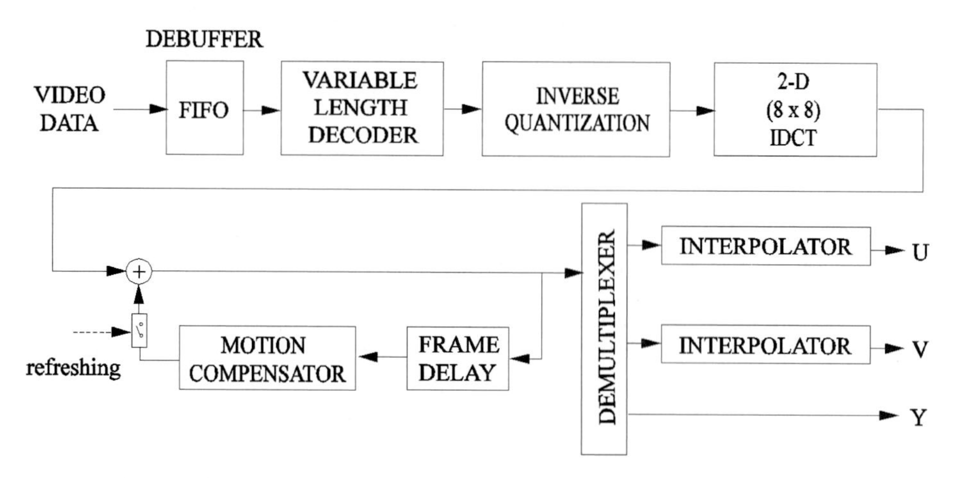

Figure 13.3 A digital video decoder block diagram in the DigiCipher system.

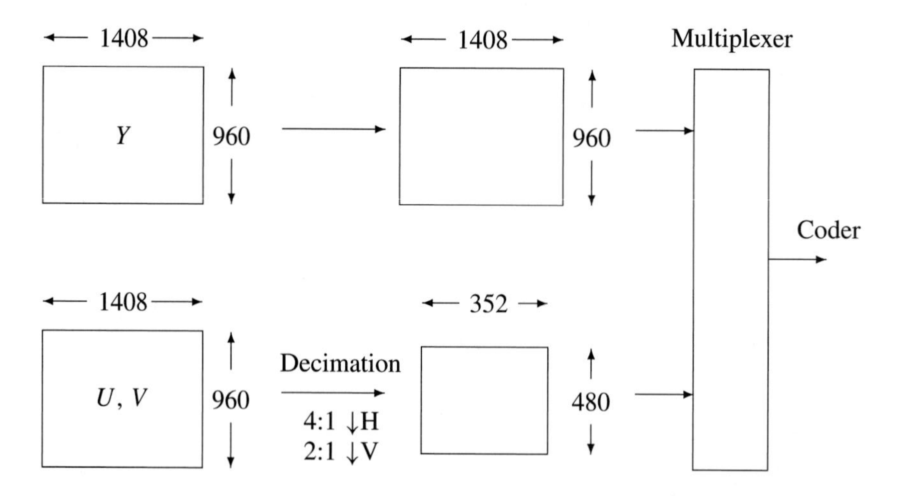

Figure 13.4 Decimation of chrominance components before video coding.

is reserved and indicates to the decoder that no data will be transmitted. The quantization scaling does not apply to the DC coefficient (always 8 bpp).

Motion estimation is performed at the encoder only and a motion vector is derived on the basis of the (32×16) Y block and is applied (appropriately scaled) to the (8×8) U and V blocks. The motion vector range is $+31/-32$ pixels horizontally and $+7/-8$ lines vertically. These limits allow the tracking of objects

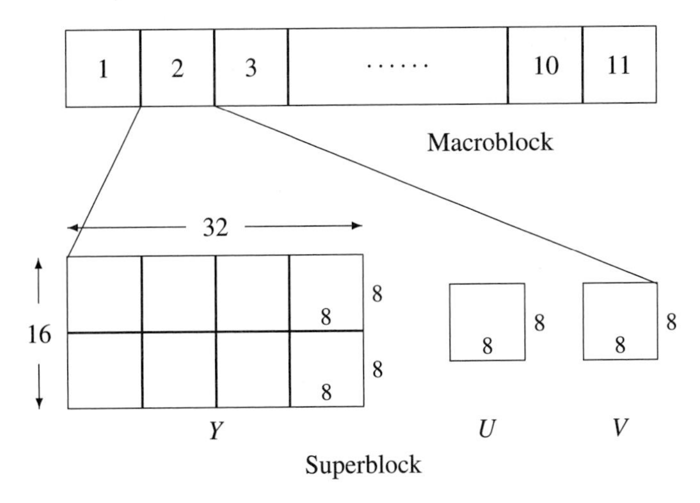

Figure 13.5 Superblock and macroblock definition in the DigiCipher HDTV system.

moving at 0.68 picture width/sec and nearly 0.25 picture height/sec. The overhead required to send a motion vector is 10 bits per superblock (0.0195 bit/pixel).

The difference between the current frame and the motion-compensated previous frame is coded and transmitted. A lower bit rate, however, is occasionally possible by direct PCM coding of a block instead of using the motion-compensated differential coding. This can happen during scene change and very fast motion. Direct PCM coding implies intraframe coding, i.e., the switches in both the encoder (Fig. 13.2) and decoder (Fig. 13.3) are open (no MC and no feedback loop). To obtain the lowest possible bit rate, the encoder determines the number of bits required for each of the two methods (direct PCM and MC/VLC) and then selects the method requiring the fewest bits on a per block basis. During scene changes, the motion compensation rate is less than 10% (number of blocks coded by interframe MC hybrid (DPCM/DCT) technique) whereas for most scenes this rate averages from 85% to 100% [833]. When the decoder is tuned to a new channel, it has no previous frame information for reconstructing the video. To mitigate this problem, one frame out of 11 frames is periodically intraframe coded (once every 0.37 second). This causes about a 9% reduction in the overall compression efficiency.

Frame processing works better than field processing when little or no motion occurs, while field processing works better than frame processing when motion occurs. In detailed moving areas, the interleaving of the even and odd fields from the interlaced video source will introduce spurious high vertical frequencies if frame processing is selected. This reduces the correlation between lines and gives a lower compression ratio. Therefore, the field/frame processing mode decision is

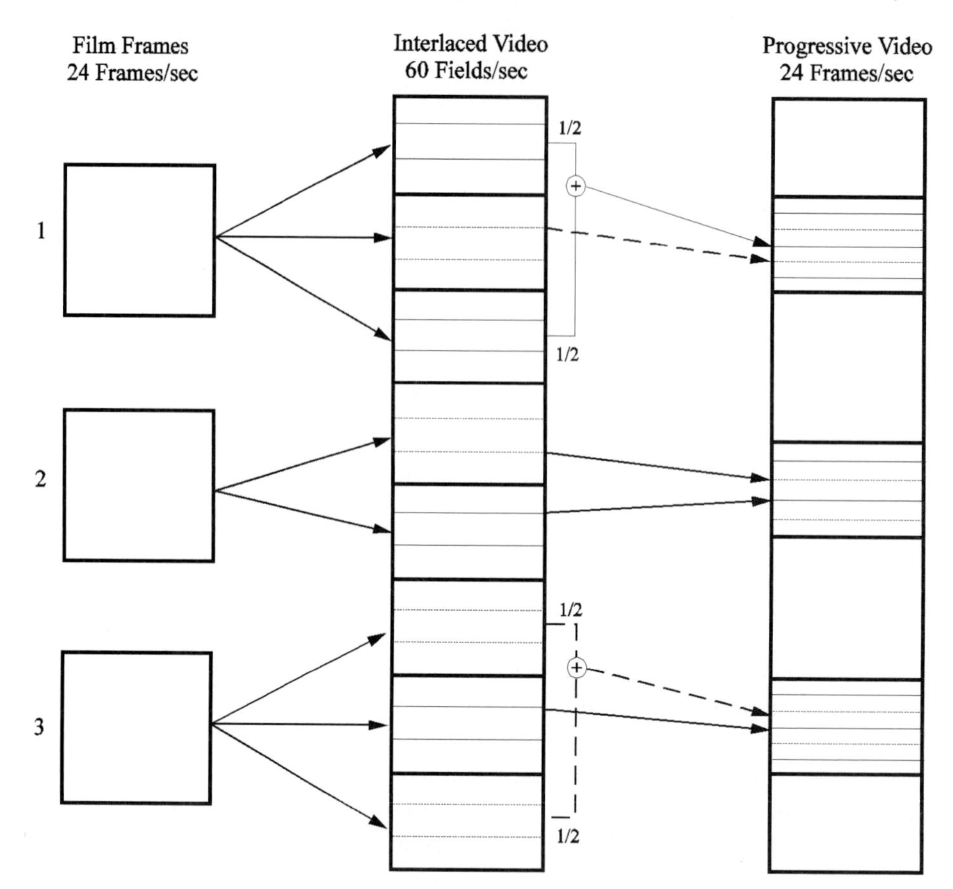

Figure 13.6 Three-two pulldown to convert 24 frames/sec film to 60 fields/sec interlaced video [833]. © 1991 GI and MIT.

performed at the encoder and one bit per superblock of overhead is included in the encoded signal.

The DigiCipher system uses the three-two pulldown process to adapt the film material at 24 frames/second. As shown in Fig. 13.6, it involves alternating between three repetitions and two repetitions of each frame of the film. The three-two pulldown process increases/decreases the number of video frames without increasing/decreasing the amount of information.

DigiCipher audio coding

The DigiCipher HDTV system uses the Dolby Laboratories AC-2 digital audio system [541, 856, 864], which is a frequency domain audio-coding scheme in a multiplicity of narrow bands to take full advantage of psychoacoustics and

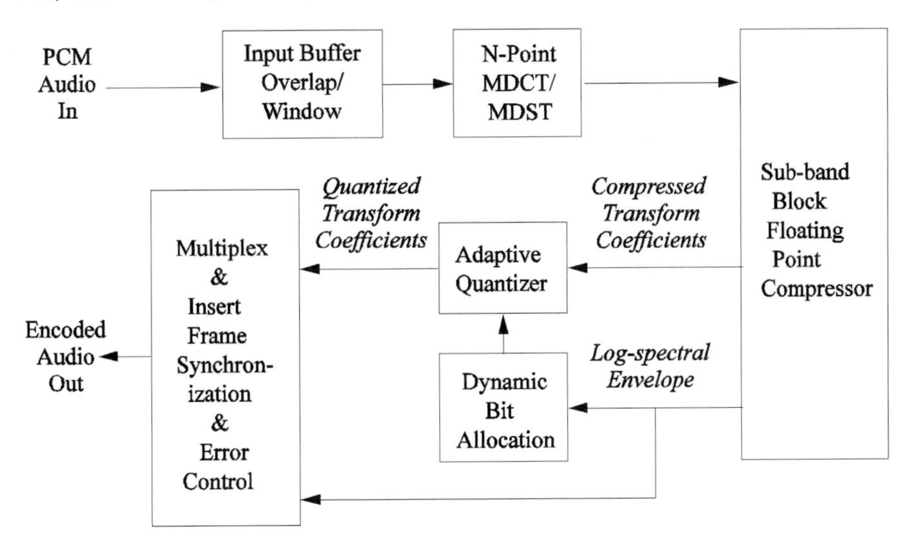

Figure 13.7 An AC-2 digital audio encoder block diagram [541]. © 1990 AES.

noise masking. The MPEG audio-coding algorithm (see Section 10.3) also exploits perceptual properties of the human ear. The AC-2 algorithm provides two high-quality data rate reduction coders: a low delay coder and a moderate delay coder. The former is designed to achieve excellent subjective and objective quality at 4:1 compression with the low delay time of less than 9 msec. The latter is designed to achieve 6:1 compression with a moderate delay time of less than 45 msec. The coding system can be used for either the 44.1 KHz or the 48 KHz sampling rate. At a frame size of $N = 512$ samples and a sampling rate of 48 KHz, a delay of approximately 45 msec is obtained, since the delay increases to about $4N$ sample time for a fully utilized processor.

Figure 13.7 shows an AC-2 digital audio encoder. Sampled audio data are buffered into frames of length N in the first stage of processing. Each new frame overlaps the previous one by a half ($N/2$ samples) as the MPEG-1 audio coder does (see Section 10.3). Next, a windowing function is applied to reduce the effect of frame boundary discontinuities on the spectral estimate provided by the transform. The windowing also improves the frequency analysis characteristics of the transform when compared to a rectangular window. A key advantage is that the half-frame overlap is achieved without increasing the required bit rate. The transformation is based on time-domain aliasing cancellation (TDAC) [855], consisting of alternating modified discrete cosine (MDCT) and modified discrete sine (MDST) transforms.

The filter bank of the coder is the primary element that allows bit-rate reduction. The auditory masking function is not constant in the frequency domain.

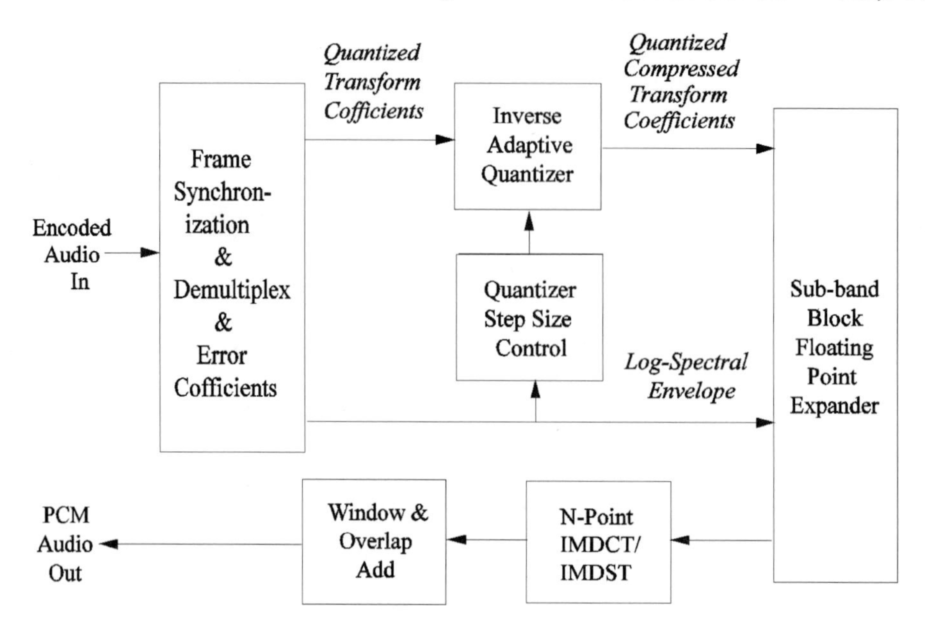

Figure 13.8 An AC-2 digital audio decoder block diagram [541] ©1990 AES.

The filter bank developed from the TDAC is of constant bandwidth, not of varying bandwidths, as in the auditory system. This can be overcome by approximating the nonuniform bandwidths of the human auditory system by grouping transform coefficients to form subbands for further decomposition and analysis. The number of coefficients per group is computed to approximate the nonuniform critical bands. Transform coefficients within one subband are then converted to a floating-point representation, with one or more mantissas per exponent, depending upon the subband center frequency. Each exponent represents the quantized peak log-amplitude and provides an estimate of the log-spectral envelope for the audio frame.

Bit allocation is performed depending upon the log-spectral envelope in estimating which subbands are most significant. Each subband mantissa is quantized to a bit resolution defined by the sum of a fixed allocation and a dynamic allocation. The total allocation is determined in approximately a 4:1 ratio. In the final stage of the encoder, the exponents are multiplexed and interleaved with the fixed and adaptive mantissa bits. Additional error correction codes (Reed Solomon error protection) are added before transmitting the bit stream.

In the AC-2 decoder, as shown in Fig. 13.8, the received log-spectral envelope is decomposed in a manner identical to the encoder bit-allocation routine to reconstruct step sizes for an adaptive inverse quantizer. The mantissas in the log-spectral envelope are combined to regenerate the transform coefficients at the floating-point expander block. These coefficients are inverse transformed and windowed again to reconstruct the digital audio output.

In the DigiCipher audio coder, four channels of analog signal are accepted. They are low pass filtered, synchronously quantized to 16-bit precision, and processed using the AC-2 algorithm at 24-bit precision. The coded data are formatted into a serial output data stream at 503 Kbps.

13.2.2 DSC HDTV system

DSC (Digital Spectrum Compatible) HDTV is also a digital simulcast system, proposed by Zenith and AT & T, that requires a single 6 MHz television transmission channel. The video source is an analog *RGB* signal with alternate 787/788 lines, progressively scanned, and a 59.94 Hz frame rate, with an aspect ratio of 16:9. Active video pixels are 720 lines by 1280 pixels per line. Chrominance signals are decimated by a factor of two both horizontally and vertically resulting in an uncompressed bit rate of 994 Mbps. Nine-bit precision is employed for all luminance and chrominance samples. The nominal bandwidth equals 33.9 MHz with the sampling rate of 75.3 MHz and a Kell factor of 0.9. The Kell factor is a usable fraction of spatial resolution when an image is displayed on some sampled display, such as a CRT monitor.

A hierarchical block-matching algorithm with a half-pixel accuracy is used for motion estimation. The motion vectors are the coarse motion vector representing a large block (32H$\times$16V) and the fine motion vector representing a DCT block (8H$\times$8V). Four digital audio channels are compressed using the Dolby Laboratories AC-2 algorithm [541, 856] as in the DigiCipher system. The compressed audio rate is 252 Kbps per pair of channels.

Time division multiplexing between 4-level and 2-level VSB (Vestigial Side Band) transmission is employed to provide improved error performance and extended coverage. A pilot carrier is positioned on 0.31 MHz above the lower edge of the 6 MHz channel. The video data rate ranges from 8.45 to 16.92 Mbps and the total transmission rate ranges from 11.14 to 21.0 Mbps. The system also provides 413 Kbps of data capacity in two separate ancillary data channels.

DSC video coding

The DSC system exploits three types of redundancy in the video compression process. Spatial redundancy is removed by the DCT spatial frequency transformation. Temporal redundancy is removed by motion compensation as in most of the video coders. Amplitude redundancy is removed by perceptual weighting that puts quantization noise in less visible areas.

For the efficient coding of motion video, the DSC encoder as shown in Fig. 13.9 uses a motion-compensated loop including 2D (8×8) DCT, adaptive quantizer, and motion-compensated prediction. In Fig. 13.10, the displaced frame

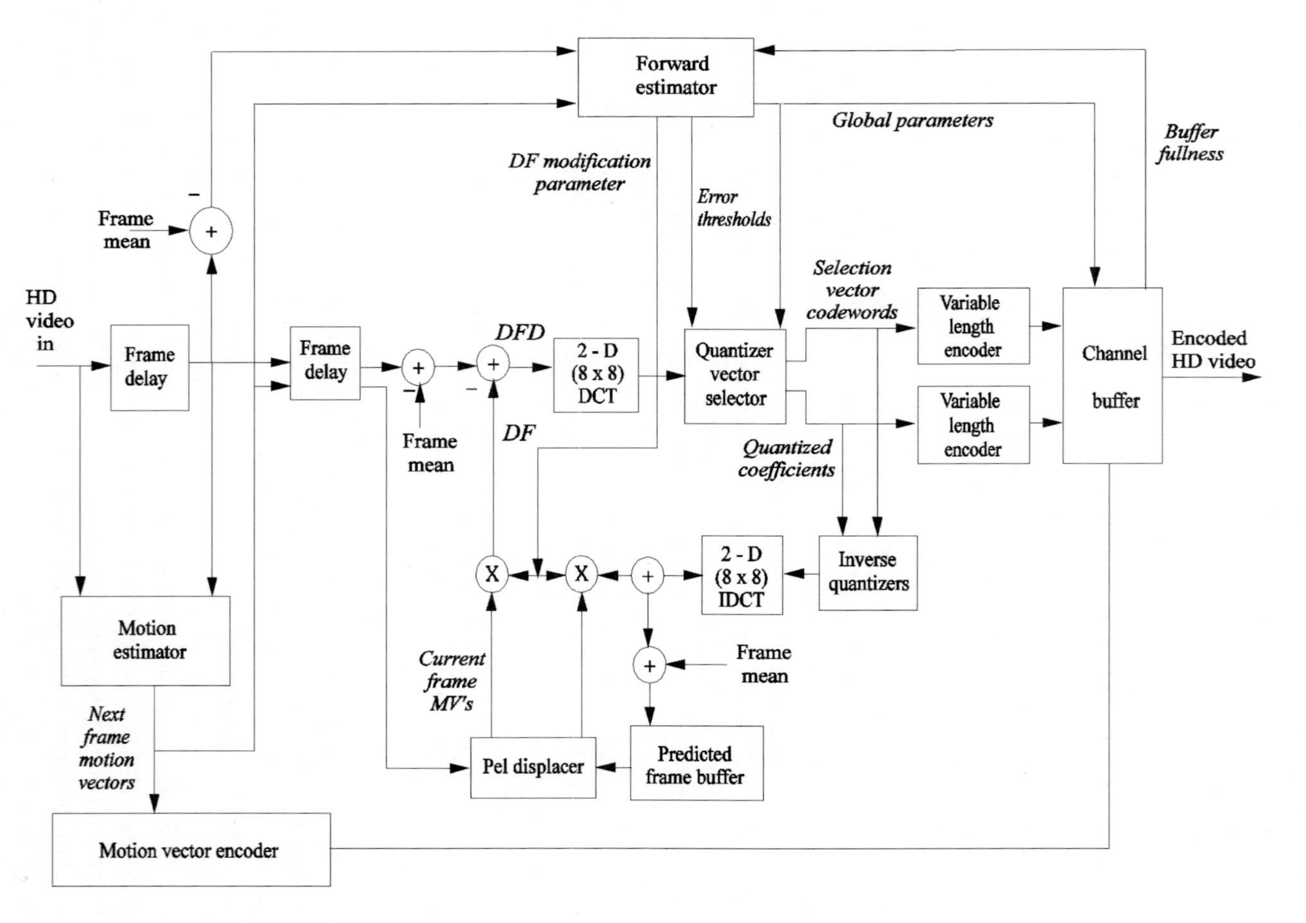

Figure 13.9 A DSC video encoder block diagram.

signals are combined with the transmitted and decoded signals (displaced frame difference) for reconstructing the decoded HD video. The main differences from the other HDTV proponents are as follows:

- Hierarchical motion estimation
- Vector quantization of quantizer selection patterns
- Perceptual quantization based on the HVS model using both image dependent and image independent parameters
- Forward analyzer to determine the coding parameters

Motion from the previous frame to the current frame is estimated using a hierarchical block-matching algorithm as shown in Fig. 13.11. At each stage the best block match is defined to be that which has the least absolute difference between the blocks. The results from one stage (coarse motion estimator) are used as a starting point for the next stage to minimize the number of block matches per image. The prediction error is defined as the sum of the absolute differences between a block of pixels in the present frame and the displaced block in the previous frame.

The hierarchical motion estimation reduces the complexity of the search, since a first-stage coarse estimation is refined by a second-stage finer estimation. A $(32H \times 16V)$ block is decimated by a factor of two both horizontally and vertically. A block size of $(16H \times 8V)$ pixels in the decimated image is used with one-pixel accuracy for motion estimation that is equivalent to two-pixel accuracy in a $(32H \times 16V)$ block. The motion vectors generated are referred in the second stage that performs a half-pixel accuracy search centered around this coarse estimate. The total search area is $(96H \times 80V)$ pixels, which provides highly accurate motion estimation.

The second stage of the motion estimator generates the prediction errors of the (8×8) blocks within the search area. The prediction errors of the coarse blocks are derived from the sums of the appropriate small block (8×8) prediction errors. At the final stage of the motion estimator, both prediction errors are compared to obtain the minimum prediction error for all blocks in every location. The motion vector selector decides whether a small block MV is to be transmitted or not. Thus two resolutions of motion vectors are sent, the first set representing the motion vectors of a large block that is unconditionally transmitted, and the second set representing the motion vectors of small block that are selected to obtain the minimum errors, i.e., the difference between the prediction error of the $(32H \times 16V)$ block and the sum of the eight prediction errors of the (8×8) blocks.

Figure 13.12 illustrates a large block (32×16) and a slice (64×48) concept in the DSC HDTV system. First, for each slice, five of the six motion vectors are

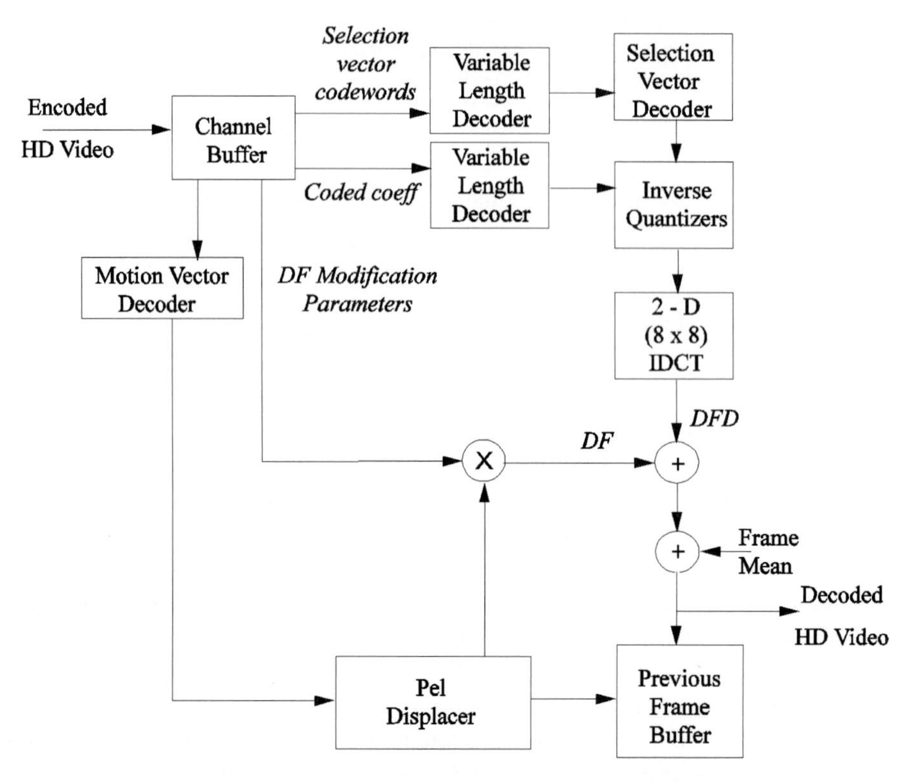

Figure 13.10 A DSC video decoder block diagram.

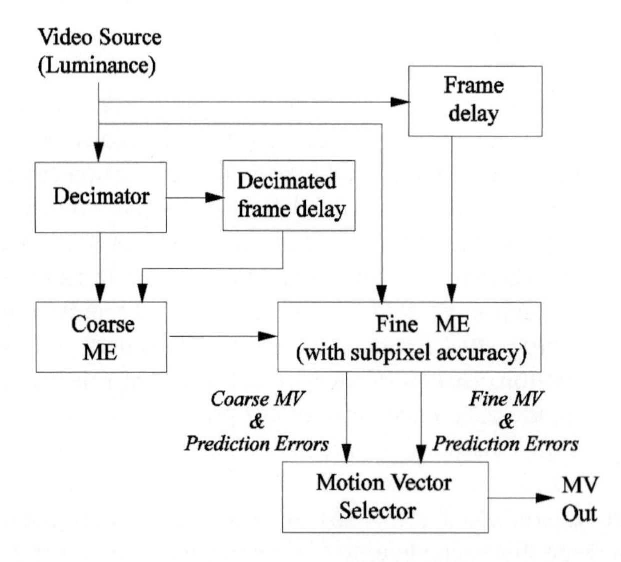

Figure 13.11 Hierarchical motion estimation and motion vector selection.

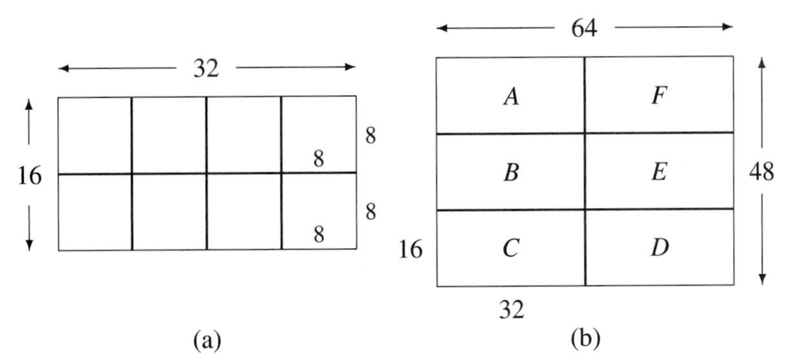

Figure 13.12 Hierarchical block for DSC video coding. (a) A large block. (b) A slice.

differentially coded. If the six motion vectors constituting the slice are numbered as shown in this figure, then the motion vectors that are coded represent the values of the motion vectors, A, $B - A$, $C - B$, $D - C$, $E - D$, and $F - E$. In addition to these motion vectors for large blocks, information is transmitted indicating which of the ($32H \times 16V$) block motion vectors have been subdivided into eight (8×8) block motion vectors. One extra bit that indicates whether any of the motion vectors are subdivided is needed to represent them.

The forward analyzer is shown in Fig. 13.13. The video input frame is transformed by 2D-DCT and analyzed by the perceptual weight calculator. The perceptual thresholds are computed by the buffer state. The coding parameters depend on the results of analysis of the transformed original frame and ideal DFD (displaced frame difference). Examples of coding parameters are the luminance and chrominance mean values, the quantizer scaling factors, DF (displaced frame) factor, buffer fullness, and frame number.

Perceptual weighting values are based on the characteristics of the human visual system. The following properties of the HVS are used: frequency sensitivity, contrast sensitivity, spatial masking, and temporal masking. Frequency sensitivity refers to the property of the HVS that tolerates more quantization errors at higher frequencies than at lower frequencies. Contrast sensitivity refers to flat field stimuli at certain areas of the luminance intensity. Spatial masking explains that the perceptual threshold is based on the amount of local texture present at each location in the input. Temporal masking refers to the increase of the perceptual threshold for high frequencies when there is motion in the scene. The luminance and chrominance perceptual thresholds are generated by human visual sensitivities. They are not transmitted but are used to optimize the transmitted information and minimize the perceptible artifacts.

A variety of quantizers is available for each DCT coefficient. Vector quantization is used to represent the possible combinations or patterns of quantizers that

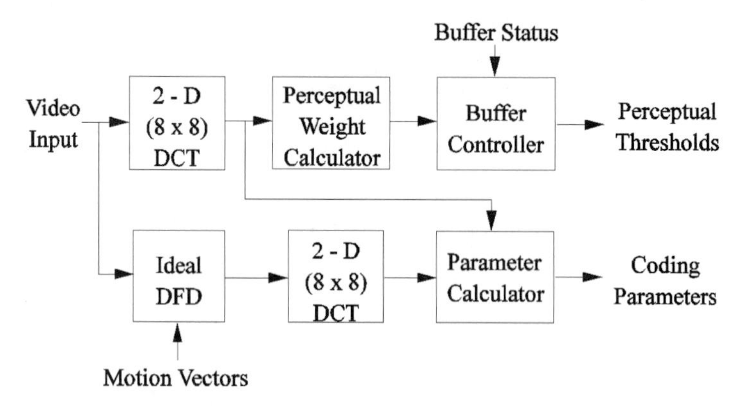

Figure 13.13 A forward analyzer block diagram.

can be applied to a given (8×8) block of coefficients. Variable-length coding is applied to the indices of these patterns. Coding efficiency is achieved by transmitting the index (variable-length coded) associated with a given quantizer selection pattern instead of the pattern itself. A quantizer vector is selected from the quantizer vector codebook. A (16×16) pixel is the domain of a given selection vector for luminance component, whereas a $(16H \times 48V)$ superblock is the domain of a given selection vector for chrominance components. The luminance codebook contains fewer than 2000 code words and each chrominance codebook contains fewer than 500 code words. The codebooks are organized differentially so that only differences between the successive quantizer vectors are computed. The quantization error is compared to a perceptual error threshold, which results in the selection error. The optimal quantizer vector is selected by considering both selection error and bit rate.

DSC audio coding

The DSC HDTV system provides four independent 125.874 Kbps channels of CD-quality audio. The Dolby AC-2 algorithm, as previously stated in the DigiCipher (Section 13.2.1), is used for compression/decompression. An audio sampling rate of 47.203 KHz (three times the NTSC horizontal scan rate) prevents low-frequency beats at the receiver. The audio channels are independently coded. This provides a main stereo pair that is carried by two-level VSB transmission. The secondary channels are carried by four-level VSB transmission.

VSB modulation

The total bit rate of 21.52 Mbps, including video, audio, and ancillary data, is combined into one data frame (525 data segments). A data segment consists

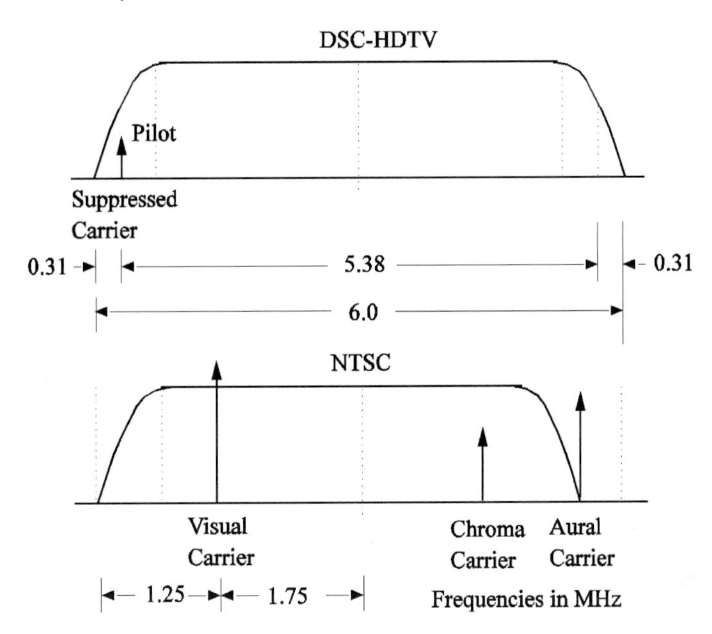

Figure 13.14 The VSB spectrum in DSC HDTV and the NTSC system.

of data segment sync (4 bytes) and data field (167 bytes) including 20 bytes parity that are set aside as RS parity bytes so that the code is characterized by RS (167, 147)$t = 10$. Twenty parity bytes can correct 10 byte errors. The DSC HDTV receiver has to operate reliably under severe cochannel interference with the NTSC channels. To reduce this interference the low end 0.31 MHz of the 6 MHz channel is used to position the suppressed carrier within the vestigial side band (VSB) modulation system.

The nominally occupied spectrum is shown in Fig. 13.14 together with an NTSC cochannel. The nominal four-level transmission signals can be represented by -3, -1, $+1$, $+3$ and the two-level signals by -2, $+2$. A pilot is added by changing these levels to -2, 0, $+2$, $+4$ (4-VSB) and to -1, $+3$, (2-VSB). The modulated signal is designated as 4-VSB when the modulating signal consists of 4-level symbols and as 2-VSB when modulated by 2-level symbols, where the data signal of maximum 21 Mbps is partially converted into 4-level symbols, 4-VSB, and partially into 2-level symbols, 2-VSB.

13.2.3 Advanced Digital HDTV

AD-HDTV [495, 832], proposed by the Advanced Television Research Consortium (ATRC), is a digital simulcast system that requires a 6 MHz television transmission channel that is used for the conventional NTSC system. The AD-HDTV video

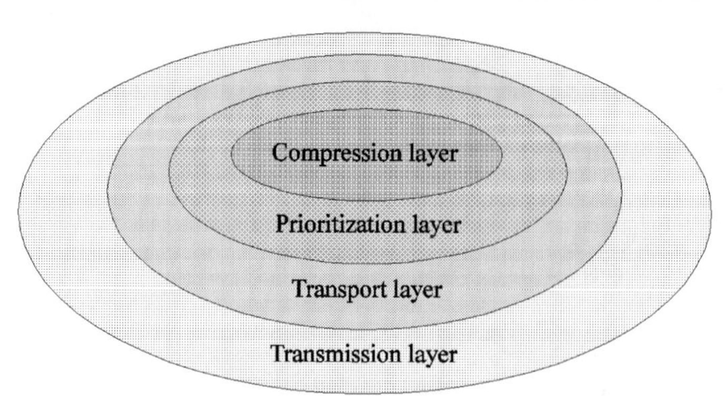

Figure 13.15 Layered structure of the AD-HDTV system.

source is an analog *RGB* signal with 1050 lines, 2:1 interlace, a 59.94 Hz field rate, and an aspect ratio of 16:9, as shown in Table 13.1.

The AD-HDTV system is a layered digital system that consists of

- MPEG++ video compression [495]
- MUSICAM audio compression at 256 Kbps per stereo pair (Section 10.3.2)
- A prioritized data transport format
- Spectrally shaped quadrature amplitude modulation(SS QAM)

In Fig. 13.15, four principal layers for AD-HDTV data encoding are shown. The compression layer performs the tasks of MPEG-1 video/audio compression/decompression. Pre- and postprocessing of input and output are also included. The prioritization layer performs the tasks of assigning priorities to video data elements at the encoder. The data transport layer is responsible for packaging MPEG++ high-priority (HP) and standard-priority (SP) data streams into separate sequences of fixed-size cells (148 bytes each). Each cell consists of a header, data, and a trailer for error detection/correction (Reed-Solomon forward error correction). The transmission layer is responsible for the tasks of modulation, channel equalization, and frequency translation. These four layers constitute the MPEG++ approach to the HDTV system.

Video compression

The MPEG++ algorithm incorporates motion-compensated predictive coding in which the motion estimation/compensation extends in both the forward and backward directions in time, i.e., a bidirectional approach, as in MPEG-1 and MPEG-2. Figure 13.16 shows the block diagram of the AD-HDTV video encoder. The analog video signal is preprocessed, digitized, and interpolated to produce

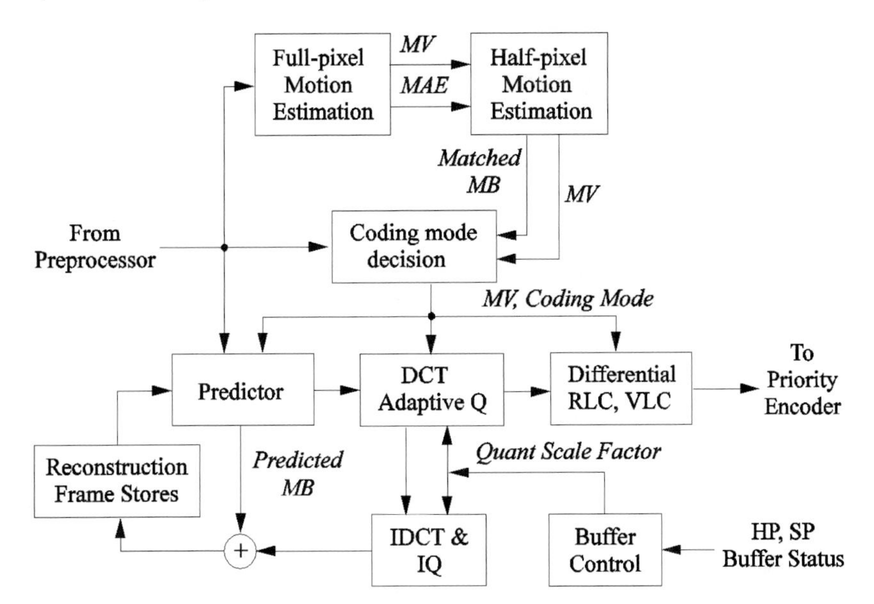

Figure 13.16 A functional block diagram of the AD-HDTV encoder.

the digital input format. The compression algorithm essentially conforms to the MPEG-1 algorithm (see Section 10.2). It is supported by a full-search block-matching motion estimation (half-pixel resolution). Motion-compensated predictive coding, 2D (8×8) DCT, adaptive quantization, and inverse DCT operation are also included in the encoder.

Video compression in AD-HDTV is based on the MPEG-1 algorithm with improvements to quantization and bit rate control. Adaptive quantization is performed on each (8×8) block of DCT coefficients. Different quantization weighting matrices are used (Table 13.2) for the quantization of intra- and interframe coded macroblocks. The same weighting matrix is used for both the luminance and chrominance blocks.

The weights for a macroblock are modified by a quantization scale factor. The scale factor is provided by the buffer control logic and is computed subject to scene complexity as well as the buffer status. The buffer contains a coded bit stream of video data with different picture types and different priorities.

The three types of pictures (I, P, and B) need different numbers of bits to be coded. I pictures require the most bits because they are intraframe coded, whereas the B pictures require the least bits, because they can be efficiently coded by the bidirectional motion-compensated predictive coding.

The buffer control subsystem is responsible for bit allocation and rate control. The bit allocation target is maintained both globally (long term) and

Table 13.2 Quantization weighting matrices: (a) intraframe coding, (b) interframe coding

(a)

8	16	19	22	26	27	29	34
16	16	22	24	27	29	34	37
19	22	26	27	29	34	34	38
22	22	26	27	29	34	37	40
22	26	27	29	32	35	40	48
26	27	29	32	35	40	48	58
26	27	39	34	38	46	56	69
27	29	35	38	46	56	69	83

(b)

16	17	18	19	20	21	22	23
17	18	19	20	21	22	23	24
18	19	20	21	22	23	24	25
19	20	21	22	23	24	25	27
20	21	22	23	24	25	27	28
21	22	23	24	25	27	28	30
22	23	24	25	27	28	30	31
23	24	25	27	28	30	31	33

locally (within a picture). The main variable for achieving the desired bit rate is the quantization weighting factor, which may be variable for each macroblock, as described in Section 10.2. The global control of the quantization factor is to control the allocation of the total number of bits used by each picture to maintain continuity of subjective quality of the video sequence.

The forward analyzer gives the improved bit allocation by analyzing image statistics on a macroblock basis. The backward analyzer keeps track of the preceding results to decide, a priori, the number of bits that will be devoted to code an image and its expected quality. Thus bit allocation and overall quality are dynamically varied to achieve high quality under the constraint of a constant bit rate. The global control sets rate/quality targets for the video sequence, while the local control refines the bit allocation within the image being coded. For a detailed description of the main compression algorithm (motion-compensated DCT and variable-length encoding) the reader is referred to Section 10.2.

The AD-HDTV decoder, as shown in Fig. 13.17, provides video decompression and error concealment. The error concealment unit estimates and replaces video data lost due to uncorrectable channel errors. When no errors are present, data pass without alteration. When there are errors in the bit stream, the decoder receives an appropriate error token from the priority subsystem, indicating the type of error event.

The error concealment procedure can be implemented as follows. The cyclic redundancy check (CRC) code detects the cell errors. For standard-priority error, the error concealment is performed on a macroblock level. For high-priority errors,

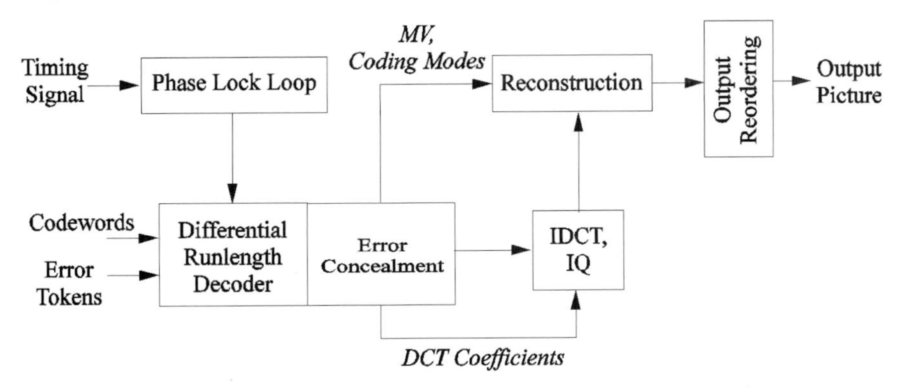

Figure 13.17 A functional block diagram of the AD-HDTV decoder.

a slice level error recovery is performed. The corresponding macroblock or slice in the received video in error is discarded. Each missing part is replaced with either a colocated one from the previous reconstructed frame, or spatial concealment in which corrupted AC coefficients of the DCT block are replaced by zero values.

Data prioritization and transport

A dynamic priority assignment algorithm takes into account the code word type (header, data, or error check), picture type (I, P, or B), and relative occupancies of high-priority (HP) and standard-priority (SP) rate buffers at the output of the encoder. The most important one has the highest priority, i.e., headers are required to initiate decoding of the received bit stream. Data in I pictures and motion vectors in P pictures are also given relatively high priority, while most of the data in B pictures can be transmitted with standard priority. The nominal HP to SP rate is 1:4, corresponding to HP and SP modem bit rates of 4.8 and 19.2 Mbps, respectively.

Table 13.3 shows a generalized rank list for the different types of pictures. The code words are stacked in an analysis buffer and their length field information is accumulated in the order specified in the table. A priority breakpoint divides the data in the analysis buffer into a high-priority portion and a standard-priority portion. The priority decoder performs the function of reassembling a single video bit stream from the received high- and standard-priority data streams. A variable-length decoder (Fig. 13.17) decodes the data and then asynchronously multiplexes the HP and the SP bit streams. The output bit stream is then decoded as a single bit stream.

The compression and prioritization layers are intended to be with a two-tier transmission system in which the HP and the SP data are transmitted with different reliability levels. Both the HP and the SP bit streams are formatted into fixed-

Table 13.3 Priority rank list for different pictures

I picture	1. Picture headers
	2. Slice headers
	3. MB addresses, types, and quant
	4. DC values
	5. Low-frequency coefficients
	6. High-frequency coefficients
P and B pictures	1. Picture headers
	2. Slice headers
	3. MB addresses, type, and quant
	4. Motion vectors
	5. DC values
	6. Low-frequency coefficients
	7. High-frequency coefficients

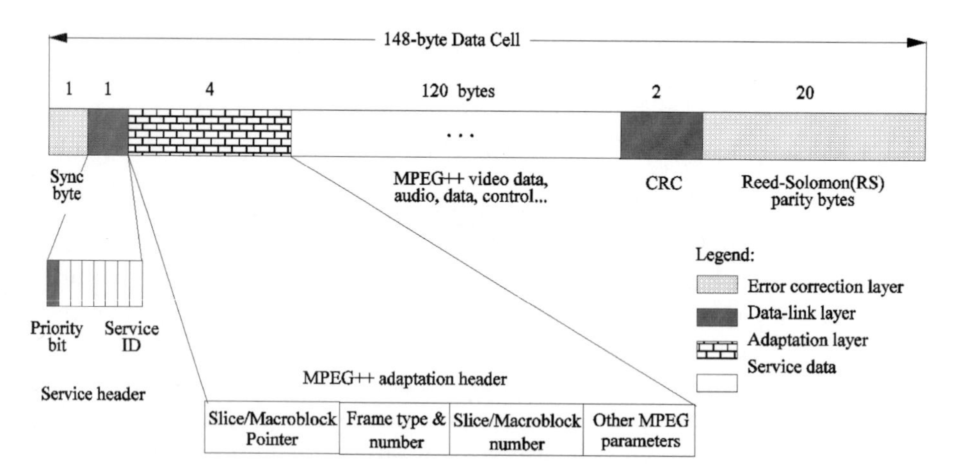

Figure 13.18 The AD-HDTV data format structure [832].

length, 148-byte cells, as shown in Fig. 13.18. This fixed-length cell structure provides good error protection and rugged data synchronization under fluctuating channel conditions. The data cells also provide flexibility and new programming opportunities by delivering a dynamic combination of video, audio, and auxiliary data to receivers. The adaptation sublayer supports rapid decoder recovery when one or more cells are received in error.

The AD-HDTV prototype for terrestrial simulcast uses a two-carrier spectrally shaped modem (SS QAM) to realize prioritized transport. A narrowband 1.125 MHz QAM carrier is positioned in the 1.25 MHz band below the NTSC picture carrier to transmit high-priority data. A second wideband 4.5 MHz carrier in the 4.75 MHz band is positioned above the picture carrier to transmit standard-priority data. This concept is illustrated in Fig. 13.19. Thus the SS QAM signal

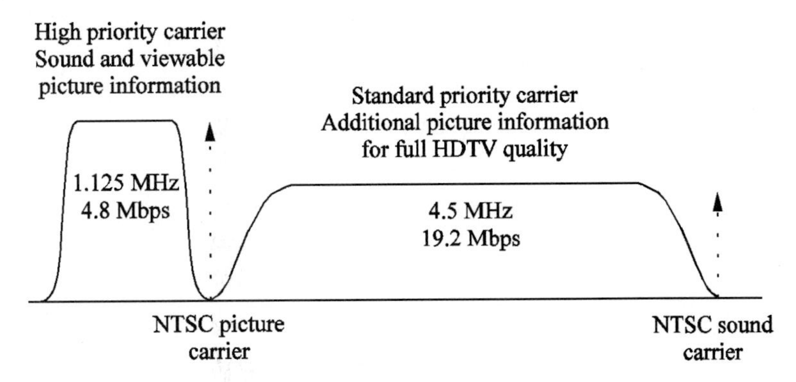

Figure 13.19 Spectrally shaped QAM concept in AD-HDTV.

is transmitted without causing cochannel interference on NTSC receivers that use filters for suppressing picture carrier on reference frequency and sound carrier on 4.5 MHz [851].

For 32-QAM operation [848, 849], the modulator provides a gross bit rate of 24 Mbps. After accounting for error correction overhead from trellis coding (at rate 9/10) and Reed-Solomon block coding (rate 127/148), a net service rate of 18.53 Mbps (3.7 and 14.8 Mbps for high- and low-priority channels) is available for audio and video data transport. More detailed specifications are shown in Table 13.1.

13.2.4 CCDC HDTV system

The CCDC (Channel Compatible DigiCipher) HDTV system, proposed by the American Television Alliance (MIT and GI Corp.), is also a digital simulcast system that requires a single 6 MHz television transmission channel. The main specifications are similar to those of the DSC system (Section 13.2.2): 787/788 lines alternate video source, 59.94 Hz frame rate, progressive scanning, $(1280H \times 720V)$ display format, 75.52 MHz video-sampling frequency. The video-compression algorithm is motion-compensated interframe hybrid (DPCM/DCT) and intraframe adaptive coding. Information related to the compressed video is entropy coded for transmission, including motion vectors and parameters for deciding intraframe or interframe coding. Two transmission modes are supported: 32 QAM, the primary transmission mode, and 16 QAM, both with a symbol rate of 5.29 Msymbol/sec. The CCDC system provides four digital audio channels using the MIT Audio Coder (MIT-AC) for compression/decompression.

CCDC video coding

The video compression algorithm is briefly discussed in this section. The main distinct features in the CCDC system are as follows:

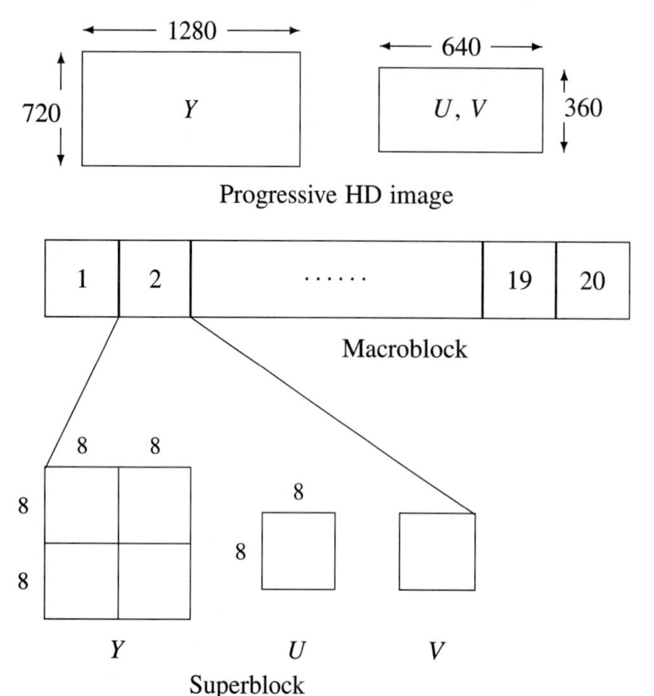

Figure 13.20 Superblock and macroblock concept in CCDC system.

- Motion estimation/compensation with adaptive block size
- Spatially adaptive inter/intra mode encoding
- Weighting by the HVS
- Vector coding of DCT coefficients

The basic unit for the 2D-DCT is an (8×8) block, as in all the previous systems. The CCDC system, however, defines particular sizes of superblock and macroblock as illustrated in Fig. 13.20. A superblock is a grouping of four luminance blocks and the associated two chrominance blocks (U and V) corresponding to a (16×16) pixel area as defined in MPEG-1 (Fig. 10.20). A macroblock is a grouping of 20 consecutive superblocks. Motion vectors are based on either an (8×8) block or a superblock adaptively.

The input *RGB* signal is converted into *YUV* components. The chrominance components U and V are low-pass filtered and subsampled by a factor of two along both the horizontal and vertical directions. The preprocessed *YUV* signal is input to the encoder as shown in Fig. 13.21.

During a scene change the prediction error from motion estimation/compensation is so large that coding the prediction error may be more difficult than coding the original image. In these cases, intraframe (with no ME/MC) processing

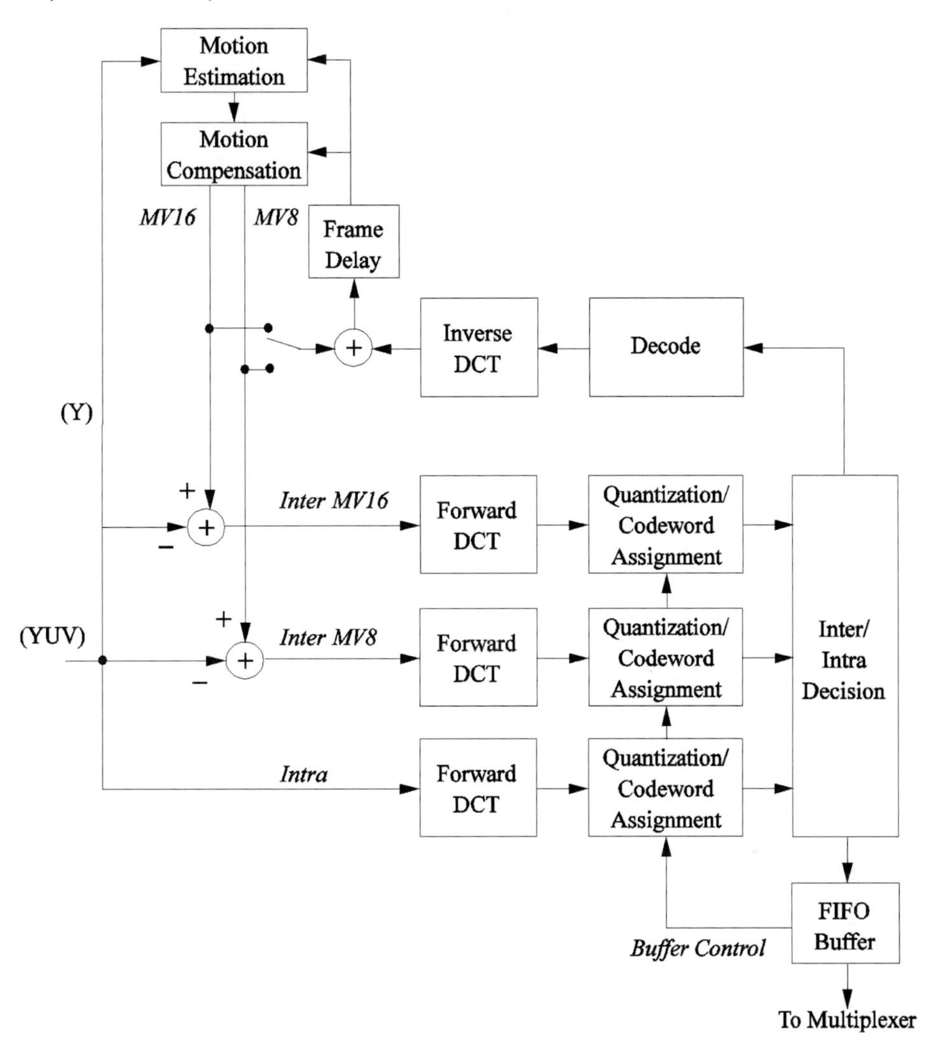

Figure 13.21 A CCDC video encoder block diagram.

is performed instead of interframe processing. The intra/inter processing mode decision is performed in a spatially adaptive manner based on local scene complexity and buffer fullness. The optimal spatial block size over which ME/MC is performed may also vary from one region to another. For processing some regions of the prediction error, a (16×16) superblock may be appropriate, while for other regions, an (8×8) block may yield better performance. There is a trade-off between large- and small-block ME/MC operations considering the number of motion vectors to be encoded.

The motion vectors are estimated based on the luminance component only and are applied to both luminance and chrominance components. For the latter the MV are appropriately scaled. The motion vector range is $+15/-16$ pixels horizontally and $+7/-8$ pixels vertically with a half-pixel accuracy. This search range will allow the encoder to track objects moving at up to 0.75 frame width and 0.67 frame height per second. This is especially useful for encoding video with large frame-to-frame displacements such as sporting events.

Each (16×16) superblock of the prediction error is encoded using one of the following three superblock encoding modes:

- Intra mode encoding of all blocks
- Intra mode encoding and inter mode encoding using MV_{16} [(16×16) block ME]
- Intra mode encoding and inter mode encoding using both MV_{16} and MV_8 [(8×8) block ME]

The decision criterion is based on the fewest number of bits required to achieve the same quality reconstructed video. This algorithm allows adaptive inter/intra mode processing and adaptive block size selection for ME/MC. To exploit the human visual system, an adaptive quantizer is used for encoding the DCT coefficients. There are 6 times 32 different sets of (8×8) weighting factors that may be applied to each (8×8) block of DCT coefficients. The quantizer is adaptive in the sense that different weighting factors are used depending upon the intraframe or interframe mode; human visual sensitivity of the luminance or chrominance component and of the spatial frequencies; and measures of the local scene activity and buffer status. The weighting factors are selected to produce coarser quantization of the high-frequency coefficients and finer quantization of the low-frequency coefficients. The DCT coefficients weighted by the HVS are uniformly quantized.

The quantized DCT block is vector coded rather than run/length encoded. In the vector-coding method, the (8×8) block of DCT coefficients is first divided into four regions, each containing 16 transform coefficients as shown in Fig. 13.22. The division is such that the first region contains the lowest-frequency coefficients and the last region contains the highest-frequency coefficients. For each region, a variable-length code word is chosen to identify the positions of all selected coefficients. Special code words are also defined in case all the coefficients in the remaining regions are zero and the selection pattern is uniquely coded. In addition, amplitudes of the selected coefficients are also statistically and adaptively vector encoded.

CCDC audio coding

The CCDC HDTV system uses the MIT-AC [831] for audio compression. Four channels of digital audio are provided. The input to and output from MIT-AC

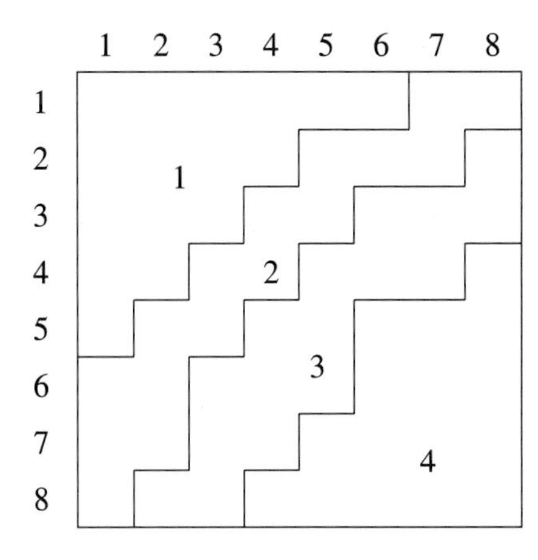

Figure 13.22 Division of an (8×8) DCT block for the vector-coding scheme.

is wideband digital audio sampled at 48 KHz, which ensures a clean digital interface to both the studio and the user. The encoder, as shown in Fig. 13.23, consists of transform analysis, spectral envelope encoding, transform coefficient encoding, and exploiting the properties of the human auditory system.

The overall algorithm of the MIT-AC system is similar to the AC-2 digital audio-coding algorithm developed by Dolby Laboratories (see Fig. 13.7). The audio signal is converted from time domain to frequency domain by a critically sampled single-sideband filterbank that is implemented with a fast transform algorithm. The data are first segmented by overlapping windows and a raised-cosine window with a duration of approximately 20 ms. For each window (512 samples) the temporal samples are transformed into a frame of spectral coefficients. To overcome the aliasing problem, the overlap-add process (time domain aliasing cancellation) is performed. The spectral envelope provides a fair approximation to the energy within each subband and can therefore serve, along with an interband masking model, as an indicator of the relative perceptual importance of each subband within the frame. According to the spectral envelope information, bits are dynamically allocated for each subband.

The transformed coefficients are then quantized within the range specified by the spectral envelope using the allocated bits for the subband. In addition, the spectral envelope is encoded and protected with error control coding. Approximately 20% of the total number of bits available for audio are used to encode the spectral envelope information.

The audio decoder performs error concealment and transform synthesis, as shown in Fig. 13.24. The spectral envelope information is first decoded. Error control encoding and decoding allows robust transmission. The spectral envelope also provides the range information for the coefficients within each subband. It is therefore required for decoding the coefficient amplitudes. Error concealment is necessary when the errors are detected but not corrected (the channel is very noisy). It is performed by repeating previous subbands of spectral coefficients.

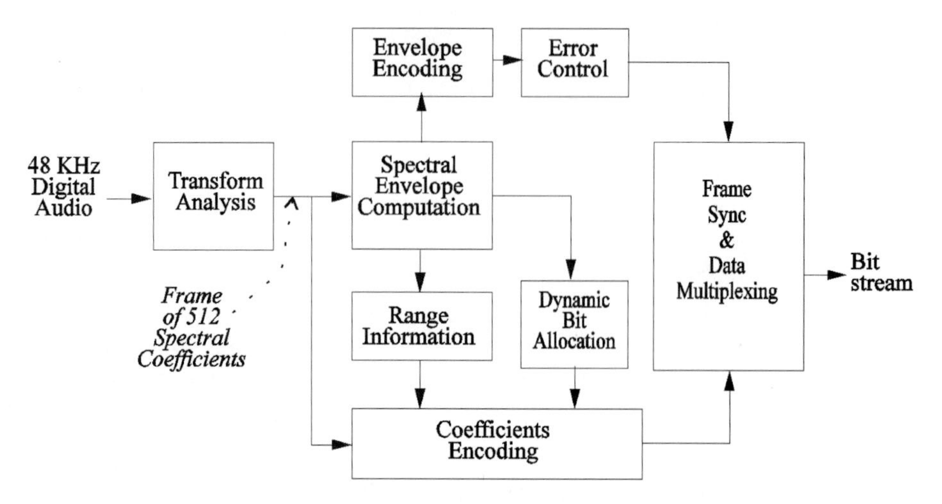

Figure 13.23 An MIT-AC audio encoder block diagram [831]. © 1992 MIT.

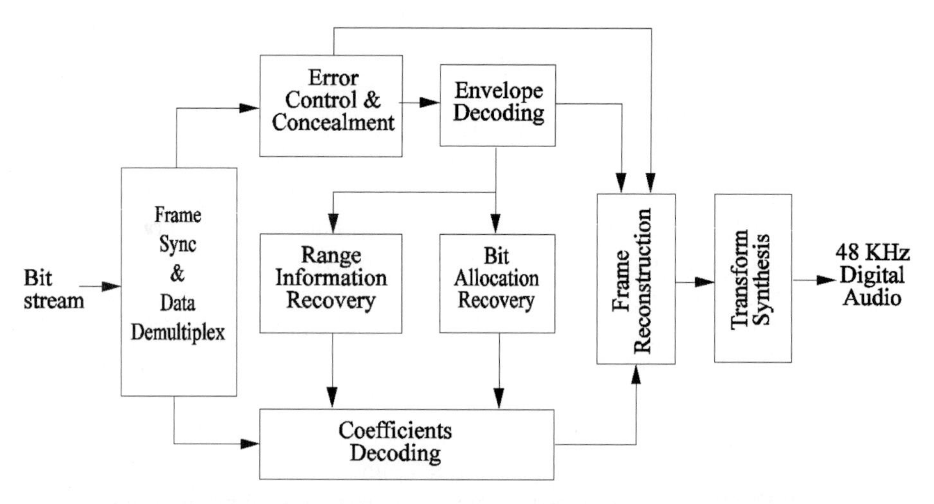

Figure 13.24 An MIT-AC audio decoder block diagram [831]. © 1992 MIT.

This method is satisfactory for maintaining undegraded audio up to a bit error rate of about 10^{-4} and maintaining pleasant audio up to a bit error rate of about 10^{-2}.

13.3 Grand Alliance System

The FCC ACATS has shown that all four proponents of the digital HDTV systems—DigiCipher, DSC-HDTV, AD-HDTV, and CCDC—have provided practical digital HDTV systems that lead the world in this technology [853]. The

Special Panel did not, however, recommend any one of the four excellent systems for adoption as a U.S. HDTV standard. Rather, the Special Panel recommended that these four finalists be authorized to implement a Grand Alliance system in May 1993.

The Special Panel examined five selection criteria: quality, transmission, scope of services and features, extensibility, and interoperability. The quality and transmission criteria were based on actual system testing, and the other three were primarily the subject of detailed analysis of the systems.

The DigiCipher and AD-HDTV systems showed an overall advantage over the other systems in video quality testing, but there were no significant differences among all the proponents. Based on the need for multichannel audio coding, audio subjective tests of the new multichannel audio systems were conducted. In view of interoperability considerations, the systems are required to deliver over alternative media (cable, satellite, packet networks), to transcode with NTSC, film, and so forth, and to integrate with computers and digital technology. For interoperability with computers, DSC-HDTV and CCDC ranked better than the others. Only AD-HDTV had its proposal for a packetized data structure and headers/descriptors as a critical enabling concept for flexibility.

For developing the best HDTV standard for the United States, the Grand Alliance system combines the advantages of all the proposed digital systems. The Grand Alliance has agreed on a goal of a 1080-line (the actual number is 1125) system. Later, another format (720 lines) was added [858, 862]. For the video compression, the Alliance reached consensus on MPEG-2 high profile (see Table 11.4). The compression algorithm is a hybrid coder including I, P, and B picture modes. For the audio compression, the Alliance recommended Dolby Lab's AC-3 multichannel algorithm [861, 863]. One of the most difficult decisions was the transmission algorithm. Following extensive paper study and field testing, Zenith's vestigial sideband (VSB) approach was selected. A packetized and prioritized data format is used to provide flexibility of services and extensibility.

The GA HDTV system is a layered digital system that uses headers/descriptors to provide flexibilities at each layer. The main characteristics of the system can be categorized into four layers as shown in Fig. 13.25 (compare with Fig. 13.15) and is summarized as follows:

- Picture layer: Multiple picture formats and frame rates are provided. The system designers can choose the best format for their particular use, as shown in Table 13.4.

- Compression layer: The MPEG-2 algorithm is used for video compression at a nominal data rate of approximately 18.4 Mbps. Dolby AC-3 is used for audio compression at a nominal data rate of 384 Kbps.

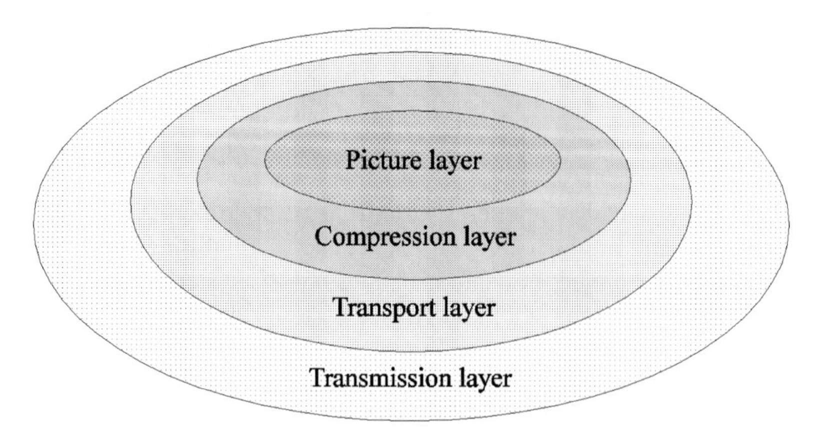

Figure 13.25 Layered structure of the GA HDTV system.

Table 13.4 Video formats in GA HDTV system

Spatial format ($H \times V$)	Frame rate (Hz)	Progressive/interlaced
1280 × 720	23.976/24	Progressive
(Square pixels)	29.97/30	Progressive
	59.94/60	Progressive
1920 × 1080	23.976/24	Progressive
(Square pixels)	29.97/30	Progressive
	59.94/60	Interlaced

- Transport layer: The MPEG-2 transport protocol is used as a packet format that provides the flexibility to deliver a wide variety of picture, sound, and data services.
- Transmission layer: A trellis-coded 8-VSB modulation technique that delivers a data rate of 19.3 Mbps in the 6 MHz terrestrial simulcast channel is used.

Table 13.5 shows the summarized GA specifications including the four layers. We discuss the GA coding algorithms, video compression, audio compression, transport layer, and transmission techniques in subsequent sections.

13.3.1 GA video coding

The GA proposal includes two main formats, 720 active lines per frame and 1080 active lines per frame, with different numbers of pixels per line. The 720-line format uses 1920 active pixels per line, yielding square pixels on the 16:9 aspect ratio. The frame rate can be 60 Hz, 30 Hz, or 24 Hz and the NTSC frame rates, 59.94 Hz, 29.97 Hz, and 23.98 Hz, are supported. Only the 60 Hz and 59.94 Hz

Table 13.5 Summary of the GA HDTV system specifications

Video parameter	Format 1	Format 2
Active pixels	1280 H × 720 V	1920 H × 1080 V
Total samples	1600 H × 787.5 V	2200 H × 1125 V
Frame rate	60 Hz progressive	60 Hz interlaced, 30 Hz progressive, 24 Hz progressive
Chrominance sampling	4:2:0	
Aspect ratio	16:9	
Data rate	10 to 45 Mbps, or Variable	
Colorimetry	SMPTE 240 M	
Picture coding types	Intra coding (I) Predictive coding (P) Bidirectionally predictive coding (B)	
Video refreshing	I picture/Progressive	
Picture structure	Frame	Frame/Field (60 Hz only)
Coefficient scanning	Zigzag	Zigzag, Alternate
DCT modes	Frame	Frame/Field (60 Hz only)
Motion compensation	Frame	Frame/Field (60 Hz only), Dual-prime (60 Hz only)
Motion vector range	Horizontal: unlimited by syntax Vertical: −128, +127.5	
MV precision	1/2 pixel	
DC coefficient precision	8 bits, 9 bits, 10 bits	
Rate control	Modified TM5 with forward analyzer	
Film mode processing	Automated 3:2 pull-down	
Max. VBV buffer size	8 Mbits	
Quantization matrices	Downloadable (scene dependent)	
VLC coding	Intra and inter run-length/amplitude VLC	
Error concealment	MC frame holding (slice level)	
Audio parameter		
Number of channels	5.1	
Audio bandwidth	20 KHz	
Sampling frequency	48 KHz	
Dynamic range	100 dB	
Compressed data rate	384 Kbps	
Transport parameter		
Multiplex technique	MPEG-2 systems layer	
Packet size	188 bytes	
Packet header	4 bytes including sync	
Number of services:		
– Conditional access	Payload scrambled on service basis	
– Error handling	4-bit continuity counter	
– Prioritization	1 bit/packet	
System multiplex	Multiple program capability	

table continues

Table 13.5 *(cont.)*

Transmission parameter	Terrestrial mode	High data rate Cable mode
Channel bandwidth	6 MHz	6 MHz
Excess bandwidth	11.5%	11.5%
Symbol rate	10.76 Msps	10.76 Msps
Bits per symbol	3	4
Trellis FEC	2/3 rate	None
Reed-Solomon FEC	(208, 188) T=10	(208, 188) T=10
Segment length	836 symbols	836 symbols
Segment sync	4 symbols/seg	4 symbols/seg
Frame sync	1 per 313 segments	1 per 313 segments
Payload data rate	19.3 Mbps	38.6 Mbps
NTSC co-channel rejection	NTSC rejection filter in receiver	N/A
Pilot power contribution	0.3 dB	0.3 dB
C/N threshold	14.9 dB	28.3 dB

variations for the 1080-line format are encoded as interlaced scanned images as shown in Table 13.4. Other formats are coded as progressively scanned. The frame rates 24 Hz and 30 Hz are included to be coded as film modes. The number of pixels per line can be reduced to 1440 samples per line for the interlaced 1080-line format only.

Analog video signal is received in *RGB* format and is digitized using 10-bit A/D converters. To compensate for the nonlinear property of the display, the digitized video is gamma-corrected as shown in Fig. 13.26. The samples are then converted to the SMPTE 240M (YP_bP_r) color space using a linear transformation as described by Eq. (2.8). Programmable FIR filters are used to shape the frequency response and to prevent aliasing during the down-sampling process. The chrominance components are down-sampled by a factor of 2, both horizontally and vertically, since the GA system uses the 4:2:0 format.

Motion estimation/compensation

The GA compression algorithm utilizes multiple prediction modes including P picture prediction, B picture prediction, and dual-prime prediction for effective motion compensation (Fig. 11.16). In P picture mode the prediction is performed in the forward direction only (prediction from the previous I or P picture), as discussed in Section 11.2.3. In B picture mode the prediction is based on both previous and future frames bidirectionally. The dual-prime mode (Fig. 11.16) is for interlaced fields only and is precluded when a sequence uses B pictures. P pictures are predicted using the most recently encoded P picture or I picture in the group of pictures (Fig. 10.8). Each macroblock within a P picture can be

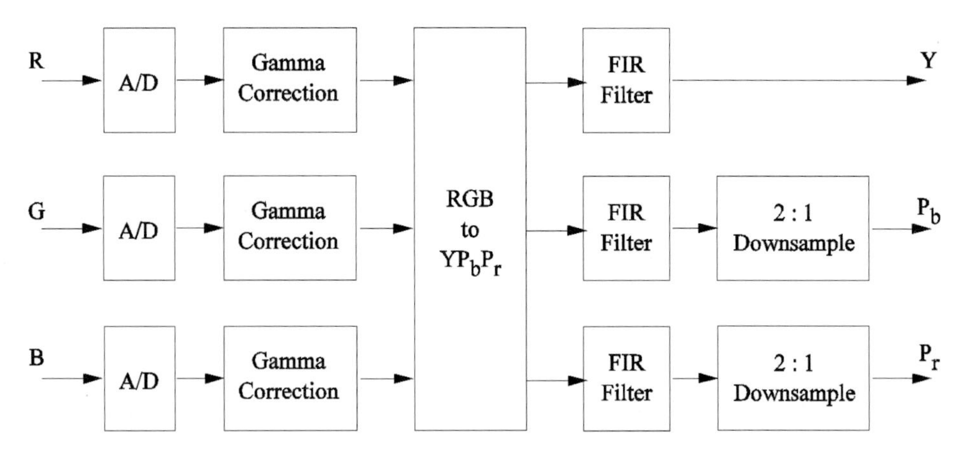

Figure 13.26 Video processor including gamma-correction, format conversion, and down-sampling.

either forward-predicted or intraframe coded. For the forward prediction, either frame mode or field mode can be used. Regardless of picture mode, field-based, frame-based, or dual-prime prediction, the predicted macroblock is subtracted from the target macroblock and only the difference values (motion-compensated residual) are coded and transmitted. Figure 13.27 shows the functional block diagram of the GA video encoder. Motion estimation/compensation is performed by the motion processor that produces motion vectors and prediction errors.

B picture prediction is useful for increasing the coding efficiency when encoding latency is not an important factor. It works well with both progressive and interlaced picture modes, at the expense of additional frame memory in the receiver. It is included in the GA system because the coding efficiency is increased especially with progressive scanning where dual-prime prediction is not available. When B picture prediction is used, the forward and backward predictors are averaged and then subtracted from the target macroblock to calculate the prediction errors. Dual-prime prediction based on field-based predictions of both fields is obtained by averaging two separate prediction errors calculated from the two nearest decoded fields. The final averaged prediction is then subtracted from the target macroblock field as discussed in Example 11.1.

Mode selection

The GA video compression system provides both intraframe coding and interframe coding. When motion-compensated prediction is quite good, the interframe coding mode is selected. For some frames, such as those at scene changes, however, the prediction is not as good (the prediction errors are large). The intraframe coding mode is then selected for (16×16) macroblock coding. For

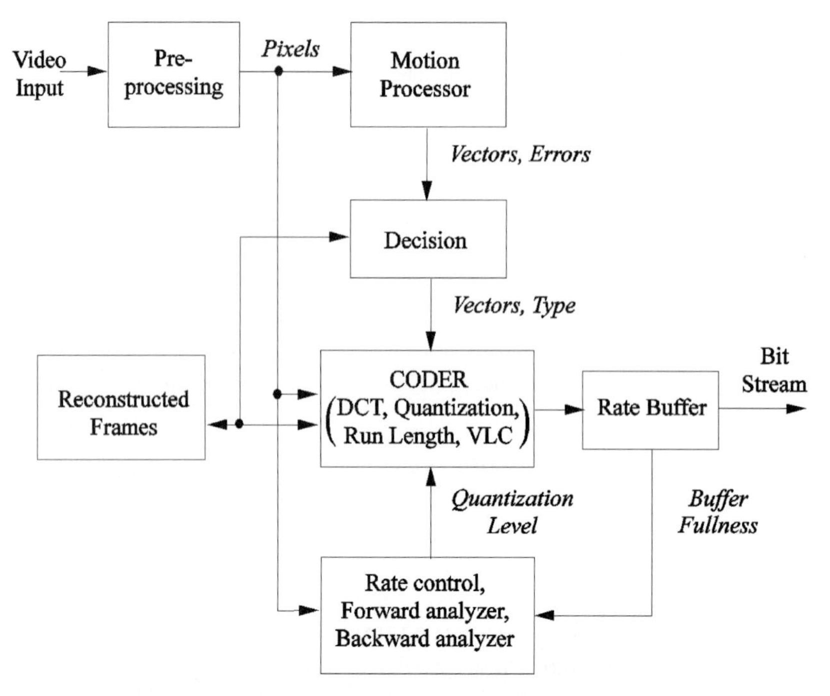

Figure 13.27 A block diagram of the GA video encoder [858]. © 1994 FCC ACATS.

motion-compensated prediction, furthermore, an initial frame is needed at the decoder to start the prediction loop. Therefore, the entire frame is encoded using the intraframe mode (I pictures). That is, the entire frame is refreshed periodically as shown in Fig. 10.8. This mechanism provides clean insertion points in the compressed bit stream at the expense of a larger buffer, increased latency, and complicated rate control.

There are two options for processing interlaced video signals. The first one is to separate each frame into its two field components and then process the two fields independently (Fig. 13.28a). The second one is to interleave the lines of field 1 and 2 (Fig. 13.28b) and then process the two fields as a single frame. Frame-based processing generally works better than field-based processing, since there is more correlation between lines in a frame than in a field. For example, pictures with little detail of any sort that are moving horizontally with little horizontal detail and that are moving vertically with little vertical detail can be coded by using frame-based processing. In detailed moving areas, however, field-to-frame conversion (interleaving the two fields) would introduce spurious high vertical frequencies. In this case, field processing works better than frame processing. The GA system combines the advantages of both frame processing and field processing

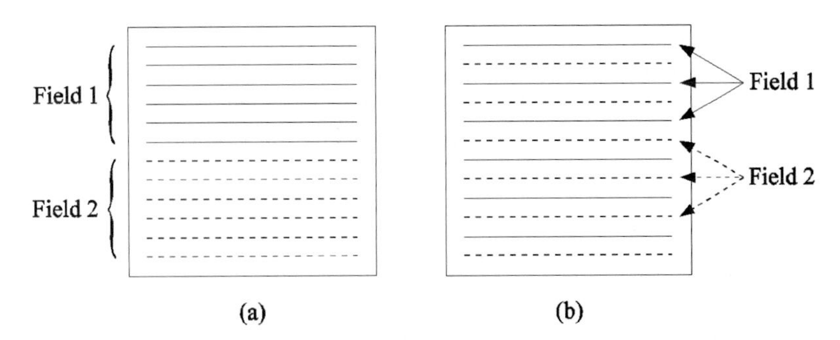

Figure 13.28 Field/frame mode processing of interlaced video signals. (a) Field processing; (b) frame processing.

by adaptively selecting one of the two modes on a block-by-block basis. Note that frame processing is always used for progressive sources and the adaptive field/frame processing is only enabled for interlaced sources.

Adaptive coding process

Two-dimensional (8×8) DCT is used not only for the motion-compensated residual but also for the intra mode macroblocks. Motion compensation reduces the temporal redundancy, whereas DCT reduces the spatial redundancy. The residual is partitioned into (8×8) blocks and then independently transformed. The transform coefficients are adaptively quantized based on their local characteristics. Low-frequency coefficients may be finely quantized and the less important high-frequency coefficients may be quantized more coarsely. The step sizes are normalized and weighted based on human visual sensitivity. Quantization matrices are determined by the perceptual weighting, and the actual video compression is achieved by using adaptive quantization such as linear or nonlinear matrices, as illustrated in Fig. 11.19, and scalar or vector quantization. The DC coefficients are coded differentially to take advantage of high spatial correlation.

The quantized coefficients are variable-length coded according to their probability distribution. At first, the 2D array of transform coefficients is scanned and prioritized into a 1D sequence as shown in Fig. 11.20. The coefficients are then efficiently represented through run-length encoding. Hence, the run length and the coefficient value can be entropy coded either separately or jointly. Here the run length indicates the run of consecutive quantized zero coefficients along the specified scan before a nonzero (quantized) coefficient appears. Huffman coding, which is used in the GA system, is one of the most common entropy coding schemes. In Huffman coding, a codebook is generated so that the entropy is minimized subject to the code word constraints of integer length and unique decodability. Events that are more likely to occur are assigned shorter-length code words while those that are less likely to occur are assigned longer-length code words. Near optimal performance can be achieved by using alternate codebooks where elements with similar statistics are grouped and coded together. Moreover, the size of each codebook can be determined by grouping events that have similar statistics.

Interoperability

The GA system provides interworking with MPEG systems and computers. MPEG-2 high- and high-1440–level decoders can decode the GA signal format. A GA decoder can not decode the MPEG-2 high- or MPEG-2 main-level bit stream, however, unless the manufacturer adds functionality to the decoding device. This does not mean that the decoding engine has to be changed, because the GA system is a superset of MPEG-2. The MPEG-2 main level is not able to decode the GA bit stream due to memory and speed constraints.

The GA system interoperates with computer displays in a similar manner, at either the transport layer or the compression layer. The computer, however, requires some display conversion of frame rate, picture aspect ratio (16:9 to 4:3), interlaced to progressive (if needed), and pel aspect ratio (if needed).

13.3.2 GA audio coding

Dolby AC-3 is a frequency domain–based audio compression algorithm that processes audio in discrete blocks representing 256 samples for each of the input channels. It is designed to accommodate a wide variety of input and output channel formats from a single mono channel up to five full bandwidth audio channels plus a low-frequency enhancement channel (referred to as 5.1 channels) at any bit rate up to 384 Kbps. A wide range of bit rates and audio coding modes are also supported.

The AC-3 is applicable for many applications including HDTV, satellite broadcasting systems, cable TV, laser discs, video cassettes, compact discs, digital audio tapes, and PC multimedia applications. The reader may compare this high-quality audio-coding system with others such as multichannel MUSICAM (Section 10.3), MPEG-2 audio (Section 11.3), the AC-2 codec adopted by DigiCipher and DSC-HDTV, and the MIT-AC codec proposed by CCDC (Section 13.2). The AC-3 was selected for the DigiCipher2 system developed by GI Corp. in the summer of 1993.

An AC-3 serial-coded audio bit stream is composed of synchronization frames (SF) containing six coded audio blocks (AB), each of which represents 256 samples, as illustrated in Fig. 13.29. A synchronization information (SI) header is included at the beginning of each frame. A bit stream information (BSI) header then follows and contains parameters for the mode of the coded audio services.

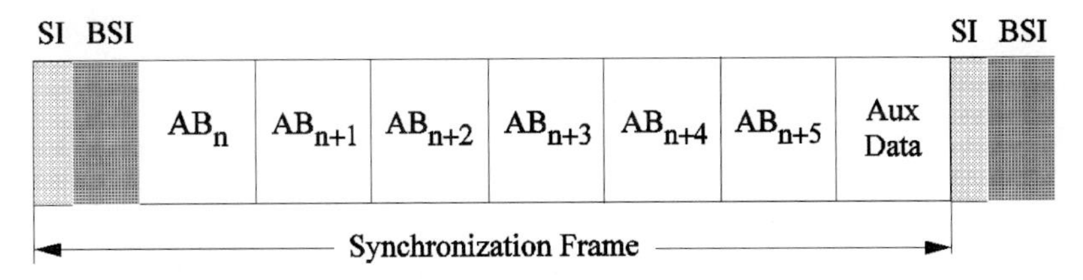

Figure 13.29 The AC-3 synchronization frame structure.

An auxiliary data field (Aux) may also be inserted. The sync frame can be verified by use of an embedded CRC check at the end of the sync frame.

AC-3 encoder/decoder

AC-3 is a transform coder using the TDAC (time domain aliasing cancellation) transform, which is represented in two transforms by means of the half overlap/add windows. The block size of 512 points is a good trade-off between frequency resolution and complexity. The 512-point transform is performed every 256 points, providing a time resolution of 5.3 ms (at 48 KHz sampling rate, calculated by 1/48 KHz $\times$ 256 samples), that means 93 Hz frequency resolution. Since the time resolution is sometimes insufficient, a finer time resolution is needed to prevent prenoise artifacts at low bit rates. The encoder can switch to a 256-point transform for a time resolution of 2.7 ms. Figure 13.30 shows the block diagram of the AC-3 encoder, representing filter bank, bit allocation, spectral envelope, and multiplexing.

Transform filter bank: Input audio samples are transformed by the MDCT/-IMDCT with discrete audio blocks of 512 samples or 256 samples determined adaptively. The adaptation occurs by switching the block length in a signal-dependent manner. The output of the 512-point TDAC transform gives 256 real-valued frequency coefficients.

Spectral envelope: The logarithm of the absolute values of the coefficients forms the spectral envelope. The frequency and time resolution of the spectral envelope can also be signal dependent and can be determined at the encoder. AC-3 uses a differential coding strategy in which the exponents (results of the logarithm) for a channel are differentially coded across frequency. The differential exponents are combined into groups in the audio block (Fig. 13.29). The main difference between the group modes is the number of differential exponents that are combined together. Each differential exponent can take on one of five values: -2, -1, 0, $+1$, $+2$, representing 6 dB steps from the previous exponent. The encoder decodes the spectral envelope to make use of it as a reference for the quantization and the bit allocation routine.

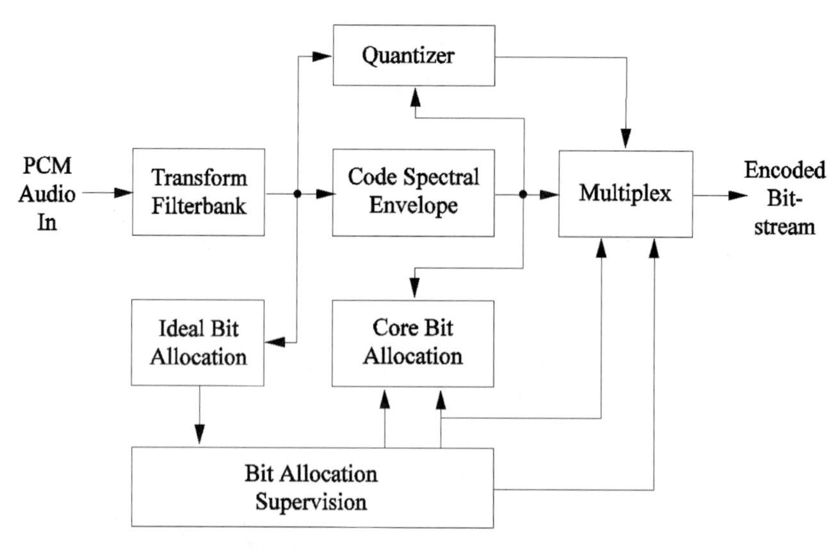

Figure 13.30 The AC-3 digital audio encoder block diagram [858]. © 1994 ACATS.

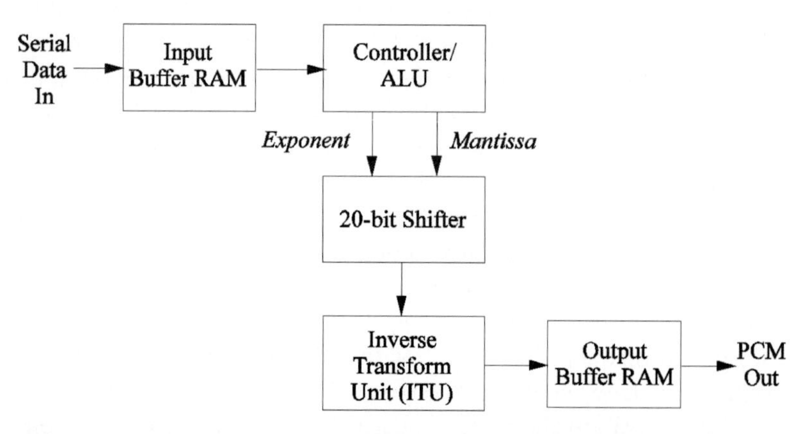

Figure 13.31 The AC-3 digital audio decoder block diagram [858]. © 1994 ACATS.

Parametric bit allocation: This routine analyzes the spectral envelope of the audio signal being coded with respect to masking effects to determine the number of bits to assign to each transform coefficient mantissa. The bits are allocated to each channel and it is possible to allocate a different number of bits for each individual channel. The bit allocation contains a parametric model of human hearing for estimating the masked noise threshold. The model parameters

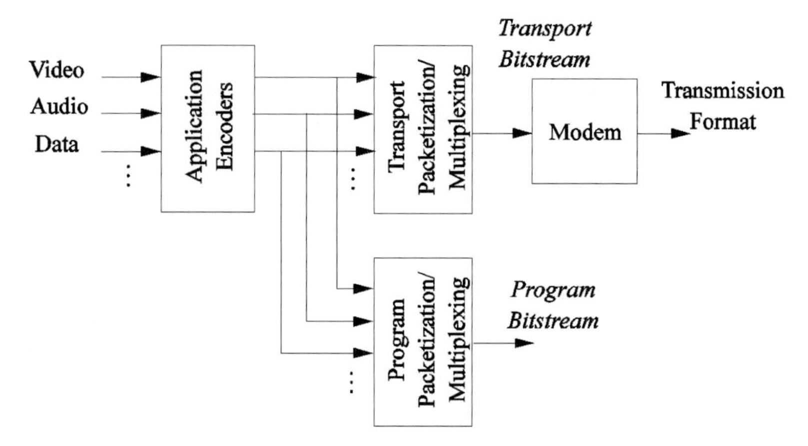

Figure 13.32 System level multiplexing approaches (transport stream and program stream).

are conveyed to the decoder as side information. The AC-3 encoder calculates an ideal bit allocation from the transform coefficients and compares the results of this allocation with those of the core bit allocation at the bit allocation supervision routine. The parameters of the psychoacoustic model can then be altered to improve subjective quality. The core bit allocation routine requires an ALU (arithmetic logic unit) with two 16-bit accumulators as shown in Fig. 13.31. The ALU requires the following instructions: add, subtract, compare, AND, OR, XOR, conditional branch, and shift, but no multiplier.

Rates and modes: The AC-3 algorithm supports the sampling rates of 48 KHz, 44.1 KHz, and 32 KHz. The GA system may not support the use of more than one sampling rate. Again, the GA system recommends the use of nominal bit rates up to 384 Kbps, whereas the AC-3 provides nineteen nominal bit rates of 32, 40, . . ., 640 Kbps. The main audio service mode can vary from monophonic (1/0) up to all combinations of multiple channels (2/1, 2/2, 3/0, 3/1, 3/2) (Fig. 11.31), with optional use of a low-frequency enhancement channel (0.1 channel).

13.3.3 Transport layer

The sources (video, audio, data, program/system control information, etc.) are independently encoded by relevant coders. The bit streams go through a PES (packetized elementary stream) layer packetization prior to the GA transport layer. The PES packetization interval is application dependent and of variable length with a maximum size of 2^{16} bytes.

The variable-length bit streams are to be packetized and multiplexed into a transport bit stream in transport layer. The two multiplexing approaches are motivated by different application scenarios, as shown in Fig. 13.32. For example,

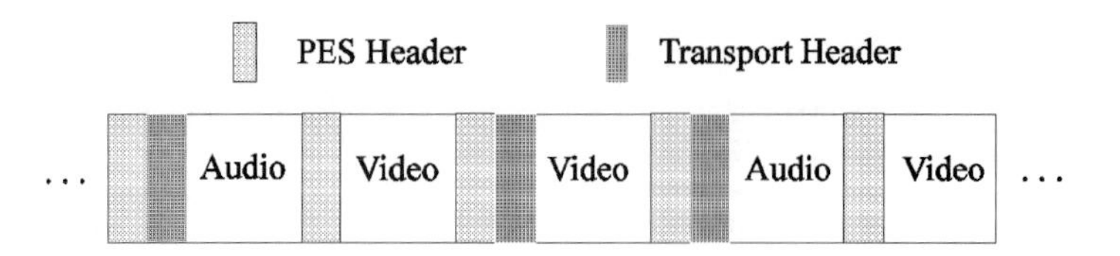

Figure 13.33 Illustration of bit stream for transport stream.

the video PES and audio PES may be packetized as transport stream or as program stream. Transport streams are defined for environments where errors and data loss events are likely, including storage applications and transmission over noisy channels. Program streams, on the other hand, are defined for relatively error-free media, e.g. CD-ROMs. The transport streams are transmitted as the payload of fixed-length transport packets, while the program streams are transmitted by multiplexing the bits for the complete PES packets in sequence.

Both approaches have been used in the MPEG-2 system. The transport stream approach of MPEG-2 has been found to support most of the functional requirements for the GA system that carries ATV, CATV, DBS, and so forth, on noisy channels. Hence, the transport stream approach forms the basis of the GA system definition. Although the GA transport system can still identify and carry the program stream of MPEG-1 video and audio services, compatibility with MPEG-1 systems is not a concern of the GA.

As illustrated in Fig. 13.33, each PES packet for a particular elementary bit stream occupies a variable number of transport packets, and data from various elementary bit streams are generally interleaved at the transport packet layer. The PES header carries various rate, timing, and data descriptive information, as set by the encoder. The PES packet includes stream identification, packet length, PES header, and data block. In essence, all the fields of the PES header are optional. All data for a PES packet, including the header, are transmitted contiguously as the payload of transport packets (Fig. 13.34). New PES packet data always start a new transport packet, and PES packets that end in the middle of a transport packet are followed by stuffing bytes for the remaining length of the transport packet.

Each transport packet consists of 188 bytes. The choice of this packet size is motivated by a few factors. It is to be large so that overhead may not be significant. It is not to be too large under inefficient error correcting conditions. It is to be interoperable with the ATM format wherein a cell consists of 53 bytes. A single GA transport packet is then transmitted in four ATM cells.

The transport header includes a fixed-length (4-byte) link header and a variable-length adaptation header (Fig. 13.34). This link and adaptation level functionality is directly used for the terrestrial link on which the GA bit stream is

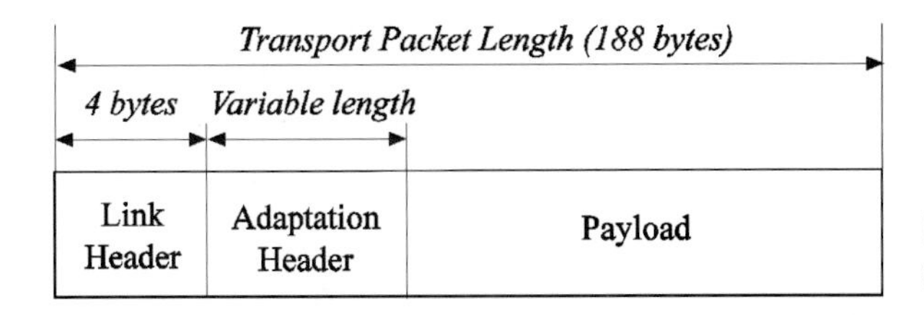

Figure 13.34 Illustration of the GA transport packet format.

transmitted. These headers can be ignored in the ATM system, however, in which the GA bit stream may just remain the payload to be carried.

The link header (4-byte) includes packet synchronization (1-byte), packet identification (13-bit PID), error indicator (error or not), packet continuity counter (4-bit), scrambling control (2-bit), priority, adaptation field control, and payload start indicator. The priority bit is reset, however, since the GA transmission system is currently not expected to support prioritization.

The adaptation header in the GA packet is a variable-length field. The functionality of this header is basically related to higher-level decoding. The header format is based on the use of flags to indicate the presence of the particular extensions to the field. The typical functions of this layer are synchronization of the decoding by transmitting a timing information sample of a 27 MHz clock in the program clock reference field (PCR), random entry (access) points for program tuning and program switching, and insertion of local programming, e.g., commercials, into a bit stream at a broadcast headend. The random entry points coincide with the start of PES packets and I frame in the group of pictures.

The GA transport system is compatible with two of the most important transport systems, MPEG-2 and ATM. It may also be compatible with other applications, such as CATV and DBS, provided that these systems are based on MPEG-2. The GA system is interoperable with MPEG-2 decoders since the GA transport layer is currently a constrained subset of the MPEG-2 transport layer. The constraints are imposed for reasons of increased performance of channel acquisition, bandwidth efficiency, and decoder complexity.

The GA system can be transferred in a link layer that supports ATM transmission. Three techniques are suggested for mapping the GA transport packet into the ATM transport packet. First, the simplest method is the null AAL (ATM adaptation layer) byte structure. Note that the ATM cell consists of two parts: a 5-byte header and a 48-byte user data field, where up to 4 of these bytes can be allocated to an adaptation layer as shown in Fig. 13.35.

The GA transport packet is partitioned into 48-byte payloads that can be included in the user data field of the ATM cell in succession. Note that the packet length of 188 bytes is not an integer multiple of the ATM payload, 48-bytes. The second method is the single AAL byte structure. The GA transport packet is partitioned into 47-byte payloads resulting in an integer multiple of the ATM

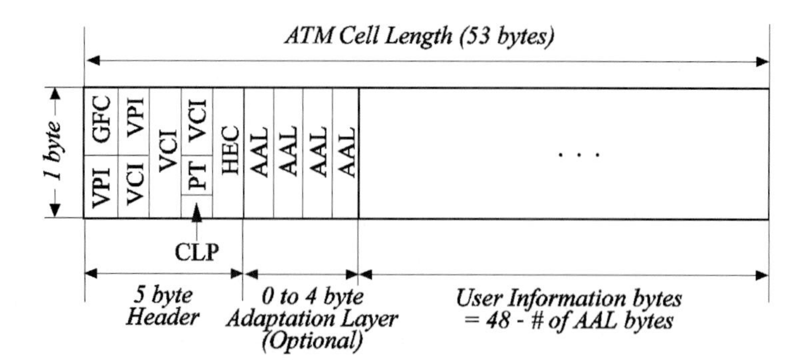

(**GFC**: 4-bit generic flow control; **VPI**: 8-bit network virtual path identifier; **VCI**: 16-bit network virtual channel identifier; **PT**: 3-bit payload type; **CLP**: 1-bit cell loss priority flag; **HEC**: 8-bit header error control field; **AAL**: user specific ATM Adaptation Layer)

Figure 13.35 Structure of the ATM cell.

payload. The transport packet of 188 bytes can fit into four ATM cells and the FEC codes may be included in the 1-byte AALs [626]. The third method is the dual AAL byte structure. The partitioned packet size is to be a 46-byte length. The PID of the transport header can be discarded, since it can be reconstructed from the ATM headers (VPI and VCI). These methods are, however, only a suggested approach and do not represent a complete design.

13.3.4 Transmission layer

The vestigial sideband (VSB) digital transmission system was selected for GA signal transmission, instead of the QAM system, because of its simplicity and better performance. Zenith's DSC-HDTV has proposed technique similar to that discussed in Section 13.2.2. The VSB system has two modes: a simulcast terrestrial broadcast mode, and a high–data rate cable mode. The terrestrial broadcast mode is designed for transmitting an ATV signal on currently unusable taboo NTSC channels (6 MHz bandwidth) with minimal interference to/from other channels. The high–data rate cable mode is designed for transmitting two ATV signals in a 6 MHz channel, since the cable does not have as severe an environment as that of the terrestrial system. The specific comparison of the two modes is shown in Table 13.5.

The VSB transmission system has the structure of frame, field, data segment, and synchronization segment, as illustrated in Fig. 13.36, for transmitting the video data on a 6 MHz channel. Originally the Zenith's VSB system was designed within the NTSC frame/field structure. The concept of frame/field of the GA VSB system, however, does not match that of the NTSC or an entire picture as a basic coding layer. Most of the other specifications are similar to those of the Zenith's

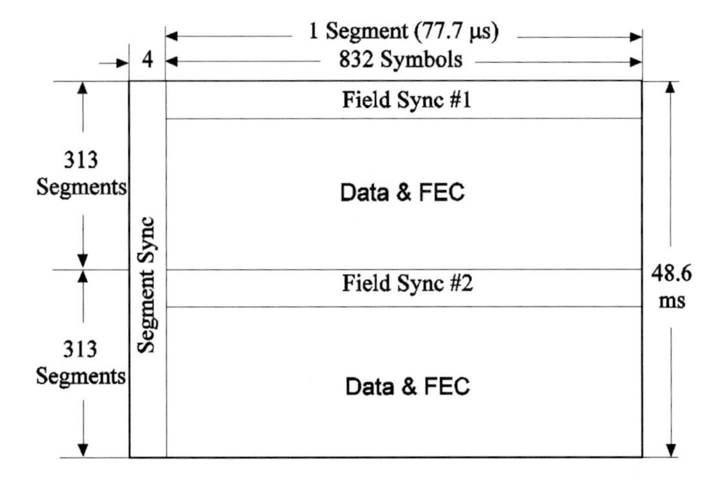

Figure 13.36 The VSB data frame structure of the GA system.

system. The exact symbol rate is 10.76 Msymbol/sec = 836 symbols/77.7 μsec. The frame is organized into segments each with 836 symbols, taking 48.6 ms (20.6 frames/sec). The frequency of a segment is calculated as follows:

$$\begin{aligned} f_{seg} &= (10.76\ \text{Msymbols/sec})/(836\ \text{symbols/seg}) \\ &= 12.87\ \text{Ksegments/sec} \end{aligned} \tag{13.1}$$

For terrestrial broadcast mode, 8-VSB is used, and the number of bits per segment is

$$3\ \text{bits/symbol} \times 832\ \text{symbols} = 2496\ \text{bits/segment} \tag{13.2}$$

For the high rate cable mode, 16-VSB (4 bits/symbol) is used, resulting in 3328 bits/segment.

The spectrum for the VSB transmission is flat, as shown in Fig. 13.14, except for the band edges where a root-raised cosine response results in 620 KHz transition regions. The symbols (data and syncs) are modulated by suppressed--carrier modulation and the lower sideband is removed before transmission.

Figure 13.37 shows the VSB terrestrial broadcast transmitter and receiver. (For the cable mode, there is no trellis code or noise-rejection filter.) A short description for each part is as follows.

Data randomizer: This is used to randomize the data payload. A 16-bit maximum length pseudo-random sequence is generated in a 16-bit shift register that has 9 feedback taps.

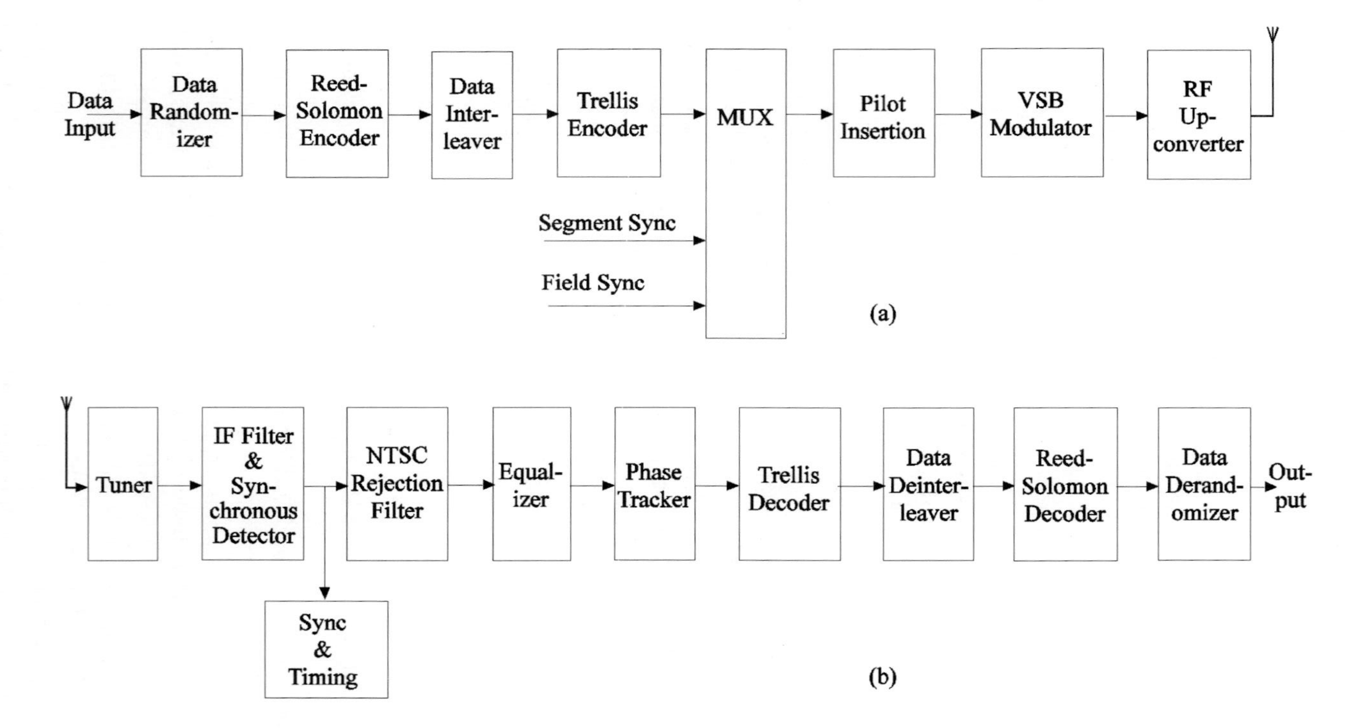

Figure 13.37 The VSB terrestrial broadcast. (a) Transmitter. (b) Receiver.

Reed-Solomon encoder/decoder: The RS code $t = 10(208, 188)$ is used, in which the data block size is 188 bytes. A total block size of 208 bytes is transmitted per data segment. Twenty RS parity bytes can correct up to 10 byte errors per block. The $(208, 188)t = 10$ RS decoder decodes the trellis-decoded byte data. Any burst errors are greatly reduced by the combination of interleaving and RS error correction.

Interleaver: The data are interleaved for further protection against burst errors. The goal of the interleaver is to spread the data bytes from the same RS block over time so that a long burst of noise is necessary to encompass more than 10 data bytes and overrun the RS error protection. The interleaver is an 87 data segment diagonal byte interleaver. Interleaving is provided to a depth of about 1/3 of a data field, and intrasegment interleaving is also performed, encoding symbols $(0, 12, 24, 36, \ldots)$ as one group, symbols $(1, 13, 25, 37, \ldots)$ as a second group, and so on (a total of 12 groups).

Trellis coder/decoder: The terrestrial broadcast mode employs a 2/3 rate trellis code so that one input bit is encoded into two output bits using a 1/2 rate convolution code while the other input bit is left unencoded. The signaling waveform used with the trellis code is an 8-level (3-bit) 1 D constellation (8-VSB). Twelve identical trellis encoders and decoders are required for coding the interleaved data.

VSB modulator: The 10.76 Msymbols/sec, 8-level trellis composite data signal with pilot and syncs is filtered for proper shaping by a linear phase raised-cosine Nyquist filter. The filter response is essentially flat across the entire band, except for the transition regions at each end of the band. The orthogonal baseband signals are converted to an analog signal and then modulated on quadrature IF (intermediate frequency) carriers (46.69 MHz) to create the VSB IF signal.

Chapter 14
CMTT Digital Broadcasting Standards

Summary

This chapter describes the algorithms recommended by the ITU for digital coding and transmission of TV signals in component form as specified in ITU-R BT.601 (4:2:2 format). Two bit rates, both for contribution quality, are specified: 34 to 45 Mbps (third digital hierarchy) for ITU-R CMTT.723, and nearly 140 Mbps (fourth digital hierarchy) for ITU-R CMTT.721-1. The former is based on the traditional hybrid scheme (MC DPCM/DCT), complemented by some adaptive features and HVS weighting, whereas the latter is based on nonadaptive DPCM.

14.1 ITU-R CMTT.723

CMTT (the Committee for Mixed Telephone and Television) is the joint study group of CCITT and CCIR. Under the new structure of the ITU, this has been transferred to the Telecommunication Standardization Sector as ITU-T Study Group 9. To constitute a common standard for the various television formats (NTSC, PAL, SECAM), digital coding and transmission of component television signals at bit rates of about 34 to 45 Mbps are specified [899] in the format defined by ITU-R BT.601 [898]. The video-coding algorithms are based on a hybrid predictive/transform scheme. The temporal redundancy is reduced by motion-compensated prediction and the spatial redundancy is reduced by 2D (8 × 8) DCT. The transform coefficients are variable-length coded. Hence, most of the coding algorithms are similar to those described in Chapters 9 through 13.

Specification of component TV coders for about 34 or 45 Mbps (the third level of the digital transmission hierarchy) is summarized as follows:

- Video input/output: 4:2:2 ITU-R BT.601 signal (525- or 625-line digital video in component form) (Table 2.1)
- Preprocessing: Samples per active line are 720 for luminance and 360 for each chrominance component. The number of active lines is 240 per field in 525-line system and 288 per field in 625-line system. The samples are converted to an 8-bit two's-complement representation.
- Coding: Three modes (intrafield, interfield, and MC interframe) are used. Two-dimensional (8 × 8) DCT is applied to blocks in a macroblock (intrafield mode), or to differential blocks obtained by difference between the current and a reference block taken in the previous field (interfield mode) or field with same parity in the previous frame (interframe mode). Quantization parameters are adapted to the buffer occupancy, the type of block (luminance/-chrominance), and the criticality of the quantized DCT coefficients.
- Buffer memory capacity: 1,572,864 bits
- Error protection: Reed-Solomon (255, 239) code
- Multiplex: 2.048 Mbps (or 1.544 Mbps) audio channels are multiplexed, but neither the coding nor the error protection of the audio channels is covered by this specification. Channels for Teletex, test line, supervision, conditional access, and time code are also combined with a video channel and audio channels.

14.1.1 Coding mode decision

Mode selection (intrafield, interfield, and interframe) is based on each macroblock (Fig. 14.1) and is indicated by an overhead bit. A macroblock is composed of an (8 × 16) luminance and two colocated (8 × 8) chrominance components. A stripe comprises eight active lines of the ITU-R 601 luminance component.

The choice of mode selection, the method of motion estimation and the actual implementation of various functions such as 2D (8 × 8) DCT and 2D (8 × 8) IDCT are not specified. One possible method of determining the mode is to calculate the macroblock difference between the current block and the block in the previous field or frame, as discussed in Sections 9.3 and 10.2.4. The order of mode choice is (i) intrafield mode, (ii) interfield mode (no motion compensation),

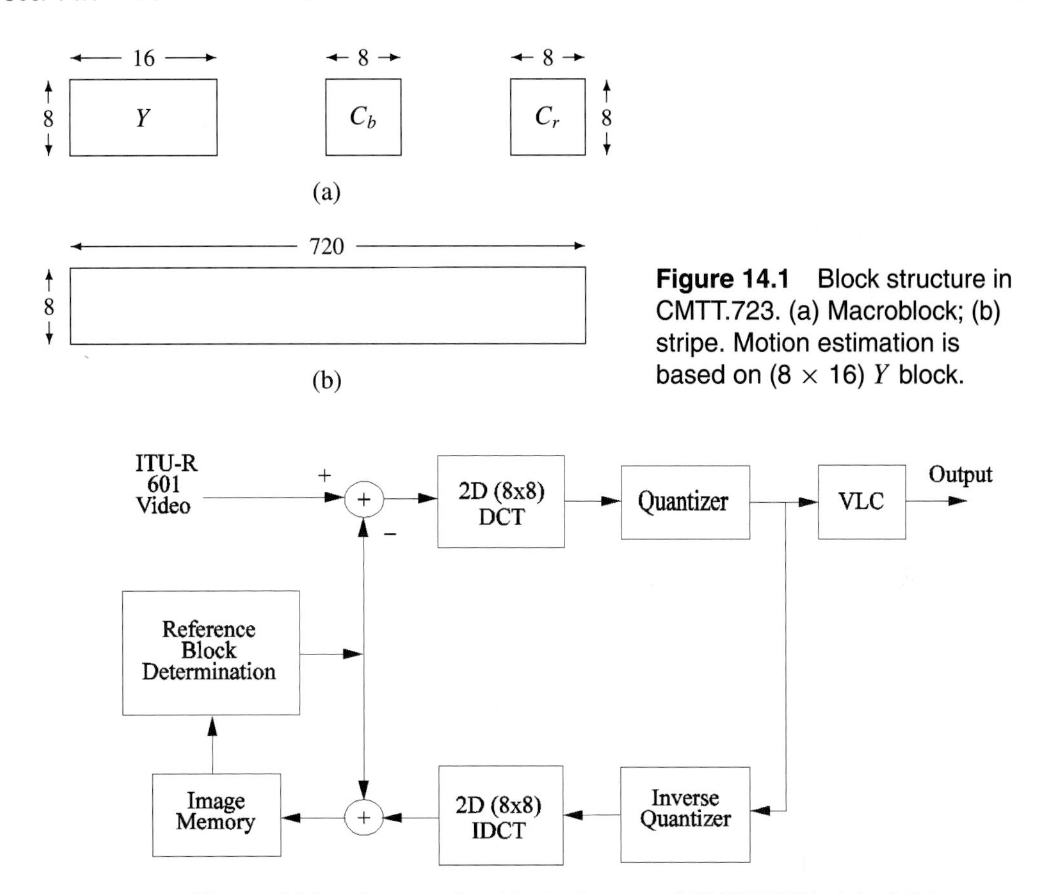

Figure 14.1 Block structure in CMTT.723. (a) Macroblock; (b) stripe. Motion estimation is based on (8×16) Y block.

Figure 14.2 An encoding block diagram of CMTT.723 in interfield and interframe modes.

and (iii) interframe mode with motion compensation. It is recommended that a forced intrafield mode be chosen periodically to avoid the temporal propagation of transmission error effects.

The encoding block diagram is shown in Fig. 14.2. The input video signal is preprocessed from ITU-R 601 as stated earlier. The pixels outside the active picture are set to zero. In intrafield mode, the input samples are transformed by 2D (8×8) DCT into the frequency domain. Inverse quantized and inverse transformed field data, however, are stored in the memory to be used as reference for coding the next image. The two last previously decoded fields are also stored to predict the current block. The reference block is determined from the previous interpolated field (interfield mode) or from the previous frame (interframe mode).

Table 14.1 Motion estimation parameters in ITU-R CMTT.723

Motion vector range	$\pm$ 14 pels and $\pm$ 7 lines
Resolution	Half-pel and half-line
Number of possible vectors	1653

Motion estimation and compensation

For the interfield mode, the reference block is computed with pixels from the previous field. Since two fields in an image frame are interleaved, the predicted block is interpolated from two neighboring lines as

$$x_{Np}(i, j) = (x_{N-1}(i, j-1) + x_{N-1}(i, j+1))/2 \qquad (14.1)$$

where N denotes field number.

For the interframe mode, the reference block is taken in the field of the previous frame with the same parity as the current field. Its position is obtained from the position of the current block by a translation given by the motion vector. The same motion vector is used for all the blocks belonging to a macroblock. For motion compensation, there is no ambiguity in the definition of the reference block when the motion vector coordinates (x, y) are integers. If one of the coordinates has a nonzero fractional part (noninteger accuracy), an interpolation scheme has to be used to build the reference block.

The parameters of the motion estimation are given in Table 14.1. The number of possible motion vectors is $1653 = \{(14+14)\times 2+1\}\times\{(7+7)\times 2+1\}$, i.e., all vectors within the search area are permitted. The motion vectors are represented with half-pel and half-line accuracy. MV differences are coded by variable-length coding ranging from -28 to $+28$ as shown in Table 14.2. The motion vectors that apply to the chrominance blocks are derived from the macroblock luminance motion vector in the following way: the vertical component is identical to that of the luminance and the horizontal component is equal to half that of the luminance.

Quantization of DCT coefficients

The quantizer parameters signaled to the decoder are the transmission factor and the criticality. The transmission factor is related to the buffer occupancy and it is provided at the stripe level (Fig. 14.1). It can vary for each stripe. At the end of a stripe of 8 lines, buffer occupancy is evaluated. The buffer occupancy is converted to an 8-bit integer for determining the transmission factor. Different values of the transmission factor may be used for the luminance and chrominance components of a stripe. The criticality factor is a macroblock classification index

Table 14.2 Code words for motion vector differences

Code word	MV_x or MV_y	Code word	MV_x or MV_y
01	0	00	−0.5
11 00	+0.5	10 01	−1
11 01	+1	10 00	−1.5
11 10 00	+1.5	10 11 01	−2
11 10 01	+2	10 11 00	−2.5
11 11 00	+2.5	10 10 01	−3
11 11 01	EOB_1	10 10 00	EOB_0
11 10 10 00	+3	10 11 11 01	−3.5
11 10 10 01	+3.5	10 11 11 00	−4
11 10 11 00	+4	10 11 10 01	−4.5
11 10 11 01	+4.5	10 11 10 00	−5
11 11 10 00	+5	10 10 11 01	−5.5
11 11 10 01	+5.5	10 10 11 00	−6
11 11 11 00	+6	10 10 10 01	−6.5
11 11 11 01	+6.5	10 10 10 00	−7
11 10 10 10 11	+7	10 11 11 11 01	−7.5
11 10 10 10 01	+7.5	10 11 11 11 00	−8
11 10 10 11 00	+8	10 11 11 10 01	−8.5
11 10 10 11 01	+8.5	10 11 11 10 11	−9
11 10 11 10 00	+9	10 11 10 11 01	−9.5
11 10 11 10 01	+9.5	10 11 10 11 00	−10
11 10 11 11 00	+10	10 11 10 10 01	−10.5
11 10 11 11 01	+10.5	10 11 10 10 00	−11
11 11 10 10 00	+11	10 10 11 11 01	−11.5
11 11 10 10 01	+11.5	10 10 11 11 00	−12
11 11 10 11 00	+12	10 10 11 10 01	−12.5
11 11 10 11 01	+12.5	10 10 11 10 00	−13
11 11 11 10 00	+13	10 10 10 11 01	−13.5
11 11 11 10 01	+13.5	10 10 10 11 00	−14
11 11 11 11 00	+14	10 10 10 10 01	−14.5
11 11 11 11 01	+14.5	10 10 10 10 00	−15
11 10 10 10 10 00	+15	10 11 11 11 11 01	−15.5
11 10 10 10 10 01	+15.5	10 11 11 11 11 00	−16
11 10 10 10 11 00	+16	10 11 11 11 10 01	−16.5
11 10 10 10 11 01	+16.5	10 11 11 11 10 00	−17
11 10 10 11 10 00	+17	10 11 11 10 11 01	−17.5
11 10 10 11 10 01	+17.5	10 11 11 10 11 00	−18
11 10 10 11 11 00	+18	10 11 11 10 10 01	−18.5
11 10 10 11 11 01	+18.5	10 11 11 10 10 00	−19
11 10 11 10 10 00	+19	10 11 10 11 11 01	−19.5
11 10 11 10 10 01	+19.5	10 11 10 11 11 00	−20
11 10 11 10 11 00	+20	10 11 10 11 10 01	−20.5
11 10 11 10 11 01	+20.5	10 11 10 11 10 00	−21
11 10 11 11 10 00	+21	10 11 10 10 11 01	−21.5
11 10 11 11 10 01	+21.5	10 11 10 10 11 00	−22
11 10 11 11 11 00	+22	10 11 10 10 10 01	−22.5
11 10 11 11 11 01	+22.5	10 11 10 10 10 00	−23
11 11 10 10 10 00	+23	10 10 11 11 11 01	Null

table continues

Table 14.2 *(cont.)*

Code word	MV_x or MV_y	Code word	MV_x or MV_y
11 11 10 10 10 01	+23.5	10 10 11 11 11 00	−23.5
11 11 10 10 11 00	+24	10 10 11 11 10 01	−24
11 11 10 10 11 01	+24.5	10 10 11 11 10 00	−24.5
11 11 10 11 10 00	+25	10 10 11 10 11 01	−25
11 11 10 11 10 01	+25.5	10 10 11 10 11 00	−25.5
11 11 10 11 11 00	+26	10 10 11 10 10 01	−26
11 11 10 11 11 01	+26.5	10 10 11 10 10 00	−26.5
11 11 11 10 10 11	+27	10 10 10 11 11 01	−27
11 11 11 10 10 01	+27.5	10 10 10 11 11 00	−27.5
11 11 11 10 11 00	+28	10 10 10 11 10 01	−28

and is transmitted for each macroblock. One possible method of determining the block criticality is based on the AC energy of the (8×8) block (Fig. 14.3).

AC coefficient quantization: A different quantization characteristic is used for each coefficient. The quantization is achieved in two steps: first, obtaining relative coefficients from transformed coefficients and second, applying a nearly uniform quantizer. The relative coefficient C_{uv} is calculated as follows:

$$C_{uv} = 2S_{uv}/T(u, v, m, f)$$

$$T(u, v, m, f) = 2^{N(u,v,m,f)/16}$$

$$N(u, v, m, f) = \text{Min}[N'(u, v, m, f), 175]$$

$$N'(u, v, m, f) = \text{Max}[Q(u, v, m, f), 0]$$

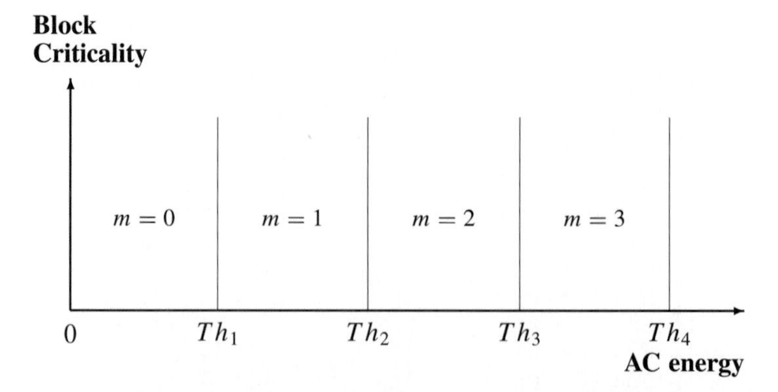

Figure 14.3 A possible method for determining the block criticality using AC energy calculation.

Table 14.3 Translation and limit values depending on the block criticality

Criticality m	Translation $Tr(m)$	Limit $Th(m)$	
		Luminance	Chrominance
0	8	No (i.e. 44 + 8)	No (i.e. 26 + 8)
1	2	No (i.e. 44 + 2)	No (i.e. 26 + 2)
2	0	34	16
3	0	24	9

$$Q(u, v, m, f) = \text{Min}[2P(u, v, m) - 48, f] + f$$

$$P(u, v, m) = \text{Min}[P_0(u, v) + Tr(m), Th(m)] \tag{14.2}$$

where $P_0(u, v)$ denotes the relative visibility matrix for the DCT coefficients as shown in Fig. 14.4, $T(u, v, m, f)$ is the transmission threshold for (u, v) coefficients, $m = 0, 1, 2, 3$, according to the block criticality factor, and f is the transmission factor, which is related to the buffer occupancy and provided at the stripe level, i.e., for all the macroblocks belonging to each group of 8 video lines. $Tr(m)$ and $Th(m)$ denote the translation value and limit value, respectively, where these values depend on the block criticality m as shown in Table 14.3. Only the criticality factor and the transmission factor are transmitted to the decoder. Ranges of the quantization parameters are

Transmission factor	$0 \sim 175$
Transmission threshold	$0 \sim 175$
Relative visibility	$0 \sim 44$

The obtained relative cofficients C_{uv} are quantized by the nearly uniform quantizer (Table 14.4). The data accuracy of 12-bit transform coefficients is not changed by relative coefficients calculation, where the nearly uniform quantizer provides 11-bit levels.

DC coefficient quantization: The same process as the AC coefficient quantization is used but the scaling factor $N(0, 0, m, f)$ of the DC coefficient is limited to the range [0, 48].

Example 14.1 *The DCT coefficients described in Eq. (8.5) are quantized using Eq. (14.2) and Table 14.4. We assume that the block criticality is zero, since the image block is not very active, and the transmission factor is* 96 *(roughly half the buffer occupancy). The quantized DCT coefficients*

$$S_{quv} = \begin{pmatrix} 155 & -4 & 1 & 0 & 0 & 0 & 0 & 0 \\ 3 & 0 & 0 & 0 & 0 & 0 & 0 & 0 \\ 1 & 0 & 0 & 0 & 0 & 0 & 0 & 0 \\ 0 & -1 & 0 & 0 & 0 & 0 & 0 & 0 \\ 0 & 0 & 0 & 0 & 0 & 0 & 0 & 0 \\ 0 & 0 & 0 & 0 & 0 & 0 & 0 & 0 \\ 0 & 0 & 0 & 0 & 0 & 0 & 0 & 0 \\ 0 & 0 & 0 & 0 & 0 & 0 & 0 & 0 \end{pmatrix} \quad (14.3)$$

can be compared with Eq. (8.6).

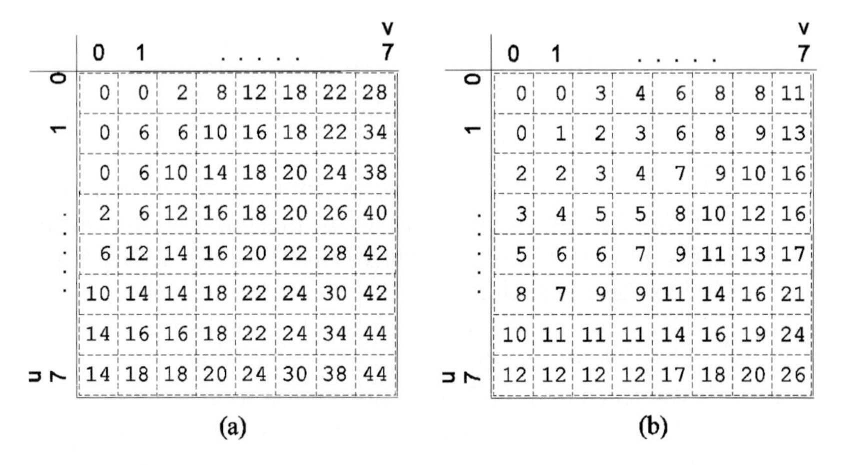

Figure 14.4 Relative visibility matrix $P_0(u, v)$ for (a) luminance and (b) chrominance.

Table 14.4 Nearly uniform symmetric quantizer (positive levels only are shown)

Relative coefficients intervals C_{uv}	Quantizer levels	Reconstruction level C'_{uv}
0	0	0
1	1	1
2	2	2
255	255	255
256:257	256	256
⋮	⋮	⋮
512:515	384	513
⋮	⋮	⋮
1024:1031	512	1027
⋮	⋮	⋮
2040:2047	639	2043

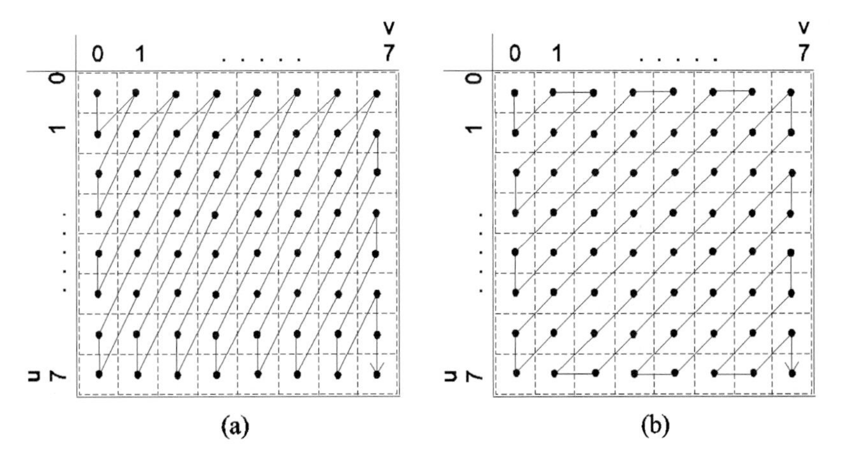

Figure 14.5 The scanning path for the DCT coefficients. (a) Luminance; (b) chrominance components.

Table 14.5 Variable-length code word structure

Word length (bits)	Word structure	No. of words
2	$0X_0$	2
4	$1X_1\ 0X_0$	4
6	$1X_2\ 1X_1\ 0X_0$	8
8		16
10		32
12	$\vdots \quad \vdots$	64
14		128
16		256
18	$1X_8 \cdots 1X_1\ 0X_0$	512
18	$1X_8 \cdots 1X_1\ 1X_0$	512

Variable-length coding

The quantized coefficients are scanned along the zigzag path for chrominance and the alternate path for luminance as shown in Fig. 14.5. The scanning path for luminance may be effective in the field-based processing (see the alternate scan shown in Fig. 11.20b for MPEG-2).

Variable-length code words in this standard are systematically generated. Each word comprises groups of a pair of bits. Table 14.5 shows the word structure with maximum word length 18 and the corresponding number of words. The first bit of each pair is the continuation bit (1 or 0). If this bit is 0, the present pair is the last one except for the word having length 18. The second bits of the pairs of bits are the information bits, which are determined by the generation rules and by

Table 14.6 Code words for run-length of zeros 29 to 63

Code word	Run of zeros	Code word	Run of zeros
11 10 10 10 10 00	0*29	10 11 11 11 11 00	0*30
11 10 10 10 10 01	0*31	10 11 11 11 11 01	0*32
11 10 10 10 11 00	0*33	10 11 11 11 10 00	0*34
11 10 10 10 11 01	0*35	10 11 11 11 10 01	0*36
11 10 10 11 10 00	0*37	10 11 11 10 11 00	0*38
11 10 10 11 10 01	0*39	10 11 11 10 11 01	0*40
11 10 10 11 11 00	0*41	10 11 11 10 10 00	0*42
11 10 10 11 11 01	0*43	10 11 11 10 10 01	0*44
11 10 11 10 10 00	0*45	10 11 10 11 11 00	0*46
11 10 11 10 10 01	0*47	10 11 10 11 11 01	0*48
11 10 11 10 11 00	0*49	10 11 10 11 10 00	0*50
11 10 11 10 11 01	0*51	10 11 10 11 10 01	0*52
11 10 11 11 10 00	0*53	10 11 10 10 11 00	0*54
11 10 11 11 10 01	0*55	10 11 10 10 11 01	0*56
11 10 11 11 11 00	0*57	10 11 10 10 10 00	0*58
11 10 11 11 11 01	0*59	10 11 10 10 10 01	0*60
11 11 10 10 10 00	0*61	10 10 11 11 11 00	0*62
11 11 10 10 10 01	0*63	10 10 11 11 11 01	Null

Table C.7 in Appendix C. The total number of words is 1534, of which 66 are used to code the run-length of zeros (the maximum run length is 63), two kinds of EOB words (EOB_0 and EOB_1), and a NULL word. In case of underflow, a number of NULL words is used for coding instead of an EOB. Code words for run-length of zeros 29 to 63 are shown in Table 14.6.

Note that 1466 code words (1534 – 66 – 2) are assigned to code the quantized levels from -733 to $+733$. The two code words (all 11 and 10 pairs) are not assigned. Code words for run-length of zeros are uniquely defined rather than mixing with the nonzero coefficients as adopted in other standards. The information bits for the positive values and the corresponding negative values are complemented. If there are one or more coefficients of $+1$ between two runs of zeros, or between a run of zeros and the EOB, one of them is not transmitted and the decoder reinserts it. This may reduce the bit rate more.

Two VLC words (EOB_0 and EOB_1) are used for the end of block event. A pseudorandom generator produces 180 sequences. [A stripe consists of 180 luminance and chrominance blocks, each of size (8×8)]. The pseudorandom generator based on the polynomial $g(x) = 1 + x^5 + x^9$ is reset to 100111000 at the beginning of each stripe (end of 180 blocks). At the end of each (8×8) block, the pseudorandom generator steps forward one bit and the output of the generator (0 or 1) determines which of the two EOB words (EOB_0 or EOB_1) is applicable.

The assignment for values of quantized levels is made in two's-complement representation using the following rules:

- $-733, \ldots, -479$: Add $+1501$, then generate nine information bits (The last continuation bit is set to one). For example, given the coefficient -733, $-733 + 1501 = 768 = (001\ 100000000)_2$ and the code word is 11 10 10 10 10 10 10 10 10.
- $-478, \ldots, -17$: Add -34, then generate up to nine information bits until the last zero bit. For example, given the coefficient -23, $-23 - 34 = -57 = (111111\ 000111)_2$ and the code word is 10 10 10 11 11 01.
- $-16, \ldots, +16$: Refer to Table C.7 in Appendix C.
- $+17, \ldots, +478$: Add $+33$, then generate up to nine information bits until the last one bit. For example, given the coefficient $+23$, $+23 + 33 = 56 = (000000\ 111000)_2$ and the code word is 11 11 11 10 10 00.
- $+479, \ldots, +733$: Add $+34$, then generate nine information bits (the last continuation bit is set to one).

Example 14.2 *An example for VLC encoding and decoding is illustrated as follows:*

Coefficients	−2	000	+1	+1	00	+1	EOB
Code words	1001	111000	01	01	1000	01	111101
Transmitted	1001	111000	01		1000		111101
Decoded	1001	111000	01	01	1000	01	111101

Here $+1$ *is not transmitted in two places. Note that in case of using the Null word,* $+1$ *must be transmitted between two Null words.*

Coding the motion vectors

Motion estimation is carried out on the basis of a macroblock (Fig. 14.1). Predictive encoding of the motion vectors is performed along a stripe of blocks. The components of the MV prediction errors along the horizontal (MV_x) and vertical (MV_y) directions are VLC coded (Table 14.2) using the same code word structure as that for the coefficients coding (Table C.7). The motion vector of a macroblock becomes a prediction for the motion vector of the next macroblock along the stripe. The motion vector prediction for the first macroblock of a stripe and for a macroblock following an intrafield or interfield macroblock is set to zero. Since the motion compensation is performed only in interframe mode, the differential motion vectors are sent if the macroblock is interframe encoded and if the MV prediction error (MV_x, MV_y) is different from (0, 0).

Error correction coding

Reed-Solomon (255, 239) code is used for protection from transmission errors. The code can correct 8 octet errors and has 2 octet interleaving. The generator polynomial of the Reed-Solomon code is given by $\prod_{i=0}^{15}(x+\alpha_i)$, where α_i is a root of the binary primitive polynomial $x^8+x^4+x^3+x^2+1$. The data stream at the output of the video encoder is arranged in a matrix of 16 rows of 239 columns. Each column corresponds to one 16-bit word of video data. The RS (255, 239) code is computed on each of the 2 rows of octets and the 16 octet error-control group is added to the corresponding row.

14.2 ITU-R CMTT.721-1

The ITU-R CMTT.721-1 [900] describes coding and transmission of digital television signals in component form for contribution-quality applications at bit rates of nearly 140 Mbps. This is the fourth level of the digital transmission hierarchy. Input digital video signals conform with Recommendation ITU-R BT. 601 (4:2:2 format). The standard mainly supports the 625-line system, while CMTT.723 in Section 14.1 provides coding algorithms for both formats of the 525-line and the 625-line systems. Since it codes and transmits at higher bit rates (small compression ratio), no subsampling, motion compensation, or discrete cosine transform is included in the coding process.

The main characteristics recommended for the coding of component-coded digital television signals at bit rates of nearly 140 Mbps are summarized as follows:

- Video input/output: 4:2:2 ITU-R BT.601 signal (625-line digital video in component form)
- Preprocessing: No subsampling/prefiltering
- Coding: Two-dimensional intrafield prediction (DPCM) for luminance and color-difference components. The second-order predictor referring to two previous pixels horizontally and vertically is designed with 8-bit accuracy. Video levels outside the active picture are set to 16 (black level) for Y and 128 for C_b, C_r to preset the initial value of the predictor in both the coder and decoder. No motion compensation or adaptive control of predictors is required. A symmetric quantizer (Table 14.7) combined with nonadaptive DPCM is used. Moreover, no variable-length coding or postprocessing is required.
- Video data rate: 124.416 Mbps (720 active pels/line $\times$ 576 active lines/frame $\times$ 6 bits/pel $\times$ 25 frames/sec)

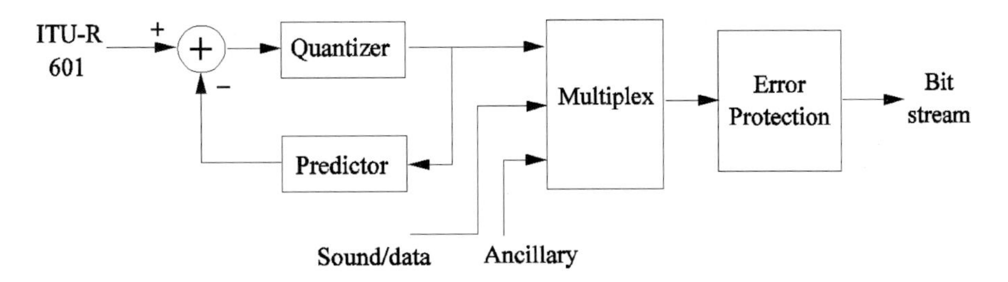

Figure 14.6 Encoding block diagram of ITU-R CMTT.721-1.

- Error correction coding: (110, 108, 3) RS code for rows, (102, 100, 3) RS code for columns
- Multiplexing: This scheme makes use of a so-called TV container with a data rate of 138.240 Mbps that is able to convey different video source signals with data rates of 135.000 Mbps and sound/data signals with data rates of 2.048 Mbps. Within the video frame data rate of 129.600 Mbps, two 2.048 Mbps channels are available for sound/data transmission, which can be used for the transmission of two stereo sound signals, e.g. according to Recommendation ITU-R CMTT.724.

The encoder is described in Fig. 14.6. The following two-dimensional predictor is used to predict the present pel X.

Line $n-1$ ———— C ★ ——

n —— A ★ ———— X ⊗ ——

$$X = \frac{A + C}{2}$$

Fixed predictors are applied for both the luminance and color-difference components. Nonadaptive DPCM in combination with symmetric quantizer forms the main elements of the coding scheme. The larger prediction errors are coarsely quanitized as shown in Table 14.7.

For multiplexing, the 6-bit structure of video data is changed to an 8-bit structure. The transmission order of the video components is C_b, Y, C_r, Y, C_b, Y, C_r, and so forth, where a group of the two color-differences and two luminances allow 3-octet capacity. Hence, there are 360 by 3 octets in a TV line and we get the bit rate of 124.416 Mbps for the 625-line system, which has 576 active lines and a 25 frame rate. Two 2.048 Mbps sound/data channels associated with the justification signal are included in the video multiplex. A data rate of 576 Kbps is available for ancillary data (e.g., teletext, test signals, etc.), which is normally

Table 14.7 Symmetric quantizer characteristics in ITU-R CMTT.721-1 scheme

Level	Decision	Reconstruction
0	0	0
1	1	1
2	2	2
⋮	⋮	⋮
9	9	9
10	$10 \sim 12$	11
11	$13 \sim 15$	14
⋮	⋮	⋮
29	$107 \sim 113$	110
30	$114 \sim 120$	117
31	$121 \sim 127$	124

transmitted in the vertical blanking interval of the conventional TV signal. The sound/data and ancillary data are interleaved between a 9×3 octets video signal. In addition, the video multiplex includes a 2-byte synchronization word at the beginning of each line. In conclusion, there are 40×27 byte video information, two 16×1 byte sound/data channels, 5×1 byte ancillary data, and 6×1 byte justification signals in a line multiplex, which result in 1125 bytes per line, which is preceded by a 2-byte justification signal. Hence, the video frame has a data rate of 129.600 Mbps. Forward error correction coding increases the bit rate by 3.89%, resulting in 135 Mbps. The output data rate in a TV container is 138.240 Mbps including TV container framing, service channels, and another sound/data channel.

The complete video multiplex is protected from transmission errors by a 2D FEC product code using a 1-octet correcting Reed-Solomon coder in each direction; (110, 108, 3) RS code for rows and (102, 100, 3) RS code for columns. The field generator polynomial is $f(\alpha) = \alpha^8 + \alpha^4 + \alpha^3 + \alpha^2 + 1$, and the code generator polynomial is $g(x) = (x + \alpha)(x + 1) = x^2 + \alpha x + x + \alpha$, where α is defined as 0000 0010.

Appendix A
Manufacturers and Vendors

There are many companies and vendors all over the world who specialize in chip sets, software, and application systems. It is almost impossible to prepare a complete list because the field is continuously changing. The list shows only those companies that are well known in open literature or in the public domain of network news. The structure of these companies is changing rapidly, i.e., some are merged, new ones have started, some have closed, and some have moved (new addresses, phone, fax, etc.). Some of the efforts by Universities and Research Institutes are directed toward research and development as opposed to immediate commercialization.

Every effort is made to make this list as exhaustive, comprehensive, and up-to-date as possible. We arrange this list as follows and focus on the three categories that would be helpful to the reader who has interest in hardware- or software-based implementations of the international standards for digital audio/video coding that we have discussed in this book.

1. Names and addresses of vendors
2. Available chips, chip sets, software, boards, and systems
3. Special features and relevant information

We have compiled this list as follows: first, video coding implementation of JPEG, H.261, MPEG algorithms applicable to still image processing, video data storage, videophone, and TV or HDTV video processing, second, audio coding implementation based on the ITU-T and MPEG algorithms; third, motion estimation chip sets and techniques; fourth, software implementation that can manipulate overall image processing functions; and finally, the vendors who provide image capture and digitizing boards.

A.1 JPEG, H.261, MPEG Video

AEG Olympia GmbH

1. Max-Stromeyer Strasse 160, D–7750 Konstanz, Germany
   Tel: (07531)818-0, Fax: (07531)818-500
2. Mike (Multifunctional ISDN Communication Equipment)
   – H.261 and G.711 compatible videophone system
   – 270 ms coding and decoding delay for CIF
   TeViFon Videophone System
   – CIF and QCIF at 48, 64, 96 Kbps

ALCATEL

1. 33, rue Emeriau, 75725 Paris Cedex 15, France
   Tel: 33(1) 40 58 58 58, Fax: 33(1) 45 77 88 61
   1225 North Alma Rd., Richardson, TX 75081-2206
   Tel: 800-ALCATEL, Fax: 214-996-5409
2. 1741 VC—High Definition Video Codec
   1250/50, 1125/60 system HDTV Codec
   DTV-45 Video Codec
   – Up to 74.25 MHz video sampling rate
   – 34 to 140 Mbps transmission bit rates
   – Parallel processing with five processors

Alpha Systems Lab

1. 2361 McGaw Ave., Irvine, CA 92714
   Tel: 714-252-0117, Fax: 714-252-0887
2. MegaMotion—Multimedia full-motion video capture (board) and playback based on motion JPEG

Apple Computer Inc.

1. 2025 Marini Ave., Cupertino, CA 95014
2. Quick Time, image compression manager
   – JPEG compression/decompression software on Macintosh

Ariel Corp.

1. 433 River Rd., Highland Park, NJ 08904
Tel: 908-249-2900, Fax: 908-249-2123

2. AT & T VCOS and Ariel PC plug-in MP 3210, JPEG encoder/decoder

Array Microsystems

1. 1420 Quail Lake Loop, Colorado Springs, CO 80906
Tel: 719-540-7900, Fax: 719-540-7950

2. Image Compression/Expansion Processor
   ICC (Image Compression Coprocessor)
   MEC (Motion Estimation Coprocessor)
   – Up to 7.2 BOPS block-matching engine
   – In part funded by Samsung Electronics Co.
   – JPEG, H.261, MPEG-1 fully compatible (programmable)
   – Multimedia PC add-in card that implements H.261, MPEG, and JPEG Videoflow chip set, a 77500 hardware evaluation kit, a 77800 software toolkit, 77900 software development tools
   – Video encoder/decoder based on Motion JPEG, H.261, MPEG-1, and transcoding

Asia Concepts Ltd

1. 20/F Kam Chung Comm. Bldg., 19-21 Hennessy Rd., Wanchai, Hong Kong
Tel: 852-2861-3615, Fax: 852-2520-2987

2. MPEG encoding solutions based on Optibase products

Atlanta Signal Processors, Inc.

1. 1375 Peachtree St., NW, Suite 690, Atlanta, GA 30309
Tel: 404-892-7265, Fax: 404-892-2512

2. Elf™ DSP Application Toolkits
   – JPEG and MPEG encoding/decoding using DSP chip (TMS320C31) on PC

AT & T Microelectronics

1. Dept. AL-500404200, 555 Union Blvd., Allentown, PA 18103
Tel: 800-372-2447, Fax: 215-778-4106

2. AVP-1300E (single-chip H.261/MPEG-1 encoder)
   AVP-1400D (single-chip H.261/MPEG-1 decoder)
   AV6110A (MPEG-2 digital video decoder)
   – 0.9/0.75 μm CMOS design, 45 MHz operating frequency
   – Single-chip solution for CIF 30 Hz (AVP-1x00)
   – Real-time decoding of MPEG-2 MP/ML (AV6110A, up to 54 Mbps)
   AVP-AV4310A (single-chip H.261 encoder)
   AVP-AV4220A (single-chip H.261 decoder)
   AVP-AV4120A (system-level protocol for H.320)
   AVP Video/Audio Processor AV4400A, single-chip video/audio processor. Supports MPEG-1, H.261, H.263, JPEG, motion JPEG. Compliant with H.320, H.324, and ISO MPEG-1.

Black Box Corp.

1. P.O. Box 12800, Pittsburgh, PA 15241
   Tel: 412-746-5105
2. Videoconferencing system
   – H.261 for video and G.711 and G.722 for audio

Canon USA, Inc.

1. One Canon Plaza, Lake Success, NY 11042-1113
   Tel: 516-488-6700, or 800-OK-Canon, Fax: 516-328-5069
2. JPEG image compression software, electronic still-video cameras

C–Cube Microsystems

1. 1778 McCarthy Blvd., Milpitas, CA 95035
   Tel: 408-944-6300, Fax: 408-944-6314, email: techpubs@c-cube.com (for literature requests)
2. CL550 and CL560 (JPEG image compression processors)
   CL450 (MPEG video decoder)
   CL480 (MPEG-1 aduio and video decoder in real time)
   CL4000 (programmable video compression processor)
   CLM4200 (H.261 video codec)
   CLM4400 (Four-chip microcode implementation for MPEG-2 encoding
   CLM4500 (MPEG-1 video encoder)
   CLM4600 (MPEG-2 video encoder)
   Video RISC MPEG encoder development station
   – Software and development kit available

– 0.8 μm CMOS, up to 60 MHz operating frequency (CL4500)
– JPEG, H.261, MPEG can be programmed to run (CL4000)
– Consumer quality (CLM4500) and broadcast quality (CLM4600)
CL9100 (multimode video decoder, MPEG-2 ML/MP, ML/SP, Digicipher II, and MPEG-1)
CL9110 (MPEG-2 transport layer demultiplexer)
CLM4100 family of multimedia accelerators
– Support encode/decode of MPEG-1, H.261, and others
MPEG-2 encoder development system

CeQuadrat (USA), Inc.

1. 5 Thomas Mellon Circle, Suite 105, San Francisco, CA 94134
   Tel: 800-330-MPEG, Fax: 415-715-5619
2. Pixelschrink, software-only MPEG-1 encoder. MPEG video/audio interleaved by multiplexer to a single White Book compatible target MPEG system

Channelmatic, Inc.

1. 821 Tavern Rd., Alpine, CA 91901
   Tel: 619-445-2691, Fax: 619-445-3293
2. Digital encoding unit: on-board MPEG compressor and broadcast-oriented ad insertion system

Cinax Designs, Inc.

1. P.O. Box 93581, Nelson Park PO, Vancouver, B.C. Canada V5Y 3A7
   Tel: 604-681-0905, Fax: 604-681-0955
2. PC-based MPEG-1 CD-ROM products

CLI, Compression Labs Inc.

1. 2860 Junction Ave., San Jose, CA 95134
   Tel: 408-435-3000, Fax: 408-922-5429
2. Cameo Personal Video System, Model 2001
   – H.261, videophone software
   – 64 kbps video and audio rates up to 15 fps
   "Magnitude" MPEG Encoding System
   – MPEG-1 encoder upgradable to MPEG-2

MPEG-2 Encoder
Rembrandt-II/VP video codec (56 to 2048 Kbps)
Consumer Videophone ($<$ 10 Kbps)
Spectrumsaver (compressed digital video broadcast system)

COMSAT Lab.

1. 22300 COMSAT Dr., Clarksburg, MD 20871
   Tel: 301-428-4681, Fax: 301-428-4534
2. MPEG-2 SCPC encoders and decoders for satellite, news gathering, backhaul, or broadcast applications

Connectware

1. 1301 E. Arapaho Rd., Richardson, TX 75081
   Tel: 214-997-4111, Fax: 214-997-4308
2. Single-slot raster video S bus. Displays full-motion video. Image freeze and capture (SUN Raster, JPEG, TIFF). JPEG/MPEG module

Coreco, Inc.

1. 6969 Trans-Canada Highway, Suite 142, St. Laurent, Quebec, Canada H4T 1V8,
   Tel: 800-361-4914 or 514-333-1301, Fax: 514-333-1388
2. OC ULVS-VCP (JPEG Image Compression and Decompression board)
   – H.261 encode/decode, MPEG-1 real-time video encode/decode, MPEG-2 real-time video decode

Creative Labs, Inc.

1. 1901 McCarthy Blvd., Milpitas, CA 95035
   Tel: 408-428-6600, Fax: 408-428-6611
2. Videoblaster RT 300
   – 320 $\times$ 240 resolution, 30 frames/sec, 6:1 compression and playback
   Videoblaster FS 200
   – Mix/match video from various sources; compression and playback

Crest National Videotape & Film Lab.

1. 1000 N. Highland Ave., Hollywood, CA 90038-2407
   Tel: 213-466-0624, Fax: 213-461-8901
2. MPEG mastering studio (encoding services)

Data Cell

1. West End Rd., Mortimer Common, Reading, Berks RG7 3TF, Canada,
   Tel: 734-33-3666, Fax: 734-33-2827
2. PC JPEG compression and video capture card
   – Implements the baseline sequential process
   – Downloadable Q-tables and Huffman tables

Diamond Multimedia Systems

1. 1130 E. Arques Ave., Sunnyvale, CA 94086-4062
   Tel: 800-4-MULTIMEDIA, Fax: 800-380-0030
2. Stealth Video VRAM, full-motion full-screen video playback, motion video player daughtercard for hardware MPEG decoding
   Stealth 64 line of PC multimedia accelerators bundled with MPEG Arcade player from Mediamatics, Inc.

Digital Equipment Corp.

1. 146 Main St., Maynard, MA 01754-2571
   Tel: 508-493-5111, Fax: 800-344-4825, WWW: http://www.dec.com
2. Mediaplex ad-server MPEG-2 codec
   Digital media studio unit: MPEG-1 and -2 encoding services

Digital Video Arts

1. 715 Twining Rd., Suite 107, Dresher, PA 19025
   Tel: 215-576-7920, Fax: 215-576-7932
2. Real-time, full-motion video playback from CD-ROM, hard disk, RAM, or digital network. Digital video file support for Indeo™ Video, PLV, RTV, and True Motion

Digital Video Computing

1. Seestrasse 7, D–82211, Herrsching, Germany
   Tel: 49-81-52-93010, Fax: 49-81-52-91331
2. Motion JPEG, H.261, MPEG, either software or hardware RAM-based video sequence recorder for HDTV, digital video, and MPEG coding

Digital Video Solutions GmbH

1. Herrsching, Germany
2. SCSI/Video: Motion JPEG hardware to compress and decompress video

DIVICOM

1. 1708 McCarthy Blvd., Milpitas, CA 95035
   Tel: 408-944-6700, Fax: 408-944-6705
2. MPEG-2 program encoder/remultiplexer
   – Encode video (MP@ML) and audio & multiplex into MPEG-2 transport bit stream
   DMC2 program encoder
   DRMX2 remultiplexer and system controller

Dornier GmbH

1. Bildverarbeitung, Max-Stromeyer Str. 160, D–7750 Kanstanz, Germany
   Tel: 49-75-95-82008, Fax: 49-75-95-88108
2. CIF and QCIF H.261 video codec at $p \times 64$ Kbps
   – Supports G.711 audio codec

Eastman Kodak Co.

1. Electronic Photography Div., 343 State St., Rochester, NY 14650
   Tel: 800-44 Kodak, Ext. 610, WWW: http://www.kodak.com
2. SV 9600 and SV 9610 still video transceiver (DCT-based compression). Transmission of compressed images over telephone or 56/64 Kbps lines

ETRI

1. Electronics and Telecommunications Research Institute, Computer Hardware Sec., 161, Kajong-dong, Yusong-gu, Taejon, 305-350, Korea
   Tel: 82-42-860-6656, Fax: 82-42-860-6645
2. JESP (JPEG Engine for Scan Layer Processing)
   – 1 μm CMOS design, 33 Mpel/sec performance
   – Applicable to real-time implementation of motion JPEG
   Combi Station (multimedia personal workstation)
   – Video: 30 frames/sec compression/decompression (MPEG-1, H.261, JPEG)
   – Audio: MPEG-1 or ADPCM compression/decompression

FAST Electronic

1. Landsberger Str. 76, 80339 München, Germany
   Tel: 49-89-50206-0, Fax: 49-89-50206-199
2. PC-based full-motion JPEG compression board

Fast Forward Video

1. 18200 West McDurmott, Irvine, CA 92714
   Tel: 714-852-8404, Fax: 714-852-1226
2. Real-time JPEG full-motion video (NTSC/PAL) encoder/decoder BANDIT: random-access digital video recorder that records and plays back full-screen, full-motion digital video streams with built-in JPEG compression

Fluent Machines Inc.

1. 1881 Worcester Rd., Framingham, MA 01701
   Tel: 508-626-2144
2. VSA-1000 single AT-compatible board
   – Captures video from different sources and formats (NTSC, PAL)

Future Tel, Inc.

1. 1092 E. Arques Ave., Sunnyvale, CA 94086
   Tel: 408-522-1400, Fax: 408-522-1404, WWW: http://www.ftelinc.com
2. Prime View II, MPEG-1 real-time PC-based video/audio encoder including ITU-R 601 to SIF
   MPEG-2 transport stream encoder: transport stream compliant multiplexer

GC Technology

1. 6F Annex Tohshin Bldg, 4-36-19 Yoyogi Shibuya-ku, Tokyo 151, Japan
   Fax: 81-3-5351-0184
2. DA7190 MPEG audio decoder chip
   – Decodes layer I or layer II audio bit stream
   – Supports 128, 192, 256, and 384 Kbps for stereo mode

GEC Plessy Semiconductors

1. Cheney Manor, Swindon, Wilshire SN2 2QW, U.K.
   Tel: 44-793-518000, Fax: 44-793-518411
2. VP2611 (video compression source coder)
   VP2615 (video reconstruction processor)
   VP510 (color space converter)
   VP520 (three channel video filter)
   VP530 (NTSC/PAL encoder)
   VP2612 (video multiplexer)
   VP2614 (video demultiplexer)
   VP101 (triple D/A converter)
   – Single-chip solution for H.261 CIF 30 Hz

General Instrument Corp.

1. Videocipher Division, 6262 Lusk Blvd., San Diego, CA 92121
   Tel: 619-455-1500, Fax: 619-535-2485, WWW: http://www.gi.com
2. DigiCipher I Digital Video Decoder
   – Upgradable to DigiCipher II and MPEG-2 standard compatibility; DigiCipher encoder.

Genoa Systems Corp.

1. 75 East Trimble Rd., San Jose, CA 95131
   Tel: 800-93-GENOA, Fax: 408-432-9090
2. GVision DX card: combo 64-bit graphics card and MPEG playback card; features 30-fps full-screen playback of MPEG, CD-i, AVI, etc.

Grass Valley

1. P.O. Box 1114, Grass Valley, CA 95945
   Tel: 916-478-3000, Fax: 916-478-3808
2. Multichannel DS3 codec based on motion JPEG

Hartman Multimedia Service

1. Dipl. Ing. Stefan Hartman, Keplerstr. 11B, 10589 Berlin, Germany
   Tel: 49-30-344-23-66, Fax: 49-30-344-92-79, email: harti@contrib.de, WWW: http://www.b-2.de.contrib.net
2. Internet MPEG CD-ROM

Hewlett-Packard Company

1. 5301 Stevens Creek Blvd., Santa Clara, CA 95052-8059
   Tel: 800-367-4788, Fax: 408-553-3905, WWW: http://www.hp.com
2. HP Broadcast Video Server. MPEG encoder/decoder boards. Storage and playback of broadcast quality video (bit rate 1.5 to 15 Mbps). A number of options including upgrades for new developments such as HDTV
   MPEG-2 Protocol Viewer: this software troubleshoots MPEG-2 implementation problems

Hitachi

1. Central Research Lab., 1-280 Higashi-koigakubo, Kokubunji, Tokyo, 185, Japan
   Tel: 81-423-23-1111
2. CA-200 and DP-200 video conferencing system
   Video and audio codec, ISDN interface
   – H.261 and G.711/G.721 compatible
   – Lip sync adjustable up to 0.5 sec
   DV 6190 (MPEG-1 Video LSI, decompression and display of video)
   H.320 and H.261 chip set
   MPEG-1 and 2 (MP/ML) encoder and decoder
   MPEG-1 HD 814103FE video decoder and HD 814102F audio decoder

Hyundai Electronics America

1. 510 Cottonwood Dr. Milpitas, CA 95035
   Tel: 408-232-8000, Fax: 408-232-8131, email: tvakay@hea.com, WWW: www.hea.com
2. HDM8211M single-chip decoder
   – Decodes MPEG-1 and MPEG-2 MP/ML bit streams
   – Demultiplexes and decodes audio/video

IBM Microelectronics Division

1. Marketing Commun., 1580 Route 52, Bldg. 504, Hopewell Junction, NY 12533-6531
   Tel: 800-IBM-0181, Ext. 1501, email: askibm@info.ibm.com, WWW: http://www.ibm.com and http://www.chips.ibm.com
2. MPEG-2 single-chip video decoder
   MPEG-2 single-chip video encoder
   – Both chips are MPEG-1 compatible

I^2M: International Interactive Media

1. Bruce J. Meek, U.S. Agent c/o Sudden Impact Corp., The Mill, 49 Richmondville Ave., Westport, CT 06880
   Tel: 203-226-8923, Fax: 203-226-4428
2. MPEG-1 video/audio decoder
   – Macintosh or PC equipped with a CD-ROM drive to play CD-i and video CD discs

IIT

1. Integrated Information Tech, Inc., 2445 Mission College Blvd., Santa Clara, CA 95054
   Tel: 408-727-1885, Fax: 408-980-0432
2. VC (vision controller), VP (vision processor)
   VCP (vision controller & processor)
   – Programmable chip set for JPEG, H.261, MPEG-1 coder/decoder and MPEG-2 decoder
   – 33 MHz, 1.0 μm CMOS design
   IIT3304 (multimedia encoder processor)
   – Single-chip authoring and playback engine; encodes and decodes MPEG-1 video including full-motion video capture

IMAGRAPH

1. 11 Elizabeth Dr., Chelmsford, MA 01824
   Tel: 508-256-4624, Fax: 508-250-9155
2. IMASCAN/JPEG, PCI board for real-time full-motion JPEG compression and playback

In-Motion Technologies, Inc.

1. 1940 Colony St., Mountain View, CA 94043
   Tel: 415-968-6363, Fax: 415-968-4488
2. Movie Perfect Pro (motion JPEG video encoder/decoder)
   Digital VCR (add-on companion to Picture Perfect Pro)

Intelligent Resources Integrated Systems

1. 3030 Salt Creek Lane, #100 Arlington Heights, IL 60005-5000
   Tel: 708-670-9388, Fax: 708-670-0585,
   WWW: http://www.IntelligentResources.com
2. MPEG authoring and playback workstation: real-time encoder, supports video CD, CD-ROM, CD-i, and other formats

Intergraph Computer Systems

1. Huntsville, AL 35894
   Tel: 800-763-0242, WWW: http://www.intergraph.com/ics.
2. TD Workstation MPEG-2 compliant, multimedia application with sound and video integrated on the motherboard

ITRI CC Labs

1. Computer & Communication Research Lab., B300, Bldg. 11, 195 Sec. 4. Chung Hsing Rd., Chutung, Hsinchu, Taiwan 31015
   Tel: 886-35-917228, Fax: 886-35-968764
2. CIC86121 (DCT/IDCT chip set)
   CIC86200 (motion estimation unit)
   – applicable to JPEG, H.261, MPEG
   – ±7 search range for motion vector

JVC Professional Products Co.

1. 41 Slater Dr., Elmwood Park, NJ 07407
   Tel: 201-794-3900, Fax: 201-523-2077, WWW: http://www.jvc.ca/jvc
2. MPEG system: MPEG-2 video server (video/audio encoder and multiplexer) and MPEG-2 decoder

LSI Logic Corp.

1. 1551 McCarthy Blvd., Milpitas, CA 95035
   Tel: 408-433-8000, Fax: 408-433-8989, WWW: http://www.lsilogic.com
2. JView Emulation Board (JPEG Video-in-a-Window)
   L64702 (single-chip JPEG compression device)
   – Data rates up to 8.25 MBytes/sec with a 3MHz clock
   – Including software that runs under MicroSoft Windows 3.1
   Video rate compression/decompression system
   L64765 (color and raster-block converter)
   L64735 (DCT processor)
   L64745 (JPEG coder)
   – Full-motion JPEG coder/decoder up to 27 MBytes/sec
   Digital TV MPEG video decoder
   L64111 (MPEG audio decoder)
   L64112 (MPEG-1 video decoder)
   L64714 (Reed-Solomon encoder/decoder)
   – Decodes MPEG-1 bit stream up to 15 Mbps
   L64000 (MPEG-2 video decoder, MP/ML)
   L64002 (integrated MPEG-2 audio/video decoder)
   – Video: MP/ML, audio: layers 1 and 2 at 32, 44.1 and 48 KHz
   JPEG video rate bit buster evaluation board
   L64007 (MPEG-2 transport decoder)
   MPEG-2 audio/video decoder evaluation system

M3 Dimensions, Inc.

1. 14618 E. Whittier Blvd., Suite B, Whittier, CA 90605
   Tel: 310-907-6590, Fax: 310-907-6563
2. MPEG-1 encoding service, Video CD, CD-i and CD-ROM—MPEG-1 audio/video decoder

MasPar Computer

1. Sunnyvale, CA
2. Dennon/MasPar MPEG video encoding system

Matrox Electronic Systems Ltd.

1. 1055 St. Regis Blvd., Dorval, Quebec, Canada H9P2T4
   Tel: 514-685-2630, Fax: 514-685-2853
2. Matrox Marvel II Multimedia Controller
   – MPEG-1 audio/video decoder

Matsushita Elec. Industrial Co.

1. Audio Video Information Tech. Lab., Corporate Product Development Div., Tokyo, Japan
   Tel: 81-6-908-1131, Fax: 81-6-090-7054
2. VDSP2 MPEG-2 video codec
   – Encoding and decoding MPEG-2 by changing program on the same chip

Mediamatics, Inc.

1. 4633 Old Ironsides Dr., Suite 328, Santa Clara, CA 95054
   Tel: 408-496-6360, Fax: 408-496-6634
2. Q-speed: motion JPEG video playback in software. MPEG-1 based proprietary technology licensed to Wester Digital Corp.

Micropolis Corp.

1. 21211 Nordhoff St., Chatsworth, CA 91311
   Tel: 818-701-2866, Fax: 818-701-2809, WWW: http://www.micropolis.com
2. Micropolis AV Server: multichannel MPEG-2 video decoder board (up to 15 Mbps per channel)

Microware Systems Corp.

1. 1866 NW 114th St., Des Moines, IA 50325
   Tel: 515-224-1929, email: info@microware.com
2. Motion Picture File Manager (MPEG Software) on OS–9 systems
   – Support for C–Cube CL450 and Motorola MCD260 decoders

Minerva Systems, Inc.

1. 2933 Bunker Hill Lane, Suite 202, Santa Clara, CA 95054
   Tel: 408-970-1780, Fax: 408-982-9877
2. Minerva Compressionist^{TM}
   Audio/video real-time MPEG-1 interactive encoder
   Video/audio capture subsystems: upgradable to MPEG-2
   Compressionist 200: MPEG-2 encoding/publishing system optimized for video-server applications

Miro Computer Products, Inc.

1. 955 Commercial St., Palo Alto, CA 94303
   Tel: 415-855-0940, Fax: 415-855-9004
2. Miro Video DC 1 tv: motion JPEG compression to hard disk, record/playback in real time to/from hard disk

Mitec, Inc.

1. Espace UN Bldg 5F, 1-238, Takabata, Nakagawa-ku, Nagoya 454, Japan
   Tel: 81-52-351-2286, Fax: 81-52-363-1911,
   email: HFGO2162@niftyserve.or.jp
2. Real-time MPEG-1 video encode/decode board (I picture only), 320 × 240 (15 fps), 160 × 120 (30 fps), wavelet image encode/decode software

Mitsubishi Electronics America, Inc.

1. Information and Communication Systems Group, Visual Telecom Division, 5757 Plaza Dr., Cypress, CA 90630
2. Series 8000 (all-in-one mobile videoconferencing system)
   – Includes MVC-8100 codec based on H.261
   MELFACE 880, H.261 codec, 56/64 – 384 Kbps
   HDTV codec, video coding similar to MPEG-2, audio coding MPEG-1 layer 2, 40 Mbps

Motorola, Inc.

1. P.O. Box 20912, Phoenix, AZ 85036
   Tel: 800-441-2446 or 602-952-3000

2. MCD250 (MPEG full-motion video decoder)
   – 16 MHz System clock
   – Up to Mbps, 30Hz SIF decoding
   MC68VDP (Video decompression processor, MPEG-2 and DigiCipherII)
   MC68ADP (AC3 audio decoder up to 448 Kbps)

NEC

1. 1-1 Miyazaki 4 Chome, Miyamae-ku, Kawasaki 216, Japan
2. Low-cost MPEG video encoder based on a single-chip DSP
   – Several H.261-based video codecs
   – DS-3 level (44.736 Mbps) broadcast TV codecs
   VisuaLink 7000
   MPEG-2 video codec, SP@ML, and MPEG-1 audio layer 2
   MPEG-2 system TS/PS (2 – 15 Mbps)

New Media Graphics

1. 780 Boston Road, Billerica, MA 01821-5925
   Tel: 508-663-0666, Fax: 508-663-6678
2. JPEG compression/decompression
   – Variable compression ratio 12:1 to 80:1
   – Supports scaling, zooming, and windowing
   – Converts image files to TIFF, GIF, WMF formats

News Datacom, Ltd.

1. Maidenhead, Berks, U.K.
2. VCS 4000 Digital Video System: multichannel system for the full range of MPEG-2 specs for video, audio, and data

North Valley Research, Inc.

1. 15232 NW Greenbrier Parkway, Beaverton, OR 97006
   Tel: 503-236-6121, Fax: 503-236-7358, email: marketing@nvr.com, WWW: http://www.nvr.com
2. MPEG compressor software
   – Portable, upgradable and modular MPEG-1 compression video and audio layers 1 and 2
   Video-CD and CD-i compatible

MPEG player
– MPEG-1 decoder
Video and audio layers 1 and 2

NTL

1. 34 A-C Parham Dr., Boyatt Wood Industrial Estate, Eastleigh, Hampshire S0504NU, U.K.
   Tel: 44-1703-498045, Fax: 44-1703-498043
2. MPEG Compressor Software
   – Portable, upgradable and modular. MPEG-1 compression video and audio layers 1 and 2
   Video-CD and CD-i compatible.
   MPEG player – MPEG-1 decoder. Video and audio layers 1 and 2.
   MPEG-2 encoder (MP@ML), modulator, multiplexer, decoder, and system controller

NTT Visual Media Lab.

1. 1-2356 Take Yokosaka-shi, Kanagawa-ken, 238-03 Japan
   Tel: 81-486-59-2800, Fax: 81-486-59-2806
2. Visual telephones
   – N-ISDN or B-ISDN available

Nuts Technologies

1. 101 Metro Dr., Suite 750, San Jose, CA 95110
   Tel: 408-441-2166, Fax: 408-441-2177
2. Desktop video conferencing
   – Includes codec board, ISDN interface board, software, camera
   – QCIF processing up to 30 frames/sec with audio

Oak Technology

1. Sunnyvale, CA
2. ProMotion Development Kit: MPEG-1 decoder (firmware/software) for Video CD 2.0, Karaoke CD, etc., including MPEG audio

Oksori Co. Ltd.

1. 355-11 Songnae-dong, Sosa-gu, Puchon, Kyonggi, Korea
   Tel: 32-664-1761, Fax: 32-666-1930
2. Oksori CD-Vision Pro (Video-CD/CD-i Player)
   – Decoding playback board using CD-ROM drive for full-motion video and CD-quality audio compressed by MPEG-1 in IBM PC (video: 30 frames/sec, audio: stereo, 74 minutes per CD title).

Optibase, Inc.

1. 5000 Quorum Dr., Suite 700, Dallas, TX 75240
   Tel: 214-774-3800 or 800-451-5101, Fax: 214-239-1273,
   email: mpeg.quality@optibase.com, WWW: http://www.optibase.com
2. JPG–1000 JPEG Compression Engine
   MPG–1000 MPEG-1 Video Codec
   – use LSI Logic DCT or C–Cube CL550 chipsets
   – MicroSoft Windows 3.1 available
   MPEG Lab^{+}, MPEG Lab Pro
   MPEG Lab^{TM} suite (Advanced MPEG-1 encoder/decoder for the PC/AT)
   MDI-5000 (MPEG digital interface board)
   PC Motion^{TM} Pro (MPEG-1 playback board for the PC/AT)
   QuadStream: VME-based 6U board that sustains 4 channel MPEG decoding for video-on-demand and pay-per-view solutions for Cable TV
   Gemini single slot EISA MPEG-2 playback board

Optical Disc Corp.

1. 12150 Mora Dr., Santa Fe Springs, CA 90670
   Tel: 310-946-3050, Fax: 310-946-6030
2. DVD–7100 (Digital Videodisc Recorder)
   Super CD up to 4 hours storage of CCIR 601 resolution digital video at 5.6 Mbps on a single side 12" disc. MPEG extended/MPEG-2 compression

OptImage

1. 7185 Vista Dr. West, Des Moines, IA 50266-9313
   Tel: 515-225-7000, Fax: 515-225-0252
2. MPEG-1 video/audio authoring system for Video CD, CD ROM, CD-i, and 3DO. Includes "real-time preview"

Optivision

1. 1477 Drew Ave., Suite 102, Davis, CA 95616
   Tel: 916-757-4850 or 800-562-8934, Fax: 916-756-1309,
   WWW: http://www.optivision.com
2. OPTIPAC Series 250/ISA
   – JPEG and CCITT G3/4 image compression board
   MPEG real-time audio/video encoder and decoder
   – Flexible user control up to 5 Mbps
   OptiVideo VGA Decoder
   – Decompresses MPEG bitstream for audio/video playback in an IBM PC compatible

Pegasus Imaging Corp.

1. 4350 W. Cypress, Suite 908, Tampa, FL 33607
   Tel: 813-875-7575, Fax: 813-875-7705
2. JPEG and proprietary compression algorithms for photo/ID system, image data base, etc.

Philips

1. Thurn-Und-Taxis Strasse 14, D–8500 Nürnberg 10, Germany
   Tel: 49-911-5260, Fax: 49-911-5262850
2. TITAN Video Codec Videophone System
   Video and Audio Codec, ISDN Interface
   AxPe640V (Programmable Video Processor)
   – H.261 and G.711, G.722 compatible
   – JPEG Baseline mode transmission optimal
   MPEG Lab Pro (Real-time MPEG-1 Video/audio Encoder)
   MPEG Lab™ VCD (Compatible Video-CD Authoring System with real-time video-CD: white book, CD-i: green book and MPEG-1 encoder)
   MPEG Lab™ Suite (Advanced MPEG-1 Encoding System)

PictureTel

1. The Tower at Northwoods, 222 Rosewood Dr., Danvers, MA 01923
   Tel: 508-762-5000, Fax: 508-762-5245
2. System 4000 – High Performance Dial-up Videoconferencing
   – H.261 Video and G.722/G.711 audio compatible
   PictureTel Live PCS 50, Windows-based desktop videoconferencing (CIF

and QCIF video, G.711 and G.728 audio)
PictureTel Live PCS 100, high-end videoconferencing

Pivotal Graphics

1. Santa Clara, CA
2. MPEG-500 PFW Workstation (Distributer). Software for video authoring and CD formatting
   90 MHz Pentium PC-based CD Station: CD multimedia authoring (includes MPEG-1 video encoding)

Radius, Inc.

1. 1710 Fortune Dr., San Jose, CA 95131
   Tel: 408-434-1010, Fax: 408-434-0127
2. VideoVision Studio – Real time hardware-accelerated JPEG compression and decompression Macintosh MPEG decoder adapter board Cinepak Toolkit: software-based video encoder/decoder

Ricoh Co. Ltd.

1. 3001 Orchard Parkway, San Jose, CA 95134-2088
   Tel: 408-432-8800, Fax: 408-432-8372
2. Three chip set that implements JPEG baseline system including color conversion, raster/block conversion, and sampling

S3, Inc.

1. 2770 San Tomas Expressway, Santa Clara, CA 95051
   Tel: 408-980-5400
2. Developer of NATIVE-MPEG, MPEG-1 software based decompression utility for Windows 3.1 system
   S3 Native-MPEG: Host-only MPEG software decoder
   Single-chip audio/video MPEG-1 decoders

Samsung Electronics Co., Ltd.

1. San#14, Nongseo-ri, Kihung-eup, Yongin-kun, Kyungki-do, Korea
   Tel: 82-331-280-9454, Fax: 82-331-280-9459

2. DV-530KV (CD/CDG/CDV/LD/Video-CD)
   – Plays Laserdiscs, Audio CD, Video CD, and CD-i FMV movies
   – Adopts C-Cube's CL450 chip for MPEG-1 video coding and TI's AV110 chip for audio coding

Scientific Atlanta, Inc.

1. 4357 Park Drive, Norcross, GA 30093-2990
   Tel: 404-698-6850
2. Video compression system. Compliance with MPEG-2 and DVB standards for digital video transmission
   PowerVu: integrated MPEG-2 compliant system

Script Media, Inc.

1. 221 Elizabeth St., Utica, NY 13501
   Tel: 315-797-4052, Fax: 315-797-1850
2. CD Motion (desktop publishing system for multimedia system for compact disc)

SGS–THOMSON Microelectronics

1. 1000 East Bell Rd., Phoenix, AZ 85022
   Tel: 602-867-6100
2. STV 3200, 3208 (DCT Processor)
   STi 3220 (Motion Estimation Processor)
   STi 3240 (MPEG, H.261 Decoder)
   STi 3430 (Single chip MPEG-1 decoder for Video CD, PC multimedia, and digital TV applications)
   STi 3500 (MPEG-2/CCIR 601 Video Decoder)
   – MPEG and H.261 video decoding up to 10 Mbps
   – Supports MPEG-2 Main Profile/Main Level (STi 3500)
   STi 3520 and STi 3520A (MPEG Audio/MPEG-2 Video Integrated Decoder)
   – Supports MPEG-2 MP/ML, Layers 1 and 2 audio, multimedia, and DVD

Sigma Designs, Inc.

1. 46501 Landing Parkway, Fremont, CA 94538
   Tel: 510-770-0100, Fax: 510-770-2918
2. Real Magic Adapter Board
   – Decodes, scales, and outputs video data (MPEG-1) and audio
   Real Magic Producer: MPEG-1 authoring system

Silicon Graphics

1. 2011 N. Shoreline Blvd., Mountain View, CA 94043
   Tel: 415-960-1980 or 800-800-7441
2. Indigo2 Video and Indy Video cards
   – Captures video to disk with JPEG compression
   CosmoCompress Motion-JPEG card: full motion real time video encoder/decoder for the firm's Indigo family of desktop workstations

Smart and Friendly

1. 16539 Saticoy Street, Van Nuys, CA 91406
   Tel: 800-366-6001
2. MPEG Creation Station Pro/Deluxe, video/audio capture, encode/decode MPEG-1 on CD-ROM

Sonic Solutions

1. 1891 East Francisco Blvd., San Rafael, CA 94901
   Tel: 415-485-4800, Fax: 415-485-4877
2. Sonic digital video playback option—Playback of JPEG video in sync with audio stored on Sonic hard disks

Sonitech Intl., Inc.

1. 14 Mica Lane, Wellesley, MA 02181
   Tel: 617-235-6824, Fax: 617-235-2531
2. JPEG compression/decompression software, JPEG-X-DOS, and JPEG-X-SUN

Sony Corp., BISCO

1. Norihisa Sayanagi, 3-3-1 Tsujido Shinmachi, Fukisawashi, Kanagawaken 251, Japan
   Tel: 81-466-30-4055, Fax: 81-466-30-4207
2. MPEG-1 video audio Layers 1 and 2 and system encoder/decoder
   MPEG-2 video MP@ML, audio Layer 2
   MPEG-2 system encoder, MPEG-2 video MP@ML decoder

Storm Technology, Inc.

1. 1861 Landings Dr., Mountain View, CA 94043
   Tel: 800-275-5734, Fax: 415-691-9825
2. Picture Press 2.5 (JPEG Compression Software)
   – Variable compression ratio and adaptive encoding

Tandberg Telecom AS

1. Veritasreinen 9, P.O. Box 346, N-1322, Hovik, Norway
   Tel: 47-2-51-60-90, Fax: 47-2-51-61-12
2. CIF and QCIF for H.261 with audio at different rates (G.711, G.722)

Tektronix, Inc.

1. Television Division MS 58-699, P.O. Box 500, Beaverton, OR 97077-0001
   Tel: 503-627-1555, Fax: 503-627-5801
2. Profile™ Professional Desk Recorder (PDR) 100/101, professional-quality video/audio manipulation and storage using motion JPEG MTS100 MPEG-2 Test System: transport stream generation and analysis product for test and evaluation

Telephoto Communications

1. 11722 Sorrento Valley Rd., Suite D, San Diego, CA 92121-1084
   Tel: 619-452-0903, Fax: 619-792-0075
2. ALICE – JPEG Image Compression Software
   – DOS, Windows, OS/2, and UNIX versions available

Televideo, Inc.

1. San Jose, CA
2. TeleMPEG Pro and TelePort: videoboards for the PC.
   Handles MPEG-1, Indeo and Cinepak algorithms

Texas Instruments

1. P.O. Box 655303, Dallas, TX 75265
   Tel: 800-336-5236 or 214-995-2201, WWW: http.//www.ti.com/dsps

2. SN74ACT6341 (JPEG Image Compression/Decompression Processor)
   – Fully JPEG Compliant, 0.8μm CMOS Technology
   TMS 320C80 (Multimedia Video Processor (MVP))
   – Software implementation of a real-time MPEG video, H.261, JPEG codec
   TMS 320AV220 (MPEG-1 Video Decoder)
   H.320 software library includes video module H.261, audio G.728, G.722, and G.711 and communication H.221, H.242, and H.230. Also H.324

Toshiba Corp.

1. Electronics Systems Dept., International Operations Information and Commun. Systems, 1-1 Shibaura, 1-Chome, Minato-ku, Tokyo 105, Japan
   Tel: 03-457-3246, Fax: 03-456-1287
2. Color videophone for ISDN
   – Pair of 64 Kbps digital line, color video (5 to 10 frames/sec)
   Video-conferencing system
   – 32 Kbps to 1.5 Mbps. Includes 16 Kbps voice codec
   MPEG-2 Video Encoder (MP@ML, NTSC/PAL)
   TC81200F (MPEG-2 MP@ML Video Decoder)

TrueVision

1. 7340 Shadeland Station, Indianapolis, IN 46256-3919
   Tel: 317-841-0332, Fax: 317-576-7700
2. Targa 2000 and Targa 2000E. NTSC/PAL video capture. JPEG/customized compression on board. Record video direct to disk (full motion/full frame)

TV/COM International

1. 16516 Via Esprillo, San Diego, CA 92127
   Tel: 619-451-1500, Fax: 619-451-1505
2. Compression Networks
   – Number of encoders/decoders based on MPEG-2 Main Profile video and Musicam audio

VideoLogic, Inc.

1. 245 First St., Suite 1403, Cambridge, MA 02142
   Tel: 617-494-0530, Fax: 617-494-0534

2. MediaSpace – Motion JPEG encoder/decoder. Microsoft video for Windows. Audio 4-bit and 8-bit ADPCM
MPEG player—MPEG-1 video/audio decoder. Compatible with several standards

Videomail, Inc.

1. 568-4 Weddell Dr., Sunnyvale, CA 94089
Tel: 408-747-0223, Fax: 408-747-0225
2. VMC-2 Integrated Multimedia Controller with JPEG/MPEG/TV options–displays real-time video at 30 frames/sec at up to 1024×768 resolution. JPEG Codec Option—real-time full motion JPEG compression and decompression to hard disk. MPEG Decoder Option
– MPEG-1 playback from Video CD or hard disk
VMC-4 Real-time MPEG Encoder – single slot real-time compressor board. MPEG-1 and H.261 compressed video and audio streams for real-time disc authoring

Viewpoint Systems

1. Two Metro Square, 2665 Villa Creek Dr., Suite 100, Dallas, TX 75234
Tel: 214-243-0634, Fax: 214-243-0635
2. Real-time full motion video codec (56 – 640 Kbps), desktop videoconferencing

Visionetics International

1. 21311 Hawthorne Blvd., Suite 300, Torrance, CA 90503
Tel: 310-316-7940, Fax: 310-316-7457
2. MPEG Master™ – Playback MPEG-1 audio/video from Video CD, Karaoke CD, and CD-i – FMV). Playback multimedia card

VistaCom Inc.

1. 20395 Pacifica Dr., Suite 109, Cupertino, CA 95014
Tel: 408-253-5165, Fax: 408-253-5170
2. VCI/CP Board (H.221/H.242 Communication Processor)
VCI/DCT Board (H.261 Video Compression Board for PC/AT bus)
VCI/NI Boards (Network interface boards for easy access to various CCITT networks)

VCI/OEM (PC based video codec for H.320 applications)
VCI-10 CL (Video Codec, 144 - 128 Kbps)
VCI-10 GL (Video Codec, 56 - 128 Kbps)
VCI-10 GLX (Video Codec, 56 - 384 Kbps)
VCI-10 CL (Video Codec, 144 - 128 Kbps)
– VCI-10 also has audio codec, video capture, overlay, etc.

Visual Circuits

1. 3989 Central Ave. NE, Suite 630, Minneapolis, MN 55421
   Tel: 612-781-2186, Fax: 612-659-6629
2. MPEG Video Producer. Software for MPEG-1 video/audio compression. Hardware for decompression and display

Vitec Multimedia–Products & Technology

1. 99 rue Pierre Semard, F-92324 Chatillon Cedex, France
   Tel: 331-46-730606, Fax: 331-46-730600
2. VM422 MPEG Encoder ASIC
   – Captures and compresses/decompresses MPEG-1 I-frame, QSIF, or MPEG-1 SIF at 25/30 frames/sec
   Video Fun (multistandard AVI capture card for PC)
   Video Maker Classic (multistandard AVI capture card for PC and MPEG-1 I-frame encoder/decoder)
   Video NT (AVI and MPEG-1 Real-time Capture, Coder/decoder)
   DRT 1 (MPEG-1 Audio/video Decoder for TV)
   MM 1 (MPEG-1 Audio/video Decoder and High Resolution MPEG-1 I-frame Decoder)
   MM 2 (MPEG-1/JPEG/H.261 Multistandard Decoder)
   Video Clip MPEG—create and edit MPEG-1 files, video files, audio files, and system files

VIVO Software, Inc.

1. 411 Waverly Oaks Rd., Waltham, MA 02154
   Tel: 617-899-8900, Fax: 617-899-1400
2. H.320 – Compliant Software Compression/decompression Communications Technology
   – Confirms to H.320, H.261, G.711, H.221, H.230, H.242, H.243, Q.921, and Q.931. Transmit/receive at 112 or 128 Kbps

VTEL Corp.

1. 108 Wild Basin Rd., Austin, TX 78746
   Tel: 512-314-2742, Fax: 512-314-2718
2. DeskMax—Media Conferencing System
   – Video and audio conferencing available
   – H.261 and G.711, G.722 compatible
   – DOS/Windows operating environment

Wegner Communications

1. 11350 Technology Circle, Dulutte, GA 30136
   Tel: 404-623-0096, Fax: 404-623-0968
2. VideoLynx MPEG-2 digital SNG encoder/decoder, input video PAL, NTSC, ITU-R 601, 2 – 20 Mbps, MPEG layers 1 and 2 audio

Winbond Electronics Corp.

1. No. 4 Creation Rd. III, Science Based Industrial Park, Hsinchu 30077, Taiwan, ROC
   Tel: 886-35-770066, Fax: 886-35-789467
2. W9910 (MPEG-1 Audio Decoder)
   W9920 (MPEG-1 Video Decoder)
   W9925 (MPEG-1 Audio/Video Combo Decoder)
   W9922 (MPEG-2 Video Decoder, Main Profile)

Xing Technology Corp.

1. 1540 West Branch St., Arroyo Grande, CA 93420
   Tel: 805-473-0145 or 800-2-XINGIT, Fax: 805-473-0147, email: xing@xingtech.com, WWW: http://www.xingtech.com
2. Picture Prowler Still Image Manager
   Scalable MPEG Software (Video)
   – Compression speed: 1 to 2 seconds per frame
   – SIF image display up to 30 frames per second on 486 PC Windows 3.1
   Xingit! real-time quarter resolution MPEG video capture and compression board

ZORAN Corp.

1. 1705 Wyatt Dr., Santa Clara, CA 95054
   Tel: 408-986-1314, Fax: 408-986-1240
2. ZR 36040 (JPEG Image Coder/Decoder)
   ZR 36020 (DCT Processor)
   ZR 34161 (16–bit Vector Signal Processor)
   ZR 36031 (Image Compression Coder/Decoder)
   – JPEG compatible up to 21 MHz operating
   – Development Board on PC 386/486 available
   ZR 36100 (single-chip MPEG-1 system and video decoder and audio/video synchronizer)
   055 Motion JPEG multimedia chip set includes ZR 36055 Motion JPEG Controller and ZR 36050 JPEG Image Compression Processor (codec)—user selectable compression ratio. JPEG Baseline and Lossless mode supported

A.2 MPEG Audio

Analog Devices

1. P.O. Box 9106, Norwood, MA 02062-9016
   Tel: 617-461-3881
2. AD 1846 and AD 1847 (Complete digital audio "systems on a chip")
   – Implements ITU-T Rec. G.728
   – Programmable multi-function PC sound card chip set

Ariel Corp.

1. 433 River Rd., Highland Park, NJ 08904
   Tel: 908-249-2900, Fax: 908-249-2123
2. Real-time MPEG audio coding/decoding
   – AT & T VCOS development platform (multimedia on PC)

Atlanta Signal Processors, Inc.

1. 1375 Peachtreet St. NW, Suite 690, Atlanta, GA 30309
   Tel: 404-892-7265, Fax: 404-892-2512
2. MPEG Audio Encoder/Decoder
   – Layers 1, 2, and 3, mono, stereo, joint stereo, and dual channel modes

British Telecom

1. 81 Newgate Street, London EC IA7AJ, U.K.
2. PC Videophone Kit, VC 8000 Multimedia Communication Card
   – Video at QCIF (H.261), Audio G.728, G.711, and G.722 compliant, up to 128 Kbps

C–Cube Microsystems

1. 1778 McCarthy Blvd., Milpitas, CA 95035
   Tel: 408-944-6300, Fax: 408-944-6314
2. MPEG Audio Encoder
   – Layers 1 and 2 real time (32–384 Kbps)
   – Input 32, 44.1, 48 KHz, 16-bit stereo, analog, or digital

CLI, Compression Labs Inc.

1. 2860 Junction Ave., San Jose, CA 95134
   Tel: 408-435-3000, Fax: 408-922-5429
2. MPEG Audio Encoder/Decoder
   – Layers 1, 2, 56–192 Kbps mono (112–384 Kbps stereo)

Crystal Semiconductor Corp.

1. P.O. Box 17847, 4210 S. Industrial Dr., Austin, TX 78744
   Tel: 512-445-7222 or 800-888-5016, Fax: 512-445-7581
2. CS 4920 Multi-standard audio decoder
   – Programmable DSP supporting industry standard and proprietry DSP algorithms
   – Software for MPEG audio layers 1 and 2 and AC-2 is available

Fraunhofer-IIS

1. Weichselgarten 3, D-91058 Erlangen, Germany
   Fax: 49-9131-776-399
2. MPEG Audio layer 3 shareware encoder/decoder

Graphics Communication Technologies

1. Column-Minami, Aoyama Bldg. 6F, 7-1-5 Minami-Aoyamo, Minatoku, Tokyo, Japan 107
   Tel: 81-3-3498-7141, Fax: 81-3-3498-7543
2. DA 7190 MPEG audio decoder chip
   – Single chip real-time decoder
   – Decodes audio data from 64–192 Kbps mono (128–384 Kbps streo)
   – Soft muting and demuting of output data

Hitachi

1. Central Research Lab., 1-280 Higashi-koigakubo, Kokubunji, Tokyo, 185, Japan
   Tel: 81-423-23-1111
2. HD 814102F MPEG-1 audio decoder, Layers 1 and 2, single, dual, stereo, and joint stereo, 32, 44.1, and 48 KHz

Kauai Media

1. 3790 Via dela Valle, Suite 204, Del Mar, CA 92014
   Tel: 619-793-3903, Fax: 619-756-4155, email: bbal@kauai.com, WWW: http://www.kauai.com
2. MPEG audio compression and decompression software

LSI Logic Corp.

1. 1151 McCarthy Blvd., Milpitas, CA 95035
   Tel: 408-433-8000, Fax: 408-434-6457
2. L64111 (MPEG Audio Decoder)
   – Layers 1 and 2 and MUSICAM
   – Supports up to 15 Mbps sustained channel data rate
   – Decodes audio data from 32 to 192 Kbps mono (64–384 Kbps streo)
   DV 003AC3 (Dolby AC-3 Audio Decoder)

Matrox Electronics Systems

1. 1055 St. Regis Blvd., Dorval, Quebec, Canada H9P 2T4
   Tel: 514-685-2630, Fax: 514-685-2853
2. Matrox Marvel II Multimedia Controller
   – MPEG-1 audio/video decoder
   – Layers 1 and 2, 32, 44.1, and 48 Kbps sampling rates

Microware Systems Corp.

1. 1900 NW, 114th Street, Des Moines, IA 50325
   Tel: 515-224-1929, Fax: 515-224-1352
2. Microware's OS-9 operating system with Motion Picture File Manager can decode MPEG audio/video files for playback in real-time

MPR Teltech, Ltd

1. Multimedia Telecommunications, 8999 Nelson Way, Burnaby, B.C., Canada V5A 4B5
   Tel: 604-293-5183, Fax: 604-293-5787
2. L2 Blue MPEG layer 2 audio codec and Capella MPEG Layer II PC audio codec

Optivision Inc.

1. 3450 Hillview Ave., Palo Alto, CA 94304
   Tel: 800-562-8934, Fax: 415-855-0222, WWW: http://www.optivision.com
2. MPEG real-time audio/video encoder and decoder (Both MPEG 1 and 2)
   – Flexible user control up to 15 Mbps
   MPEG-2 Encoding workstation
   MPEG Video Transmission Systems
   Multichannel MPEG-2 Decoder

Philips Semiconductors

1. 811 E. Arques Ave., P.O. Box 3409, Sunnyvale, CA 94088
   Tel: 818-880-6304
2. SAA 2500 MPEG Audio Decoder
   – Demultiplexes MPEG audio bitstream for layers 1 and 2 and decodes
   – Broadband audio output signal at 16, 18, 20, or 22 bits

SGS-THOMSON Microelectronics

1. 17 Ave. des Martyrs, BP 217, 38019 Grenoble, France
   Tel: 76-58-56-10
2. STi 4500/4510 (MPEG-1 Audio Decoder)
   – Layers 1 and 2 at all sampling rates, all modes (mono, dual, stereo, and joint stereo), and all bit-rates

Sigma Design, Inc.

1. 47900 Bayside Parkway, Fremont, CA 94538
   Tel: 510-770-0100, Fax: 510-770-2640
2. MPEG Layers 1 and 2 Decoder
   – 44.1 KHz sampling rate

Texas Instruments

1. P.O. Box 655303, Dallas, Texas 75265
   Tel: 800-336-5236 or 214-995-2201
2. TMS320AV110, MPEG Audio Decoder
   – Single-chip ISO MPEG audio (layers 1 and 2) decoder

VideoLogic, Inc.

1. 245 First Street, Suite 1403, Cambridge, MA 02142
   Tel: 617-494-0530, Fax: 617-494-0534
2. MediaSpace—Motion JPEG encoder/decoder. Microsoft video for Windows. Audio 4-bit and 8-bit ADPCM
   MPEG player—MPEG-1 video/audio decoder. Compatible with several standards

VistaCom Inc.

1. 20395 Pacifica Dr., Suite 109, Cupertino, CA 95014
   Tel: 408-253-5165, Fax: 408-253-5170
2. MPEG Audio Encoder
   – Layers 1 and 2 real time (32–384 Kbps)
   – Input 32, 44.1, 48 KHz, 16-bit stereo, analog or digital

Winbond Electronics Corp.

1. No.4 Creation Road III, Science Based Industrial Park, Hsinchu 30077, Taiwan, ROC
   Tel: 886-35-770066, Fax: 886-35-789467
2. W9910 (MPEG-1 Audio Decoder)
   – Layers 1 and 2 (32, 44.1, and 48 KHz)

Xing Technology Corp.

1. 1540 West Branch St., Arroyo Grande, CA 93420
   Tel: 805-473-0145 or 800-2-XINGIT, Fax: 805-473-0147
2. XingSound: Real-time Audio Compression Software
   – As much as 12:1 compression ratio by using MPEG-1 algorithms on 386/486 PC
   XingCD: MPEG Audio Compression/Decompression CD
   – Compresses/decompresses full-motion MPEG audio/video streams

ZORAN Corp.

1. 1705 Wyatt Dr., Santa Clara, CA 95054
   Tel: (408)986-1314, Fax: (408)986-1240
2. ZR 38000 DSP Processor Family
   – Two-channel MPEG-1 audio decoder (ZR 38511, up to 448 Kbps)
   – Two-channel Dolby AC-3 audio decoder (ZR 38501, 320 Kbps)
   – Six-channel Dolby AC-3 audio decoder (ZR 38500, 320 Kbps)

A.3 Motion Estimation

Array Microsystems

1. 1420 Quail Lake Loop, Colorado Springs, CO 80906
   Tel: (719)540-7900, Fax: (719)540-7950
2. MEC (Motion Estimation Coprocessor)
   – BMA 8×8, or 16×16 block motion estimation
   – Full-pixel and quarter-pixel resolution

Bellcore

1. 331 Newman Springs Rd., Red Bank, NJ 07701, and 445 South St., Morristown, NJ 07960
2. Full-search BMA
   – Various block sizes (8 × 8, 16 × 16, 32 × 32, . . .)
   – 30 MHz speed, −8 to +7 pels-lines/frame MV range

Graphics Communication Technologies

1. Column-Minami, Aoyama Bldg 6F, 7-1-5 Minami-Aoyamo, Minatoku, Tokyo, Japan 107
   Tel: 81-3-3498-7141, Fax: 81-3-3498-7543
2. Full-search BMA
   – Various block sizes (8 × 8, 16 × 16, 32 × 32)
   – 30 MHz speed, −8 to +7 pels-lines/frame MV range

HHI fuer Nachrichtentechnik Berlin

1. Einsteinufer 37, D–1000, Berlin 10, Germany
   or TU-Berlin, Inst. fuer Microelectronik, Jebensstr, D–1000, Berlin 12, Germany
2. Two chip set design for BMA/ME
   – Real-time TV and HDTV application

ITRI CC Labs

1. Computer & Communication Research Lab., B300, Bldg. 11, 195 Sec. 4. Chung Hsing Rd., Chutung, Hsinchu, Taiwan 31015
   Tel: 886-35-917228, Fax: 886-35-968764
2. CIC86200 (Motion Estimation Unit)
   – 15 frames/sec at 25 MHz clock rate
   – 16 × 16 block size, ±7 search range

LSI Logic Corp.

1. 1151 McCarthy Blvd., Milpitas, CA 95035
   Tel: 408-433-8000, Fax: 408-434-6457

2. L64720 Full Search BMA
   – 8 × 8, 16 × 16 block sizes, 8 × 8, 16 × 16 search window
   – Clock rate up to 40 MHz

SGS-THOMSON Microelectronics

1. 17 Ave. des Martyrs, BP 217, 38019 Grenoble, France
   Tel: 76-58-56-10
2. STi3220 Full Search BMA
   – Various block sizes (8 × 8, 16 × 16)
   – 13.5 MHz speed, −8 to +7 pels-lines/frame MV range

Siemens AG

1. Corporate R&D, Otto-Hahn-Ring 6, D–8000 Munich 83, Germany
   Tel: 089-636-46242
2. Full-search BMA
   – Line scan mode and block scan mode

Telecom Paris University

1. 46 rue Barrault, 75634 Paris, France
2. Full Search BMA
   – Block size and MV range are variable
   – 18 MHz speed, ±15 pels-lines/frame maximum MV range

Telettra

1. via Trento 30, 20059 Vimercate, Italy 039/66551
   Tel: 039-66551
2. Full Search BMA
   – Motion vector for 16 × 8 block with a 1/2 pel resolution

University of Hannover

1. Appelstr. 9A, D–3000, Hannover 1, Germany
2. Full-search BMA
   – Block size 8 × 8, ±8 pels-lines/frame MV range
   – Systolic array processors

VistaCom, Inc.

1. 20395 Pacifica Dr., Suite 109, Cupertino, CA 95014
   Tel: 408-253-5165, Fax: 408-253-5170
2. VCI/MA Programmable Motion Accelerator
   – Motion estimation for CIF at 30 Hz, up to ± 15 pels/frame
   – Full-pixel and half-pixel resolution

A.4 Image Processing Software

Alacron

1. 71 Spit Brook Rd., Suite 204, Nashua, NH 03060
   Tel: 603-891-2750, Fax: 603-891-2745
2. RIPL—Real-Time Image Processing Library, SIPL—Scientific Image Processing Library
   – Basic image-processing software to operate Alacron's multiprocessor boards

Datacube

1. 300 Rosewood Dr., Danvers, MA 01923-4505
   Tel: 508-777-4200
2. ImageFlow: Software for Datacubes' Max Video 200, Max Video 20 and Max Video image processor devices
   MD1/8GB Digital Image Recoder
   Wit Flow and DIGIT: interactive Graphical Imaging Tools

Data Translation

1. 100 Locke Dr., Marlboro, MA 01752-1192
   Tel: 508-481-3700 or 800-525-8528
2. Color frame grabber
   – Real-time image capture, storage and display, plug-in processor boards, RGB ⇔ HSI conversion, image processing software

Mathworks

1. 24 Prime Park Way, Natick, MA 01760
   Fax: 508-653-6284, email: info@mathworks.com

2. Image Processing Toolbox
   – Basic image processing operations, including 2D transforms

Matrox Electronic Systems Ltd.

1. 1055 St. Regis Blvd., Dorval, Quebec, Canada H9P2T4
   Tel: 514-685-2630, Fax: 514-685-2853
2. Matrox tools
   – Software/hardware various image processing operations

Media Cybernetics

1. 8484 Georgia Ave., Silver Spring, MD 20910
   Tel: 800-992-HALO or +1 301-495-3305, Fax: +1 301-495-5964
2. Image-Pro Plus, for Microsoft windows
   – Image analysis software for gray scale and true color image. Windows interface, Image management. Image acquisition and enhancement

Precision Digital Images Corp.

1. 8730, 148th Ave. N.E., Redmond, WA 98052
   Tel: 206-882-0218, Fax: 206-867-9177
2. Media Station 5000, A real-time multimedia workstation for MPEG, JPEG, and P×64
   – Various image processing algorithms, based on TMS320C80 chip (MVP)

A.5 Image Processing Systems

Abekas Video Systems, Inc.

1. 101 Galveston Dr., Redwood City, CA 94063
   Tel: 415-369-5111

Analogic Corp.

1. 8 Centennial Dr., Peabody, MA 01960
   Tel: 508-977-3000, Fax: 508-977-6813
2. DASM-VIP (Video Image Processor)

Colorado Video

1. Box 928, Boulder, CO 80306

Coreco, Inc.

1. 6969 Trans-Canada Highway, Suite 113, St. Laurent, Quebec, Canada H4T 1V8
   Tel: 514-333-1301, Fax: 514-333-1388

Cyber Research, Inc.

1. 25 Business Park Dr., Branford, CT 06405
   Tel: 203-483-8815 or 800-341-2525, Fax: 203-483-9024

Data Translation

1. 100 Locke Dr., Marlboro, MA 01752
   Tel: 508-481-3700

Datron/Transco Inc.

1. 2071 Concourse Dr., San Jose, CA 95131

Digital Video Computing

1. Seestrasse 7, D–82211, Herrsching, Germany
   Tel: 49-81-52-93010, Fax: 49-81-52-91331
   Also: 400 Oyster Point Blvd., Suite 57, San Francisco, CA 94080
   Tel: 415-875-7100

DVS GMBH

1. Krepenstrasse 8, 3000 Hannover 1, Germany
   Tel: 4951163-0077, Fax: 4951163-0077

EPIX, Inc.

1. 381 Lexington Drive, Buffalo Grove, IL, 60089
   Tel: 708-465-1819, Fax: 708-465-1919

Gould Electronics

1. Imaging and Graphics Division, 46360 Fremont Blvd., Fremont, CA 94538
   Tel: 415-498-3200

Graftek Imaging

1. 240 Oral School Rd., Mystic, CT 06355
   Tel: 800-959-3011, Fax: 203-572-7328

ImageNation Corp.

1. P.O. Box 276, Beaverton, OR 97075
   Tel: 503-641-7408, Fax: 503-643-2458

Imaging Technology Inc.

1. 55 Middlesex Turnpike, Bedford, MA 01730-1421
   Tel: 617-275-2700, Fax: 617-275-9590

Intel Corporation

1. 3065 Bowers Ave., Santa Clara, CA 95051
   Tel: 408-765-8080
2. Action Media 750 Capture Board and 750 Delivery Board (DVI Technology)

Intelligent Systems Intl.

1. Liuderstraat 9 B-3210 Linden, Belgium
   Tel: 32-1662-1585, Fax: 32-1662-1584, email: virtuoso@blx.com

K^2T, Inc.

1. One Library Place, Duquesne, PA 15110
   Tel: 412-469-3150, Fax: 412-469-8120

Loughborough Sound Images Plc.

1. Loughborough Park, Ashby Road, Loughborouh, Leicestershire, LE 113 NE, England
   Tel: 44-1509-634444, Fax: 44-1509-634450

Noesis

1. 6800 Cote de Liesse, Suite 200, St. Laurent, Quebec H4T 2A7, Canada
   Tel: 514-345-1400, Fax: 514-345-1575, email: noesis@cam.org

Optimas Corp.

1. 190 West Dayton St., Edmonds, WA 98020

PC Video Conversion Corp.

1. 1340 Tully Rd., Suite 309, San Jose, CA 95122
   Tel: 408-279-2442, Fax: 408-279-6105

Pentek Innovative DSP Systems

1. 55 Walnut St., Norwood, NJ 07648
   Tel: 201-767-7100, Fax: 201-767-3994

Philips Semiconductor

1. 2001 West Blue Heron Blvd., P.O. Box 10330, Riviera Beach, FL 33404
   Tel: 407-881-3200, 800-447-3762, Fax: 407-881-3300

Raytheon Semiconductor

1. 350 Ellis St., P.O. Box 7016, Mountain View, CA 94039-7016
   Tel: 415-968-9211, Literature: 800-722-7074, Consumer Service: 408-522-7051

Recognition Concepts

1. 342 Ski Way, P.O. Box 8510, Incline Village, NV 89450
   Tel: 702-831-0473

Sonitech Intl

1. 14 Mica Lane, Wellesley, MA 02181
   Tel: 617-235-6824, Fax: 617-235-2531, email: info@sonitech.com

Toshiba

1. I. E. OEM Division, 9740 Irving Blvd., Irving, CA 92718
   Tel: 714-583-3180
2. Toshiba IP9506 Image Processor

Transtech Parallel Systems Corp.

1. 20 Thornwood Dr., Ithaca, NY 14850-1263
   Tel: 607-257-6502, Fax: 607-257-3980, email: transtech@transtech.com

TraQuair Data Systems, Inc.

1. Tower Bldg., 112 Prospect Street, Suite 114, Ithaca, NY 14850
   Tel: 607-272-4417, Fax: 607-272-6211
2. Toshiba IP9506 Image Processor

Truevision

1. 7340 Shadeland Station, Indianapolis, IN 46256-3919
   Tel: 317-841-0332, Fax: 317-576-7700

TRW LSI Products, Inc.

1. 4243 Campus Point Ct., San Diego, CA 92121
   Tel: 619-457-1000, Fax: 619-455-6314

Univision Technologies, Inc.

1. Three Burlington Woods, Burlington, MA 01803
   Tel: 617-221-6700, Fax: 617-221-6777

VITEC, Visual Information Technologies Inc.

1. 3460 Lotus, Plano, TX 75075
   Tel: 214-985-2210

Zoran Corp.

1. 1705 Wyatt Dr., Santa Clara, CA 95054
   Tel: 408-986-1314, Fax: 408-986-1240

A.6 Image Capture/Digitizer Boards

Analogic Corp.

1. 8 Centennial Dr., Peabody, MA 01960
   Tel: 508-977-3000, Fax: 508-977-6813

Aspro Technologies Ltd.

1. 5805 White Rd., Suite 202, Mississauga, Ontario, L4Z 2J1, Canada
   Tel: 905-712-2131, Fax: 905-712-1887

AT & T Electronic

1. Photography and Imaging Center, 2002 Wellesley Blvd., Indianapolis, IN 46219
   Tel: 317-352-6120

Chorus Data Systems

1. P.O. Box 370, 6 Continental Blvd., Merrimack, NH 03054
   Tel: 603-424-2900

Coreco, Inc.

1. 6969 Trans-Canada Highway, Suite 113, St. Laurent, Quebec, Canada H4T 1V8
   Tel: 514-333-1301, Fax: 514-333-1388

Datacube

1. 300 Rosewood Dr., Danvers, MA 01923-4505
   Tel: 508-777-4200

Data Translation

1. 100 Locke Dr., Marlboro, MA 01752
   Tel: 508-481-3700

Dipix

1. 1051 Baxter Rd., Ottawa, Ontario, Canada K2C 3P1
   Tel: 613-596-4942, Fax: 613-596-4914

Epix, Inc.

1. 381 Lexington Dr., Buffalo Grove, IL 60089
   Tel: 708-465-1818, Fax: 708-465-1919

Imaging Technology Inc.

1. 55 Middlesex Turnpike, Bedford, MA 01730-1421
   Tel: 617-275-2700, Fax: 617-275-9590

Imagraph

1. 11 Elizabeth Dr., Chelmsford, MA 01824
   Tel: 508-256-4624, Fax: 508-250-9155

IMAVISION

1. Roy Ball Associated Ltd., 1750 Courtwood Crescent, Suite 300, Ottawa, Ontario K2C 2B5 Canada
   Tel: 613-226-7890

Intel Corporation

1. 3065 Bowers Ave., Santa Clara, CA 95051
   Tel: 408-765-8080

Matrox Electronic System Ltd.

1. 1025 St., Regis Blvd., Dorval, Quebec, Canada H9P 2T4
   Tel: 514-685-2630, Fax: 514-685-2853

MRT

1. MRT micro as. Stromsv 74N-2010, Strommen, Norway
   Tel: 4763892020, Fax: 4763801212, (USA, Tel: 603-465-2830, Fax: 603-465-2680)

MuTech

1. 800 West Cummings Park, Suite 3800, Woburn, MA 01801
   Tel: 617-935-1770, Fax: 617-935-3054

Precision Digital Images Corp.

1. 8711 148th Ave., Redmond, WA 98052
   Tel: 206-882-0218, Fax: 206-867-9177

TECON INC.

1. 14613 N.E. 87th St., Redmond, WA 98052
   Tel: 206-881-7551

Transtech Parallel Systems Corp.

1. 20 Thornwood Dr., Ithaca, NY 14850-1263
   Tel: 607-257-6502, Fax: 607-257-3980, email: transtech@transtech.com

TRUE VISION

1. 7340 Shadeland Station, Indianapolis, IN 46256
   Tel: 317-841-0332

Zytronix, Inc.

1. 1208 Apollo Way, Suite 504, Sunnyvale, CA 94086
   Tel: 408-749-1326, Fax: 408-749-1329

Appendix B
Information on the Internet

This list will be a guide for the reader who has an interest in implementing the international standards (JPEG, H.261, MPEG-1 and 2, HDTV, CMTT, etc.). Most of the sites can be accessed by Internet anonymous ftp or www. They also provide source code (generally written in C), image/video viewer, and simulation data. Source codes for UNIX machines have to be properly installed with the user-specific parameters, and those for IBM PC or Macintosh run on the computer without many parameter changes. The list is referred by the news groups as "comp.compression," "comp.dsp," "comp.graphics," etc. The concerning FAQs (frequently asked questions) from the various news groups are available by anonymous ftp at rtfm.mit.edu: /pub/usenet/news.answers.

JPEG

- Source code for most any machine:
  ftp.uu.net:/graphics/jpeg/jpegsrc.v5.tar.gz
  nic.funet.fi:/pub/graphics/packages/jpeg/jpegsrc.v5*.tar.gz
  Contact: jpeg-info@uunet.uu.net (Independent JPEG Group)
- havefun.stanford.edu:pub/jpeg/JPEGv1.2.tar.Z (supports lossless mode), Contact: Andy Hung <achung@cs.stanford.edu>
- ftp.cs.cornell.edu:/pub/multimed/ljpg.tar.Z (lossless jpeg)
- xv, an image viewer that can read JPEG pictures, is available in ftp.cis.upenn.edu:/pub/xv/xv-3.00a.tar.Z
- etro.vub.ac.be:/pub/DCT_ALGORITHMS/* (DCT algorithms)
  Contact: Charilos Christopoulos <chchrist@etro2.vub.ac.be>
- JPEG Extensions: Contact Dr. Daniel T. Lee, email: dlee@hplabs.hp.com

H.261 and H.263

- havefun.stanford.edu:pub/p64/P64v1.2.tar.Z
  Contact: Andy Hung <achung@cs.stanford.edu>
- zenon.inria.fr:/rodeo/ivs/last_version/ivs*-src.tar.gz (Inria videoconference system)
  Contact: Thierry Turletti <turletti@sophia.inria.fr>
- Source code for H.263 TMN5 is available from bonde.nta.no:
  /pub/tmn/software/tmn-1.6b.tar.gz or tmndec-1.6b.tar.gz
- ftp.std.com File:/vendors/Picture Tel/h324/h324 info.txt
  Telenor R & D, H.263 test model simulation software, velearci.6,
  ftp://bonde.no/pub/tun/software, Dec. 1995

MPEG

- http://www.crsq.it/~luigi/MPEG/mpegcompanies.html Browse under MPEG companies in NETSCAPE for MPEG-related products.
- havefun.stanford.edu:/pub/mpeg/MPEGv1.2.1.tar.Z
  Contact: Andy Hung <achung@cs.stanford.edu>
- s2k-ftp.cs.berkeley.edu: /pub/multimedia/mpeg/mpeg_play-2.0.tar.Z
  s2k-ftp.cs.berkeley.edu: /pub/multimedia/mpeg/mpeg_encode-1.3.tar.Z
  Contact: mpeg-bugs@cs.berkeley.edu
- flash.bu.edu: /pub/code/mpeg_system/mpeg_system_source_v1.0.tar.gz
  (MPEG-1 Multi-Stream System Layer encoder/player; includes an enhanced version of mpeg_play)
  Contact: Jim Boucher <jboucher@spiderman.bu.edu> or Ziv Yaar <zyaar@bu.edu>
- ftp.mni.mcgill.ca:/pub/mpeg/mpeg_lib-1.1.tar.gz [MPEG library]
  Contact: Gregory Ward <greg@pet.mni.mcgill.ca>
- ftp.netcom.com:/pub/cfogg/mpeg/vmpeg12a.zip
  Contact: Stefan Eckart <stefan@lis.e.technik.tu-muenchen.de>
  decel.ecel.uwa.edu.au:/users/michael/mpegw32f.zip
  (for Windows and NT)
- nvr.com:/pub/NVR-software/Product-1.0.4.tar.Z (192.82.231.50)
  (free demo copy of NVR's software toolkit for SPARCstations)
  Contact: Todd Brunhoff <toddb@nvr.com>
- ftp.netcom.com:/pub/cfogg/mpeg2/* (MPEG-2 encoder and decoder)
  Contact: MPEG-L@netcom.com (MPEG Software Simulation Group)
- ftp.informatik.tu-muenchen.de [131.159.0.198]
  /pub/comp/graphics/mpex/mplex/mplex-1.0a_beta.tar.gz

(MPEG-1 systems multiplexer in ANSI-C)
/pub/comp/graphics/mpeg/mplex/mpeg_systems_paper_0.99.ps.gz
(PostScript paper discussing MPEG-1 systems)

- wuarchive.wustl.edu: graphics/x313/pub/video_verification/paris
  (109 MBytes of verification bit sreams and picture reconstructions)
  graphics/x313/pub/video_verification/samples
  (13 Mbytes of sample MPEG-2 and MPEG-1 streams)
- ftp.microsoft.com: /developers/DEVTOOLS/WIN32SDK
  (Win32s patch executables and drivers that allow 32-bit programs to be executed on 386, 486, and Pentiums from within the traditional 16-bit Windows (e.g., version 3.1) environment)
- atum.ee.upenn.edu: /dist/pkg/MPEGTool/ (MPEGTool and utilties, Word Wide Web (WWW) Graphics Archive) WWW browser: Anonymous to ftp.rahul.net:/pub/bryanw/wwwfaq.zip

MPEG Audio

- A lossless compressor for 8-bit and 16-bit audio data (.au) is available by anonymous ftp at
  svr-ftp.eng.cam.ac.uk:/pub/comp.speech/sources/shorten.tar.gz
  Contact: Tony Robinson <ajr@eng.cam.ac.uk>.
- Source code for a *fast* G.721 32 kbps ADPCM (lossy) is available by ftp from ftp.cwi.nl as /pub/adpcm.shar. You can get a G.721/722/723 package by email to teledoc@itu.arcom.ch, with GET ITU-3022
- Source code for the MPEG-1 audio decoder layers 1, 2, and 3 is available on fhginfo.fhg.de (153.96.1.4) in /pub/layer3/public_c. There are two files: mpeg1_iis.tar.Z (Unix: lines separated by line feed only), mpeg1iis.zip (PC: lines separated by carriage return and line feed).
  Contact: Harald Popp <popp@iis.fhg.de>
- The MPEG conformance audio test bit streams are available from
  ftp.tnt.uni-hannover.de: /pub/MPEG/audio/mpeg-testbitstreams.tar.Z
- Audio software is available on sunsite.unc.edu in subdirectories of /pub/electronic-publications/IUMA/audio_utils:
  – An MPEG audio Layer-2 decoder is in
  mpeg_players/Workstations/maplay1_2.tar.Z
  – The sources of the XING MPEG audio player for Windows is in mpeg_players/Windows/mpgaudio.zip.
  – Audio codec source code is in converters/source/mpegaudio.tar.Z.
- CELP C code for Sun SPARCs is available for anonymous ftp in
  ftp.super.org:/pub/speech/celp_3.2a.tar.Z

- GSM 06.10 13 kbps RPE/LTP speech compression Version 1.0 of the implementation is available for anonymous ftp from tub.cs.tu-berlin.de as /pub/tubmik/gsm-1.0.tar.Z. Questions and bug reports should be directed to toast@tub.cs.tu-berlin.de.
- The QCELP process is documented in the Common Air Interface (CAI), which is available for anonymous ftp from lorien.qualcomm.com in /pub/cdma, as PostScript files.
- You can find the comp.dsp FAQ by ftp on rtfm.mit.edu in /pub/usenet/news.answers/audio-fmts/part1.
- For G.723 C source code, contact Ron Richter <rrichter@dspg.com> Tel: 408-986-4469, Fax: 408-986-4490. Also from ITU.

JBIG

- JBIG software and the JBIG specification are available on nic.funet.fi:/pub/graphics/misc/test-images/jbig.tar.gz.
  You can find a description of JBIG in ISO/IEC CD 11544, contained in document ISO/IEC JTC1/SC2/N2285. The only way to get it is to ask your national standards body for a copy. In the United States, call ANSI at 212-642-4900.
- JBIG Extensions contact P. G. Howard <pgh@research.att.com>

Fractals

- A fractal image compression program is available by ftp in lyapunov.ucsd.edu: /pub/young-fractal/unifs10.zip or yuvpak20.zip. A fractal image decompression program is available on the same ftp site in /pub/inls-ucsd/fractal-2.0.tar (file size 1.3 MB).
- Articles, some C codes, and the SIGGRAPH '92 course notes from Yuval Fisher <yfisher@ucsd.edu> are available by ftp to legendre.ucsd.edu: pub/Research/Fisher.
  The source code for the program published in the Oct. '93 issue of *Byte* is in ftp.uu.net:/published/byte/93oct/fractal.exe.
  Several papers on fractal image compression are available by ftp on ftp.informatik.uni-freiburg.de:/documents/papers/fractal. A bibliography is in schmance.uwaterloo.ca:/pub/Fractal/fractal.biblio.ps.Z.
- Another fractal compression program is available by ftp in vision.auc.dk:/pub/Limbo/Limbo*.tar.Z.

Wavelets

- Preprints and software are available by anonymous ftp from ceres.math.yale.edu: /pub/wavelets and pub/software.
- epic: (pyramid wavelet coder) is available from whitechapel.media.mit.edu:/pub/epic.tar.Z [18.85.0.125]
  Contact: Eero P. Simoncelli <eero@media.mit.edu>
- hcompress: (wavelet image compression) is available from stsci.edu:/software/hcompress/hcompress.tar.Z
- wavethresh: (wavelet software for the language S) is available from gdr.bath.ac.uk:/pub/masgpn/wavethresh2.2.Z
  Contact: gpn@maths.bath.ac.uk
- rice-wlet: (wavelet software) is available from cml.rice.edu:/pub/dsp/software/rice-wlet-tools.tar.Z
  Contact: Ramesh Gopinath <ramesh@rice.edu>
- scalable: (two- and three-dimensional subband transformation) is available from robotics.eecs.berkeley.edu:/pub/multimedia/scalable2.tar.Z
  Contact: scalable@robotics.eecs.berkeley.edu
- AccuPress, wavelet image compression software,
  Aware, Inc., 1 Oak Park, Bedford, MA 01730
  Tel: 617-276-4000, email: sales@aware.com,
  WWW: http://www.aware.com
  Houston Advanced Research Center (HARC),
  Woodland, TX, "HARC-C" wavelet compression
- WWW: http://saigon.ece.wisc.edu/~waveweb/QAMF.html

Vector Quantization

- Some vector quantization software for data analysis is available from cochlea.hut.fi (130.233.168.48) in the /pub directory. One package is lvq_pak and one is som_pak (som_pak generates Kohonen maps of data using lvq to cluster it).

Test Data/Images

- The ftp site eedsp.gatech.edu (130.207.226.2) has a bunch of standard images (lenna, baboon, cameraman, crowd, moon, etc.) in directory /database/images.
- The site ipl.rpi.edu (128.113.14.50) has standard images in two directories: /pub/image/still/usc and /pub/image/still/canon. Sequence images (table tennis, flower garden, football, etc.) are in the directory /pub/image/sequence as Sun Raster RGB format.

- All the images are in Sun rasterfile format. You can use the pbm utilities to convert them to whatever format is most convenient. [pbm is available in ftp.ee.lbl.gov:/pbmplus*.tar.Z.] Questions about the ipl archive should be sent to help@ipl.rpi.edu. The original lenna image is 512×512, 8 bits/pel, red, green, and blue fields. Gray-scale versions of lenna have been obtained in two different ways from the original: (1) Using the green field as a gray-scale image, and (2) doing an RGB-to-YUV transformation and saving the Y component. Method 1 makes it easier to compare different groups' results since everyone's version should be the same using that method. Method 2 produces a more correct image. The editorial in the January 1992 issue of *Optical Engineering* (vol. 31, no. 1) details how *Playboy* has finally caught on to the fact that their copyright on Lenna Sjooblom's photo is being widely infringed. Essentially, you may have to get permission from *Playboy* to publish it.
- There are few gray-scale still images and some raw data of test results available on nic.funet.fi:/pub/graphics/misc/test-images. There are lots of .gif images in nic.funet.fi:/pub/pics.
- Medical images can be found in decaf.stanford.edu: /pub/images/medical/mri, eedsp.gatech.edu: /database/images/wchung/medical, and omicron.cs.unc.edu: /pub/softlab/CHVRTD.
- The CCITT test images are available on nic.funet.fi in directory pub/graphics/misc/test-images, files ccitt1.tif to ccitt8.tif. Alternate versions are in http://www.cs.waikato.ac.nz/~singlis/ccitt.html.

Documents

- The official specification of ISO/IEC standards can be ordered from the national standards bodies: for example, in the United States, ANSI, 11 West 42nd St., New York, NY 10036 [Tel: (212) 642-4900], in Germany, DIN-Beuth Verlag, Auslandsnormen, Burggrafenstr. 6, D-10772 Berlin, Germany, Tel: 030-2601-2757, Fax: 030-2601-1231. The official ITU address is ITU, Place des Nations, CH-1211, Geneve 20, Switzerland. The official ISO address is 1 rue de Verambe, CH-1211, Geneve 20, Switzerland.
- The ITU-T documents are available on src.doc.ic.ac.uk, but the directory name keeps changing. Check one of the following:
  /computing/ccitt/ccitt-standards/ccitt/
  /computing/ccitt/standards/ccitt, or /doc/ccitt-standards/ccitt
  For example, the V.42bis and the H.261 standards are in *standards/ccitt/1992/v/v42bis.asc.Z and /computing/ccitt/standards/ccitt/1992/h/h261.doc.Z, or h261.rtf.Z. You can also check ipl.rpi.edu: /image-archive/bitmap/ccitt.

- A mail server for ITU-T documents is available at teledoc@itu.arcom.ch or itudoc@itu.ch.
- A Gopher server is also available:
  Name=International Telecommunications Union (ITU)
  Host=info.itu.ch
  Port=70
- The latest versions of H.263 and MPEG-2 documents are available from http://www.nta.no/brukere/DVC/.
- A copy of the Grand Alliance HDTV standard may be obtained at an ftp site: ga-doc.sarnoff.com, or else send an email message to grand_alliance@sarnoff.com. The full ATSC AC-3 standard document specification is provided by the United States Advanced Television Systems Committee, Standard A/52, Nov. 1994. Also via Internet: ftp.atsc.org/pub/standards/ WWW: http:/www.atsc.org
- MPEG-4 Documents. contact Dr. Cliff Reader, email: creader@samsung.com or Dr. Thomas Sikora, email: sikora@mailsam.HHI.DE

Appendix C
VLC Tables for Coding the DCT Coefficients

Summary

The tables from ISO/IEC 10918:1994, ISO/IEC 11172:1993 and ISO/IEC 13818:1994 have been reproduced with the permission of the International Organization for Standardization, ISO, and the International Electrotechnical Commission, IEC. The complete standards can be obtained from any ISO or IEC member or from the ISO or IEC Central Offices, Case Postal, 1211 Geneva 20, Switzerland. Copyright remains with ISO and IEC.

The tables from ITU-T H.261, ITU-T H.263, and ITU-R CMTT.723 have been reproduced with the permission of the International Telecommunication Union, Telecommunication Standardization Sector and Radiocommunication Sector. Copyright remains with ITU.

C.1 VLC Table for Luminance AC Coefficients in JPEG

Run	Size	Code word	Run	Size	Code word
0	0	1010 (EOB)	3	5	1111 1111 1001 0000
0	1	00	3	6	1111 1111 1001 0001
0	2	01	3	7	1111 1111 1001 0010
0	3	100	3	8	1111 1111 1001 0011
0	4	1011	3	9	1111 1111 1001 0100
0	5	1101 0	3	A	1111 1111 1001 0101
0	6	1111 000	4	1	1110 11
0	7	1111 1000	4	2	1111 1110 00
0	8	1111 1101 10	4	3	1111 1111 1001 0110
0	9	1111 1111 1000 0010	4	4	1111 1111 1001 0111
0	A	1111 1111 1000 0011	4	5	1111 1111 1001 1000
1	1	1100	4	6	1111 1111 1001 1001
1	2	1101 1	4	7	1111 1111 1001 1010
1	3	1111 001	4	8	1111 1111 1001 1011
1	4	1111 1011 0	4	9	1111 1111 1001 1100
1	5	1111 1110 110	4	A	1111 1111 1001 1101
1	6	1111 1111 1000 0100	5	1	1111 010
1	7	1111 1111 1000 0101	5	2	1111 1110 111
1	8	1111 1111 1000 0110	5	3	1111 1111 1001 1110
1	9	1111 1111 1000 0111	5	4	1111 1111 1001 1111
1	A	1111 1111 1000 1000	5	5	1111 1111 1010 0000
2	1	1110 0	5	6	1111 1111 1010 0001
2	2	1111 1001	5	7	1111 1111 1010 0010
2	3	1111 1101 11	5	8	1111 1111 1010 0011
2	4	1111 1111 0100	5	9	1111 1111 1010 0100
2	5	1111 1111 1000 1001	5	A	1111 1111 1010 0101
2	6	1111 1111 1000 1010	6	1	1111 011
2	7	1111 1111 1000 1011	6	2	1111 1111 0110
2	8	1111 1111 1000 1100	6	3	1111 1111 1010 0110
2	9	1111 1111 1000 1101	6	4	1111 1111 1010 0111
2	A	1111 1111 1000 1110	6	5	1111 1111 1010 1000
3	1	1110 10	6	6	1111 1111 1010 1001
3	2	1111 1011 1	6	7	1111 1111 1010 1010
3	3	1111 1111 0101	6	8	1111 1111 1010 1011
3	4	1111 1111 1000 1111	6	9	1111 1111 1010 1100
6	A	1111 1111 1010 1101	A	6	1111 1111 1100 1011
7	1	1111 1010	A	7	1111 1111 1100 1100
7	2	1111 1111 0111	A	8	1111 1111 1100 1101
7	3	1111 1111 1010 1110	A	9	1111 1111 1100 1110
7	4	1111 1111 1010 1111	A	A	1111 1111 1100 1111
7	5	1111 1111 1011 0000	B	1	1111 1110 01
7	6	1111 1111 1011 0001	B	2	1111 1111 1101 0000

Appendix C.1 *(cont.)*

Run	Size	Code word	Run	Size	Code word
7	7	1111 1111 1011 0010	B	3	1111 1111 1101 0001
7	8	1111 1111 1011 0011	B	4	1111 1111 1101 0010
7	9	1111 1111 1011 0100	B	5	1111 1111 1101 0011
7	A	1111 1111 1011 0101	B	6	1111 1111 1101 0100
8	1	1111 1100 0	B	7	1111 1111 1101 0101
8	2	1111 1111 1000 000	B	8	1111 1111 1101 0110
8	3	1111 1111 1011 0110	B	9	1111 1111 1101 0111
8	4	1111 1111 1011 0111	B	A	1111 1111 1101 1000
8	5	1111 1111 1011 1000	C	1	1111 1110 10
8	6	1111 1111 1011 1001	C	2	1111 1111 1101 1001
8	7	1111 1111 1011 1010	C	3	1111 1111 1101 1010
8	8	1111 1111 1011 1011	C	4	1111 1111 1101 1011
8	9	1111 1111 1011 1100	C	5	1111 1111 1101 1100
8	A	1111 1111 1011 1101	C	6	1111 1111 1101 1101
9	1	1111 1100 1	C	7	1111 1111 1101 1110
9	2	1111 1111 1011 1110	C	8	1111 1111 1101 1111
9	3	1111 1111 1011 1111	C	9	1111 1111 1110 0000
9	4	1111 1111 1100 0000	C	A	1111 1111 1110 0001
9	5	1111 1111 1100 0001	D	1	1111 1111 000
9	6	1111 1111 1100 0010	D	2	1111 1111 1110 0010
9	7	1111 1111 1100 0011	D	3	1111 1111 1110 0011
9	8	1111 1111 1100 0100	D	4	1111 1111 1110 0100
9	9	1111 1111 1100 0101	D	5	1111 1111 1110 0101
9	A	1111 1111 1100 0110	D	6	1111 1111 1110 0110
A	1	1111 1101 0	D	7	1111 1111 1110 0111
A	2	1111 1111 1100 0111	D	8	1111 1111 1110 1000
A	3	1111 1111 1100 1000	D	9	1111 1111 1110 1001
A	4	1111 1111 1100 1001	D	A	1111 1111 1110 1010
A	5	1111 1111 1100 1010	E	1	1111 1111 1110 1011
E	2	1111 1111 1110 1100	F	1	1111 1111 1111 0101
E	3	1111 1111 1110 1101	F	2	1111 1111 1111 0110
E	4	1111 1111 1110 1110	F	3	1111 1111 1111 0111
E	5	1111 1111 1110 1111	F	4	1111 1111 1111 1000
E	6	1111 1111 1111 0000	F	5	1111 1111 1111 1001
E	7	1111 1111 1111 0001	F	6	1111 1111 1111 1010
E	8	1111 1111 1111 0010	F	7	1111 1111 1111 1011
E	9	1111 1111 1111 0011	F	8	1111 1111 1111 1100
E	A	1111 1111 1111 0100	F	9	1111 1111 1111 1101
F	0	1111 1111 001 (ZRL)	F	A	1111 1111 1111 1110

C.2 VLC Table for Chrominance AC Coefficients in JPEG

Run	Size	Code word	Run	Size	Code word
0	0	00 (EOB)	1	6	1111 1111 0101
0	1	01	1	7	1111 1111 1000 1000
0	2	100	1	8	1111 1111 1000 1001
0	3	1010	1	9	1111 1111 1000 1010
0	4	1100 0	1	A	1111 1111 1000 1011
0	5	1100 1	2	1	1101 0
0	6	1110 00	2	2	1111 0111
0	7	1111 000	2	3	1111 1101 11
0	8	1111 1010 0	2	4	1111 1111 0110
0	9	1111 1101 10	2	5	1111 1111 1000 010
0	A	1111 1111 0100	2	6	1111 1111 1000 1100
1	1	1011	2	7	1111 1111 1000 1101
1	2	1110 01	2	8	1111 1111 1000 1110
1	3	1111 0110	2	9	1111 1111 1000 1111
1	4	1111 1010 1	2	A	1111 1111 1001 0000
1	5	1111 1110 110	3	1	1101 1
3	2	1111 1000	6	8	1111 1111 1010 1100
3	3	1111 1110 00	6	9	1111 1111 1010 1101
3	4	1111 1111 0111	6	A	1111 1111 1010 1110
3	5	1111 1111 1001 0001	7	1	1111 010
3	6	1111 1111 1001 0010	7	2	1111 1111 000
3	7	1111 1111 1001 0011	7	3	1111 1111 1010 1111
3	8	1111 1111 1001 0100	7	4	1111 1111 1011 0000
3	9	1111 1111 1001 0101	7	5	1111 1111 1011 0001
3	A	1111 1111 1001 0110	7	6	1111 1111 1011 0010
4	1	1110 10	7	7	1111 1111 1011 0011
4	2	1111 1011 0	7	8	1111 1111 1011 0100
4	3	1111 1111 1001 0111	7	9	1111 1111 1011 0101
4	4	1111 1111 1001 1000	7	A	1111 1111 1011 0110
4	5	1111 1111 1001 1001	8	1	1111 1001
4	6	1111 1111 1001 1010	8	2	1111 1111 1011 0111
4	7	1111 1111 1001 1011	8	3	1111 1111 1011 1000
4	8	1111 1111 1001 1100	8	4	1111 1111 1011 1001
4	9	1111 1111 1001 1101	8	5	1111 1111 1011 1010
4	A	1111 1111 1001 1110	8	6	1111 1111 1011 1011
5	1	1110 11	8	7	1111 1111 1011 1100
5	2	1111 1110 01	8	8	1111 1111 1011 1101
5	3	1111 1111 1001 1111	8	9	1111 1111 1011 1110
5	4	1111 1111 1010 0000	8	A	1111 1111 1011 1111
5	5	1111 1111 1010 0001	9	1	1111 1011 1
5	6	1111 1111 1010 0010	9	2	1111 1111 1100 0000
5	7	1111 1111 1010 0011	9	3	1111 1111 1100 0001
5	8	1111 1111 1010 0100	9	4	1111 1111 1100 0010

Appendix C.2 *(cont.)*

Run	Size	Code word	Run	Size	Code word
5	9	1111 1111 1010 0101	9	5	1111 1111 1100 0011
5	A	1111 1111 1010 0110	9	6	1111 1111 1100 0100
6	1	1111 001	9	7	1111 1111 1100 0101
6	2	1111 1110 111	9	8	1111 1111 1100 0110
6	3	1111 1111 1010 0111	9	9	1111 1111 1100 0111
6	4	1111 1111 1010 1000	9	A	1111 1111 1100 1000
6	5	1111 1111 1010 1001	A	1	1111 1100 0
6	6	1111 1111 1010 1010	A	2	1111 1111 1100 1001
6	7	1111 1111 1010 1011	A	3	1111 1111 1100 1010
A	4	1111 1111 1100 1011	D	A	1111 1111 1110 1100
A	5	1111 1111 1100 1100	E	1	1111 1111 1000 00
A	6	1111 1111 1100 1101	E	2	1111 1111 1110 1101
A	7	1111 1111 1100 1110	E	3	1111 1111 1110 1110
A	8	1111 1111 1100 1111	E	4	1111 1111 1110 1111
A	9	1111 1111 1101 0000	E	5	1111 1111 1111 0000
A	A	1111 1111 1101 0001	E	6	1111 1111 1111 0001
B	1	1111 1100 1	E	7	1111 1111 1111 0010
B	2	1111 1111 1101 0010	E	8	1111 1111 1111 0011
B	3	1111 1111 1101 0011	E	9	1111 1111 1111 0100
B	4	1111 1111 1101 0100	E	A	1111 1111 1111 0101
B	5	1111 1111 1101 0101	F	0	1111 1110 10 (ZRL)
B	6	1111 1111 1101 0110	F	1	1111 1111 1000 011
B	7	1111 1111 1101 0111	F	2	1111 1111 1111 0110
B	8	1111 1111 1101 1000	F	3	1111 1111 1111 0111
B	9	1111 1111 1101 1001	F	4	1111 1111 1111 1000
B	A	1111 1111 1101 1010	F	5	1111 1111 1111 1001
C	1	1111 1101 0	F	6	1111 1111 1111 1010
C	2	1111 1111 1101 1011	F	7	1111 1111 1111 1011
C	3	1111 1111 1101 1100	F	8	1111 1111 1111 1100
C	4	1111 1111 1101 1101	F	9	1111 1111 1111 1101
C	5	1111 1111 1101 1110	F	A	1111 1111 1111 1110
C	6	1111 1111 1101 1111			
C	7	1111 1111 1110 0000			
C	8	1111 1111 1110 0001			
C	9	1111 1111 1110 0010			
C	A	1111 1111 1110 0011			
D	1	1111 1111 001			
D	2	1111 1111 1110 0100			
D	3	1111 1111 1110 0101			
D	4	1111 1111 1110 0110			
D	5	1111 1111 1110 0111			
D	6	1111 1111 1110 1000			
D	7	1111 1111 1110 1001			
D	8	1111 1111 1110 1010			
D	9	1111 1111 1110 1011			

C.3 VLC Table for DCT Coefficients in H.261

Run	Level	Code word	Run	Level	Code word
EOB		10	3	4	0000 0000 1001 1s
0	1	1s	4	1	0011 0s
		(If first coefficient)	4	2	0000 0011 11s
0	1	11s (If not	4	3	0000 0001 0010 s
		first coefficient)	5	1	0001 11s
0	2	0100 s	5	2	0000 0010 01s
0	3	0010 1s	5	3	0000 0000 1001 0s
0	4	0000 110s	6	1	0001 01s
0	5	0010 0110 s	6	2	0000 0001 1110 s
0	6	0010 0001 s	7	1	0001 00s
0	7	0000 0010 10s	7	2	0000 0001 0101 s
0	8	0000 0001 1101 s	8	1	0000 111s
0	9	0000 0001 1000 s	8	2	0000 0001 0001 s
0	10	0000 0001 0011 s	9	1	0000 101s
0	11	0000 0001 0000 s	9	2	0000 0000 1000 1s
0	12	0000 0000 1101 0s	10	1	0010 0111 s
0	13	0000 0000 1100 1s	10	2	0000 0000 1000 0s
0	14	0000 0000 1100 0s	11	1	0010 0011 s
0	15	0000 0000 1011 1s	12	1	0010 0010 s
1	1	011s	13	1	0010 0000 s
1	2	0001 10s	14	1	0000 0011 10s
1	3	0010 0101 s	15	1	0000 0011 01s
1	4	0000 0011 00s	16	1	0000 0010 00s
1	5	0000 0001 1011 s	17	1	0000 0001 1111 s
1	6	0000 0000 1011 0s	18	1	0000 0001 1010 s
1	7	0000 0000 1010 1s	19	1	0000 0001 1001 s
2	1	0101 s	20	1	0000 0001 0111 s
2	2	0000 100s	21	1	0000 0001 0110 s
2	3	0000 0010 11s	22	1	0000 0000 1111 1s
2	4	0000 0001 0100 s	23	1	0000 0000 1111 0s
2	5	0000 0000 1010 0s	24	1	0000 0000 1110 1s
3	1	0011 1s	25	1	0000 0000 1110 0s
3	2	0010 0100 s	26	1	0000 0000 1101 1s
3	3	0000 0001 1100 s	ESC		0000 01

Note: The last bit 's' denotes the sign of the level, 0 for positive, 1 for negative. EOB: end-of-block code, ESC: escape code.

C.4 VLC Table for DCT Coefficients in MPEG-1 (Table Zero in MPEG-2)

Run	Level	Code word	Run	Level	Code word
EOB		10	0	30	0000 0000 0100 01s
0	1	1s	0	31	0000 0000 0100 00s
		(If first coefficient)	0	32	0000 0000 0011 000s
0	1	11s (If not	0	33	0000 0000 0010 111s
		first coefficient	0	34	0000 0000 0010 110s
0	2	0100 s	0	35	0000 0000 0010 101s
0	3	0010 1s	0	36	0000 0000 0010 100s
0	4	0000 110s	0	37	0000 0000 0010 011s
0	5	0010 0110 s	0	38	0000 0000 0010 010s
0	6	0010 0001 s	0	39	0000 0000 0010 001s
0	7	0000 0010 10s	0	0	0000 0000 0010 000s
0	8	0000 0001 1101 s	1	1	011s
0	9	0000 0001 1000 s	1	2	0001 10s
0	10	0000 0001 0011 s	1	3	0010 0101 s
0	11	0000 0001 0000 s	1	4	0000 0011 00s
0	12	0000 0000 1101 0s	1	5	0000 0001 1011 s
0	13	0000 0000 1100 1s	1	6	0000 0000 1011 0s
0	14	0000 0000 1100 0s	1	7	0000 0000 1010 1s
0	15	0000 0000 1011 1s	1	8	0000 0000 0011 111s
0	16	0000 0000 0111 11s	1	9	0000 0000 0011 110s
0	17	0000 0000 0111 10s	1	10	0000 0000 0011 101s
0	18	0000 0000 0111 01s	1	11	0000 0000 0011 100s
0	19	0000 0000 0111 00s	1	12	0000 0000 0011 011s
0	20	0000 0000 0110 11s	1	13	0000 0000 0011 010s
0	21	0000 0000 0110 10s	1	14	0000 0000 0011 001s
0	22	0000 0000 0110 01s	1	15	0000 0000 0001 0011 s
0	23	0000 0000 0110 00s	1	16	0000 0000 0001 0010 s
0	24	0000 0000 0101 11s	1	17	0000 0000 0001 0001 s
0	25	0000 0000 0101 10s	1	18	0000 0000 0001 0000 s
0	26	0000 0000 0101 01s	2	1	0101 s
0	27	0000 0000 0101 00s	2	2	0000 100s
0	28	0000 0000 0100 11s	2	3	0000 0010 11s
0	29	0000 0000 0100 10s	2	4	0000 0001 0100 s
2	5	0000 0000 1010 0s	18	1	0000 0001 1010 s
3	1	0011 1s	19	1	0000 0001 1001 s
3	2	0010 0100 s	20	1	0000 0001 0111 s
3	3	0000 0001 1100 s	21	1	0000 0001 0110 s
3	4	0000 0000 1001 1s	22	1	0000 0000 1111 1s
4	1	0011 0s	23	1	0000 0000 1111 0s
4	2	0000 0011 11s	24	1	0000 0000 1110 1s
4	3	0000 0001 0010 s	25	1	0000 0000 1110 0s
5	1	0001 11s	26	1	0000 0000 1101 1s

Appendix C.4 *(cont.)*

Run	Size	Code word	Run	Size	Code word
5	2	0000 0010 01s	27	1	0000 0000 0001 1111 s
5	3	0000 0000 1001 0s	28	1	0000 0000 0001 1110 s
6	1	0001 01s	29	1	0000 0000 0001 1101 s
6	2	0000 0001 1110 s	30	1	0000 0000 0001 1100 s
6	3	0000 0000 0001 0100 s	31	1	0000 0000 0001 1011 s
7	1	0001 00s	ESC		0000 01
7	2	0000 0001 0101 s			
8	1	0000 111s			
8	2	0000 0001 0001 s			
9	1	0000 101s			
9	2	0000 0000 1000 1s			
10	1	0010 0111 s			
10	2	0000 0000 1000 0s			
11	1	0010 0011 s			
11	2	0000 0000 0001 1010 s			
12	1	0010 0010 s			
12	2	0000 0000 0001 1001 s			
13	1	0010 0000 s			
13	2	0000 0000 0001 1000 s			
14	1	0000 0011 10s			
14	2	0000 0000 0001 0111 s			
15	1	0000 0011 01s			
15	2	0000 0000 0001 0110 s			
16	1	0000 0010 00s			
16	2	0000 0000 0001 0101 s			
17	1	0000 0001 1111 s			

Note: The last bit 's' denotes the sign of the level, 0 for positive, 1 for negative.
EOB: end-of-block; ESC: escape.

C.5 VLC Table for DCT Coefficients in MPEG-2 (Table 1 for Intra Coding)

Run	Level	Code word	Run	Level	Code word
EOB		0110	0	8	1111 011s
0	1	10s	0	9	1111 100s
1	1	010 s	0	10	0010 0011 s
0	2	110 s	0	11	0010 0010 s
2	1	0010 1s	1	5	0010 0000 s
0	3	0111 s	2	4	0000 0011 00s
3	1	0011 1s	3	3	0000 0001 1100 s
4	1	0001 10s	4	3	0000 0001 0010 s
1	2	0011 0s	6	2	0000 0001 1110 s
5	1	0001 11s	7	2	0000 0001 0101 s
6	1	0000 110s	8	2	0000 0001 0001 s
7	1	0000 100s	17	1	0000 0001 1111 s
0	4	1110 0s	18	1	0000 0001 1010 s
2	2	0000 111s	19	1	0000 0001 1001 s
8	1	0000 101s	20	1	0000 0001 0111 s
9	1	1111 000s	21	1	0000 0001 0110 s
ESC		0000 01	0	12	1111 1010 s
0	5	1110 1s	0	13	1111 1011 s
0	6	0001 01s	0	14	1111 1110 s
1	3	1111 001s	0	15	1111 1111 s
3	2	0010 0110 s	1	6	0000 0000 1011 0s
10	1	1111 010s	1	7	0000 0000 1010 1s
11	1	0010 0001 s	2	5	0000 0000 1010 0s
12	1	0010 0101 s	3	4	0000 0000 1001 1s
13	1	0010 0100 s	5	3	0000 0000 1001 0s
0	7	0001 00s	9	2	0000 0000 1000 1s
1	4	0010 0111 s	10	2	0000 0000 1000 0s
2	3	1111 1100 s	22	1	0000 0000 1111 1s
4	2	1111 1101 s	23	1	0000 0000 1111 0s
5	2	0000 0010 0s	24	1	0000 0000 1110 1s
14	1	0000 0010 1s	25	1	0000 0000 1110 0s
15	1	0000 0011 1s	26	1	0000 0000 1101 1s
16	1	0000 0011 01s	0	16	0000 0000 0111 11s
0	17	0000 0000 0111 10s	11	2	0000 0000 0001 1010 s
0	18	0000 0000 0111 01s	12	2	0000 0000 0001 1001 s
0	19	0000 0000 0111 00s	13	2	0000 0000 0001 1000 s
0	20	0000 0000 0110 11s	14	2	0000 0000 0001 0111 s
0	21	0000 0000 0110 10s	15	2	0000 0000 0001 0110 s
0	22	0000 0000 0110 01s	16	2	0000 0000 0001 0101 s
0	23	0000 0000 0110 00s	27	1	0000 0000 0001 1111 s
0	24	0000 0000 0101 11s	28	1	0000 0000 0001 1110 s
0	25	0000 0000 0101 10s	29	1	0000 0000 0001 1101 s

Appendix C.5 *(cont.)*

Run	Size	Code word	Run	Size	Code word
0	26	0000 0000 0101 01s	30	1	0000 0000 0001 1100 s
0	27	0000 0000 0101 00s	31	1	0000 0000 0001 1011 s
0	28	0000 0000 0100 11s			
0	29	0000 0000 0100 10s			
0	30	0000 0000 0100 01s			
0	31	0000 0000 0100 00s			
0	32	0000 0000 0011 000s			
0	33	0000 0000 0010 111s			
0	34	0000 0000 0010 110s			
0	35	0000 0000 0010 101s			
0	36	0000 0000 0010 100s			
0	37	0000 0000 0010 011s			
0	38	0000 0000 0010 010s			
0	39	0000 0000 0010 001s			
0	40	0000 0000 0010 000s			
1	8	0000 0000 0011 111s			
1	9	0000 0000 0011 110s			
1	10	0000 0000 0011 101s			
1	11	0000 0000 0011 100s			
1	12	0000 0000 0011 011s			
1	13	0000 0000 0011 010s			
1	14	0000 0000 0011 001s			
1	15	0000 0000 0001 0011 s			
1	16	0000 0000 0001 0010 s			
1	17	0000 0000 0001 0001 s			
1	18	0000 0000 0001 0000 s			
6	3	0000 0000 0001 0100 s			

Note: The last bit 's' denotes the sign of the level, 0 for positive, 1 for negative.
EOB: end-of-block, ESC: escape.

C.6 3D VLC Table for DCT Coefficients in H.263

Last	Run	Level	Bits	Code word
0	0	1	3	10s
0	0	2	5	1111s
0	0	3	7	0101 01s
0	0	4	8	0010 111s
0	0	5	9	0001 1111s
0	0	6	10	0001 0010 1s
0	0	7	10	0001 0010 0s
0	0	8	11	0000 1000 01s
0	0	9	11	0000 1000 00s
0	0	10	12	0000 0000 111s
0	0	11	12	0000 0000 110s
0	0	12	12	0000 0100 00s
0	1	1	4	110s
0	1	2	7	0101 00s
0	1	3	9	0001 1110s
0	1	4	11	0000 0011 11s
0	1	5	12	0000 0100 001s
0	1	6	13	0000 0101 0000s
0	2	1	5	1110s
0	2	2	9	0001 1101s
0	2	3	11	0000 0011 10s
0	2	4	13	0000 0101 0001s
0	3	1	6	0110 1s
0	3	2	10	0001 0001 1s
0	3	3	11	0000 0011 01s
0	4	1	6	0110 0s
0	4	2	10	0001 0001 0s
0	4	3	13	0000 0101 0010s
0	5	1	6	0101 1s
0	5	2	11	0000 0011 00s
0	5	3	13	0000 0101 0011s
0	6	1	7	0100 11s
0	6	2	11	0000 0010 11s
0	6	3	13	0000 0101 0100s
0	7	1	7	0100 10s
0	7	2	11	0000 0010 10s
0	8	1	7	0100 01s
0	8	2	11	0000 0010 01s
0	9	1	7	0100 00s
0	9	2	11	0000 0010 00s
0	10	1	8	0010 110s
0	10	2	13	0000 0101 0101 s
0	11	1	8	0010 101s
0	12	1	8	0010 100s

Appendix C.6 *(cont.)*

Last	Run	Level	Bits	Code word
0	13	1	9	0001 1100s
0	14	1	9	0001 1011s
0	15	1	10	0001 0000 1s
0	16	1	10	0001 0000 0s
0	17	1	10	0000 1111 1s
0	18	1	10	0000 1111 0s
0	19	1	10	0000 1110 1s
0	20	1	10	0000 1110 0s
0	21	1	10	0000 1101 1s
0	22	1	10	0000 1101 0s
0	23	1	12	0000 0100 010s
0	24	1	12	0000 0100 011s
0	25	1	13	0000 0101 0110 s
0	26	1	13	0000 0101 0111 s
1	0	1	5	0111s
1	0	2	10	0000 1100 1s
1	0	3	12	0000 0000 101s
1	1	1	7	0011 11s
1	1	2	12	0000 0000 100s
1	2	1	7	0011 10s
1	3	1	7	0011 01s
1	4	1	7	0011 00s
1	5	1	8	0010 011s
1	6	1	8	0010 010s
1	7	1	8	0010 001s
1	8	1	8	0010 000s
1	9	1	9	0001 1010s
1	10	1	9	0001 1001s
1	11	1	9	0001 1000s
1	12	1	9	0001 0111s
1	13	1	9	0001 0110s
1	14	1	9	0001 0101s
1	15	1	9	0001 0100s
1	16	1	9	0001 0011s
1	17	1	10	0000 1100 0s
1	18	1	10	0000 1011 1s
1	19	1	10	0000 1011 0s
1	20	1	10	0000 1010 1s
1	21	1	10	0000 1010 0s
1	22	1	10	0000 1001 1s
1	23	1	10	0000 1001 0s
1	24	1	10	0000 1000 1s
1	25	1	11	0000 0001 11s
1	26	1	11	0000 0001 10s
1	27	1	11	0000 0001 01s

Appendix C.6 *(cont.)*

Last	Run	Level	Bits	Code word
1	28	1	11	0000 0001 00s
1	29	1	12	0000 0100 100s
1	30	1	12	0000 0100 101s
1	31	1	12	0000 0100 110s
1	32	1	12	0000 0100 111s
1	33	1	13	0000 0101 1000s
1	34	1	13	0000 0101 1001s
1	35	1	13	0000 0101 1010s
1	36	1	13	0000 0101 1011s
1	37	1	13	0000 0101 1100s
1	38	1	13	0000 0101 1101s
1	39	1	13	0000 0101 1110s
1	40	1	13	0000 0101 1111s
ESC			7	0000 011

Note: The last bit 's' denotes the sign of the level, 0 for positive 1 for negative. There is no end-of-Block code. ESC denotes the escape code.

C.7 VLC Table for DCT Coefficients in ITU-R CMTT.723

Level or zero-run			
Luminance	Common	Chrominance	Code word
	1		01
	2		11 00
	0∗1		11 01
	0∗3		11 10 00
3		0∗5	11 10 01
4		0∗7	11 11 00
	EOB_1		11 11 01
0∗5		3	11 10 10 00
0∗7		4	11 10 10 01
	0∗9		11 10 11 00
	0∗11		11 10 11 01
	5		11 11 10 00
6		0∗13	11 11 10 01
7		0∗15	11 11 11 00
8		0∗17	11 11 11 01
0∗13		6	11 10 10 10 00
0∗15		7	11 10 10 10 01
0∗17		8	11 10 10 11 00
	0∗19		11 10 10 11 01

Appendix C.7 *(cont.)*

Level or zero-run			
Luminance	Common	Chrominance	Code word
	0*21		11 10 11 10 00
	0*23		11 10 11 10 01
	0*25		11 10 11 11 00
	0*27		11 10 11 11 01
	9		11 11 10 10 00
	10		11 11 10 10 01
	11		11 11 10 11 00
	12		11 11 10 11 01
	13		11 11 11 10 00
	14		11 11 11 10 01
	15		11 11 11 11 00
	16		11 11 11 11 01
	−1		00
	−2		10 01
	0*2		10 00
	0*4		10 11 01
−3		0*6	10 11 00
−4		0*8	10 10 01
	EOB_0		10 10 00
0*6		−3	10 11 11 01
0*8		−4	10 11 11 00
	0*10		10 11 10 01
	0*12		10 11 10 00
	−5		10 10 11 01
−6		0*14	10 10 11 00
−7		0*16	10 10 10 01
−8		0*18	10 10 10 00
0*14		−6	10 11 11 11 01
0*16		−7	10 11 11 11 00
0*18		−8	10 11 11 10 01
	0*20		10 11 11 10 00
	0*22		10 11 10 11 01
	0*24		10 11 10 11 00
	0*26		10 11 10 10 01
	0*28		10 11 10 10 00
	−9		10 10 11 11 01
	−10		10 10 11 11 00
	−11		10 10 11 10 01
	−12		10 10 11 10 00
	−13		10 10 10 11 01
	−14		10 10 10 11 00
	−15		10 10 10 10 01
	−16		10 10 10 10 00

Bibliography

Chapter 1 Introduction

[1] P. Clarkson and H. Stak (Eds.), *Signal processing methods for audio, images and telecommunications*, Orlando, FL: Academic Press, 1995.

[2] Special issue on still and video/image compression, *J. of Visual Commun. and Image Representation*, vol. 5, Dec. 1994.

[3] Special issue on image sequence compression, *IEEE Trans. Image Process.*, vol. 3, Sept. 1994.

[4] Special issue on advances in image and video compression, *Proc. IEEE*, vol. 83, pp. 135–340, Feb. 1995.

[5] Special issue on digital television, *Proc. IEEE*, vol. 83, April 1995.

[6] Conference on Wavelet Applications (part of 1995 Intl. Symp. on Aerospace/Defense, Sensing and Dual-use Photonics), Orlando, FL, April 1995.

[7] H. M. Hang and J. W. Woods, *Handbook of visual communications*, Orlando, FL: Academic Press, 1995.

[8] K. Sayood, *Introduction to data compression*, San Francisco, CA: Morgan Kaufmann Publishers, 1995.

[9] J. Watkinson, *Compression in video and audio*, Boston, MA: Focal Press, UK/Butterworth-Heineman, 1995.

[10] V. Bhaskaran and K. Konstantinides, *Image and video compression standards*, Hingham, MA: Kluwer Academic, 1995.

[11] W. Kou, *Digital image compression—algorithms and standards*, Hingham, MA: Kluwer Academic, 1995.

[12] B. Furht, S. W. Smoliar, and H. Zhang, *Video and image processing in multimedia systems*, Hingham, MA: Kluwer Academic, 1995.

Chapter 2 Color formats

[13] K. Jack, *Video demystified*, Solana Beach, CA: Hightext Pub. Inc., 1993.

[14] J. Slater, *Modern television systems to HDTV and beyond*, London, UK: Pitman Pub., 1991.

[15] ITU-R Recommendation BT.601, *Encoding parameters of digital television for studios*, 1982.

[16] SMPTE standard for television, SMPTE 240M-1988, *Signal parameters for 1125/60 high-definition production system*, Mar. 14, 1988.

[17] A. K. Jain, *Fundamentals of digital image processing*, Englewood Cliffs, NJ: Prentice Hall, 1984.

[18] J. Watkinson, *The art of digital video*, London, UK: Focal Press, 1990.

Chapter 3 Quantization

[19] A. Gersho, "Quantization," *IEEE Commun. Magazine*, vol. 15, pp. 16–29, Sept. 1977.

[20] P. A. Wintz and A. J. Kurtenbach, "Waveform error control in PCM telemetry," *IEEE Trans. Inform. Theory*, vol. IT-14, pp. 650–661, Sept. 1968.

[21] P. Pirsch, "Design of DPCM quantizers for video signals using subjective tests," *IEEE Trans. Commun.*, vol. COM-29, pp. 990–1000, July 1981.

[22] S. A. Kassam, "Quantization based on the mean-absolute-error criterion," *IEEE Trans. Commun.*, vol. COM-26, pp. 267–270, Feb. 1978.

[23] F. S. Lu and G. L. Wise, "A further investigation of Max's algorithm for optimum quantization," *IEEE Trans. Commun.*, vol. COM-33, pp. 746–750, July 1985.

[24] A. J. Kurtenbach and P. A. Wintz, "Quantizing for noisy channels," *IEEE Trans. Commun.*, vol. COM-17, pp. 291–302, Apr. 1969.

[25] P. F. Panter and W. Dite, "Quantization distortion in pulse-count modulation with nonuniform spacing of levels," *Proc. of the IRE*, pp. 44–48, vol. 39, Jan. 1951.

[26] J. J. Y. Huang and P. M. Schultheiss, "Block quantization of correlated Gaussian random variables," *IEEE Trans. Commun. System*, vol. CS-11, pp. 289–296, Sept. 1963.

[27] J. Max, "Quantizing for minimum distortion," *IRE Trans. Inform. Theory*, vol. IT-6, pp. 16–21, Mar. 1960.

[28] H. Gish and J. N. Pierce, "Asymptotically efficient quantizing," *IEEE Trans. Inform. Theory*, vol. IT-14, pp. 676–683, 1968.

[29] A. Gersho, "Asymptotically optimum block quantization," *IEEE Trans. Inform. Theory*, vol. IT-25, pp. 373–380, July 1979.

[30] Y. Yamada, S. Tazaki, and R. M. Gray, "Asymptotic performance of block quantizers with different distortion measures," *IEEE Trans. Inform. Theory*, vol. IT-26, pp. 6–14, Jan. 1980.
[31] R. M. Gray, J. C. Kieffer and Y. Linde, "Locally optimal block quantizer design," *Inform. Contr.*, vol. 45, pp. 178–198, May 1980.
[32] R. M. Gray and Y. Linde, "Vector quantizers and predictive quantizers for Gauss Markov sources," *IEEE Trans. Commun.*, vol. COM-30, pp. 391–389, Feb. 1982.
[33] S. P. Lloyd, "Least square quantization in PCM," *IEEE Trans. Inform. Theory*, vol. IT-28, pp. 129–137, 1982.
[34] M. D. Paez and T. H. Glisson, "Minimum mean-square error quantization in speech PCM and DPCM systems," *IEEE Trans. Commun.*, vol. COM-20, pp. 225–230, April 1972.
[35] P. Kabal, "Quantizers for the gamma distribution and other symmetrical distributions," *IEEE Trans. Acoust., Speech, and Signal Process.*, vol. ASSP-32, pp. 836–841, 1984.
[36] W. C. Adams, Jr., and C. E. Giesler, "Quantizing characteristics for signals having Laplacian amplitude probability density function," *IEEE Trans. Commun.*, vol. COM-26, pp. 1295–1297, Aug. 1978.
[37] K. N. Ngan and K. S. Leong, "Fast convergence method for Lloyd-Max quantizer design," *Electronics Letters*, vol. 22, pp. 844–846, July 1986.
[38] K. H. Tzou, "Embedded Max quantization," *ICASSP '86*, pp. 505–508, Tokyo, Apr. 1986.
[39] P. F. Swaszek and J. B. Thomas, "Design of quantizers from histograms," *IEEE Trans. Commun.*, vol. COM-32, pp. 240–245, Mar. 1984.
[40] K. H. Tzou, "A fast computational approach to the design of block quantization," *IEEE Trans. Acoust, Speech, and Signal Process.*, vol. ASSP-35, pp. 235–237, Feb. 1987.
[41] V. R. Algazi, "Useful approximations to optimal quantization," *IEEE Trans. Commun. Tech.*, vol. COM-14, pp. 297–301, June 1966.
[42] W. Mauersberger, "Experimental results on the performance of mismatched quantizers," *IEEE Trans. Inform. Theory*, vol. IT-25, pp. 381–386, July 1979.
[43] W. G. Bath and V. D. Vandlinde, "Robust memoryless quantization for minimum signal distortion," *IEEE Trans. Inform. Theory*, vol. IT-28, pp. 296–306, Mar. 1982.
[44] K. Kroschel, "A comparison of quantizers optimized for corrupted and uncorrupted input signals," *Signal Process.*, vol. 12, pp. 169–176, Mar. 1987.
[45] L. Wang and M. Goldberg, "Progressive image transmission by multistage transform coefficient quantization," *ICC '86*, pp. 419–423, Toronto, Canada, June 1986.

[46] Z. Orbach, "New types of adaptive quantizers," *IEEE Trans. Acoust., Speech, and Signal Process.*, vol. ASSP-32, pp. 1006–1013, Oct. 1984.

[47] R. Schafer, "Design of adaptive and nonadaptive quantizers using subjective criteria," *Signal Process.*, vol. 5, pp. 333–345, 1983.

[48] B. Girod *et al.*, "A subjective evaluation of noise-shaping quantization for adaptive intra-/interframe DPCM coding of color television signals," *IEEE Trans. Commun.*, vol. 36, pp. 332–346, Mar. 1988.

[49] A. K. Jain, *Fundamentals of digital image processing*, Englewood Cliffs, NJ: Prentice Hall, 1984.

[50] N. S. Jayant and P. Noll, *Digital coding of waveforms, principles and applications to speech and video*, Englewood Cliffs, NJ: Prentice Hall, 1984.

[51] R. Reininger and J. Gibson, "Distribution of the two dimensional DCT coefficients for images," *IEEE Trans. Commun.*, vol. COM-31, pp. 835–839, June 1983.

[52] K. N. Ngan, "Adaptive transform coding of video signals," *Proc. IEE*, vol. 129, pp. 28–40, Feb. 1982.

Chapter 4 Predictive coding

[53] R. C. Brainard and J. H. Othmer, "VLSI implementation of a DPCM compression algorithm for digital TV," *IEEE Trans. Commun.*, vol. COM-35, pp. 854–856, Aug. 1987.

[54] R. C. Brainard and A. Puri, "Compact coder for component color television," *IEEE Trans. Commun.*, vol. COM-38, pp. 223–232, Feb. 1990.

[55] X. Song, T. Viero, and Y. Neuvo, "Interframe DPCM with robust median-based predictors for transmission of image sequences over noisy channels," *submitted to IEEE Trans. Image Process.*, vol. 3, p. 244, Mar. 1994.

[56] T. Ishiguro et al., "Composite interframe coding of NTSC color TV signals," *NTC*, pp. 6.4-1–6.4-4, Dallas, TX, 1976.

[57] P. Tischer, "Optimal predictors for image compression," *submitted to IEEE Trans. Image Process.*, vol. 3, p. 225, Mar. 1994.

[58] D. G. Daut, D. Zhao, and J. C. Wu, "Double predictor differential pulse code modulation algorithm for image data compression," *Optical Engineering*, vol. 32, pp. 1514–1523, July 1993.

[59] N. S. Jayant and P. Noll, *Digital coding of waveforms—principles and applications to speech and video*, Englewood Cliffs, NJ: Prentice Hall, 1984.

[60] J. B. O'Neal Jr., "Differential pulse code modulation with entropy coding," *IEEE Trans. Inform. Theory*, vol. IT-21, pp. 169–174, Mar. 1976.

[61] B. S. Atal and M. R. Schroeder, "Predictive coding of speech signals and subjective error criteria," *IEEE Trans. Acoust., Speech, and Signal Process.*, vol. ASSP-27, pp. 247–254, June 1979.

[62] KDD R&D Labs, Proposal submitted to T1Y1.1, *Specification of 45 Mbps universal digital TV codec*, 1990.

[63] S. Sabri and D. Lemay, Northern Telecom, Proposal submitted to T1Y1.1, *Digital encoding of system M-NTSC television signals for broadcast quality transmission at the DS3 rate*, 1989.

[64] Proposal submitted to T1Y1.1, *ABL Engineering algorithm disclosure*, 1990.

[65] P. Pirsch, "Design of DPCM quantizers for video signals using subjective tests," *IEEE Trans. Commun.*, vol. COM-29, pp. 990–1000, July 1981.

[66] T. Koga, NEC proposal submitted to T1Y1.1, *Algorithms for broadcast quality encoding of NTSC television for transmission at 44.736 Mb/s (DS3)*, 1989.

[67] J. O. Limb and C. B. Rubinstein, "On the design of quantizers for DPCM coders, A functional relationship between visibility, probability and masking," *IEEE Trans. Commun.*, vol. COM-26, pp. 573–578, May 1978.

[68] W. J. Ho and W. T. Chang, "Multiresolution interpolative DPCM for data compression," *SPIE/VCIP*, vol. 2501, pp. 524–539, Taipei, Taiwan, May 1995.

[69] C. C. Lien, C. L. Huang and I. C. Chang, "New ADPCM image coder using frequency weighted directional filters," *SPIE/VCIP*, vol. 2501, pp. 647–657, Taipei, Taiwan, May 1995.

[70] C. H. Chang and J. J. Leon, "Detection and correction of transmission errors in DPCM images," *IEEE Trans. CSVT*, vol. 5, pp. 166–171, Apr. 1995.

[71] F. Arp, "BIGCHAIR-DPCM, A new method for visually irrelevant coding of pictorial information," *ISCAS '88*, Helsinki, Finland, pp. 231–234, June 1988.

[72] J. J. Hwang, M. H. Lee, and K. R. Rao, "Transmission noise eliminations in BDPCM image," *IEEE Trans. Consumer Electronics*, vol. 39, pp. 151–158, Aug. 1993.

[73] B. S. Atal, "Predictive coding of speech at low bit rates," *IEEE Trans. Commun.*, vol. COM-30, pp. 600–614, Apr. 1982.

[74] B. Girod et al., "A subjective evaluation of noise shaping quantization for adaptive intra/interframe DPCM coding of colour television signals," *IEEE Trans. Commun.*, vol. 36, pp. 332–346, Mar. 1988.

Chapter 5 Transform coding

[75] P. Lee and F. Y. Huang, "Restructured recursive DCT and DST algorithms," *IEEE Trans. Signal Process.*, vol. 42, pp. 1600–1609, July 1994.

[76] A. N. Skodras, "Fast discrete cosine transform pruning," *IEEE Trans. Signal Process.*, vol 42, pp. 1833–1836, July 1994.

[77] N. Ahmed, T. Natarajan and K. R. Rao, "Discrete cosine transform," *IEEE Trans. Comput.*, vol. C-23, pp. 90–93, Jan. 1974.

[78] K. R. Rao and P. Yip, *Discrete cosine transform: Algorithms, Advantages, Applications*, New York, NY: Academic Press, 1990.

[79] Y. S. Hsu *et al.*, "Pattern recognition experiments in the Mandala/cosine domain," *IEEE Trans. Pattern Anal. Machine Intell.*, vol. PAMI-5, pp. 512–520, Sept. 1983.

[80] F. Arguello and E. L. Zapata, "Fast cosine transform based on successive doubling method," *Electronics Letters*, vol. 26, pp. 1616–1618, Sept. 1990.

[81] S. C. Chan and K. L. Ho, "A new two-dimensional fast cosine transform algorithm," *IEEE Trans. Signal Process.*, vol. 39, pp. 481–485, Feb. 1991.

[82] Z. Wang, "Pruning the fast discrete cosine transform," *IEEE Trans. Commun.*, vol. COM-39, pp. 640–643, May 1991.

[83] A. N. Skodras and A. G. Constantinides, "Efficient input-reordering algorithms for fast DCT," *Electronics Letters*, vol. 27, pp. 1973–1975, Oct. 1991.

[84] N. I. Cho and S. U. Lee, "Fast algorithm and implementation of 2D discrete cosine transform," *IEEE Trans. Circuits, Systems*, vol. 38, pp. 297–305, Mar. 1991.

[85] N. I. Cho, I. D. Yun, and S. U. Lee, "On the regular structure for the fast 2-D DCT algorithm," *IEEE Trans. Circuits, Systems*, vol. 40, pp. 259–266, Apr. 1993.

[86] Z. Cvetkovic and M. V. Popovic, "New fast recursive algorithms for the computation of discrete cosine and sine transforms," *IEEE Trans. Signal Process.*, vol. 40, pp. 2083–2086, Aug. 1992.

[87] E. Feig and S. Winograd, "Fast algorithms for the discrete cosine transform," *IEEE Trans. Signal Process.*, vol. 40, pp. 2174–2193, Sept. 1992.

[88] M. A. Haque, "A two-dimensional fast cosine transform," *IEEE Trans. Acoust., Speech, and Signal Process.*, vol. ASSP-33, pp. 1532–1539, Dec. 1985.

[89] M. T. Heideman, "Computation of an odd-length DCT from a real-valued DFT of the same length," *IEEE Trans. Signal Process.*, vol. 40, pp. 54–61, Jan. 1992.

[90] W. Li, "A new algorithm to compute the DCT and its inverse," *IEEE Trans. Signal Process.*, vol. 39, pp. 1305–1313, June 1991.

[91] K. J. R. Liu et al., "Optimal unified architectures for the real-time computation of time-recursive discrete sinusoidal transforms," *IEEE Trans. CSVT*, vol. 4, pp. 168–180, Apr. 1994.

[92] Y. F. Jang, "A 0.8μ 100-MHz 2-D DCT core processor," *ICCE '94*, pp. 330–331, Chicago, IL, June 1994.

[93] J. C. Corlach, P. Penard, and J. L. Sicre, "TCAD: a 27-MHz (8×8) discrete cosine transform chip," *ICASSP '89*, pp. 2429–2432, 1989.

[94] S. Uramoto et al., "A 100-MHz 2-D discrete cosine transform processor," *IEEE J. of Solid State Circuits*, vol. 27, pp. 492–498, Apr. 1992.

[95] C. T. Chiu and K. J. R. Liu, "Parallel implementation of transform based DCT filter bank for video communications," *IEEE Intl. Conf. on Consumer Electronics*, pp. 152–153, Chicago, Il, June 1994.

[96] C. T. Chiu and K. J. R. Liu, "Real-time parallel and fully-pipelined two-dimensional DCT lattice structures with application to HDTV systems," *IEEE Trans. CSVT*, vol. 2, pp. 25–37, Mar. 1992.

[97] J. Canaris, "A VLSI architecture for the real time computation of discrete trigonometric transforms," *J. of VLSI Signal Process.*, vol. 5, pp. 95–104, 1993.

[98] A. Hiregange, E. Subramaniyan, and N. Srinivasa, "1-D FFT and 2-D DCT routines for the Motorola DSP 56100 family," *Intl. Conf. on SP Applications & Technology*, pp. 797–801, Dallas, TX, Oct. 1994.

[99] M. El. Sharkawy and W. Eshmawy, "Pruned discrete cosine transform algorithms for image compression," *Intl. Conf. on SP Applications & Technology*, pp. 980–985, Dallas, TX, Oct. 1994.

[100] N. Subramani and T. Ogunfunmi, "VLSI design and implementation of a DCT chip for video compression using synthesis tools," *37th Midwest Symp. Circuits and Systems*, Lafayette, LA, Aug. 1994.

[101] C. W. Kok, "Fast algorithm for computing discrete cosine transform," *IEEE Workshop on Visual Signal Process. and Commun.*, Piscataway, NJ, Sept. 1994.

[102] A. Jalali and K. R. Rao, "An architecture for hybrid coding of NTSC TV signals," *Comput. and Elec. Engrg.*, vol. 9, pp. 45–51, 1982.

[103] M. J. Narasimha and A. M. Peterson, "On the computation of the discrete cosine transform," *IEEE Trans. Commun.*, vol. COM-26, pp. 934–936, June 1978.

[104] B. G. Lee, "A new algorithm to compute the discrete cosine transform," *IEEE Trans. Acoust., Speech, and Signal Processing*, vol. ASSP-32, pp. 1243–1245, Dec. 1984.

[105] B. G. Lee, "Input and output index mappings for a prime-factor--decomposed computation of discrete cosine transform," *IEEE Trans. Acoust., Speech, and Signal Processing*, vol. ASSP-37, pp. 237–244, Feb. 1989.

[106] H. S. Hou, "A fast recursive algorithm for computing the discrete cosine transform," *IEEE Trans. Acoust., Speech, and Signal Processing*, vol. ASSP-35, pp. 1455–1461, Oct. 1987.

[107] F. A. Kamangar and K. R. Rao, "Fast algorithms for the 2-D discrete cosine transform," *IEEE Trans. Comput.*, vol. C-31, pp. 899–906, Sept. 1982.

[108] P. Duhamel and C. Guillemot, "Polynomial transform computation of 2-D DCT," *ICASSP '90*, pp. 1515–1518, Apr. 1990.

[109] M. Vetterli, "Fast 2D discrete cosine transform," *ICASSP '85*, pp. 1538–1541, Tampa, FL, Mar. 1985.,

[110] P. Lee and F. Y. Huang, "An efficient prime-factor algorithm for the discrete cosine transform and its hardware implementations," *IEEE Trans. Signal Process.*, vol. 42, pp. 1996–2005, Aug. 1994.

[111] Z. J. Mau and F. Jutand, "A high-speed low-cost DCT architecture for HDTV applications," *ICASSP '91*, pp. 1153–1156, Toronto, Canada, May 1991.

[112] K. Chau and I. F. Wang, "VLSI implementation of a 2-D DCT in a compiler," *ICASSP '91*, pp. 1233–1236, Toronto, Canada, May 1991.

[113] Y. Huang, N. P. Galatsanos, and H. M. Dreizen, "Priority DCT coding for image sequences," *ICASSP '91*, pp. 2629–2632, Toronto, Canada, May 1991.

[114] N. I. Cho, I. D. Yun, and S. U. Lee, "A fast algorithm for 2-D DCT," *ICASSP '91*, pp. 2197–2200, Toronto, Canada, May 1991.

[115] E. Linzer and E. Feig, "New scaled DCT algorithms for fused multiply/add architectures," *ICASSP '91*, pp. 2201–2204, Toronto, Canada, May 1991.

[116] R. Gluth, "Regular FFT-related transform kernels for DCT/DST-based polyphase filter banks," *ICASSP '91*, pp. 2205–2208, Toronto, Canada, May 1991.

[117] S. Cucchi and M. Fratti, "A novel architecture for VLSI implementation of the 2-D DCT/IDCT," *ICASSP '92*, vol. 5, pp. 693–696, San Francisco, CA, Mar. 1992.

[118] T. Eude et al., "On the distribution of the DCT coefficients," *ICASSP '94*, vol. 5, pp. 365–368, Adelaide, Australia, Apr. 1994.

[119] L. P. Chau and W. C. Siu, "Efficient formulation for the realization of discrete cosine transform using recursive structure," ICASSP '94, vol. 3, pp. 437–440, Adelaide, Australia, Apr. 1994.

[120] D. P. Lun, "On efficient software realization of the prime factor discrete cosine transform," *ICASSP '94*, vol. 3, pp. 465–468, Adelaide, Australia, Apr. 1994.

[121] A. W. Johnson, M. H. Chan and J. Princess, "Frequency scalable video coding using the MDCT," *ICASSP '94*, vol. 5, pp. 477–480, Adelaide, Australia, Apr. 1994.

[122] F. M. Wang and S. Liu, "Hybrid video coding for low bit-rate applications," *ICASSP '94*, vol. 5, pp. 481–484, Adelaide, Australia, Apr. 1994.

[123] J. H. Kim and G. M. Park, "Two layered DCT based coding scheme for a new digital HD-VCR," *ICCE '94*, pp. 24–25, Chicago, IL, June 1994.

[124] H. Bessalah, "VLSI architectures for fast orthogonal transforms on-line computation," *ICSPAT '93*, pp. 1607–1618, Santa Clara, CA, Sept./Oct. 1993.

[125] C. C. Chuang, "An adaptive image compression scheme based on wavelet transform and hybrid DCT/VQ/SQ," *ICSPAT '94*, pp. 904–909, Dallas, TX, Oct. 1994.

[126] D. K. Mitrakos and A. G. Constantinides, "Layered DCT video coding for embedded data transmission over BISDN," *ICASSP '91*, pp. 2841–2844, Toronto, Canada, May 1991.

[127] A. N. Skodras, "A fast input reordering algorithm for the discrete cosine transform," *ICASSP '92*, vol. 5, pp. 53–56, San Francisco, CA, Mar. 1992.

[128] Y. H. Chan and W. C. Siu, "A cyclic correlated structure for the realization of discrete cosine transform," *IEEE Trans. Circuits and Systems II: Analog and Digital Signal Process.*, vol. 39, pp. 109–113, Feb. 1992.

[129] J. L. Wu, S. H. Hsu, and W. J. Duh, "A novel two-stage algorithm for DCT and IDCT," *IEEE Trans. Acoust., Speech, and Signal Process.*, vol. 40, pp. 1610–1612, June 1992.

[130] R. J. Chen and B. C. Chieu, "A fully adaptive DCT based color image sequence coder," *Signal Processing: Image Communication*, vol. 6, pp. 289–301, Aug. 1994.

[131] A. Neri, G. Russo and P. Talone, "Interblock filtering and downsampling in DCT domain," *Signal Process.: Image Commun.*, vol. 6, pp. 303–317, Aug. 1994.

[132] S. Sridharan, E. Dawson, and B. Goldburg, "Speech encryption in the transform domain," *Electronics Letters*, vol. 26, pp. 655–656, 10 May 1990.

[133] B. Goldburg, S. Sridharan, and E. Dawson, "Design of a discrete cosine transform based speech scrambler," *Electronics Letters*, vol. 27, pp. 613–614, 28 Mar. 1991.

[134] B. Goldburg, S. Sridharan, and E. Dawson, "Design and cryptanalysis of transform-based analog speech scramblers," *IEEE J. Selected Areas in Commun.*, vol. 11, pp. 735–743, June 1993.

[135] B. Goldburg, S. Sridharan, and E. Dawson, "A secure analog speech scrambler using the discrete cosine transform," *Abstracts of Asiacrypt '91*, Japan, pp. 179–183, Nov. 1991.

[136] S. Sridhraran, B. Goldburg, and E. Dawson, "Discrete cosine transform speech encryption system," *Intl. Conf. on Speech Science and Technology*, pp. 472–476, Nov. 1990.

[137] B. Chitprasert and K. R. Rao, "Discrete cosine transform filtering," *ICASSP '90*, pp. 1281–1284, Albuquerque, NM, Apr. 1990.

[138] B. Chitprasert and K. R. Rao, "Discrete cosine transform filtering, *Signal Process.*, vol. 19, pp. 233–245, 1990.

[139] K. R. Rao, "Theory and the applications of the discrete cosine transform," *Jordan Intl. Electrical and Electronic Engrg. Conf.*, pp. 259–264, Amman, Jordan, Apr.–May 1985.

[140] M. Miran and K. R. Rao, "Fast progressive reconstruction of images using the DCT," *Circuits, Systems, and Signal Process.*, vol. 11, pp. 377–385, 1992.

[141] J. Y. Nam and K. R. Rao, "Image coding using a classified DCT/VQ based on two channel conjugate vector quatization," *IEEE Trans. Circuits and Systems for Video Technology*, vol. 1, pp. 325–336, Dec. 1991.

[142] B. Chitprasert and K. R. Rao, "Human visual weighted progressive image transmission," *IEEE Trans. Commun.*, vol. 38, pp. 1040–1044, July 1990.

[143] K. N. Ngan, "Image display techiques using the cosine transform," *IEEE Trans. Acoust., Speech, and Signal Process.*, vol. ASSP-32, pp. 173–177, Feb. 1984.

[144] N. C. Griswold, "Perceptual coding in the cosine transform domain," *Optical Engineering*, vol. 19, pp. 306–311, May–June 1980.

[145] N. B. Nill, "A visual model weighted cosine transform for image compression and quality assessment," *IEEE Trans. Commun.*, vol. COM-33, pp. 551–557, June 1985.

[146] K. N. Ngan, K. S. Leong, and H. Singh, "Cosine transfrom coding incorporating human visual system model," *SPIE Fiber '86*, pp. 165–171, Cambridge, MA, Sept. 1986.

[147] K. N. Ngan and R. J. Clarke, "Lowpass filtering in the cosine transform domain," *ICC '80*, pp. 31.7.1–31.7.5, Seattle, WA, June 1980.

[148] W. H. Chen and S. C. Fralick, "Image enhancement using cosine transform filtering," *Image Science, Math. Symp.*, Monterey, CA, Nov. 1976.

[149] J. I. Guo, C. M. Liu, and C. W. Jen, "A memory based approach to design and implement systolic arrays for DFT and DCT," *ICASSP '92*, vol. 5, pp. 621–624, San Francisco, CA, Mar. 1992.

[150] K. N. Ngan, "Experiments on two-dimensional decimation in time and orthogonal transform domain," *Signal Process.*, vol. 11, pp. 249–263, Oct. 1986.

[151] A. Palau and G. Mirchandani, "Image coding with discrete cosine transforms using efficient energy-based adptive zonal filtering," *ICASSP '94*, vol. 5, pp. 337–340, Adelaide, Australia, Mar. 1994.

[152] J. H. Kim and M. G. Park, "Two-layered DCT based coding scheme for recording digital HDTV Signals," *ICASSP '94*, vol. 5, pp. 425–428, Adelaide, Australia, Mar. 1994.

[153] A. W. Johnson, J. Princen, and M. H. Chan, "Frequency scalable video coding using the MDCT," *ICASSP '94*, vol. 5, pp. 477–480, Adelaide, Australia, Apr. 1994.

[154] F. M. Wang and S. Liu, "Hybrid video coding for low bit rate applications," *ICASSP '94*, vol. 5, pp. 481–484, Adelaide, Australia, Apr. 1994.

[155] K. H. Tzou et al., "Compatible HDTV coding for broadband ISDN," *GLOBECOM '88*, pp. 743–749, Miami, FL, Dec. 1988.

[156] L. Crutcher and J. Grinham, "The networked video jukebox," *IEEE Trans. CSVT*, vol. 4, pp. 105–120, Apr. 1994.

[157] T. K. Tan, K. K. Pang, and K. N. Ngan, "A frequency scalable coding scheme employing pyramid and subband techniques," *IEEE Trans. CSVT*, vol. 4, pp. 203–207, Apr. 1994.

[158] Y. F. Jang, J. N. Kao, and P. C. Huang, "VLSI architecture and implementation of a high speed 2-D DCT core processor," *Intl. Workshop on HDTV '94*, Torino, Italy, Oct. 1994.

[159] M. H. Lee, "On computing 2-D systolic algorithm for discrete cosine transform," *IEEE Trans. Circuits, Systems*, vol. 37, pp. 1321–1323, Oct. 1990.

[160] M. H. Lee and Y. Yasuda, "New 2D systolic array algorithm for DCT/DST," *Electronics Letters*, vol. 25, pp. 1702–1703, July 1989.

[161] C. S. Burrus et al., *Computer based exercises for signal processing using MATLAB*, Englewood Cliffs, NJ: Prentice Hall, 1994.

[162] V. Srinivasan and K. J. R. Liu, "Full custom VLSI implementation of high-speed 2-D DCT/IDCT chip," *ICIP '94*, pp. III. 606–610, Austin, TX, Nov. 1994.

[163] U. V. Koc and K. J. R. Liu, "Discrete-cosine/sine transform based motion estimation," *ICIP '94*, pp. III. 771–775, Austin, TX, Nov.1994.

[164] M. Barazande-Pour and J. W. Mark, "Adaptive MHDCT coding of images," *ICIP '94*, pp. I. 90–94, Austin, TX, Nov. 1994.

[165] A. B. Watson, "Perceptual optimization of DCT color quantization matrices," *ICIP '94*, pp. I. 100–104, Austin, TX, Nov. 1994.

[166] L. V. Oliveira, "Identification of dominant coefficients in DCT image coders using weighted vector quantization," *ICIP '94*, pp. I. 110–113, Austin, TX, Nov. 1994.

[167] D. W. Kang et al., "Sequential vector quantization of directionally decomposed DCT coefficients," *ICIP '94*, pp. I. 114–118, Austin, TX, Nov. 1994.

[168] J. A. Small and K. J. Parker, "Multitoning using generalized projections and the DCT," *ICIP '94*, pp. II. 1027–1031, Austin, TX, Nov. 1994.

[169] C. L. Wang and C. Y. Chen, "A linear systolic array for the 2-D discrete cosine transform," *IEEE Asia-Pacific Conf. on Circuits and Systems*, pp. 4c.2.1–4c.2.6, Taipei, Taiwan, Dec. 1994.

[170] L. Wang, "Error accumulation in hybrid DPCM/DCT video coding," *SPIE/VCIP*, vol. 2308, pp. 343–352, Chicago, IL, Sept. 1994.

[171] G. D. Mandyam, N. Ahmed, and N. Magotra, "DCT-based scheme for lossless image compression," *SPIE/IS&T*, vol. 2419, pp. 474–478, San Jose, CA, Feb. 1995.

[172] L. Kasperovich, "Multiplication free scaled 8×8 DCT algorithm with 530 additions," *SPIE/IS&T*, vol. 2419, pp. 499–504, San Jose, CA, Feb. 1995.

[173] B. Feher, "New inner product algorithm of the 2D DCT," *SPIE/IS&T*, vol. 2419, pp. 436–444, San Jose, CA, Feb. 1995.

[174] V. Ratnakar et al., "Runlength encoding of quantized DCT coefficients," *SPIE/IS&T*, vol. 2419, pp. 398–406, San Jose, CA, Feb. 1995.

[175] H. De Perthuis et al., "Fast VLSI architecture for 8×8 2D DCT," *SPIE/IS&T*, vol. 2419, pp. 429–435, San Jose, CA, Feb. 1995.

[176] N. I. Cho and S. U. Lee, "DCT algorithms for VLSI implementations," *IEEE Trans. Acoust. Speech, and Signal Process.*, vol. ASSP-38, pp. 121–127, Jan. 1990.

[177] M. T. Sun, T. C. Chen, and A. M. Gottlieb, "VLSI implementation of a 16×16 discrete cosine transform," *IEEE Trans. Circuits, Systems*, vol. CAS-36, pp. 610–617, Apr. 1989.

[178] *IEEE Standard specifications for the implementations of 8×8 inverse discrete cosine transform*, IEEE Standard 1180–1990, Mar. 18, 1991.

[179] A. W. Johnson et al., "Filters for drift reduction in frequency scalable video coding schemes," *Electronics Letters*, vol. 30, pp. 471–472, Mar. 17, 1994.

[180] R. Mokry and D. Anastassiou, "Minimal error drift in frequency scalability for motion compensated DCT coding," *IEEE Trans. Circuits and Systems for Video Technology*, vol. 4, pp. 392–406, 1994.

[181] K. N. Ngan et al., "Frequency scalability experiments for MPEG-2 standard," *Asia-Pacific Conf. on Commun.*, pp. 298–301, Taejon, Korea, Aug. 1993.

[182] T. Sikora, T. K. Tan, and K. N. Ngan, "A performance comparison of frequency domain pyramid scalable coding schemes within the MPEG framework," *Picture Coding Symp.*, pp. 16.1–16.2, Lausanne, Switzerland, Mar. 1993.

[183] T. K. Tan, K. K. Pang, and K. N. Ngan, "A frequency scalable coding scheme employing pyramid and subband techniques," *IEEE Trans. Circuits and Systems for Video Technology*, vol. 4, pp. 203–207, Apr. 1994.

[184] Y. S. Tan and K. N. Ngan, "A flexible three layer frequency/SNR scalable video coder," *Australian Telecommun. Networks & Applications Conf.*, Melbourne, Australia, Dec. 1994.

[185] T. K. Tan, *Efficient frequency scalable coders*, Ph.D. Thesis, Monash University, Clayton, Victoria, Australia, 1994.

[186] F. Bellifemine and R. Picco, "Video signal coding with DCT and vector quantization," *IEEE Trans. Commun.*, vol. 42, pp. 200–207, Feb. 1994.

[187] R. de Queiroz and K. R. Rao, "HVS weighted progressive image transmission using the LOT," *J. of Electronic Imaging*, vol. 1, pp. 328–338, July 1992.

[188] W. H. Chen, C. H. Smith, and S. C. Fralick, "A fast computational algorithm for the discrete cosine transform," *IEEE Trans. Commun.*, vol. COM-25, pp. 1004–1009, Sept. 1977.

[189] N. Suehiro and M. Hatori, "Fast algorithms for the DFT and other sinusoidal transforms," *IEEE Trans. Acoust., Speech, and Signal Processing*, vol. ASSP-34, pp. 642–644, June 1986.

[190] C. Loeffler, A. Ligtenberg, and G. S. Moschytz, "Practical fast 1-D DCT algorithm with 11 multiplications," *Intl. Conf. on Acoust., Speech, and Signal Process.*, pp. 988–991, Glasgow, Scotland, May 1989.

[191] S. H. Tan, K. K. Pang, and K. N. Ngan, "Classified perceptual coding with adaptive quantization," *IEEE Trans. Circuits and Systems for Video Technology*, vol. 6, 1996.

[192] J. J. Hwang, G. H. Yang, and M. H. Lee, "Visual weighted DCT coding for monochrome still images," *J. KITE*, vol. 29-B, pp. 93–101, Nov. 1992.

[193] E. Feig and E. Linzer, "Scaled DCTs on input sizes that are composite," *IEEE Trans. Signal Process.*, vol. 43, pp. 43–50, Jan. 1995.

[194] Y. H. Hu and Z. Wu, "An efficient CORDIC array structure for the implementation of discrete cosine transform," *IEEE Trans. Signal Process.*, vol. 43, pp. 331–336, Jan 1995.

[195] E. Scopa et al., "A 2D-DCT low-power architecture for H.261 coders," *ICASSP '95*, pp. 3271–3274, Detroit, MI, May 1995.

[196] J. Bruguera and T. Lang, "2-D DCT using on-line arithmetic," *ICASSP '95*, pp. 3275–3278, Detroit, MI, May 1995.

[197] K. A. Birney and T. R. Fischer, "On the modeling of DCT and subband image data for compression," *IEEE Trans. Signal Process.*, vol. 4, pp. 186–193, Feb. 1995.

[198] J. F. Yang, B. L. Bai, and S. C. Hsia, "An efficient two-dimensional inverse discrete cosine transform algorithm for HDTV receivers," *IEEE Trans. Circuits and Systems for Video Technology*, vol. 5, pp. 25–30, Feb. 1995.

[199] C. L. Wang and C. Y. Chen, "High throughput VLSI architectures for the 1-D and 2-D discrete cosine transforms," *IEEE Trans. Circuits and Systems for Video Technology*, vol. 5, pp. 31–40, Feb. 1995.

[200] Y. Takahashi, T. Nomura, and S. Sakai, "Shape-adaptive DCT for generic coding of video," *IEEE Trans. Circuits and Systems for Video Technology*, vol. 5, pp. 59–62, Feb. 1995.

[201] W. C. Siu, Y. H. Chan, and L. P. Chau, "New 2^N discrete cosine transform algorithm using recursive filter structure," *ICASSP '95*, pp. 1169–1172, Detroit, MI, May 1995.

[202] R. J. Clarke, *Transform coding of images*, New York, NY: Academic Press, 1985.

[203] N. Ahmed and K. R. Rao, *Orthogonal transforms for digital signal processing*, Berlin, Germany: Springer-Verlag, 1975.

[204] M. Rabbani and P. W. Jones, *Digital image compression techniques*, Bellingham, WA: SPIE Optical Engineering Press, 1991.

[205] J. Jeong and J. M. Jo, "Adaptive Huffman coding of 2-D DCT coefficients for image sequence compression," *Signal Process.: Image Commun.*, vol. 7, pp. 1–11, Mar. 1995.

[206] R. Horng and A. J. Ahumada Jr., "Fast DCT block smoothing algorithm," *SPIE/VCIP*, vol. 2501, pp. 28–39, Taipei, Taiwan, May 1995.

[207] B. Deknuydt et al., "Description and evaluation of a non-DCT-based codec," *SPIE Visual Commun. and Image Process.*, vol. 2501, pp. 40–52, Taipei, Taiwan, May 1995.

[208] M. Yuen, H. R. Wu, and K. R. Rao, "Performance evaluation of projection onto convex sets (POCS) loop filtering in generic MC/DPCM/DCT video coding," *SPIE/VCIP*, vol. 2501, pp. 65–75, Taipei, Taiwan, May 1995.

[209] A. Puri, R. V. Kollarits, and B. G. Haskell, "Stereoscopic video compression using temporal scalability," *SPIE/VCIP*, vol. 2501, pp. 745–756, Taipei, Taiwan, May 1995.

[210] G. Deng and L. Cahill, "Isotropic quadratic filter design using the discrete-cosine transform,"*IEEE Intl. Symp. on Circuits and Systems*, vol. 2, pp. 873–876, Seattle, WA, Apr./May 1995.

[211] C. Dre et al., "Alternative architectures for the 2-D DCT algorithm," *IEEE Intl. Symp. on Circuits and Systems*, vol. 3, pp. 2156–2159, Seattle, WA, Apr./May 1995.

[212] C. H. Hsieh, "DCT-based codebook design for vector quantization of images," *IEEE Trans. CSVT*, vol. 2, pp. 401–409, Dec. 1992.

[213] Y. T. Chang and C. L. Wang, "New systolic array implementation of the 2-D discrete cosine transform and its inverse," *IEEE Trans. CSVT*, vol. 5, pp. 150–157, Apr. 1995.

[214] A. Madisetti and A. N. Willson Jr. "A 100 MHz 2-D 8×8 DCT/IDCT processor for HDTV applications," *IEEE Trans. CSVT*, vol. 5, pp. 158–165, Apr. 1995.

[215] M. Zhou, "Vector radix IDCT implementation for MPEG decoding," *IEEE Intl. ASIC Conf.*, Austin, TX, Sept. 1995.

[216] R. Neogi, "Real-time integrated video compression architecture for HDTV and advanced multimedia applications," *IEEE Intl. ASIC Conf.*, Austin, TX, Sept. 1995.

[217] M. Potkonjak and A. Chandrakasan, "Synthesis and selection of DCT algorithms using behavioral systhesis-based algorithm space exploration," *ICIP '95*, vol. 1, pp. 65–68, Washington, DC, Oct. 1995.

[218] V. Bhaskaranand and B. K. Natarajan, "A fast approximate algorithm for scaling down digital images in the DCT domain," *ICIP '95*, vol 2, pp. 241–244, Washington, DC, Oct. 1995.

[219] S. A. Martucci, "Image resizing in the discrete cosine transform domain," *ICIP '95*, vol. 2, pp. 244–247, Washington, D.C., Oct. 1995.

Chapter 6 Hybrid Coding and Motion Compensation

[220] L. D. Vos, "VLSI - architecture for the hierarchial block matching algorithm for HDTV applications," *SPIE/VCIP*, vol. 1360, pp. 398–409, Cambridge, MA, Nov. 1990.

[221] L. D. Vos and M. Stegherr, "Parameterized VLSI architectures for the full search block matching algorithm," *IEEE Trans. Circuits, Systems*, vol. 36, pp. 1309–1316, Oct. 1989.

[222] L. D. Vos, M. Stegherr and T. G. Noll, "VLSI architectures for the full search block matching algorithm," *ICASSP '89*, pp. 1687–1690, Glasgow, Scotland, May 1989.

[223] C. V. Reventlow, "System considerations and the system level design of a chip set for real time TV and HDTV motion estimation," *J. of VLSI Signal Process.*, vol. 5, pp. 237–248, 1993.

[224] D. Bailey et al., "Programmable vision processor/controller," *IEEE Micro*, Programmable (Microcode) BMA-ME (IIT Inc.), pp. 33–39, Oct. 1992.

[225] T. Koga et al., "Motion compensated interframe coding for video conferencing," *NTC '81, National Telecommun. Conf.*, pp. G5.3.1–G5.3.5, New Orleans, LA, Nov.–Dec. 1981.

[226] W. Lee et al., "Real time MPEG video on a single chip multiprocessor," *SPIE/IS&T, Symp. on Electronic Imaging*, vol. 2187, pp. 32–42, San Jose, CA, Feb. 1994.

[227] M. J. Chen et al., "VLSI implementation of real time motion estimation for video compression," *4th VLSI DESIGN/CAD Workshop*, pp. 129–133, Taipei, Taiwan, Aug. 1993.

[228] I. Tamitani et al., "An encoder/decoder chip set for the MPEG video standard," *ICASSP '92*, vol. 5, pp. 661–664, San Francisco, CA, 1992.

[229] B. M. Wang, "An efficient VLSI architecture of hierarchial block matching algorithm in HDTV applications," *Intl. Workshop on HDTV*, Ottawa, Canada, Oct. 1993.

[230] K. Muller et al., "A flexible real time HDTV motion vector estimation chip set based on phase correlation," *Intl. Workshop on HDTV*, Ottawa, Canada, Oct. 1993.

[231] T. Yoshino et al., "A 54 MHz motion estimation engine for real-time MPEG video encoding," *ICCE '94*, pp. 76–77, Chicago, IL, June 1994.

[232] J. Wiseman, "HDTV motion vector decoding with a TMS 320 C30," *Intl. Conf. on SP Applications & Technology*, pp. 55–60, Dallas, TX, Oct. 1994.

[233] Z. He and M. L. Liou, "A new array architecture for motion estimation," *IEEE Workshop Visual Signal Process. and Commun.*, Piscataway, NJ, Sept. 1994.

[234] L. A. de Barros and N. Demassieux, "A VLSI architecture for motion compensated temporal interpolation," *IEEE Workshop Visual Signal Process. and Commun.*, Piscataway, NJ, Sept. 1994.

[235] B. M. Wang, J. C. Yen, and S. Chang, "Zero waiting cycle hierarchical block matching algorithm and its array architectures," *IEEE Trans. CSVT*, vol. 4, pp. 18–28, Feb. 1994.

[236] T. Komarek and P. Pirsch, "Array architectures for block matching algorithms," *IEEE Trans. Circuits, Systems*, vol. 36, pp. 1301–1308, Oct. 1989.

[237] S. K. Azim, "A low cost application specific video codec for consumer videophone," *IEEE Custom Integrated Circuits Conf.*, San Diego, CA, pp. 115–118, May 1994.

[238] K. M. Yang, M. T. Sun, and L. Wu, "A family of VLSI designs for the motion compensation block-matching algorithm," *IEEE Trans. Circuits, Systems*, vol. 36, pp. 1317–1325, Oct. 1989.

[239] C. H. Hsieh and T. P. Lin, "VISI architecture for block-matching motion estimation algorithm," *IEEE Trans. CSVT*, vol. 2, pp. 169–175, 1992.

[240] E. Ogura et al., "A cost effective estimation processor LSI using a simple and efficient algorithm," *IEEE, Intl. Conf. on Consumer Electronics*, pp. 248–249, Chicago, IL, June 1995.

[241] C. V. Reventlow et al., "Chipset for real time HDTV motion vector estimation using the phase-correlation," *Intl. Symp. on Fiber Optic Networks and Video Commun.*, SPIE, Berlin, Germany, Apr. 1993.

[242] F. Scalise, A. Zuccaro and A. Cremonesi, "Motion estimation on very high speed video signals. A flexible and cascadable block matching processor," *Intl. Workshop on HDTV '94*, Torino, Italy, Oct. 1994.

[243] F. M. Yang and R. Laur, "VLSI implementation of the PDC algorithm for full-search block-matching HDTV motion detection," *Intl. Workshop on HDTV '94*, Torino, Italy, Oct. 1994.

[244] S. Kumar et al., "A simple FPGA-based conjugate search motion estimator," *IEEE Asia-Pacific Conf. on Circuits and Systems*, pp. 109–114, Taipei, Taiwan, Dec. 1994.

[245] G. Tziritas and C. Labit, *Motion analysis for image sequence coding*, Amsterdam, Netherlands: Elsevier Science, B.V. 1994.

[246] M. J. Chen, L. G. Chen, and T. D. Chiueh, "One dimensional full-search motion estimation algorithm for video coding," *IEEE Trans. Circuits and Systems for Video Technology*, vol. 4, pp. 504–509, Oct. 1994.

[247] M. I. Sezan and R. L. Lagendijk, *Motion analysis and image sequence processing*, Hingham, MA: Kluwer Academic, 1993.

[248] R. W. Young and N. G. Kingsbury, "Video compression using lapped transforms for motion estimation/compensation and coding," *Optical Engineering*, vol. 7, pp. 1451–1463, July 1993.

[249] R. W. Young and N. G. Kingsbury, "Frequency domain motion estimation using a complex lapped transform," *IEEE Trans. Image Process.*, vol. 2, pp. 2–17, Jan. 1993.

[250] J. R. Jain and A. K. Jain, "Displacement measurement and its application in interframe image coding," *IEEE Trans. Commun.*, vol. COM-29, pp. 1799–1808, Dec. 1981.

[251] K. A. Prabhu and A. N. Netravali, "Pel recursive motion compensated color coding," *Proc. Intl. Conf. Commun.*, pp. 2G.8.1–2G.8.5, Philadelphia, PA, June 1982.

[252] T. Koga et al., "A 1.5 Mb/s interframe codec with motion compensation," *Proc. Intl. Conf. Commun.*, pp. D.8.71–D.8.75, Boston, MA, June 1983.

[253] R. Srinivasan and K. R. Rao, "Predictive coding based on efficient motion estimation," *IEEE Trans. Commun.*, vol. COM-33, pp. 888–896, Aug. 1985.

[254] R. Srinivasan and K. R. Rao, "Motion compensated coder for videoconferencing," *IEEE Trans. Commun.*, vol. COM-35, pp. 297–304, Mar. 1987.

[255] S. Kappagantula and K. R. Rao, "Motion compensated interframe image prediction," *IEEE Trans. Commun.*, vol. COM-33, pp. 1011–1015, Sept. 1985.

[256] Y. Ninomiya and Y. Ohtsuka, "A motion compensated interframe coding scheme for television pictures," *IEEE Trans. Commun.*, Vol. COM-30, pp. 201–211, Jan. 1982.

[257] Y. Ninomiya and Y. Ohtsuka, "A motion compensated interframe coding scheme for television signals," *IEEE Trans. Commun.*, vol. COM-32, pp. 328–334, Mar. 1984.

[258] M. Bierling, "Displacement estimation by hierarchial block matching," *SPIE/VCIP*, vol. 1001, pp. 942–951, 1988.

[259] D. R. Walker and K. R. Rao, "Improved pel recursive motion compensation," *IEEE Trans. Commun.*, vol. COM-33, pp. 1011–1015, Sept. 1985.

[260] H. G. Musmann, P. Pirsch, and H. J. Grallert, "Advances in picture coding," *Proc. IEEE*, vol. 73, pp. 523–548, Apr. 1985.

[261] V. Seferidis and M. Ghanbari, "General approach to block matching motion estimation," *Optical Engineering*, vol. 32, pp. 1464–1474, July 1993.

[262] H. D. Luke, *Model-based motion estimation in image sequences*, PhD Thesis, Aachen University of Technology, Aachen, Germany, 1994.

[263] W. Li and E. Salari, "Successive estimation algorithm for motion estimation," *IEEE Trans. Signal Process.*, vol. 4, pp. 105–107, Jan. 1995.

[264] Y. Huang and X. Zhuang, "An adaptively refined block matching algorithm for motion compensated video coding," *IEEE Trans. Circuits and Systems for Video Technology*, vol. 5, pp. 56–59, Feb. 1995.

[265] L. Wang, "Error accumulation in hybrid DPCM/DCT video coding," *Signal Process.: Image Commun.*, vol. 7, pp. 93–104, Mar. 1995.

[266] J. W. Kim and S. U. Lee, "Rate-distortion optimization between the hierarchical variable block size motion estimation and motion sequence coding," *SPIE/VCIP*, vol. 2501, pp. 822–833, Taipei, Taiwan, May 1995.

[267] S. Kim, J. Chalidabhongse, and C. C. J. Kuo, "Fast motion vector estimation by using spatiotemporal correlation of motion field," *SPIE/VCIP*, vol. 2501, pp. 810–821, Taipei, Taiwan, May 1995.

[268] C. K. Wong and O. C. Au, "Fast motion compensated temporal interpolation for video," *SPIE/VCIP*, vol. 2501, pp. 1108–1115, Taipei, Taiwan, May 1995.

[269] C. Hsin, "Estimation of image motion through mathematical modeling," *SPIE/VCIP*, vol. 2501, pp. 1129–1140, Taipei, Taiwan, May 1995.

[270] C. H. Lin and J. L. Wu, "Fast motion estimation algorithm with adjustable search area," *SPIE/VCIP*, vol. 2501, pp. 1328–1336, Taipei, Taiwan, May 1995.

[271] S. A. Seyedin, "Motion estimation using the Radon transform in dynamic scenes," *SPIE/VCIP*, vol. 2501, pp. 1337–1348, Taipei, Taiwan, May 1995.

[272] S. Kwon and J. Kim, "Simple motion-compensated up-conversion method for the TV/HDTV compatible video coding," *SPIE/VCIP*, vol. 2501, pp. 1349–1357, Taipei, Taiwan, May 1995.

[273] J. Li, P. Y. Cheng, and C. C. J. Kuo, "On the improvements of embedded zerotree wavelet (EZW) coding," *SPIE/VCIP*, vol. 2501, pp. 1490–1501, Taipei, Taiwan, May 1995.

[274] I. K. Kim and R. H. Park, "Block matching algorithm using a genetic algorithm," *SPIE/VCIP*, vol. 2501, pp. 1545–1552, Taipei, Taiwan, May 1995.

[275] P. Nasiopoulos and R. Ward, "A hybrid coding method for digital HDTV signals," *IEEE Intl. Symp. on Circuits and Systems*, vol. 2, pp. 769–772, Seattle, WA, April/May, 1995.

[276] Y. Kim, C. Rim, and B. Min, "A block matching algorithm with 16:1 subsampling and its hardware design," *IEEE Intl. Symp. on Circuits and Systems*, vol. 1, pp. 613–616, Seattle, WA, April/May, 1995.

[277] M. Chen and A. Willson, "A high accuracy predictive logarithmic motion estimation algorithm for video coding," *IEEE Intl. Symp. on Circuits and Systems*, vol. 1, pp. 617–620, Seattle, WA, April/May, 1995.

[278] S. Honken, F. Yang, and R. Laur, "A HDTV-suited architecture for a fast full search block matching algorithm," *IEEE Intl. Symp. on Circuits and Systems*, vol. 1, pp. 621–624, Seattle, WA, April/May, 1995.

[279] A. Ohtani et al., "A motion estimation processor for MPEG2 video real time encoding at wide search range," *IEEE Custom Integrated Circuits Conf.*, pp. 405–408, Santa Clara, CA, May 1995.

[280] N. Hayashi, "A compact motion estimator with a simplified vector search strategy maintaining encoded picture quality," *IEEE Custom Integrated Circuits Conf.*, pp. 409–412, Santa Clara, CA, May 1995.

[281] J. M. M. Anderson and G. B. Giannakis, "Image motion estimation algorithms using cumulants," *IEEE Trans. Image Process.*, vol. 4, pp. 346–357, Mar. 1995.

[282] Y. Y Lee and J. W. Woods, "Motion vector quantization for video coding," *IEEE Trans. Image Process.*, vol. 4, pp. 378–382, Mar. 1995.

Chapter 7 Vector Quantization and Subband Coding

[283] Y. Linde, A. Buzo, and R. M. Gray, "An algorithm for vector quantizer design," *IEEE Trans. Commun.*, vol. COM-28, pp. 84–95, Jan. 1980.

[284] W. H. Equitz, "A new vector quantization clustering algorithm," *IEEE Trans. Acoust., Speech, and Signal Processing*, vol. 37, pp. 1568–1575, Oct. 1989.

[285] R. F. Chang and W. T. Chen, "A fast finite-state codebook design algorithm for vector quantization," *SPIE/VCIP*, vol. 1605, pp. 172–178, Boston, MA, Nov. 1991.

[286] C. H. Hsieh, "DCT-based codebook design for vector quantization of images," *IEEE Trans. CSVT*, vol. 2, pp. 401–409, Dec. 1992.

[287] B. Marangelli, "A vector quantizer with minimum visible distortion," *IEEE Trans. Signal Process.*, vol. 39, pp. 2718–2721, Dec. 1991.

[288] A. Gersho and R. M. Gray, *Vector quantization and signal compression*, Norwell, MA: Kluwer Academic, 1992.

[289] N. Moayeri, K. L. Neuhoff, W. E. Stark, "Fine-coarse vector quantization," *IEEE Trans. Acoust., Speech, and Signal Processing*, vol. 39, pp. 1503–1515, July 1991.

[290] C. H. Hsieh et al., "A fast codebook design algorithm for vector quantization of images," *IEEE Intl., Symp. Circuits and Systems, IS-CAS'91*, pp. 288–291, June 1991.

[291] E.Yair, K. Zeger, and A. Gersho, "Competitive learning and soft competition for vector quantizer design," *IEEE Trans. Signal Process.*, vol. 40, pp. 294–309, Feb. 1992.

[292] K. Zeger, J. Vaisey, and A. Gersho, "Globally optimal vector quantizer design by stochastic relaxation," *IEEE Trans. Signal Process.*, vol. 40, pp. 310–322, Feb. 1992.

[293] N. Akrout et al., "A fast algorithm for vector quantization: application to codebook generation in image subband coding," *EUSIPCO '92*, Brussels, Belgium, Aug. 1992.

[294] T. Kim, "Side match and overlap match vector quantizers for images," *IEEE Trans. Image Process.*, vol. 1, pp. 170–185, Apr. 1992.

[295] V. J. Mathews, "Multiplication free vector quantization using L1 distortion measure and its variants," *IEEE Trans. Image Process.*, vol. 1, pp. 11–17, Jan. 1992.

[296] J. W. Modestino and Y. H. Kim, "Adaptive entropy-coded predictive vector quantization of images," *IEEE Trans. Signal Process.*, vol. 40, pp. 633–644, Mar. 1992.

[297] A. Madisetti et al., "Architectures for integrated circuits for real time vector quantization of images," *ICASSP '92*, pp. 677–680, San Francisco, CA, Mar. 1992.

[298] C. K. Chan and C. K. Ma, "A fast image codevector generation method," *IEEE Workshop on Visual Signal Process. and Commun.*, pp. 68–73, Raleigh, NC, Sept. 1992.

[299] S. C. Chan et al., "Codebook generation and search algorithm for vector quantization," *IEEE Workshop on Visual Signal Process. and Commun.*, pp. 74–79, Raleigh, NC, Sept. 1992.

[300] M. T. Orchard, "A fast nearest-neighbor search algorithm," *ICASSP '91*, pp. 2297–2300, Toronto, Canada, May 1991.

[301] C. K. Chan and L. M. Po, "A complexity reduction technique for image vector quantization," *IEEE Trans. Image Process.*, vol. 1, pp. 312–321, July 1992.

[302] C. M. Huang et al., "Fast full search equivalent encoding algorithms for image compression using vector quantization," *IEEE Trans. Image Process.*, vol. 1, pp. 413–416, July 1992.

[303] K. N. Ngan and H. C. Koh, "Predictive classified vector quantization," *IEEE Trans. Image Process.*, vol. 1, pp 269–280, July 1992.

[304] R. K. Kolagotla, S. S. Yu, and J. F. Jaja, "VLSI implementation of a tree searched vector quantization," *IEEE Trans. Signal Process.*, vol. 41, pp. 901–905, Feb. 1993.

[305] C. M. Huang and R. W. Harris, "A comparison of several vector quantization codebook generation approaches," *IEEE Trans. Image Process.*, vol. 2, pp. 108–112, Jan. 1993.

[306] S. W. Wu and A. Gersho, "Lapped vector quantization of images," *Optical Engineering*, vol. 32, pp. 1489–1495, July 1993.

[307] Special issue on "Vector Quantization," *IEEE Trans. Signal Process.*, Fall 1995.

[308] A. Kumar, "Vector quantization of images using input-dependent weighted square error destortion," *IEEE Trans. Circuits and Systems for Video Technology*, vol. 3, pp. 435–439, Dec. 1993.

[309] X. Wu and K. Zhang, "A subjective distortion measure for vector quantization," *DCC '94, Data Compression Conference*, pp. XIV+549, 22–31, Snowbird, UT, Mar. 1994.

[310] L. Torres and J. Huguet, "An improvement on codebook search for vector quantization," *IEEE Trans. Commun.*, vol. 42, pp. 208–210, Feb. 1994.

[311] C. K. Chan and C. K. Ma, "A fast method of designing better codebooks for image vector quantization," *IEEE Trans. Commun.*, vol. COM-42, pp. 237–242, Feb. 1994.

[312] C. K. Ma and C. K. Chan, "Maximum descent method for image vector quantization," *Electronics Letters*, vol. 27, pp. 1772–1773, Sept. 1991.

[313] L. M. Po and C. K. Chan, "Novel subspace distortion measurement for efficient implementation of image vector quantizer," *Electronics Letters*, vol. 26, pp. 480–482, Mar. 1990.

[314] M. R. Soleymani and S. D. Morgera, "A fast MMSE encoding technique for vector quantization," *IEEE Trans. Commun.*, vol. COM-37, pp. 656–659, June 1989.

[315] X. Wu and L. Guan, "Acceleration of the LBG algorithm," *IEEE Trans. Commun.*, vol. 42, pp. 1518–1523, Apr. 1994.

[316] N. M. Nasrabadi and R. A. King, "Image coding using vector quantization: A review," *IEEE Trans. Commun.*, vol. COM-36, pp. 957–971, Aug. 1988.

[317] R. M. Gray, "Vector quantization," *IEEE Acoust., Speech, and Signal Process. Magazine*, vol. 1, pp. 4–29, Apr. 1984.

[318] P. C. Chang and R. M. Gray, "Gradient algorithms for designing predictive vector quantizers," *IEEE Trans. Acoust., Speech, and Signal Process.*, vol. ASSP-34, pp. 679–690, Aug. 1986.

[319] K. Zeger, "Corrections to gradient algorithms for designing predictive vector quantizers," *IEEE Trans. Signal Process.*, vol. 39, pp. 764–765, Mar. 1991.

[320] N. Mohsenian, S. A. Rizvi, and N. M. Nasrabadi, "Predictive vector quantization using a neural network approach," *Optical Engineering*, vol. 32, pp. 1503–1513, July 1993.

[321] S. A. Rizvi and N. M. Nasrabadi, "Predictive vector quantizer using constrained optimization," *IEEE Signal Process. Letters*, vol. 1, pp. 15–18, Jan. 1994.

[322] V. Cuperman and A. Gersho, "Vector predictive coding of speech at 16Kb/s," *IEEE Trans. Commun.*, vol. COM-33, pp. 685–696, July 1985.

[323] H. Abut (Editor), *Vector quantization*, New York, NY: IEEE Press, 1990.

[324] M. H. Lee and G. Crebbin, "Classified vector quantization with variable block-size DCT models," *IEEE Proc. Vis. Image Signal Process.*, vol. 141, pp. 39–48, Feb. 1994.

[325] D. G. Jeong and J. D. Gibson, "Image coding with uniform and piecewise uniform vector quantizers," *IEEE Trans. Image Process.*, vol. 4, pp. 125–139, Feb. 1995.

[326] E. Riskin and R. M. Gray, "A greedy tree growing algorithm for the design of variable rate vector quantizers," *IEEE Trans. Signal Process.*, vol. 39, pp. 2500–2507, Nov. 1991.

[327] J. E. Fowler et al., "Real-time video compression using differential vector quantization," *IEEE Trans. Circuits and Systems for Video Technology*, vol. 5, pp. 14–24, Feb. 1995.

[328] C. H. Hsieh and J. S. Shue, "Frame adaptive finite-state vector quantization for image sequence coding," *Signal Processing: Image Communication*, vol. 7, pp. 13–26, Mar. 1995.

[329] J. S. Lee, R. C. Kim, and S. U. Lee, "On the transformed entropy-constrained vector quantizers employing Mandala block for image coding," *Signal Process.: Image Commun.*, vol. 7, pp. 75–92, Mar. 1995.

[330] S. A. Rizvi, N. M. Nasrabadi, and L. C. Wang, "Variable-rate predictive residual vector quantizer," *SPIE/VCIP*, vol. 2501, pp. 500–511, Taipei, Taiwan, May 1995.

[331] S. A. Rizvi, L. C. Wang, and N. M. Nasrabadi, "Finite-state residual vector quantization," *SPIE/VCIP*, vol. 2501, pp. 512–523, Taipei, Taiwan, May 1995.

[332] F. J. Jové et al., "Minimization of the mosaic effect in hierarchical multirate vector quantization for image coding," *SPIE/VCIP*, vol. 2501, pp. 555–561, Taipei, Taiwan, May 1995.

[333] C. H. Lee and L. H. Chen, "A fast search algorithm for vector quantization using means and variances of codewords," *SPIE/VCIP*, vol. 2501, pp. 619–628, Taipei, Taiwan, May 1995.

[334] W. Chen, Z. Zhang, and E. H. Yang, "A hybrid adaptive vector quantizer for image compression via gold-washing mechanism," *SPIE/VCIP*, vol. 2501, pp. 635–646, Taipei, Taiwan, May 1995.

[335] R. F. Chang and W. J. Kuo, "Two-pass side-match finite-state vector quantization," *SPIE/VCIP*, vol. 2501, pp. 658–667, Taipei, Taiwan, May 1995.

[336] K. T. Lo and J. Feng, "Improved minimum distortion encoding algorithm for image vector quantization," *SPIE/VCIP*, vol. 2501, pp. 668–675, Taipei, Taiwan, May 1995.

[337] H. A. Monawer and O. G. Zumburidze, "Intra- and inter-frame image coding via vector quantization," *SPIE/VCIP*, vol. 2501, pp. 717–732, Taipei, Taiwan, May 1995.

[338] C. C. Lai and S. C. Tai, "An efficient codebook search algorithm for vector quantization," *SPIE/VCIP*, vol. 2501, pp. 1290–1298, Taipei, Taiwan, May 1995.

[339] V. Ricordel and C. Labit, "Vector quantization by hierarchical packing of embedded truncated lattices," *SPIE/VCIP*, vol. 2501, pp. 1431–1440, Taipei, Taiwan, May 1995.

[340] Y. Huh, K. Panusopone, and K. R. Rao, "Progressive image transmission using variable block coding with classified vector quantization," *SPIE/VCIP*, vol. 2501, pp. 1533–1544, Taipei, Taiwan, May 1995.

[341] P. D. Alessandro and R. Lancini, "Video coding scheme using DCT-pyramid vector quantization," *IEEE Trans. Image Process.*, vol. 4, pp. 309–319, Mar. 1995.

[342] Y. Huh, J. J. Hwang, and K. R. Rao, "Classified wavelet transform coding of images using two-channel conjugate vector quantization," *ICIP '94*, Austin, TX, pp. 363–367, Nov. 1994.

[343] M. Vetterli, "Multidimensional subband coding: Some theory and algorithms," *Signal Process.*, vol. 6, pp. 97–112, Feb. 1984.

[344] J. M. Woods and S. D. O'Neil, "Subband coding of images," *IEEE Trans. Acoust., Speech, and Signal Process.*, vol. ASSP-34, pp. 1278–1288, Oct. 1986.

[345] P. P. Vaidyanathan, "QMF banks, M band extensions and perfect reconstruction techniques," *IEEE Acoust., Speech, and Signal Process. Magazine*, vol. 4, pp. 4–20, July 1987.

[346] J. W. Woods (Editor), *Subband image coding*, Norwell, MA: Kluwer Academic, 1991.

[347] M. J. T. Smith and T. P. Barnwell III, "Exact reconstruction techniques for tree-structured subband coders," *IEEE Trans. Acoust., Speech, and Signal Process.*, vol. ASSP-34, pp. 434–441, June 1986.

[348] P. H. Westerink et al., "Subband coding of images using vector quantization," *IEEE Trans. Commun.*, vol. COM-36, pp. 713–719, June 1988.

[349] E. B. Richardson and N. S. Jayant, "Subband coding with adaptive prediction for 56 Kbps audio," *IEEE Trans. Acoust., Speech, and Signal Process.*, vol. ASSP-34, pp. 691–695, Aug. 1986.

[350] P. P. Vaidyanathan, "Multirate digital filters, filter banks, polyphase networks, and applications: a tutorial," *Proc. IEEE*, vol. 78, pp. 56–93, Jan. 1990.

[351] G. Karlsson and M. Vetterli, "Theory of two-dimensional multirate filter banks," *IEEE Trans. Acoust., Speech, and Signal Process.*, vol. ASSP-38, pp. 925–937, June 1990.

[352] H. Gharavi and A. Tabatabai, "Subband coding of monochrome and color images," *IEEE Trans. Circuits Systems*, vol. 35, pp. 207–214, Feb. 1988.
[353] M. Vetterli, "A theory of multirate filter banks," *IEEE Trans. Acoust., Speech, and Signal Process.*, vol. ASSP-35, pp. 356–372, Mar. 1987.
[354] P. P. Vaidyanathan, *Multirate systems and filter banks*, Englewood Cliffs, NJ: Prentice Hall, 1993.
[355] O. Egger, W. Li, and M. Kunt, "High compression image coding using an adaptive morphological subband decomposition," *Proc. IEEE*, vol. 83, pp. 272–287, Feb. 1995.
[356] A. Fernandez et al., "HDTV subband DCT coding: Analysis of system complexity," *ICC '90*, vol. 4, pp. 1602–1606, Atlanta, GA, Apr. 1990.
[357] Y. Okumura, K. Irie, and R. Kishimoto, "High quality transmission system design for HDTV signals," *ICC '90*, vol. 3, pp. 1049–1053, Atlanta, GA, Apr. 1990.
[358] M. Vetterli and J. Kovacevic, *Wavelets and subband coding*, Englewood Cliffs, NJ: Prentice Hall, 1995.
[359] P. E. Fleischer, C. Lan, and M. Lucas, "Digital transport of HDTV on optical fiber," *IEEE Commun. Magazine*, vol. 29, pp. 36–41, Aug. 1991.
[360] C. I. Podilchuk, N. S. Jayant, and N. Farvardin, "Three-dimensional subband coding of video," *IEEE Trans. Signal Process.*, vol. 4, pp. 125–139, Feb. 1995.
[361] Y. H. Kim and J. Modestino, "Adaptive entropy-coded subband coding of images," *IEEE Trans. Image Process.*, vol. 1, pp. 31–48, Jan. 1992.
[362] K. Komatsu and K. Sezaki, "Reversible subband coding of images," *SPIE/VCIP*, vol. 2501, pp. 676–684, Taipei, Taiwan, May 1995.
[363] K. Sawada and T. Kinoshita, "Subband-based scalable coding schemes with motion-compensated prediction," *SPIE/VCIP*, vol. 2501, pp. 1470–1477, Taipei, Taiwan, May 1995.
[364] C. H. Lin and J. L. Wu, "Segmentation-based subband video coder," *SPIE/VCIP*, vol. 2501, pp. 1514–1524, Taipei, Taiwan, May 1995.
[365] Y. S. Ho, "Digital coding of NTSC sequences with a subband-VQ scheme," *SPIE/VCIP*, vol. 2501, pp. 1525–1532, Taipei, Taiwan, May 1995.
[366] A. Saadane, H. Senane, and D. Barba, "An entirely psychovisual based subband image coding scheme," *SPIE/VCIP*, vol. 2501, pp. 1702–1712, Taipei, Taiwan, May 1995.

Chapter 8 JPEG

[367] ISO/IEC JTC1 10918-1 } ITU-T Rec. T.81, *Information technology – Digital compression and coding of continous-tone still images: Requirements and guidelines*, 1994.

[368] ISO/IEC 10918-3 } ITU-T Rec. T.84, *Information technology – Digital compression and coding of continous-tone still images: Extensions*, 1996.

[369] ISO/IEC 10918-2 } ITU-T Rec. T.83, *Information technology – Digital compression and coding of continous-tone still images: Compliance testing*, 1995.

[370] Y. W. Lei and M. Ouhyoung, "Software-based motion JPEG with progressive refinement for computer animation," *IEEE Trans. Consumer Electronics*, vol. 40, pp. 557–562, Aug. 1994.

[371] G. P. Hudson, H. Yasuda, and I. Sebestyen, "The international standardization of a still picture compression technique," *GLOBECOM '88*, Hollywood, FL, pp. 1016–1021, 1988.

[372] G. K. Wallace, "The JPEG still picture compression standard," *Commun. of the ACM*, vol. 34, pp. 31–44, Apr. 1991.

[373] A. Leger, T. Omachi, and G.K. Wallace, "JPEG still picture compression algorithm," *Optical Engineering*, vol. 30, pp. 949–954, July 1991.

[374] W. B. Pennebaker and J. L. Mitchell, *JPEG still image data compression standard*, New York, NY: Van Nostrand Reinhold, 1993.

[375] G. K. Wallace, "The JPEG still picture compression standard," *IEEE Trans. on Consumer Electronics*, vol. 38, pp. 18–34, Feb. 1992.

[376] M. Rabbani and P. W. Jones, *Digital image compression techniques*, SPIE Press, vol. TT7, Bellingham, WA, 1992.

[377] S. W. Wu and A. Gersho, "Rate-constrained picture-adaptive quantization for JPEG baseline coders," *ICASSP '93*, Minneapolis, MN, pp. V.389–392, Apr. 1993.

[378] O. Baudin, A. Baskurt, and R. Goutte, "JPEG compression: An evaluation," *Intl. Symp. on Fiber Optic Networks and Video Commun.*, vol. 1997, Berlin, Germany, Apr. 1993.

[379] C. Hwang, S. Venkatraman, and K. R. Rao, "Human visual system weighted progressive image transmission using lapped orthogonal transform/classified vector quantization," *Optical Engineering*, vol. 32, pp. 1524–1530, July 1993.

[380] W. B. Pennebaker et al., "An overview of the basic principles of the Q-Coder adaptive binary arithmetic coder," *IBM Journal of Research and Development*, vol. 32, pp. 717–726, Nov. 1988.

[381] Y. P. Chen and Y. Yasuda, "The performance improvement of JPEG algorithm using a modified arithmetic code," *IEEE Workshop on Visual Signal Process. and Commun.*, pp. 206–209, Raleigh, NC, Sept. 1992.

[382] B. Zeng and A. N. Venetsanopoulos, "A JPEG-based interpolative coding scheme," *ICASSP '93*, Minneapolis, MN, pp. V. 393–396, Apr. 1993.

[383] J. J. Hwang and G. H. Yang, "Optimal variable quantization for JPEG extensions," *IEEE Intl. Conf. on Consumer Electronics*, Chicago, IL, pp. 266–267, June 1995.

[384] K. R. Rao and P. Yip, *Discrete cosine transform: algorithms, advantages, applications.* San Diego, CA: Academic Press, 1990.

[385] C-Cube Microsystems, *CL550A JPEG Image Compression Processor, Preliminary Data Book*, San Jose, CA, Feb. 1990.

[386] C-Cube, *CL550™ JPEG image compression processor, prelimainary data book*, Aug. 1991.

[387] LSI Logic Inc., *JPEG still image compression chip set–technical note*, Feb. 1991.

[388] LSI Logic Inc., *JPEG coprocessor technical manual*, preliminary ed., May 15, 1993.

[389] Integrated Information Technology Inc., *Using the IIT vision processor in JPEG applications, preliminary application note*, Sept. 1991.

[390] Integrated Information Technology Inc., *Single chip video codec and multimedia communications processor, IIT VCP, preliminary datasheet*, Oct. 1993.

[391] Texas Instruments, *JPEG image compression/decompression processor, technical specification*, Jan. 1993.

[392] ZORAN Corp., *JPEG image compression chip set*, May 1993.

[393] D. Bursky, "Image-processing chip set handles full-motion video," *Electronic Design*, vol. 41, pp. 117–120, May 1993.

[394] M. Uchiyama et al., "A digital still camera," *IEEE Trans. on Consumer Electronics*, vol. 38, pp. 698–702, Aug. 1992.

[395] K. Ogawa et al., "A single chip compression/decompression LSI based on JPEG," *IEEE Trans. on Consumer Electronics*, vol. 38, pp. 703–710, Aug. 1992.

[396] M. Nakagawa et al., "DCT-based still image compression ICs with bit-rate control," *IEEE Trans. on Consumer Electronics*, vol. 38, pp. 711–717, Aug. 1992.

[397] S. K. Jo and Y. H. Lee, "Statistical feedforward/feedback buffer control for transmission of digital video signal compressed by the JPEG algorithm," *Asia-Pacific Conf. on Commun.*, Taejon, Korea, pp. 406–409, Aug. 1993.

[398] J. Golston et al., "VLSI implementations of image and video compression algorithms," *TI Technical J.*, pp. 42-51, Sept./Oct. 1991.

[399] A. Razavi et al., "VLSI implementation of an image compression algorithm with a new bit rate control capability," *ICASSP '92*, pp. 699–673, San Francisco, CA, Mar. 1992.

[400] ISO/IEC JTC1/SC2/WG9, *CD11544, Progressive bi-level image compression, revision 4.1*, Sept. 16, 1992.

[401] H. Hampel et al., "Technical features of the JBIG standard for progressive bi-level image compression," *Signal Process.: Image Commun.*, vol. 4, pp. 103–111, April 1992.

[402] P. A. Ruetz et al., "A video-rate JPEG chip set," *VLSI video/image signal processing*, Edited by T. Nishitani, P.H. Ang, and F. Catthoor Hingham, MA: Kluwer Academic, 1993.

[403] A. Razavi et al., "High-performance JPEG image compression chip set for multimedia applications," *SPIE/IS&T Symp. on Electronic Imaging: Science and Technology*, vol. 1903, San Jose, CA, Jan. 31 - Feb. 4, 1993.

[404] M. Vetterli, "Filter banks allowing perfect reconstruction," *Signal Process.*, vol. 10, pp. 219–244, 1986.

[405] U. Wittenberg, "Application of the JPEG standard in a medical environment," *SPIE, Video Commun. and PACS for Medical Appl.*, Berlin, Germany, vol. 1977, pp. 121–129, Apr. 1993.

[406] X. Zhu and J. Wen, "Improving JPEG compression at complex image areas with FTC: a novel hybrid image compression algorithm," *ICSPAT '95*, Boston, MA, Oct. 1995.

[407] G. G. Langdon and A. Zandi, "Applications of universal context modeling to lossless compression of gray-scale images," *1995 Asilomar Conf. on Circuits, Systems, and Comput.*, Pacific Grove, CA, Oct. 1995.

Chapter 9 H.261

[408] M. Carr et al., "Motion video coding in CCITT SG XV—the video multiplex and transmission coding," *GLOBECOM '88*, Hollywood, FL, pp. 1005–1010, Dec. 1988.

[409] R. C. Nicol and N. Mukawa, "Motion video coding in CCITT SG XV—the coded picture format," *GLOBECOM '88*, Hollywood, FL, pp. 992–996, Dec. 1988.

[410] J. Guichard et al., "Motion video coding in CCITT SG XV—hardware trials," *GLOBECOM '88*, Hollywood, FL, pp. 37–42, Dec. 1988.

[411] R. Plompen, "Motion video coding in CCITT SG XV—the video source coding," *GLOBECOM '88*, Hollywood, FL, pp. 997–1004, Dec. 1988.

[412] T. Murakami et al., "A DSP architectural design for low bit-rate motion video codec," *IEEE Trans. on Circuits, Systems*, vol. 36, pp. 1267–1274, Oct. 1989.

[413] S. Nishimura et al., "NTSC-CIF mutual conversion processor," *SPIE, Visual Commun. and Image Proc. IV*, vol. 1199, pp. 885–894, Nov. 1989.

[414] R. Plompen, *Motion video coding for visual telephony*, PTT Research Neher Laboratories, Leidschendam, the Netherlands, 1989.

[415] G. Giunta, T. R. Reed, and M. Kunt, "Image sequence coding using oriented edges," *Signal Process.: Image Commun.*, vol. 2, pp. 429–440, Dec. 1990.

[416] C. Herpel, D. Hepper, and D. Westerkamp, "Adaptation and improvement of CCITT Reference Model 8 video coding for digital storage media

applications," *Signal Process.: Image Commun.*, vol. 2, pp. 171–185, Aug. 1990.

[417] M. Hötter, "Object-oriented analysis-synthesis coding based on moving two-dimensional objects," *Signal Process.: Image Commun.*, vol. 2, pp. 409–428, Dec. 1990.

[418] M. L. Liou, "Visual telephony as an ISDN application," *IEEE Commun. Magazine*, vol. 28, pp. 30–38, Feb. 1990.

[419] R. Plompen et al., "Motion video coding; an universal coding approach," *1990 SPIE/SPSE Symp. on Electronic Imaging Science* & Technology, vol.1244, pp. 389–405, Feb. 1990.

[420] A. Tabatabai, M. Mills, and M. L. Liou, "A review of CCITT $p \times 64$ kbps video coding and related standards," *Intl. Electronic Imaging Exposition and Conf.*, pp. 58–61, Oct. 1990.

[421] P. H. Westerink, J. Biemond, and F. Muller, "Subband coding of image sequences at low bit rates," *Signal Process.: Image Commun.*, vol. 2, pp. 441–448, Dec. 1990.

[422] N. Diehl, "Object-oriented motion estimation and segmentation in image sequences," *Signal Process.: Image Commun.*, vol. 3, pp. 23–56, Feb. 1991.

[423] Y. Endo et al., "Development of CCITT standard video codec: Visuallink 5000," *NEC Res. and Develop.*, vol. 32, pp. 557–568, Oct. 1991.

[424] M. H. A. Fadzil and T. J. Dennis, "Video subband VQ coding at 64 kbit/s using short-kernel filter banks with an improved motion estimation technique," *Signal Process.: Image Commun.*, vol. 3, pp. 3–22, Feb. 1991.

[425] H. Fukuchi, K. Shomura, and H. Fujiwara, "VLSI implementation of decoding system manager chip for motion picture decoder," *Proc. IEEE Workshop on Visual Signal Process. and Commun.*, Hsinchu, Taiwan, ROC, pp. 173–176, June 1991.

[426] T. Tajiri et al., "Single board video codec for ISDN visual telephone," *ICASSP '91*, pp. 2853–2856, Toronto, Canada, May 1991.

[427] Y. Takishima, M. Wada, and H. Murakami, "An analysis of optimum frame rate in low bit rate video coding," *SPIE/VCIP: Visual Commun.*, Boston, MA, vol. 1605, pp. 635–645, Nov. 1991.

[428] R. Saito et al., "VLSI imlplementation of a variable-length coding processor for real time video," *Proc. IEEE Workshop on Visual Signal Process. and Commun.*, Hsinchu, Taiwan, ROC, pp. 87–90, June 1991.

[429] T. Sikora, T. K. Tan, and K. K. Pang, "A two layer pyramid image coding scheme for interworking of video services in ATM," *SPIE/VCIP: Visual Commun.*, Boston, MA, vol. 1605, pp. 624–634, Nov. 1991.

[430] S. Okubo et al., "Hardware trials for verifying recommendation H.261 on $p \times 64$ kbit/s video codec," *Signal Process.: Image Commun.*, vol. 3, pp. 71–78, Feb. 1991.

[431] M. L. Liou, "Overview of $p \times 64$ kbit/s video coding standard," *Commun. of the ACM*, vol. 34, pp. 60–63, Apr., 1991.

[432] R. P. Loos et al., "Hybrid coding with pre-buffering and pre-analysis in a software-based codec environment," *Signal Process.: Image Commun.*, vol. 3, pp. 57–69, Feb. 1991.

[433] D. W. Lin, M. L. Liou, and K. N. Ngan, "Improvement of low bit rate video coding performance," *Proc. IEEE Workshop on Visual Signal Process. and Commun.*, Hsinchu, Taiwan, ROC, pp. 1–4, June 1991.

[434] T. Araki et al., "The architecture of a vector digital signal processor for video coding," *ICASSP '92*, pp. V.681–684, San Francisco, CA, Mar. 1992.

[435] F. Dufaux and M. Kunt, "Multigrid block matching motion estimation with an adaptive local mesh refinement," *SPIE/VCIP*, Boston, MA, vol. 1818, pp. 97–109, Nov. 1992.

[436] H. Fujiwara et al., "An all-ASIC implementation of a low bit-rate video codec," *IEEE Trans. CSVT*, vol. 2, pp. 123–134, June 1992.

[437] P. C. Jain, W. Schlenk, and M. Riegel, "VLSI implementation of two-dimensional DCT processor in real time video codec," *IEEE Trans. Consumer Electronics*, vol. 38, pp. 537–544, Aug. 1992.

[438] A. C. P. Loui, A. T. Ogielski, and M. L. Liou, "A parallel implementation of the H.261 video coding algorithm," *IEEE Workshop on Visual Signal Process. and Commun.*, Raleigh, NC, pp. 80–85, Sept. 1992.

[439] P. A. Ruetz et al., "A high-performance full-motion video compression chip set," *IEEE Trans. CSVT*, vol. 2, pp. 111–122, June 1992.

[440] B. Girod, "Motion compensation: visual aspects, accuracy, and fundamental limits," *in Motion analysis and image sequence processing*, edited by M.I. Sezan and R.L. Lagendijk, Hingham, MA: Kluwer Academic, 1993.

[441] M. Buck and N. Diehl, "Model-based image sequence coding," *Motion analysis and image sequence processing*, edited by M.I. Sezan and R.L. Lagendijk, Hingham, MA: Kluwer Academic, 1993.

[442] M. T. Sun et al., "Coding and interworking for videotelephony," *IEEE Intl. Symp. on Circuits and Systems*, pp. 20–23, Chicago, IL, June 1993.

[443] SGS-Thomson Microelectronics, *STi3223 motion estimation controller, product preview V.1.1*, Jan. 1993.

[444] T. Ebrahimi and F. Dufaux, "Efficient hybrid coding of video for low bitrate applications," *ICC '93*, Geneva, Switzerland, pp. 522–526, May 1993.

[445] A. C. P. Loui and M. L. Liou, "High-resolution still-image transmission based on CCITT H.261 codec," *IEEE Trans. CSVT*, vol. 3, pp. 164–169, April 1993.

[446] CCITT SG XV WP/1/Q4 Specialist Group on Coding for Visual Telephony, *Improvement of Reference Model 5 by a noise reduction filter*, Document 376, 1988.

[447] CCITT SG XV WP/1/Q4 Specialist Group on Coding for Visual Telephony, *Description of Ref. Model 6 (RM6)*, Document 396, Oct. 1988.

[448] *IEEE standard specification for the implementations of* 8×8 *Inverse Discrete Cosine Transform*, IEEE Standard 1180–1990, Mar. 18, 1991.

[449] "Draft revision of recommendation H.261: Video Codec for audiovisual services at $p \times 64$ kbit/s," *Signal Process.: Image Commun.*, vol. 2, pp. 221–239, Aug. 1990.

[450] Recommendations of the H–Series, CCITT Study Group XV, Report R37, Aug. 1990.

[451] CCITT SG XV WP/1/Q4 Specialist Group on Coding for Visual Telephony, *Description of Reference Model 8 (RM8)*, Document 525, June 1989.

[452] CCITT SG XV WP/1/Q4 Specialist Group on Coding for Visual Telephony, *Temporal pre-filtering applied to the reference model 3*, Document 199, Mar. 1987.

[453] R. Natarajan and K. R. Rao, "Design of a 64 kbps coder for teleconferencing," *SPIE 28th Annual Intl. Technical Symp.*, San Diego, CA, vol. 504, pp. 406–443, Aug. 1984.

[454] F. M. Pereira, D. Cortez, and P. Nunes, "Mobile videotelephone communications : the CCITT H.261 chances," *SPIE Video Commun. and PACS for Medical Appl.*, Berlin, Germany, vol. 1977, pp. 168–179, Apr. 1993.

[455] M. M. de Sequeira and F. M. Pereira, "Global motion compensation and motion vector smoothing in an extended H.261 recommendation," *SPIE Video Commun. and PACS for Medical Appl.*, Berlin, Germany, vol. 1977, pp. 226–237, Apr. 1993.

[456] E. Morimatsu et al., "Development of a VLSI chip set for H.261/-MPEG-1 video codec," *SPIE/VCIP*, Cambridge, MA, vol. 2094, pp. 422–433, Nov. 1993.

[457] Y. Rasse, "An H.261 single-chip low bit rate video codec," *Intl. Conf. on SP Applications and Technology*, pp. 22–31, Santa Clara, CA, Sept. 28–Oct. 1, 1993.

[458] A. Tirso and M. Luo, "Subband coding of videoconference sequence at 384 Kbps," *Intl. Conf. on SP Applications and Technology*, pp. 688–693, Santa Clara, CA, Sept 28–Oct. 1, 1993.

[459] X. Zhang, J. F. Arnold, and M. C. Cavenor, "A study of motion compensated adaptive quadtree schemes for coding videophone sequences on the broadband ISDN," *IEEE Workshop on Visual Signal Process. and Commun.*, pp. 81–84, Melbourne, Australia, Sept. 21-22, 1993.

[460] W. Blohm, "Compensation for non-uniform illumination in videotelephone images," *IEEE Workshop on Visual Signal Process. and Commun.*, Melbourne, Australia, Sept 21–22, 1993.

[461] A Suwa et al., "A video quality improvement technique for videophone/videoconferencing terminal," *IEEE Workshop on Visual Signal Process. and Commun.*, pp. 315–318, Melbourne, Australia, Sept. 21–22, 1993.

[462] M. A. Wondrow et al., "Role of standard compressed video teleconference codecs in the transmission of medical image data," *SPIE/IS&T Symp. on Electronic Imaging*, vol. 2188, pp. 243–244 (Abstract), San Jose, CA, Feb. 1994.

[463] S. Azim et al., "A low cost application specific video codec for consumer video phone," *IEEE Custom Integrated Circuits Conf.*, pp. 115–118, San Diego, CA, May 1994.

[464] E. Harborg, "A real-time wideband CELP coder for a videophone application," *ICASSP '94*, vol. 2, pp. 121–124, Adelaide, Australia, April 1994.

[465] K. Pang, H. G. Lim, and S. C. Hall, "A low complexity H.261 - compatible software video decoder," *Australian Telecommun. Networks & Applications Conf.*, Melbourne, Australia, Dec. 1994.

[466] H. Wu et al., "Real-time H.261 software based codec on the Power Mac," *SPIE/IS&T*, vol. 2419, pp. 492–498, San Jose, CA, Feb. 1995.

[467] H. Ibaraki et al., "Design and evaluation of programmable video codec board," *ICSPAT '94*, pp. 922–927, Dallas, TX, Oct. 1994.

[468] G. S. Yu and M. K. Liu, "Temporal and spatial interleaving of H.261 compression for lossy transmission," *SPIE/VCIP*, vol. 2501, pp. 1478–1485, Taipei, Taiwan, May 1995.

[469] E. Scopa et al., "A 2D-DCT low-power architecture for H.261 coders," *ICASSP '95*, Detroit, MI, May 1995.

[470] M. Ghanbari and V. Seferidis, "Efficient H.261-based two-layered video codecs for ATM networks," *IEEE Trans./CSVT*, vol. 5, pp. 171–175, April 1995.

[471] S. Ammon, "A complete audio solution for an H.320 videoconferencing system," *ICSPAT '95*, Boston, MA, Oct. 1995.

[472] A. Eleftheriadis and A. Jacquin, "Low bit rate model-assisted H.261 compatible coding of video," *ICIP '95*, vol. 2, pp.418–421, Washington, DC, Oct. 1995.

Chapter 10 MPEG-1

[473] K. M. Yang and D. J. Le Gall, "Hardware design of a motion video decoder for 1-1.5 Mbps rate applications," *Signal Process.: Image Commun.*, vol 2, pp. 117–126, Aug. 1990.

[474] K. M. Yang et al., “Design of a multi-function video decoder based on a motion compensated predictive-interpolative coder,” *SPIE/VCIP*, Lausanne, Switzerland, vol. 1360, pp. 1530–1539, Oct. 1990.

[475] B. Hürtgen, M. Gilge, and W. Guse, “Coding of moving video at 1Mbit/s–movies on CD,” *SPIE/VCIP*, Lausanne, Switzerland, vol. 1360, pp. 1092–1103, Oct. 1990.

[476] A. Puri, B. G. Haskell, and R. Leonardi, “Video coding with motion-compensated interpolation for CD-ROM applications,” *Signal Process.: Image Commun.*, vol. 2, pp. 127–144, Aug. 1990.

[477] A. Puri and R. Aravind, “On comparing motion-interpolation structures for video coding,” *SPIE/VCIP*, Lausanne, Switzerland, vol. 1360, pp. 1560–1571, Oct. 1990.

[478] E. Viscito and C. A. Gonzales, “Encoding of motion video sequences for the MPEG environment using arithmetic coding,” *SPIE/VCIP*, vol. 1360, Lausanne, Switzerland, pp. 1572–1576, Oct. 1990.

[479] A. Nagata et al., “Moving picture coding system for digital storage media using hybrid coding,” *Signal Process.: Image Commun.*, vol. 2, pp. 109–116, Aug. 1990.

[480] C. A. Gonzales et al., “DCT coding for motion video storage using adaptive arithmetic coding,” *Signal Process.: Image Commun.*, vol. 2, pp. 145–154, Aug. 1990.

[481] M. Haghiri and P. Denoyelle, “A low bit rate coding algorithm for full motion video signal,” *Signal Process.: Image Commun.*, vol. 2, pp. 187–199, Aug. 1990.

[482] F. Pereira et al., “A CCITT compatible coding algorithm for digital recording of moving images,” *Signal Process.: Image Commun.*, vol. 2, pp. 155–169, Aug. 1990.

[483] F. Sijstermans and J. van der Meer, “CD–full motion video encoding on a parallel computer,” *Commun. of the ACM*, vol. 34, pp. 82–91, April 1991.

[484] D. J. Le Gall, “MPEG: A video compression standard for multimedia applications,” *Commun. of the ACM*, vol. 34, pp. 47–58, April 1991.

[485] T. Hidaka and K. Ozawa, “Subjective assessment of redundancy-reduced moving images for interactive applications: Test methodology and report,” *Signal Process.: Image Commun.*, vol. 2, pp. 201–219, Aug. 1990.

[486] ISO/IEC JTC1/SC2/WG11, *A proposal for MPEG video report, by Brian Astle, coded representation of picture and audio information*, April 15, 1991.

[487] *ISO/IEC 11172 Information Technology: coding of moving pictures and associated audio for digital storage media at up to about 1.5 Mbit/s*, Part 1: Systems; Part 2: Video; Part 3: Audio; Part 4: Conformance Testing, 1993.

[488] ISO/IEC JTC1/SC2/WG11, *MPEG Video Simulation Model Three (SM3)*, MPEG 90/041, July 1990.

[489] I. Tamitani et al., "An encoder/decoder chip set for the MPEG video standard," *ICASSP '92*, pp. V.661–664, San Francisco, CA, Mar. 1992.

[490] W. Lynch, "Bidirectional motion estimation based on P frame from motion vectors and area overlap," *ICASSP '92*, pp. III.445–448, San Francisco, CA, Mar. 1992.

[491] S. L. Iu, "New configurations of a field-oriented MPEG system for coding HDTV image sequences," *IEEE Workshop on Visual Signal Process. and Commun.*, pp. 115–120, Raleigh, NC, Sept. 1992.

[492] D. Raychaudhuri, H. Sun, and R. S. Girons, "ATM transport and cell loss concealment for MPEG video," *ICASSP '93*, Minneapolis, MN, pp. I.117–120, Apr. 1993.

[493] D. Reininger and D. Raychaudhuri, "Bit-rate characteristics of a VBR MPEG video encoder for ATM networks," *ICC '93*, pp. 517–521, Geneva, Switzerland, May 1993.

[494] S. Okubo, "Requirements for high quality video coding standards," *Signal Process.: Image Commun.*, vol. 4, pp. 141–151, April 1992.

[495] K. Joseph et al., "MPEG++: A robust compression and transport system for digital HDTV," *Signal Process.: Image Commun.*, vol. 4, pp. 307–323, Aug. 1992.

[496] D. J. Le Gall, "The MPEG video compression algorithm," *Signal Process.: Image Commun.*, vol. 4, pp. 129–140, Apr. 1992.

[497] A. G. MacInnis, "The MPEG systems coding specification," *Signal Process.: Image Commun.*, vol. 4, pp. 153–159, Apr. 1992.

[498] K. Ramchandran, A. Ortega, and M. Vetterli, "Bit allocation for dependent quantization with applications to MPEG video coders," *ICASSP '93*, Minneapolis, MN, pp. V.381–384, Apr. 1993.

[499] K. Kamikura and H. Watanabe, "Video coding for digital storage media using hierarhical intraframe scheme," *SPIE/VCIP*, Lausanne, Switzerland, vol. 1360, pp. 1540–1550, Oct. 1990.

[500] M. Konoshima, O. Kawi, and K. Matsuda, "Principal devices and hardware volume estimation for moving picture decoder for digital storage media," *SPIE/VCIP*, Lausanne, Switzerland, vol. 1360, pp. 1551–1559, Oct. 1990.

[501] Y. T. Tse and R. L. Baker, "Camera zoom/pan estimation and compensation for video compression," *SPIE/VCIP*, vol. 1452, pp. 468–479, 1991.

[502] L. Chiariglione, "Standardization of moving picture coding for interactive applications," *GLOBECOM '89*, pp. 559–563, Nov. 1989.

[503] K. W. Chun et al., "An adaptive perceptual quantization algorithm for video coding," *IEEE Trans. Consumer Electronics*, vol. 39, pp. 555–558, Aug. 1993.

[504] H. Sun et al., "Hierarchical decoder for MPEG compressed video data," *IEEE Trans. Consumer Electronics*, vol. 39, pp. 559–564, Aug. 1993.

[505] E. Chan and S. Panchanathan, "Motion estimation architecture for video compression," *IEEE Trans. Consumer Electronics*, vol. 39, pp. 292–297, Aug. 1993.

[506] M. Hötter, "Differential estimation of the global motion parameters zoom and pan," *Signal Process.*, vol. 16, pp. 249–265, Mar. 1989.

[507] A. N. Netravali and B. G. Haskell, *Digital pictures, representation and compression*, New York, NY: Plenum Press, 1988.

[508] A. K. Jain, *Fundamentals of digital image processing*, Englewood Cliffs, NJ: Prentice Hall, 1989.

[509] S. Bose, S. Purcell, and T. Chiang, "A single chip multistandard video codec," *IEEE Custom Integrated Circuits Conf.*, San Diego, CA, pp. 11.4.1–11.4.4, May 1993.

[510] L. W. Lee et al., "On the error distribution and scene change for the bit rate control of MPEG," *IEEE Trans. Consumer Electronics*, vol. 39, pp. 545–554, Aug. 1993.

[511] M. Roser et al., "Extrapolation of a MPEG-1 video-coding scheme for low-bit-rate applications," *SPIE Video Commun. and PACS for Medical Appl.*, Berlin, Germany, vol. 1977, pp. 180–187, Apr. 1993.

[512] L. Bergonzi et al., "DIVA: a MPEG-1 video decoder for interactive applications," *SPIE Video Commun. and PACS for Medical Appl.*, Berlin, Germany, vol. 1977, pp. 409–416, Apr. 1993.

[513] C. Bouville et al., "Real-time testbed for MPEG-1 video compression," *SPIE/VCIP*, Cambridge, MA, vol. 2094, pp. 205–212, Nov. 1993.

[514] ISO/IEC JTC1/SC2/WG11, *The SR report on the MPEG/audio subjective listening tests*, Stockholm, Sweden, MPEG 91/010, June 1991.

[515] ISO/IEC JTC1/SC2/WG11, *Preliminary text for MPEG audio coding standard*, MPEG 90/265, Sept. 13, 1990.

[516] Y. Mahieux and J. P. Petit, "Transform coding of audio signals at 64 Kbit/s," *GLOBECOM '90*, San Diego, CA, pp. 518–522, Dec. 1990.

[517] N. S. Jayant, "High-quality coding of telephone speech and wideband audio," *IEEE Commun. Magazine*, vol. 28, pp. 10–20, Jan. 1990.

[518] J. D. Johnston, "Perceptual transform coding of wideband stereo signals," *ICASSP '89*, Glasgow, Scotland, pp. 1993–1996, May 1989.

[519] Y. C. Sung and J. F. Yang, "An audio compression system using modified transform coding and dynamic bit allocation," *IEEE Trans. Consumer Electronics*, vol. 39, pp. 285–259, Aug. 1993.

[520] K. Brandenburg, "OCF–A new coding algorithm for high quality sound signals," *ICASSP '87*, Dallas, TX, pp. 141–144, Apr. 1987.

[521] H. G. Musmann, "The ISO audio coding standard," *GLOBECOM '90*, San Diego, CA, pp. 551–517, Dec. 1990.

[522] LSI Logic Corp., *L64111 MPEG audio decoder*, Apr. 9, 1992.

[523] Texas Instruments, *SN74ACT6350 MPEG audio decoder*, Sept. 16, 1992.

[524] G. Stoll and Y. F. Dehery, "High quality audio bit–rate reduction system family for different applications," *Supercomm, ICC '90*, Atlanta, GA, Apr. 9, 1990.

[525] J. D. Johnston, "Transform coding of audio signals using perceptual noise criteria, *IEEE J. Selected Areas in Commun.*, vol. 6, pp. 314–323, Feb. 1988.

[526] J. D. Johnston, "Estimation of perceptual entropy using noise masking criteria, *ICASSP '88*, New York, NY, pp. 2524–2527, Apr. 1988.

[527] X. Maitre, "7 KHz audio coding within 64 kbit/s," *IEEE J. Selected Areas in Commun.*, vol. 6, pp. 283–298, Feb. 1988.

[528] M. Taka, S. Shimada, and T. Hoyama, "Multimedia multipoint teleconference system using the 7 KHz audio coding standard at 64 Kbit/s," *IEEE J. Selected Areas in Commun.*, vol. 6, pp. 299–313, Feb. 1988.

[529] R. N. J. Veldhuis, "Bit rates in audio source coding," *IEEE J. Selected Areas in Commun.*, vol. 10, pp. 86–96, Jan. 1992.

[530] P. H. N. de With et al., "Data compression systems for home-use digital video recording," *IEEE J. Selected Areas in Commun.*, vol. 10, pp. 97–121, Jan. 1992.

[531] *CCITT Recommendation G.711: Pulse code modulation (PCM) of voice frequencies*, Geneva, ITU, 1972.

[532] *CCITT Recommendation G.722: 7 KHz audio coding within 64 kbit/s*, SG XVIII Rep. R26(C), Aug. 1986.

[533] M. Iwadare and T. Nishitani, "64 Kbit/s audio signal transmission approaches using 32 Kbit/s ADPCM channel banks," *IEEE Trans. Consumer Electronics*, vol. 6, pp. 307–313, Feb. 1988.

[534] ISO/IEC JTC1/SC2/WG8, *ASPEC*, MPEG 89/205, 1989.

[535] Y. F. Dehery, "Real time software processing approach for digital sound broadcasting," *Advanced digital techniques for UHF satellite sound broadcasting, EBU*, Geneva, pp. 95–99, Aug. 1988.

[536] Y. F. Dehery et al., "Digital sound broadcasting for mobile reception," *Proc. ITU-COM '89*, Geneva, pp. 53–57, 1989.

[537] D. Y. Pan, "Digital audio compression," *Digital Technical J.*, vol. 5, pp. 28–40, Spring 1993.

[538] M. Iwadare et al., "A 128 kb/s hi–fi audio codec based on adaptive transform coding with adaptive block size MDCT," *IEEE J. Selected Areas in Commun.*, vol. 10, pp. 138–144, Jan. 1992.

[539] G. Maturi, "Single chip MPEG audio decoder," *IEEE Trans. Consumer Electronics*, vol. 38, pp. 348–356, Aug. 1992.

[540] K. Brandenburg et al., "Transform coding of high-quality digital audio at low bit rates : algorithm and implementation," *ICC '93*, Atlanta, GA, pp. 932–936, Apr. 1990.

[541] G. Davidson, L. Fielder, and M. Antill, "Low-complexity transform coder for satellite link applications," *AES 89th Convention*, Los Angeles, CA, Sept. 21–25, 1990.

[542] B. Scharf, "Critical bands," *Foundations of modern auditory theory*, edited by J. V. Tobias, Orlando, FL: Academic Press, 1970.

[543] N. S. Jayant and P. Noll, *Digital coding of waveforms, principles and applications to speech and video*, Englewood Cliffs, NJ: Prentice Hall, 1984.

[544] J. J. Zwislocki, "Masking: experimental and theoretical aspects of simultaneous, forward, backward, and central masking," *Handbook of perception*, edited by E. C. Carterette and M. P. Friedman, Orlando, FL: Academic Press, 1978.

[545] E. Eberlein, H. Gerhäuser, and S. Krägeloh, "Audio codec for 64 Kbit/sec (ISDN channel) – requirements and results," *ICASSP '90*, Albuquerque, NM, pp. 1105–1108, Apr. 1990.

[546] A. Sugiyama et al., "Adaptive transform coding with an adaptive block size (ATC–ABS)," *ICASSP '90*, Albuquerque, NM, pp. 1093–1096, Apr. 1990.

[547] M. R. Schroeder, B. S. Atal, and J. L. Hall, "Optimizing digital speech coders by exploiting masking properties of the human ear," *J. Acoust. Soc. Am.*, pp. 1647–1652, 1979.

[548] K. Konstantinides, "Fast subband filtering in MPEG audio coding," *IEEE Signal Process. Letters*, vol. 1, pp. 26–28, Feb. 1994.

[549] K. Patel, B. C. Smith, and L. A. Rowe, "Performance of a software MPEG video decoder," *ACM SIGGRAPH Multimedia '93*, Anaheim, CA, Aug. 1993.

[550] L. Yan, "Noise reduction for MPEG-1 type codec," *ICASSP '94*,vol. 5, pp. 429–432, Adelaide, Australia, Apr. 1994.

[551] R. J. Gove, "The MVP: A single-chip multiprocessor for image & video applications," *Society for Information Display, 1994 Intl. Symp. Seminar, Exhibition*, San Jose, CA, June 1994.

[552] R. J. Gove, "The MVP: A highly-integrated video compression chip," *IEEE Data Compression Conf.*, pp. XIV + 549, 215–224, Snowbird, UT, March 1994.

[553] J. B. Cheng, H. M. Hang, and D. W. Lin, "An image compression scheme for digital video cassette recording," *ICCE '94*, pp. 26–27, Chicago, IL, June 1994.

[554] C. L. McCarthy, "A low-cost audio/video decoder solution for MPEG system streams," *ICCE '94*, pp. 342–343, Chicago, IL, June 1994.

[555] Y. Honjo et al., "Karaoke CD grows to video CD," *ICCE '94*, pp. 136–137, Chicago, IL, June 1994.

[556] P. Pancha and M. E. Zarki, "MPEG coding for variable bit rate video transmission," *IEEE Commun. Magazine*, vol. 32, pp. 54–66, May 1994.

[557] T. Urabe et al., "MPEG tool: An X window-based MPEG encoder and statistics tool," *Multimedia Systems*, May 1994. (For information, email to: mpegtool@ee.upenn.edu.)

[558] C. Eddy, "The MPEG compression algorithm," *PC Graphics and Video*, vol. 3, pp. 52–55, July 1994.

[559] K. Shen, L. A. Rowe, and E. J. Delp, "Parallel implementation of an MPEG-1 encoder, faster than real time!," *SPIE/IS&T*, vol. 2419, pp. 407–418, San Jose, CA, Feb. 1995.

[560] S. Eckart, "High-performance software MPEG video player for PCs," *SPIE/IS&T*, vol. 2419, pp. 446–454, San Jose, CA, Feb. 1995.

[561] V. Bhaskaran and K. Konstantinides, "Real-time MPEG-1 software decoding on HP workstations," *SPIE/IS&T*, vol. 2419, pp. 466–473, San Jose, CA, Feb. 1995.

[562] L. Chiariglone, "The development of an integrated audiovisual coding standard: MPEG," *Proc. IEEE*, pp. 151–157, Feb. 1995.

[563] J. Shiu et al., "A low-cost cutter for MPEG-1 system streams," *ICCE '95*, pp. 172–173, Chicago, IL, June 1995.

[564] G. Fleming, "An economical solution—Video-CD," *ICCE '95*, pp. 218–219, Chicago, IL, June 1995.

[565] R. W. J. J. Saeijs et al., "An experimental digital consumer recorder for MPEG-coded video signals," *ICCE '95*, pp. 232–233, Chicago, IL, June 1995.

[566] K. Kawahara et al., "A single chip MPEG-1 decoder," *ICCE '95*, pp. 254–255, Chicago, IL, June 1995.

[567] Y. H. Wang et al., "Performance of MPEG-1 video on multiaccess integrated network," *2nd Asia-Pacific Conf. on Commun.*, pp. 628–631, Osaka, Japan, June 1995.

[568] A. Makivirta et al., "Error performance and error concealment strategies for MPEG audio coding," *Australian Telecommun., Networks & Appl. Conf.*, Melbourne, Australia, Dec. 1994.

[569] MPEG audio in cyberspace IUMA (Internet Underground Music Archives) at sunsite.unc.edu (http://sunsite.unc.edu).

[570] A. S. Spanias, "Speech coding: A tutorial review," *Proc. IEEE*, vol. 82, pp. 1541–1582, Oct. 1994.

[571] J. Watkinson, *The art of digital audio* (2nd ed.), Oxford, UK: Focal Press, 1994.

[572] L. Bergher, et al., "MPEG audio decoder for consumer applications," *IEEE Custom Integrated Circuits Conf.*, Santa Clara, CA, May 1995.

[573] A. M. Kondoz, *Digital speech: coding for low bit rate communications systems*, New York, NY: John Wiley, 1994.

[574] J. Watkinson, *Digital compression in video and audio*, Focal Press, UK: Butterworth-Heinemann, Boston, 1995.

[575] Luc Baert et al. (Eds.), *Digital audio and compact disc technology*, Focal Press, UK: Butterworth-Heinemann, Oxford, England, 1995.

[576] F. R. Jean, H. I. Lin, and H. C. Wang, "Near transparent audio coding at low bit-rate based on minimum noise loudness criterion," *ICCE '95*, pp. 360–361, Chicago, IL, June 1995.

[577] H. Jung et al., "Digital audio processor for consumer-use digital VCR," *ICCE '95*, pp. 414–415, Chicago, IL, June 1995.

[578] A. C. Hung, *PVRG-MPEG Codec 1.1*, Portable Video Research Group, Stanford University, June 1993.

Chapter 11 MPEG-2

[579] ISO/IEC JTC1/SC29/WG11, *Test Model 4*, Doc. AVC-445, Jan. 1993.

[580] Special issue on Video Coding for 10 Mbit/s, *Signal Process.: Image Commun.*, vol. 5, Feb. 1993.

[581] M. R. Civanlar and A. Puri, "Scalable Video Coding in Frequency Domain," *SPIE/VCIP*, vol. 1818, pp. 1124–1134, Boston, MA, Nov. 1992.

[582] T. Hidaka and K., Ozawa, "ISO/IEC JTC1 SC29/WG11: Report on MPEG-2 Subjective assessment at Kurihama," *Signal Process.: Image Commun.*, vol. 5, pp. 127–158, Feb. 1993.

[583] ISO/IEC JTC1/SC29/WG11, *Test Model 5*, MPEG 93/457, Document AVC-491, April 1993.

[584] S. H. Lee, S. H. Jang, and J. S. Koh, "Selective protection of coded bit-stream and error concealment for ATM transmission of MPEG video," *IEEE Workshop on Visual Signal Process. and Commun.*, pp. 307–310, Melbourne, Australia, Sept. 21-22, 1993.

[585] ISO/IEC JTC1/SC29/WG11, *Preliminary Working Draft*, MPEG 92/086, Document AVC-212, Mar. 1992.

[586] Y. S. Tan and K. K. Pang, "Optimum loop/interpolation filters in layered hybrid coders," *IEEE Workshop on Visual Signal Process. and Commun.*, pp. 259–262, Melbourne, Australia, Sept. 21–22, 1993. See also K. K. Pang and T. K. Tan, "Optimum loop filter in hybrid coders," *IEEE Trans. Circuits and Systems for Video Technology*, vol. 4, pp. 158–167, Apr. 1994.

[587] A. Puri and R. Aravind, "Motion-compensated video coding with adaptive perceptual quantization," *IEEE Trans. Circuits and Systems for Video Technology*, vol. 1, pp. 351–361, Dec. 1991.

[588] G. A. Gonzales and E. Viscito, "Motion video adaptive quantization in the transform domain," *IEEE Trans. Circuits and Systems for Video Technology*, vol. 1, pp, 374–378, Dec. 1991.
[589] A. W. Johnson and J. Princen, "Drift minimization in frequency scalable coders using block filtering, *IEEE Workshop on Visual Signal Process. and Commun.*, Melbourne, Australia, pp. 231–234, Sept. 1993.
[590] K. N. Ngan et al., "Frequency scalability experiments for MPEG-2 standard," *Asia-Pacific Conf. on Commun.*, Taejon, Korea, pp. 298–301, Aug. 1993.
[591] ISO/IEC JTC1/SC29/WG11, *CD 13818, Generic coding of moving pictures and associated audio*, Nov. 1993.
[592] ISO/IEC 13818-2—ITU-T Rec. H.262, *Generic coding of moving pictures and associated audio information: Video*, 1995.
[593] S. Okubo, K. McCann, and A. Lippman, "MPEG-2 requirements, profiles and performance verification," *Intl. Workshop on HDTV '93*, Ottawa, Canada, Oct. 26–28, 1993.
[594] C. Gonzales and E. Viscito, "Flexibly scalable digital video coding," *Signal Process.: Image Commun.*, vol. 5, pp. 5–20, Feb. 1993.
[595] C. T. Chen et al., "Hybrid extended MPEG video coding algorithm for general video applications," *Signal Process.: Image Commun.*, vol. 5, pp. 21–37, Feb. 1993.
[596] A. Puri, R. Aravind, and B. Haskell, "Adaptive frame/field motion compensated video coding," *Signal Process.: Image Commun.*, vol. 5, pp. 39–58, Feb. 1993.
[597] R. ter Horst, A. Koster, and K. Rijkse, "MUPCOS: a multi-purpose coding scheme," *Signal Process.: Image Commun.*, vol. 5, pp. 75–89, Feb. 1993.
[598] G. Morrison and I. Parke, "COSMIC: a compatible scheme for moving image coding," *Signal Process.: Image Commun.*, vol. 5, pp. 91–103, Feb. 1993.
[599] G. Schamel and J. De Lameillieure, "Subband based TV coding," *Signal Process.: Image Commun.*, vol. 5, pp. 105–118, Feb. 1993.
[600] P. Delogne, B. V. Caillie, and O. Poncin, "Video coding algorithm up to 10 Mbit/s," *Signal Process.: Image Commun.*, vol. 5, pp. 119–125, Feb. 1993.
[601] L. Hui and T. Kogure, "An adaptive hybrid DPCM/DCT method for video coding," *Signal Process.: Image Commun.*, vol. 5, pp. 199–208, Feb. 1993.
[602] T. Hanamura, W. Kameyama, and H. Tominaga, "Hierarchical coding scheme of video signal with scalability and compatibility," *Signal Process.: Image Commun.*, vol. 5, pp. 159–184, Feb. 1993.
[603] Y. Yu and D. Anastassiou, "Interlaced video coding with field-based multiresolution representation," *Signal Process.: Image Commun.*, vol. 5, pp. 185–198, Feb. 1993.

[604] Y. Ho, C. Basile, and A. Miron, "MPEG-based video coding for digital simulcasting," *Intl. Workshop on HDTV '92*, Kawasaki, Japan, 1992.

[605] G. Schamel, "Graceful degradation and scalability in digital coding for terrestrial transmission," *Intl. Workshop on HDTV '92*, Kawasaki, Japan, 1992.

[606] F. Boucherok and J. Vial, "Compatible multi-resolution coding scheme," *Intl. Workshop on HDTV '92*, Kawasaki, Japan, 1992.

[607] ISO/IEC JTC1/SC2/WG11, *Subjective test results of MPEG-2, Kurihama meeting*, MPEG 91/346, Nov. 1991.

[608] ISO/IEC JTC1/SC29/WG11, *Comparison of CD 13818 with specified requirements for MPEG-2 work*, MPEG 93/606, Nov. 1993.

[609] ITU-T SG15, Experts Group for ATM Video Coding, *Report of the eleventh meeting in Sydney and Melbourne*, Doc. AVC-496R, Apr. 7, 1993.

[610] ITU-T SG15, Experts Group for ATM Video Coding, *Report of the thirteenth experts group meeting in Brussels*, Doc. AVC-578R, Sept. 10, 1993.

[611] B. Girod et al., "A subjective evaluation of noise-shaping quantization for adaptive intra-/interframe DPCM coding of color television signals," *IEEE Trans. Commun.*, vol. 36, pp. 332–346, Mar. 1988.

[612] N. G. Panagiotidis and S. D. Kollias, "DCT/subband coding techniques in frequency scalable video coding," *SPIE, Video Commun. and PACS for Medical Appl.*, Berlin, FRG, vol. 1977, pp. 344–353, Apr. 1993.

[613] S. K. Chan and A. Leon–Garcia, "Block loss for ATM video," *SPIE/VCIP*, Cambridge, MA, vol. 2094, pp. 213–222, Nov. 1993.

[614] D. W. Lin, M. H. Wang, and J. J. Chen, "Optimal delayed-coding of video sequences subject to a buffer-size constraint," *SPIE/VCIP*, Cambridge, MA, vol. 2094, pp. 223–234, Nov. 1993.

[615] F. Moscheni, F. Dufaux, and H. Nicolas,"Entropy criterion for optimal bit allocation between motion and prediction error information," *SPIE/VCIP*, Cambridge, MA, vol. 2094, pp. 235–243, Nov. 1993.

[616] J. H. Jeon and J. K. Kim, "Hierarchical edge-based block motion estimation for video subband coding at low bit-rates," *SPIE/VCIP*, Cambridge, MA, vol. 2094, pp. 337–349, Nov. 1993.

[617] N. Haddadi and C. J. Kuo, "Fast computation of motion vectors for MPEG," *SPIE/VCIP*, Cambridge, MA, vol. 2094, pp. 350–361, Nov. 1993.

[618] R. E. H. Franich, R. L. Lagendijk, and J. Biemond, "Stereo-enhanced displacement estimation by generic block matching," *SPIE/VCIP*, Cambridge, MA, vol. 2094, pp. 362–371, Nov. 1993.

[619] A. K. Rao et al., "Low-complexity field-based processing for high-quality video compression," *SPIE/VCIP*, Cambridge, MA, vol. 2094, pp. 456–462, Nov. 1993.

[620] R. Gandhi et al., "MPEG-like pyramidal video coder," *SPIE/VCIP*, Cambridge, MA, vol. 2094, pp. 706–717, Nov. 1993.

[621] A. Puri and A. H. Wong, "Spatial-domain resolution-scalable video coding," *SPIE/VCIP*, Cambridge, MA, vol. 2094, pp. 718–729, Nov. 1993.

[622] A. Puri, "Video coding using the MPEG-2 compression standard," *SPIE/VCIP*, Cambridge, MA, vol. 2094, pp. 1701–1713, Nov. 1993.

[623] A. H. Wong and C. T. Chen, "Comparison of ISO MPEG1 and MPEG2 video coding standards," *SPIE/VCIP*, Cambridge, MA, vol. 2094, pp. 1436–1448, Nov. 1993.

[624] E. Fisch, "Scan conversion between 1050 2:1 60 Hz and 525 1:1 30 Hz U and V color components," *IEEE Trans. Consumer Electronics*, vol. 39, pp. 210–218, Aug. 1993.

[625] D. H. Lee, J. S. Park, and Y. G. Kim, "Video format conversions between HDTV systems," *IEEE Trans. Consumer Electronics*, vol. 39, pp. 285–259, Aug. 1993.

[626] S. Lei, "Forward error correction codes for MPEG2 over ATM," *IEEE Trans. Circuits and Systems for Video Technology*, vol. 4, pp. 200–203, Aug. 1993.

[627] T. Akiyama et al., "MPEG2 video codec using image compression DSP," *ICCE '94*, Chicago, IL, pp. 150–151, June 1994.

[628] SGS Thomson Microelectronics, "An integrated MPEG-1 and MPEG-2 decoder," *ICCE '94*, Chicago, IL, pp. 324–325, June 1994.

[629] L. M. van de Kerkhof et al., "MPEG1 and MPEG2 audio coding algorithms and implementation," *ICCE '94*, Chicago, IL, pp. 236–237, June 1994.

[630] T. Yoshino et al., "A 54 MHz motion estimation engine for real-time MPEG video encoding," *ICCE '94*, Chicago, IL, pp. 76–77, June 1994.

[631] T. Kopet, "Designing a multimedia PC add-in card that implements the H.261, MPEG, and JPEG video compression standards," *DSP Exposition and Symposium*, San Francisco, CA, June 1994.

[632] M. Kato, M. Tanaka, and T. Iizuka, "Chipmakers sample MPEG–2 decoders for set-top box, LAN, PC, game," *Nikkei Electronics Asia*, pp. 40–47, June 1994.

[633] R. Yates, S. Evans, and P. A. Ivey, "A 1.2 billion operations per second video signal processing chip," *ICIP '94*, Austin, TX, pp. 596–600, Nov. 1994.

[634] D. Birks and M. Isnardi, "MPEG-2 video encoder based on VLSI chip set," *IEEE Intl. Conf. on Consumer Electronics*, Chicago, IL, pp. 164–165, June 1995.

[635] S. W. Wu and A. Gersho, "Joint estimation of forward and backward motion vectors for interpolative prediction of video," *IEEE Trans. Image Process.*, vol. 3, pp. 684–687, Sept. 1994.

[636] Y. S. Tan and K. N. Ngan, "A flexible three layer frequency/SNR scalable video coder," *Australian Telecommun. Networks & Applications Conf.*, vol. 1, pp. 119–124, Melbourne, Australia, Dec. 1994.

[637] J. Feng, K. T. Lo, and H. Mehrpour, "MPEG compatible adaptive block matching algorithm for motion estimation," *Australian Telecommun. Networks & Applications Conf.*, vol. 1, pp. 131–134, Melbourne, Australia, Dec. 1994.

[638] R. Mathews and J. F. Arnold, "Drift correction in layered coders based on data partitioning," *Australian Telecommun. Networks & Applications Conf.*, vol. 1, pp. 147–152, Melbourne, Australia, Dec. 1994.

[639] Y. Wang, J. F. Arnold, and M. C. Cavenor, "SNR—scalable coding of digital HDTV based on MPEG-2," *Australian Telecommun. Networks & Applications Conf.*, vol. 2, pp. 493–498, Melbourne, Australia, Dec. 1994.

[640] M. Pickering, J. Arnold, and M. Cavenor, "A VBR rate control algorithm for MPEG-2 video coders," *Australian Telecommun. Networks & Applications Conf.*, vol. 2, pp. 521–526, Melbourne, Australia, Dec. 1994.

[641] S. Dunstan, "MPEG systems and ATM network adaptation," *Australian Telecommun. Networks & Applications Conf.*, vol. 2, pp. 431-436, Melbourne, Australia, Dec. 1994.

[642] T. Onoye et al., "Design of inverse DCT unit and motion compensator for MPEG2 HDTV decoding," *IEEE Asia-Pacific Conf. on Circuits and Systems*, pp. 608–615, Taipei, Taiwan, Dec. 1994.

[643] Y. Juan and S. Chang, "Scene change detection in a MPEG-compressed video," *SPIE/IS&T*, vol. 2419, pp. 14–25, San Jose, CA, Feb. 1995.

[644] H. H. Liu and G. L. Zick, "Scene decomposition of MPEG compressed video," *SPIE/IS&T*, vol. 2419, pp. 26–37, San Jose, CA, Feb. 1995.

[645] S. Eckert and C. E. Fogg, "ISO/IEC MPEG-2 software video codec," *SPIE/IS&T*, vol. 2419, pp. 100–109, San Jose, CA, Feb. 1995.

[646] A. T. Erdem and M. I. Sezan, "Compression of 10–bit video using the tools of MPEG-2," *Signal Process.: Image Commun.*, vol. 7, pp. 27–56, March 1995.

[647] W. Kwok, H. Sun, and J. Ju, "Obtaining an upper bound in MPEG coding performance from jointly optimizing coding mode decisions and rate control," *SPIE/VCIP*, vol. 2501, pp. 2–10, Taipei, Taiwan, May 1995.

[648] M. Yuen, H. R. Wu, and K. R. Rao, "Performance evaluation of POCS loop filtering in generic MC/DPCM/DCT video coding," *SPIE/VCIP*, vol. 2501, pp. 65–75, Taiwan, May 1995.

[649] E. D. Frimout, J. Biemond and R. L. Lagendijk, "Extraction of a dedicated fast playback MPEG bit stream," *SPIE/VCIP*, vol. 2501, pp. 76–87, Taipei, Taiwan, May 1995.

[650] M. Chien, H. Sun, and W. Kwok, "Temporal and spatial POCS-based error concealment algorithm for the MPEG encoded video sequence," *SPIE/VCIP*, vol. 2501, pp. 168–174, Taipei, Taiwan, May 1995.

[651] W. Luo and M. E. Zarki, "Analysis of error concealment schemes for MPEG-2 video transmission over ATM-based B-ISDN," *SPIE/VCIP*, vol. 2501, pp. 1358–1368, Taipei, Taiwan, May 1995.

[652] I. E. G. Richardson and M. J. Riley, "Intelligent packetizing of MPEG video data," *SPIE/VCIP*, vol. 2501, pp. 1388–1395, Taipei, Taiwan, May 1995.

[653] L. J. Lin, A. Ortega, and C. C. J. Kuo, "Gradient-based buffer control technique for MPEG," *SPIE/VCIP*, vol. 2501, pp. 1502–1513, Taipei, Taiwan, May 1995.

[654] J. L. Mitchell, W. B. Pennebaker, and D. J. Le Gall, "The MPEG digital video compression standard," New York, NY: Van Nostrand Reinhold, 1996.

[655] L. Wang, "Rate control for MPEG video coding," *SPIE/VCIP*, vol. 2501, pp. 53–64, Taipei, Taiwan, May 1995.

[656] J. Feng, H. Mehrpaur, and K. Lo, "Two-layer MPEG video coding algorithm for ATM networks," *IEEE Intl. Symp. on Circuits and Systems*, Seattle, WA, April/May 1995.

[657] M. Winzker, P. Pirsch, and R. Reimers, "Architecture and memory requirements for stand-alone and hierarchical MPEG-2 HDTV-decoders with synchronous DRAMs," *IEEE Intl. Symp. on Circuits and Systems*, Settle, WA, April/May 1995.

[658] T. Matsumura et al., "A chip set architecture for programmable real-time MPEG-2 video encoder," *IEEE Custom Integrated Circuits Conf.*, Santa Clara, CA, May 1995.

[659] S. Nakagawa et al., "A single chip, 5 GOPS, macroblock-level pixel processor for MPEG-2 real-time encoding," *IEEE Custom Integrated Circuits Conf.*, Santa Clara, CA, May 1995.

[660] J. Armer et al., "A chip set for MPEG-2 video encoding," *IEEE Custom Integrated Circuits Conf.*, Santa Clara, CA, May 1995.

[661] S. Yokono et al., "MPEG-2 data rate recordable 3.5 inch optical disk," *ICCE '95*, pp. 90–91, Chicago, IL, June 1995.

[662] S. Ueda et al., "Development of an MPEG-2 decoder for magneto-optical disk video player," *ICCE '95*, pp. 92–93, Chicago, IL, June 1995.

[663] N. Hurst and P. Meehan, "In-loop coring: Robust MPEG coding for noisy images," *ICCE '95*, pp. 166–167, Chicago, IL, June 1995.

[664] L. Hua et al., "Experimental investigaton on MPEG-2 based video coding at 22 Mbps," *ICCE '95*, pp. 168–169, Chicago, IL, June 1995.
[665] P. Meehan et al., "MPEG compliance bitstream design," *ICCE '95*, pp. 174–175, Chicago, IL, June 1995.
[666] J. Goel, D. Chan, and P. Mandl, "Pre-processing for MPEG compression using adaptive spatial filtering," *ICCE '95*, pp.246–247, Chicago, IL, June 1995.
[667] M. Froidevaux and J.-M. Gentit, "MPEG1 and MPEG2 system layer implementation trade-off between micro-coded and FSM architecture," *ICCE '95*, pp. 250–251, Chicago, IL, June 1995.
[668] C. Hanna et al., "Demultiplexer IC for MPEG2 transport streams," *ICCE '95*, pp. 252–253, Chicago, IL, June 1995.
[669] Y. Mogi et al., "A low power MPEG2 decoder chipset for set-top box," *ICCE '95*, pp. 256–257, Chicago, IL, June 1995.
[670] Y.-P. Lee, L.-G. Chen, and C.-W. Ku, "Architecture design of MPEG-2 decoder system," *ICCE '95*, pp. 258–259, Chicago, IL, June 1995.
[671] J. van den Hurk and P. Frencken, "A concept for source decoding in digital video broadcast applications," *ICCE '95*, pp. 260–261, Chicago, IL, June 1995.
[672] Y. Kanai et al., "The MPEG2 3D player system," *ICCE '95*, pp. 268–269, Chicago, IL, June 1995.
[673] I. E. G. Richardson and M. J. Riley, "MPEG coding for error-resilient transmission," *3rd ICIP '93*, pp. 559–563, Edinburgh, Scotland, July 1993.
[674] Y. S. Saw, P. M. Grant, and J. Hannah, "Reduced storage transmission buffer designs for MPEG video coder," *3rd ICIP '93*, pp. 608–612, Edinburgh, Scotland, July 1993.
[675] J. H. Kim and B. G. Lee, "Transfer of MPEG-2 stream over ATM networks," *2nd Asia-Pacific Conf. on Commun.*, pp. 672–676, Osaka, Japan, June 1995.
[676] Y. Q. Zhang and X. Lee, "Performance of MPEG codecs in the presence of errors," *J. of Visual Commun. and Image Representation*, vol. 5, pp. 379–387, Dec. 1994.
[677] Y. C. Jeung, "A 4-channel sub-band filter for the multi-channel extension of MPEG-2 audio decoder," *ICSPAT '95*, Boston, MA, Oct. 1995.
[678] A. Jalali, "MPEG-2 and JPEG on ATM," *Photonics East, SPIE*, vol. CR60, Philadelphia, PA, Oct. 1995.
[679] W. Lee and Y. Kim, "Multiprocessor architectures and algorithms for real time MPEG-2 video coding," *Photonics East, SPIE*, vol. CR60, Philadelphia, PA, Oct. 1995.
[680] T. Naveen et al., "MPEG 4:2:2 profile: high-quality video for studio applications," *Photonics East, SPIE*, vol. CR60, Philadelphia, PA, Oct. 1995.

[681] A. Puri et al., "Compression of stereoscopic video using MPEG-2," *Photonics East, SPIE*, vol. CR60, Philadelphia, PA, Oct. 1995.

[682] H. Yasuda, "MPEG impact to industries and standardization fields," *Photonics East, SPIE*, vol. CR60, Philadelphia, PA, Oct. 1995.

[683] W. Luo and M. E. Zarki, "Adaptive data portitioning for MPEG-2 video transmission over ATM based networks," *ICIP '95*, vol. 1, pp. 17–20, Washington, DC, Oct. 1995

[684] H. Sun, W. Kwok, and J. Zdepski, "Architectures for MPEG compressed bitstream scaling," *ICIP '95*, vol. 1, pp. 81–84, Washington, DC, Oct. 1995.

[685] G. Keesman et al., "Study of the subjective performance of a range of MPEG-2 encoders," *ICIP '95*, vol. 2, pp. 543–546 Washington, DC, Oct. 1995

[686] F. H. Lin, "An optimization of MPEG to maximize subjective quality," *ICIP '95*, vol. 2, pp. 547–550, Washington, DC, Oct. 1995

[687] J. B. Cheng and H. M. Hang, "Adaptive piecewise linear bits estimation model for MPEG based video coding," *ICIP '95*, vol. 2, pp. 551–554, Washington, DC, Oct. 1995.

[688] J. Katto and M. Ohta, "Mathematical analysis of MPEG compression capability and its application to rate control," *ICIP '95*, vol. 2, pp. 555–558, Washington, DC, Oct. 1995.

[689] L. J. Lin and A. Ortega, "A gradient-based rate control algorithm with applications to MPEG video," *ICIP '95*, vol. 3, pp. 392–395, Washington, DC, Oct. 1995.

[690] Y. Nakajima, H. Jori, and T. Kanoh, "Rate conversion of MPEG coded video by requantization process," *ICIP '95*, vol. 3, pp. 408–411, Washington, DC, Oct. 1995.

[691] E. Morimatsu et al., "Development of a VLSI chip for real time MPEG-2 video decoder," *ICIP '95*, vol. 3, pp. 456–459, Washington, DC, Oct. 1995.

[692] K. Shen and E. J. Delp, "A fast algorithm for video parsing using MPEG compressed sequences," *ICIP '95*, vol. 2, pp. 252–255, Washington, DC, Oct. 1995.

[693] B. L. Yeo and B. Liu, "On the extraction of DC sequence from MPEG compressed video," *ICIP '95*, vol. 2, pp. 260–263, Washington, DC, Oct. 1995.

Chapter 12 MPEG-4/H.263

[694] ISO/IEC JTC1/SC29/WG11, *MPEG-4 requirements document—Second draft*, Doc. N0711r2, July 1994.

[695] Workshop on "Very low bit rate video compression," University of Illinois, Urbana, IL, May 1, 1993.

[696] International workshop on "Coding Techniques for very low bit rate video," University of Essex, Colchester, UK, April 7–8, 1994.

[697] Special issue on very low–bit-rate video coding, *IEEE Trans. CSVT*, vol. 4, June 1994.

[698] R. Schaphorst and C. Reader, "Status of ITU and ISO/MPEG-4 coding standards at very low bit rates," *SPIE/IS&T Symp. on Electronic Imaging Science & Technology*, vol. 2187, pp. 280–289, San Jose, CA, Feb. 1994.

[699] B. Girod, "Motion estimation and very low bitrate video compression," workshop on "Very low bit rate video compression," University of Illinois, Urbana, IL, May 1, 1993.

[700] H. Jozawa and H. Kotera, "Low bit-rate video coding using segment-based motion compression." *Intl. Workshop on Mobile Multimedia Commun.*, B.1.2, Dec. 1993.

[701] Papers presented at the seminar on MPEG-4 meeting in Paris, March 1994.

[702] Papers presented at the seminar on MPEG-4 meeting in Grimstad, Norway, July 1994.

[703] Papers presented at the seminar on MPEG-4 meeting in Singapore, Nov. 1994.

[704] I. Corset, S. Jeannin, and L. Bouchard, "MPEG-4: very low bit rate coding for multimedia applications," *SPIE/VCIP*, vol. 2308, Chicago, IL, pp. 1065–1073, Sept. 1994.

[705] D. Anastassiou, "Current status of the MPEG-4 standardization effort," *SPIE/VCIP*, vol. 2308, pp. 16–24, Chicago, IL, Sept. 1994.

[706] R. Rimpela and J. Narmie, "Real time implementation of a very low bit rate video codec," *IEEE Workshop on Visual Signal Process. and Commun.*, Piscataway, NJ, Sept. 1994.

[707] ISO/IEC JTC1/SC29/WG11, *MPEG-4 call for proposals*, Doc. N0820, Nov. 1994.

[708] ITU-T SG 15 WP 15/1, *Draft Recommendation H.263 (Video coding for low bitrate communication)*, Doc. LBC-95-251, Oct. 1995.

[709] ITU-T SG 15 WP 15/1, *Video codec test model, TMN4 Rev 1*, Oct. 25, 1994.

[710] ISO/IEC JTC1/SC29/WG11, *Requirements for the MPEG-4 Syntactic Description Language (Draft Revision 2)*, N1022, MPEG95, July 1995.

[711] ISO/IEC JTC1/SC29/WG11, *MPEG-4 Proposal Package Description (PPD) Revision 3*, N998, MPEG95, July 1995.

[712] ISO/IEC JTC1/SC29/WG11, *MPEG-4 Call for Proposals* N997, MPEG95, July 1995.

[713] ISO/IEC JTC1/SC29/WG11, *MPEG-4 Testing and Evaluation Procedures*, Doc. N999, MPEG95, July 1995.

[714] A. Puri, "Status and direction of MPEG-4 standard," *Symp. on Multimedia Commun. and Video Coding*, New York, NY, Oct. 1995.

[715] R. Talluri, "MPEG-4 status and directions," *Photonics East, SPIE*, vol. CR60, Philadelphia, PA, Oct. 1995.

[716] T. Wiegand et al., "A rate-constrained encoding strategy for H.263 video compression," *Symp. on Multimedia Commun. and Video Coding*, New York, NY, Oct. 1995.

[717] H. Liu and M. E. Zark, "Data partitioning and unbalanced error protection for H.263 video transmission over wireless channels," *Symp. on Multimedia Commun. and Video Coding*, New York, NY, Oct. 1995.

[718] C. Reader, "Coding with embedded functionality MPEG-4 based coding," *ICIP '95*, Washington, D.C., Oct. 1995.

[719] D. Lindberg and H. S. Malvar, "Multimedia teleconferencing unit with H.324," *Photonics East, SPIE*, vol. CR60, Philadelphia, PA, Oct. 1995.

[720] B. Girod, "Comparison of the H.263 and H.261 video compression standards," *Photonics East, SPIE*, vol. CR60, Philadelphia, PA, Oct. 1995.

Waveform-Based Coder for MPEG-4

[721] Y. Zhang and C. Zafar, "Motion-compensated wavelet transform coding for color video compression," *IEEE Trans. CSVT*, vol, 2, pp. 285–296, Sept. 1992.

[722] P. Sriram and M. W. Marcellin, "Image coding using wavelet transform and entropy-constrained trellis coded quantization," *ICASSP '93*, Minneapolis, MN, pp. V.554–557, Apr. 1993.

[723] C. K. Cheong et al., "Motion estimation with wavelet transform and the application to motion compensated interpolation" *ICASSP '93*, Minneapolis, MN, pp. V.217–220, Apr. 1993.

[724] Z. Xiong, V. P. Galatsanos, and M. T. Orchard, "Marginal analysis prioritization for image compression based on a hierarchical wavelet decomposition," *ICASSP '93*, Minneapolis, MN, pp. V.546–549, Apr. 1993.

[725] J. M. Shapiro, "Application of the embedded wavelet hierarchical image coder to very low bit rate image coding," *ICASSP '93*, Minneapolis, MN, pp. V.558–561, Apr. 1993.

[726] A. Said and W. A. Pearlman, "Image compression using the spatial-orientation tree," *ISCAS '93*, pp. 279–282, 1993.

[727] S. Zhong, Q. Shi, and M. Cheng, "High compression ratio image compression," *ISCAS '93*, pp. 275–278, 1993.

[728] J. R. Casas and L. Torres, "Coding of details in very low bit-rate video systems," *IEEE Trans. CSVT*, vol. 4, pp. 317–327, June 1994.
[729] J. Katto et al., "A wavelet codec with overlapped motion compensation for very low bit-rate environment, *IEEE Trans. CSVT*, vol. 4, pp. 328–338, June 1994.
[730] Y. Nakaya and H. Harashima, "Motion compensation based on spatial transformations," *IEEE Trans. CSVT*, vol. 4, pp. 339–356, June 1994.
[731] P. Cicconi and H. Nicolas, "Efficient region-based motion estimation and symmetry oriented segmentation for image sequence coding," *IEEE Trans. CSVT*, vol. 4, pp. 357–364, June 1994.
[732] C. Labit and J. Leduc, "Very low bit rate (VLBR) coding schemes: a new algorithmic challenge," *SPIE/VCIP*, vol. 2308, Chicago, IL, pp. 25–38, Sept. 1994.
[733] S. A. Rajala, "Impact of human visual perception of color on very low bit-rate coding," *SPIE/VCIP*, vol. 2308, Chicago, IL, pp. 39–46, Sept. 1994.
[734] R. A. Neff and A. Zakhor, "Very low bit-rate coding using matching pursuits," *SPIE/VCIP*, vol. 2308, Chicago, IL, pp. 47–60, Sept. 1994.
[735] A. Yamada, M. Ohta, and T. Nishitani,"Video sequence quantizer for a very low bit-rate moving picture coding," *SPIE/VCIP*, vol. 2308, Chicago, IL, pp. 61–72, Sept. 1994.
[736] J. R. Casas and L. Torres, "Coding of significant features in very low bit-rate video systems," *SPIE/VCIP*, vol. 2308, Chicago, IL, pp. 73–85, Sept. 1994.
[737] J. Benois, L. Wu, and D. Barba, "Joint contour-based and motion-based image sequence segmentation for TV image coding at very low bit-rate," *SPIE/VCIP*, vol. 2308, Chicago, IL, pp. 1074–1085, Sept. 1994.
[738] K. N. Ngan and W. L. Chooi, "Very low bit rate video coding using 3D subband approach," *IEEE Trans. CSVT*, vol. 4, pp. 309–316, June 1994.
[739] H. Schiller and B. B. Chaudhuri, "Efficient coding of side information in a low bitrate hybrid image coder," *Signal Process.*, vol. 19, pp. 61–73, Jan. 1990.
[740] R. M. Mersereau et al., "Methods for low bit-rate video compression: some issues and answers," *SPIE/VCIP*, vol. 2308, pp. 2–13, Chicago, IL, Sept. 1994.
[741] R. W. Young and N. G. Kingsbury, "Frequency-domain motion estimation using a complex lapped transform," *IEEE Trans. Image Process.*, vol. 2, pp. 2–17, Jan. 1993.
[742] H. S. Malvar and D. H. Staelin, "The LOT: transform coding without blocking effects," *IEEE Trans. Acoust., Speech, and Signal Process.*, vol. 37, pp. 553–559, Apr. 1989.
[743] H. S. Malvar, *Signal processing with lapped transform*, Norwood, MA: Artech House, 1992.

[744] B. Belzer, J. Liao, and J. D. Villasenor, "Adaptive video coding for mobile networks," *ICIP '94*, Austin, TX, pp. 972–976, Nov. 1994.

[745] Y. Huh et al., "Classified wavelet transform coding of images using vector quantization," *SPIE/VCIP*, pp. 207–217, Chicago, IL, Sept. 1994.

[746] Y. Huh, J. J. Hwang, and K. R. Rao, "Block wavelet transform coding of images using classified vector quantization," *IEEE Trans. Circuits and Systems for Video Technology*, vol. 5, pp. 63–67, Feb. 1995.

[747] E. A. B. da Silva and M. Ghanbari, "On the performance of linear phase wavelet transforms in low bitrate image coding", *IEEE Trans. Image Process.*, vol. 3, pp. 222–223, Mar. 1994.

[748] G. Eude and J. C. Schmitt, "Optimized hybrid transform coding for very low bit rate: videotelephony communication on PC," *SPIE/IS&T Symp. on Electronic Imaging Science & Technology*, vol. 2187, pp. 290–300, San Jose, CA, Feb. 1994.

[749] S. W. Kim and H. K. Lee, "Video coding with wavelet transform on the very low bit-rate communication channel," *SPIE/IS&T Symposium on Electronic Imaging Science & Technology*, vol. 2187, pp. 309–320, San Jose, CA, Feb. 1994.

[750] K. Grotz, J. Mayer, and G. Süußmeier, "A 64 Kbit/s videophone codec with forward analysis and control," *Signal Process.: Image Commun.*, vol. 1, pp. 103–116, Oct. 1989.

[751] K. K. Pang and T. K. Tan, "Optimum loop filter in hybrid coders," *IEEE Trans. CSVT*, vol. 4, pp. 158–167, Apr. 1994.

Object-Oriented Coder for MPEG-4

[752] J. Ostermann, "Modelling of 3D moving objects for an analysis-synthesis coder," *SPIE, Symp. on Sensing and Reconst. of 3D Objects and Scenes*, vol. 1260, pp. 240–250, 1990.

[753] H. G. Musmann, M. Hötter, and J. Ostermann, "Object-oriented analysis-synthesis coding of moving images." *Signal Process.: Image Commun.*, vol. 1, pp. 117–138, Oct. 1989.

[754] M. Hötter, "Optimization and efficiency of an object oriented analysis--synthesis coder," *IEEE Trans. Circuts and Systems for Video Technology*, vol. 4, pp. 181–194, Apr. 1994.

[755] M. Hötter, "Object-oriented analysis-synthesis coding based on moving two-dimensional objects," *Signal Process.: Image Commun.*, vol. 2, pp. 409–428, Dec. 1990.

[756] M. Hötter, "Object-oriented analysis-synthesis coding based on the model of flexible 2D-objects," *Picture Coding Symp.*, Boston, MA, 1990.

[757] M. Hötter and R. Thoma, "Image segmentation based on object-oriented mapping parameter estimation," *Signal Process.*, vol. 15, pp. 315–334, 1988.

[758] M. Hötter and J. Ostermann, "Analysis synthesis coding based on planar rigid moving objects," *Intl. Workshop on 64 Kbps Coding of Moving Video*, Hannover, Germany, 1988.

[759] J. Ostermann, "Modelling of 2D moving objects for an analysis-synthesis coder," *SPIE Symp. on Sensing and Reconst. of 3D Objects and Scenes*, vol. 1260, pp. 240–249, Santa Clara, CA, Feb. 1990.

[760] P. Gerken, "Object-based analysis-synthesis coding of image sequences at very low bit rates," *IEEE Trans. CSVT*, vol. 4, pp. 228–235, June 1994.

Model-Based Coder for MPEG-4

[761] K. Aizawa, H. Harashima, and T. Saito, "Model-based synthetic image coding system," *Picture Coding Symp.*, Stockholm, Sweden, June 1987.

[762] K. Aizawa, H. Harashima, and T. Saito, "Model-based synthetic image coding for a person's face," *Image Commun.*, vol. 1, no. 2, pp. 139–152, 1989.

[763] C. Liedtke, H. Busch, and R. Koch, "Automatic modelling of 3D moving objects from a TV image sequence," *SPIE, Symp. on Sensing and Reconst. of 3D Objects and Scenes*, vol. 1260, pp. 230–239, 1990.

[764] K. Aizawa and T. S. Huang, "Model-based synthetic image coding," *Paper summaries of Dynamic Scene Understanding, 1991 IJCAI Workshop*, Sydney, Australia, 1991.

[765] K. Aizawa, "Model-based image coding," *SPIE/VCIP*, vol. 2308, pp. 1035–1049, Chicago, IL, Sept. 1994.

[766] C. S. Choi et al., "Analysis and synthesis of facial expressions in model-based image coding," *Picture Coding Symp.*, Cambridge, MA, Mar. 1990.

[767] R. Frochheimer and T. Kronander, "Image coding – from waveforms to animation," *IEEE Trans. ASSP*, vol. 37, pp. 2008–2023, Dec. 1989.

[768] T. Fukuhara, K. Asai, and T. Murakami, "Hierarchical division of 2-d wire-frame model and vector quantization in a model-based coding of facial images," *Picture Coding Symp.*, Cambridge, MA, Mar. 1990.

[769] M. Kaneko, A. Koike, and Y. Hatori, "Real-time analysis and synthesis of moving facial images applied to model-based image coding," *Picture Coding Symp.*, Tokyo, Japan, 1991.

[770] H. Li, P. Roivainen, and R. Forchheimer, "Recursive estimation of facial expression and movement," *ICASSP '92*, pp. 593–596, San Francisco, CA, 1992.

[771] T. Minami et al., "Knowledge-based coding of facial images," *Picture Coding Symp.*, Cambridge, MA, Mar. 1990.

[772] H. Morikawa and H. Harashima, “3-d structure extraction coding of image sequence,” *ICASSP ’90*, Albuquerque, NM, pp. 1969–1972, 1990.

[773] Y. Nakaya, Y. C. Chuah, and H. Harashima, “Model-based/waveform hybrid coding for videotelephone images,” *ICASSP ’91*, pp. 2741–2744, Toronto, Canada, 1991.

[774] W. J. Welsh, “Model-based coding of videotelephone images using an analysis synthesis method,” *Picture Coding Symp.*, Torino, Italy, Sept. 1988.

[775] W. J. Welsh, “Model-based coding of moving images at very low bit rate,” *Picture Coding Symp.*, Stockholm, Sweden, June 1987.

[776] S. Morishima, K. Aizawa, and H. Harashima, “A real-time facial action image synthesis system driven by speech and text,” *SPIE/VCIP*, vol. 1360, pp. 1151–1158, Lausanne, Switzerland, Oct. 1990.

[777] C. S. Choi et al., “Analysis and synthesis of facial image sequences in model-based image coding,” *IEEE Trans. CSVT*, vol. 4, pp. 257–275, June 1994.

[778] D. Cai, “Model-based facial image coding using deformable 3D wire frame model for very low bit transmission,” *SPIE Intl. Symp. on Optics, Imaging, and Instrumentation*, vol. 2298, pp. 751–762, San Diego, CA, July 1994.

[779] K. Aizawa, H. Harashima, and Y. Saito, “Model-based analysis synthesis image coding system for very low-rate image transmission,” *Picture Coding Symp.*, Torino, Italy, Sept. 1988.

[780] K. Aizawa and T. S. Huang, “Model-based image coding: Advanced video coding techniques for very low bit-rate applications,” *Proc. IEEE*, vol. 83, pp. 259–271, Feb. 1995.

Fractal and Morphological Coder for MPEG-4

[781] M. F. Barnsley and A. D. Sloan, “A better way to compress images,” *Byte*, 13(1), pp. 215–224, 1988.

[782] A. E. Jacquin, “A novel fractal block coding technique for digital images,” *Proc. IEEE ICASSP*, pp. 2225–2228, 1990.

[783] Y. Fisher, “Fractal image compression,” *SIGGRAPH ’92 Course Notes*, 1992.

[784] D. M. Monro and F. Dudbridge, “Fractal block coding of images,” *IEEE Electronics Letters*, 28(11), pp. 1053–1054, 1992.

[785] M. Novak, H. Nautsch, and N. Wadstromer, “Fractal coding of images,” *Proc. Symp. on Image Anal.*, Uppsala, Sweden, pp. 101–104, Mar. 1992.

[786] J. M. Beaumont, “Image data compression using fractal techniques,” *British Telecommun. Tech. J.*, 9(4), Oct. 1991.

[787] M. F. Barnsley, *Fractals everywhere*, London, UK: Academic Press, 1988.

[788] D. S. Mazel and M. H. Hayes, "Using iterated function systems to model discrete sequences," *IEEE Trans. Signal Process.*, vol. 40, pp. 1724–1734, July 1992.

[789] H. Li, M. Novak, and R. Forchheimer, "Fractal-based image sequence compression scheme," *Optical Engineering*, vol. 32, pp. 1588–1595, July 1993.

[790] M. F. Barnsley and L. P. Hurd, *Fractal image compression*, Wellesley, MA: A. K. Peters Ltd., 1992.

[791] J. Y. Lee, J. H. Suh, and H. Y. Kwon, "An image compression algorithm based on fractal coding and VQ," *Asia-pacific Conf. on Commun.*, Taejon, Korea, pp. 2D3.1–3.5, Aug. 1993.

[792] G. Lu, "Fractal image compression," *Signal Process.: Image Commun.*, vol. 5, pp. 327–343, Oct. 1993.

[793] A. Jacquin, "Fractal image coding based on a theory of iterated contractive image transformations," *SPIE/VCIP*, vol. 1360, pp. 227–239, Oct. 1990. (The best paper that explains the concept in a simple way.)

[794] A. Jacquin, "Image coding based on a fractal theory of iterated contractive image transformations" *IEEE Trans. Image Process.*, vol. 1, pp. 18–30, Jan. 1992.

[795] G. E. Oien, S. Lepsory, and T. A. Ramstad, "An inner product space approach to image coding by contractive transformations," *ICASSP '91*, pp. 2773–2776, 1991.

[796] J. Stark, "Iterated function systems as neural networks," *Neural Networks*, vol. 4, pp. 679–690, Pergamon Press, 1991.

[797] S. Graf, "Barnsley's scheme for the fractal encoding of images," *J. Complexity*, vol. 8, pp. 72–78, 1992.

[798] D. M. Monro, "A hybrid fractal transform," *ICASSP '93*, pp. V.169–172, 1993.

[799] D. M. Monro and F. Dudbridge, "Fractal approximation of image blocks," *ICASSP '92*, pp. III.485–488, 1992.

[800] D. M. Monro, D. Wilson, and J. A. Nicholls, "High speed image coding with the Bath Fractal Transform," *IEEE Intl. Symp. Multimedia Technologies*, Southampton, UK, Apr. 1993.

[801] E. W. Jacobs, Y. Fisher, and R. D. Boss, "Image compression: a study of the iterated transform method," *Signal Process.*, vol. 29, pp. 25–263, 1992.

[802] E. R. Vrscay, "Iterated function systems: theory, applications and the inverse problem," *Fractal geometry and analysis*, edited by J. Belair and S. Dubuc, Kluwer Academic, pp. 405–468, 1991.

[803] J. Domaszewicz and V. A. Vaishampayan, "Iterative collage coding for fractal compression," *ICIP '94*, Austin, TX, pp. 127–131, Nov. 1994.

[804] D. W. Lin, "Fractal image coding as generalized predictive coding," *ICIP '94*, Austin, TX, pp. 117–121, Nov. 1994.

[805] T. Naemura and H. Harashima, "Fractal coding of a multi-view 3-D image," *ICIP '94*, Austin, TX, pp. 107–111, Nov. 1994.

[806] A. Bogdan, "Multiscale (inter/intra-frame) fractal video coding," *ICIP '94*, Austin, TX, pp. 760–764, Nov. 1994.

[807] B. Paul and M. H. Hayes, "Fractal-based compression of motion video sequences," *ICIP '94*, Austin, TX, pp. 755–769, Nov. 1994.

[808] J. R. Casas and L. Torres, "Morphological filter for lossless image subsampling," *ICIP '94*, Austin, TX, pp. 903–907, Nov. 1994.

[809] O. Egger and W. Li, "Very low bit rate image coding using morphological operators and adaptive decompositions," *ICIP '94*, Austin, TX, pp. 326–330, Nov. 1994.

[810] C. Gu and M. Kunt, "Very low bit-rate video coding using multi-criterion segmentation," *ICIP '94*, Austin, TX, pp. 418–422, Nov. 1994.

[811] P. Salembier et al., "Very low bit rate video coding using morphological segmentation and contour/texture motion compensation," *12th Intl. Conference on Pattern Recognition*, Jerusalem, Israel, Oct. 1994.

[812] O. Egger, W. Li, and M. Kunt, "High compression image coding using an adaptive morphological subband decomposition," *Proc. IEEE*, vol. 83, pp. 272–287, Feb. 1995.

MPEG-4 Audio

[813] D. J. Goodman, "Trends in cellular and cordless communications," *IEEE Commun. Magazine*, pp. 31–40, June 1991.

[814] J. P. Campbell Jr., T. E. Tremain, and V. C. Welch, "The DoD 4.8 Kbps standard (proposed federal standard 1016)," *Advances in speech coding*, edited by B. S. Atal, V. Cuperman, and A. Gersho, Hingham, MA: Kluwer Academic, 1991.

[815] M. R. Schroeder and B. S. Atal, "Rate distortion theory and predictive coding," *ICASSP '81*, pp. 201–204, Atlanta, GA, Mar. 1981.

[816] M. R. Schroeder and B. S. Atal, "Code-excited linear prediction (CELP): High-quality speech at very low bit rates," *ICASSP '85*, pp. 937–940, Tampa, FL, Mar. 1985.

[817] G. Davidson and A. Gersho, "Complexity reduction methods for vector excitation coding," *ICASSP '86*, Tokyo, Japan pp. 3055–3058, Apr. 1986.

[818] V. Cuperman et al., "Low-delay vector excitation coding of speech at 16 Kb/s," *IEEE Trans. Commun.*, vol. 40, pp. 129–139, Jan. 1992.

[819] T. V. Ramabadran and C. D. Lueck, "Complexity reduction of CELP speech coders through the use of phase information," *IEEE Trans. Commun.*, vol. 42, pp. 248–251, Feb./Mar./Apr. 1994.

[820] I. A. Gerson and M. A. Jasiuk, "Vector sum excited linear prediction (VSELP) speech coding at 8 Kbps," *ICASSP '90*, Albuquerque, NM, pp. 461–464, Apr. 1990.

[821] S. Dimolitsas, C. Ravishankar, and G. Schröder, "Current objectives in 4-kb/s wireline-quality speech coding standardization," *IEEE Signal Process. Letters*, vol. 1, pp. 157–159, Nov. 1994.

[822] J. Chen et al., "A low-delay CELP coder for the CCITT 16 kb/s speech coding standard," *IEEE J. Selected Areas in Commun.*, vol. 10, pp. 830–849, June 1992.

[823] B. S. Atal, V. Cuperman, and A. Gersho (Eds.), *Advances in speech coding*, Hingham, MA: Kluwer Academic, July 1991.

[824] ITU-T SG 15 WP 15/1, *Report of the Rapporteur for very low bitrate visual telephony*, Geneva, Sept. 1993.

[825] R. D. De Iacovo, CSELT, Contributions on RPE-LTP and GSM half-rate coders.

[826] Electronic Industries Association, "Mobile-standard land station compatibility specification," *American National Standard EIA/TIA 553-1989*, Apr. 1989.

[827] Y. Naito, Mitsubishi Electric Corp., Contributions on JDC VSELP and JDC half-rate coder.

[828] S. Dimolitsas, COMSAT Labs, Additional information on the performance of the IMBE Inmarsat coder.

[829] A. Moyler, Marconi Electronics Ltd., Contributions on Marconi 5 Kbps speech coder.

[830] ITU-T SG 15, *Draft Recommendation G.723 – Dual rate speech coder for multimedia telecommunication transmitting at 5.3 & 6.3 Kbit/s*, March 1995.

Chapter 13 HDTV

[831] Massachusetts Institute of Technology, *Channel compatible DigiCipher HDTV system*, Apr. 1992.

[832] Advanced Television Research Consortium, *Advanced digital television system description*, Sept. 1992.

[833] General Instrument Corp., *DigiCipher^{TM} HDTV system description*, Aug. 1991. See also W. Paik, "DigiCipher^{TM} - all digital channel compatible HDTV broadcast system," *IEEE Trans. Broadcasting*, vol. 36, pp. 245–254, Dec. 1990.

[834] Zenith and AT & T, *Digital Spectrum Compatible, technical details*, Sept. 1991. See also A. Netravali et al., "A codec for HDTV," *IEEE Trans. Consumer Electronics*, vol. 38, pp. 325–340, Aug. 1992.

[835] N. Jayant, "Signal compression: technology targets and research directions," *IEEE J. Selected Areas in Commun.*, vol. 10, pp.796–818, June 1992.

[836] J. Kraus, J. Reimers, and K. Grüger, "A VLSI chip set for DPCM coding of HDTV signals," *IEEE Trans. Circuits and Systems for Video Technology*, vol. 3, pp. 302–308, Aug. 1993.

[837] T. Kinoshita and T. Nakahashi, "A 130 Mb/s compact HDTV codec based on a motion-adaptive DCT algorithm," *IEEE J. Selected Areas in Commun.*, vol. 10, pp. 122–129, Jan. 1992.

[838] J. M. Martinez et al., "Hierachical multiresolution coding for HDTV and TV," *IEEE VSPC '92*, Raleigh, NC, pp. 228–234, Sept. 1992.

[839] K. M. Uz, K. Ramchandran, and M. Vetterli, "Combined multiresolution source coding and modulation for digital broadcast of HDTV," *Signal Process.: Image Commun.*, vol. 4, pp. 283–292, Aug. 1992.

[840] P. Fockens and A. Netravali, "The digital spectrum-compatible HDTV system," *Signal Process.: Image Commun.*, vol. 4, pp. 293–305, Aug. 1992.

[841] K. Sawada, Y. Yashima, and H. Sakai, "An HDTV bit-rate reduction codec at the STM-1 rate of SDH," *Signal Process.: Image Commun.*, vol. 4, pp. 345–358, Aug. 1992.

[842] W. Jaß, W. Tengler, and E. Hundt, "A versatile DCT coding system for HDTV with interlaced and progressive scanning," *Signal Process.: Image Commun.*, vol. 4, pp. 389–399, Aug. 1992.

[843] S. Ono, N. Ohta, and T. Aoyama, "All-digital super high definition images," *Signal Process.: Image Commun.*, vol. 4, pp. 429–444, Aug. 1992.

[844] C. Guillemot et al., "Flexible bit rate coding for satellite transmission of HDTV," *Intl. Workshop on HDTV '92*, Kawasaki, Japan, paper no. 67, 1992.

[845] B. Marti et al., "Problems and perspectives of digital terrestrial television in Europe," *SMPTE J.*, pp. 703–711, Aug. 1993.

[846] M. Barbero and M. Stroppiana, "Video compression techniques and multilevel approaches," *SMPTE J.*, pp. 335–344, May 1994.

[847] R. J. Siracusa, "Flexible and robust packet transport for digital HDTV," *IEEE J. on Selected Areas in Commun.*, vol. 11, pp. 88–98, Jan. 1993.

[848] C. Heegard, S. A. Levy, and W. H. Paik, "Practical coding for QAM transmission of HDTV," *IEEE J. on Selected Areas in Commun.*, vol. 11, pp. 111–118, Jan. 1993.

[849] S. N. Hulyalkar et al., "Advanced digital HDTV transmission system for terrestrial video simulcasting," *IEEE J. on Selected Areas in Commun.*, vol. 11, pp. 119–126, Jan. 1993.

[850] A. N. Netravali and B. Prasada, "Adaptive quantization of picture signals using spatial masking," *Proc. IEEE*, vol. 65, pp. 536–548, Apr. 1977.

[851] N. Coppisetti, K. Challapali, and Y. Ho, "Performance analysis of the advanced digital HDTV video coding system," *IEEE Trans. Consumer Electronics*, vol. 39, pp. 779–788, Nov. 1993.

[852] E. Petajan, "Digital video coding techniques for US high-definition TV," *IEEE Micro*, pp. 13–21, Oct. 1992.

[853] FCC, "ATV system recommendation," *IEEE Trans. Broadcasting*, vol. 39, Mar. 1993.

[854] J. P. Princen, A. W. Johnson, and A. B. Bradley, "Subband/transform coding using filter bank designs based on time domain aliasing cancellation," *ICASSP '87*, Dallas, TX, pp. 2161–2164, Apr. 1987.

[855] J. P. Princen and A. B. Bradley, "Analsis/synthesis filter bank design based on time-domain aliasing cancellation," *IEEE Trans. Acoust., Speech, and Signal Process.*, vol. ASSP-34, pp. 1153–1161, Oct. 1986.

[856] L. D. Fielder and G. A. Davidson, "AC-2 family of low complexity transform based music coders," *AES 10th Intl. Conf.*, London, UK, Sept. 1991.

[857] Y. Wu and B. Caron, "Digital television terrestrial broadcasting," *IEEE Commun. Magazine*, vol. 32, pp. 46–52, May 1994.

[858] ACATS Technical Subgroup, *Grand Alliance HDTV system specification*, Version 1.0, Apr. 14, 1994.

[859] R. Hopkins, "Digital terrestrial HDTV for North America: *IEEE Trans. Consumer Electronics*, vol. 40, pp. 185–198, Aug. 1994.

[860] M. Lodman et al., "A single chip stereo AC-3 audio decoder," *ICCE '94*, Chicago, IL, pp. 234–235, June 1994.

[861] M. Davis, "The AC-3 multichannel coder," *AES 95th Convention*, Preprints no. 3774, New York, Oct. 1993.

[862] J. Mailhot and E. D. Petajan, "Grand Alliance HDTV system," *SPIE/IS&T Symp. on Electronic Imaging Science & Technology*, vol. 2186, San Jose, CA, Feb. 1994.

[863] Dolby Laboratories Inc., *Dolby AC-3, multichannel digital audio compression system*, Revision 1.12, Feb. 22, 1994.

[864] A. G. Elder and S. G. Turner, "A real-time PC based implementation of AC-2 digital audio compression," *AES 95th Convention*, Preprints no. 3773, New York, Oct. 1993.

[865] M. Bosi and S. E. Forshay, "High quality audio coding for HDTV: An overview of AC-3," *Intl. Workshop on HDTV '94*, Turin, Italy, Oct. 1994.

[866] C. Baside et al., "The U.S. HDTV standard. The Grand Alliance," *IEEE Spectrum*, vol. 32, pp. 36–45, Apr. 1995.

[867] T. E. Bell, "The HDTV 'test kitchens,'" *IEEE Spectrum*, vol. 32, pp. 46–49, Apr. 1995.

[868] MC68ADP AC-3 Dolby AC-3 audio decoder up to 448 Kbps, (44.1 and 48 KHz), mono, stereo, surround sound up to 5.1 channels. Motorola, 1994.

[869] V. Tawil, "Analysis of the grand alliance HDTV transmission subsystem field test measurement in Charlotte, NC," *ICCE '95*, pp. 12–13, Chicago, IL, June 1995.

[870] J. N. Mailhot and H. Derovanessian, "The grand alliance HDTV video encoder," *ICCE '95*, pp. 300–301, Chicago, IL, June 1995.

[871] A. Cugnini and R. Shen, "MPEG-2 video decoder for the digital HDTV grand alliance system," *ICCE '95*, pp. 302–303, Chicago, IL, June 1995.

[872] S. Vernon, "Design and implementation of AC-3 coders," *ICCE '95*, pp. 304–305, Chicago, IL, June 1995.

[873] B. Bhatt et al., "HDTV scan converter," *ICCE '95*, pp. 306–307, Chicago, IL, June 1995.

[874] P. W. Lyons, "Grand alliance prototype transport stream encoder design and implementation," *ICCE '95*, pp. 308–309, Chicago, IL, June 1995.

[875] K. Bridgewater and M. S. Deiss, "Grand alliance transport stream decoder design and implementation," *ICCE '95*, pp. 310–311, Chicago, IL, June 1995.

[876] W. Bretl and P. Snopko, "VSB modem subsystem design for grand alliance digital television receivers," *ICCE '95*, pp. 312–313, Chicago, IL, June 1995.

[877] J. G. Kim et al., "Revised record/playback interface systems for GA HDTV, HD-VCR and V3 VTR," *ICCE '95*, pp. 408–409, Chicago, IL, June 1995.

[878] A. K. Al Asmari, "A hybrid coding scheme for HDTV sequences," *ICCE '95*, pp. 140–141, Chicago, IL, June 1995.

[879] T. Nakai et al., "Development of HDTV digital transmission system through satellite," *ICCE '95*, pp. 158–159, Chicago, IL, June 1995.

[880] B. Evans, *Understanding digital TV: the route to HDTV*, New York, NY: IEEE Press, 1995.

[881] Y. Ninomiya, "High definition television systems," *Proc. IEEE*, vol. 83, pp. 1086–1093, July 1995.

[882] E. Petajan, "The HDTV grand alliance system," *Proc. IEEE*, vol. 83, pp. 1094–1105, July 1995.

[883] E. Petajan, "Digital video coding techniques for US high definition TV," *IEEE Micro.*, pp. 13–21, Oct. 1992.

[884] E. Petajan, "The path of least resistance: HDTV on the PC," *Intl. J. Imaging Systems and Technology*, vol. 5, Winter 1994.

[885] X. Lebegue, D. Mclaren, and R. Saint-Girons, "Video compression for the grand alliance: An historical perspective," *Intl. J. Imaging Systems and Technology*, vol. 5, pp. 253–258, Winter 1994.

[886] J. N. Mailhot, "Architecture and implementation of the grand alliance HDTV video encoder," *Intl. J. Imaging Systems and Technology*, vol. 5, pp. 259–262, Winter 1994.

[887] K. Challapali et al., "Grand alliance MPEG-2 based video decoder with parallel processing architecture," *Intl. J. Imaging Systems and Technology*, vol. 5, pp. 263–267, Winter 1994.

[888] J. S. McVeigh and S. W. Wu, "Partial closed-loop versus open-loop motion estimation for HDTV compression," *Intl. J. Imaging Systems and Technology*, vol. 5, pp. 268–275, Winter 1994.

[889] J. Kuriacose et al., "The grand alliance transport system for terrestrial HDTV transmission," *Intl. J. Imaging Systems and Technology*, vol. 5, Winter 1994.

[890] G. Davidson, L. Fielder, and M. Antill, "Low complexity transform coder for satellite link applications," *89th AES Convention*, Preprint No. 2966, Los Angeles, CA, Sept. 1990.

[891] A. G. Elder and S. G. Turner, "A real-time PC based implementation of AC-2 digital audio compression," *89th AES Convention*, Preprint No. 3773, New York, NY, Oct. 1993.

[892] P. Noll, "Digital audio coding for video communications," *Proc. IEEE*, vol. 83, pp. 925–943, June 1995.

[893] C. Todd, "AC-3 audio coder," *Photonics East, SPIE*, vol. CR60, Philadelphia, PA, Oct. 1995.

[894] E. D. Petajan, "Grand alliance HDTV," *Photonics East, SPIE*, vol. CR60, Philadelphia, PA, Oct. 1995.

[895] C. Ahn et al., "DTV/HDTV standard activities in Korea," *Photonics East, SPIE*, vol. CR60, Philadelphia, PA, Oct. 1995.

[896] Y. Ninomiya, "Development of systems and standards of HDTV," *Photonics East, SPIE*, vol. CR60, Philadelphia, PA, Oct. 1995.

[897] K. Terry, "A multiprocessor DSP architecture for AC-3 audio encoding," *ICSPAT '95*, Boston, MA, Oct. 1995.

Chapter 14 CMTT

[898] ITU-T Recommendation 601, *Encoding parameters of digital television for studios*, 1982.

[899] Draft revision of Recommendation ITU-R CMTT. 723, *Transmission of component-coded digital television signals for contribution-quality applications at the third hierarchical level of recommendation ITU-T G.702*, Document CMTT/BL/9-E, 14 Apr. 1993.

[900] Draft revision of Recommendation ITU-R CMTT.721-1, *Transmission of component-coded digital television signals for contribution-quality*

applications at bit rates near 140 Mbit/s, Document CMTT/BL/8-E, 26 Apr. 1993.

[901] M. Barbero et al., "A twin hybrid DCT pyramidal coding system," *Intl. Workshop on HDTV '93*, Ottawa, Canada, Oct. 1993.

[902] M. Muratori, M. Stroppiana, and Y. Nishida, "Coding efficiency of systems adopting progressive, deinterlaced and interlaced formats," *Intl. Workshop on HDTV '93*, Ottawa, Canada, Oct. 1993.

[903] P. Delogne et al., "Compatible coding of digital interlaced HDTV," *IEEE J. on Selected Areas in Commun.*, vol. 11, pp. 146–152, Jan. 1993.

Ghost Cancellation

[904] R. K. Jurgen, "Good-bye to TV ghosts," *IEEE Spectrum*, vol. 29, pp. 50–52, July 1992.

[905] J. H. Winters et al., "Ghost cancellation of analog TV signals: with applications to IDTV, EDTV and HDTV," *IEEE Trans. Circuits and Systems for Video Technology*, vol. 1, pp. 136–146, March 1991.

[906] V. Tawil and L. D. Claudy, "Field testing of a ghost cancelling systems for NTSC television broadcasting," *IEEE Trans. Broadcasting*, vol. 36, pp. 255–261, Dec. 1990.

[907] National Association of Broadcasters, "Field tests of ghost cancelling systems for NTSC television broadcasting," 1771 N. St., N. W., Washington, D.C. 20006; 202-429-5346, Jan. 31, 1992.

[908] Advanced Television System Committee (ATSC), "Ghost cancelling reference signals," 1776 K St. NW, Washington, DC 20006; 202-828-3130, Mar. 20, 1992.

[909] Cable Television Laboratories, "Field and laboratory tests of NTSC ghost cancelling system," 1050 Walnut St., Suite 500, Boulder, CO 80302; 303-939-8500, Mar. 10, 1992.

[910] Advanced Television System Committee (ATSC), "Computer simulations and laboratory tests of proposed ghost cancelling system," 1776 K St. NW, Washington, DC 20006; 202-828-3130, Mar. 1992.

[911] N. Komiya, "Ghost reduction by reproduction," *IEEE Trans. Consumer Electronics*, vol. 38, pp. 195–199, Aug. 1992.

[912] V. Tawil and L. D. Claudy, "Performance of television ghost cancelling system under field test conditions," *IEEE Trans. Consumer Electronics*, vol. 38, pp. 18–25, Aug. 1992.

[913] B. Caron, "Video ghost cancelling: evaluation by computer simulation and laboratory testing," *IEEE Trans. Consumer Electronics*, vol. 38, pp. 26–33, Aug. 1992.

[914] T. Kume, "A digital processing IC for ghost canceller," *IEEE Trans. Consumer Electronics*, vol. 38, pp. 127–134, Aug. 1992.

[915] L. D. Claudy and S. Herman, "Ghost cancelling: a new standard for NTSC broadcast television." *IEEE Trans. Consumer Electronics*, vol. 38, pp. 227–228, Dec. 1992.

[916] W. Thomas and G. Sgrignoli. "A tutorial on ghost cancelling in television systems," *IEEE Trans. Consumer Electronics*, vol. 25, pp. 9–44, Feb. 1979.

[917] Several papers in the session "Ghost Cancellation," *International Conference on Consumer Electronics '94*, Chicago, IL, June 1994.

[918] J. Lee and G. K. Ma, "On the fast RLS adaptive IIR ghost canceller," *Intl. Workshop on HDTV*, Ottawa, Canada, Oct. 1993.

[919] TV Ghost Canceller chip set one ZR 33072, 72-tap and two 288-tap ZR 33288 video rate digital filters, by ZORAN Corp, 1705 Wyatt Dr. Santa Clara, CA 95054, Phone: 408-986-1314, Fax: 408-986-1240.

[920] R. A. Peloso, "NTSC ghost cancelling," *(Panasonic) ICSPAT, DSP World Expo*, Santa Clara, CA, Sep. 28–Oct. 1, 1993.

[921] K. S. Yang et al., "A new GCR signal and its applications" *IEEE Trans. Consumer Electronics*, vol. 40, pp. 852–860, Nov. 1994.

[922] E. Abreu, S. K. Mitra, and R. Marchesani, "A new method of television signal deghosting," *ICIP '94*, pp. 283–287, Austin, TX, Nov. 1994.

[923] S. McNay, "Ghost cancellation implementation in television and VCR," *ICCE '95*, pp. 68–69, Chicago, IL, June 1995.

[924] P. Appelhans and H. Schröder "Ghost cancelling for mobile television," *ICCE '95*, pp. 70–71, Chicago, IL, June 1995.

[925] C. P. Markhauser and A. Yong, "Multighost cancellation technique for HDTV systems, using a derivation of the OLS learning algorithm," *ICCE '95*, pp. 72–73, Chicago, IL, June 1995.

Subjective Assessment of Video

[926] T. Hidada and K. Ozawa, "MPEG subjective assessment test report," ISO/IEC JTC1/SC2/WG8 MPEG 90/062, Mar. 8, 1990.

[927] T1Y1.1 Ad Hoc Experts' Group for Broadcast Quality Video Coding, R. C. Brainard, Chair, "Test program for selecting codec standard for broadcast quality NTSC television at DS3," Committee T1Y1.1, Document Number: T1Y1.1/90-502.

[928] G. Wallace, R. Vivian, and H. Poulsen, "Subjective testing results for still picture compression algorithms for international standardization," *IEEE Global Commun. Conf.*, pp. 1022–1027, Nov.–Dec. 1988.

[929] CCIR Recommendation 500-3, *Method for subjective assessment of the quality of television pictures*, 1986.

[930] CCIR Study Groups Period 1990-1994, *Subjective assessments to evaluate the Eureka-256 HDTV codec at various bit-rates*, Doc. TG CMTT-2, Feb. 1991.

[931] *Subjective assessments of HDTV based on a 4:2:2 scaled version*, Doc. IWP CMTT/2-321, IWP 11/7-321 (Spain-Italy).

[932] *Report AJ/11: Test pictures and sequences for subjective assessments of digital codecs*, Doc. CCIR 11/1085, Jan. 1990.

[933] T. Hidaka and K. Ozawa, "Subjective assessments of redundancy-reduced moving images for interactive application: test methodology and report," *Signal Process.: Image Commun.*, vol. 2, pp. 201–219, Aug. 1990.

[934] G. G. Kuperman et al., "Objective and subjective assessment of image-compression algorithm," *Society for Information Display, 1991 Intl. Symposium*, Anaheim, CA, May 6–10, 1991.

[935] CCIR Recommendation 710, *Subjective assessment methods for image quality in high definition television*, 1990.

[936] *MPEG Phase 2: simulated picture quality evaluation test*, Doc. CMTT/2-SRG/036. Dec. 10, 1991.

[937] T. Hidaka and K. Ozawa, "ISO/IECJTC 1SC29/WG 11; Report on MPEG-2, Subjective assessment at Kurihama," *Signal Process.: Image Commun.*, vol. 5, pp. 127–158, Feb. 1993.

[938] J. T. Kim, H. J. Lee, and J. S. Choi, "Subband coding using human visual characteristics for image signals," *IEEE J. Selected Areas in Commun.*, vol. 11, pp. 59–64, Jan. 1993.

[939] B. Macq, "Weighted optimum bit allocations to orthogonal transforms for picture coding," *IEEE J. Selected Areas in Commun*, vol. 10, pp. 875–883, June 1992.

[940] C. P. Cressy and G. W. Beakley, "Computer-based testing of digital video quality," *SPIE/IS&T Symposium on Electronic Imaging Science & Technology*, vol. 2187, pp. 68–78, San Jose, CA, Feb. 1994.

[941] L. Stelmach and W. J. Tam, "Viewing-based coding for advanced television systems," *Intl. Workshop on HDTV*, Ottawa, Canada, Oct. 1993.

[942] ISO/IEC JTC1/SC29/WG11, *Results of MP@ML video quality verification tests*, No. 633, MPEG 94/Mar. 1994.

[943] W. Xu and G. Hauske, "Picture quality evaluation based on error segmentation," *SPIE/VCIP*, vol. 2308, pp. 1454–1465, Chicago, IL, Sep. 1994.

[944] M. Hasegawa, Y. Kosugi, and H. Shimizu, "Development and picture quality evaluation of prototype Hi-Vision coding system for facility monitoring" *SPIE/VCIP*, vol. 2308, pp. 1884–1896, Chicago, IL, Sep. 1994.

[945] M. Miyahara, K. Kotani, and V. R. Algazi, "Objective picture quality scale (PQS) for image coding," *Society for Information Display Symposium for Image Display '92*, pp. 859–862, Boston, MA, 1992.

[946] S. A. Karunasekera and N. G. Kingsbury, "A distortion measure for image artifacts based on human visual sensitivity," *ICASSP '94*, vol. 5, pp. 117–120, Adelaide, Australia, April 1994.

[947] V. R. Algazi et al., "Comparison of image coding techniques with a picture quality scale," *SPIE, Applications of Digital Image Process. XV*, vol. 1771, 1992.

[948] T. Weyers and J. J. D. van Schalkwyk, "Picture quality assessment model," *Electronics Letters*, vol. 29, pp. 287–288, Feb. 1993.

[949] M. Miyahara, "Qality assessments for visual service," *IEEE Commun. Magazine*, vol. 26, pp. 51–60, Oct. 1988.

[950] N. Chaddha and T. H. Y. Meng, "Physcho-visual based distortion measure for monochrome image compression," *SPIE/VCIP*, vol. 2094, pp. 1680–1690, Cambridge, MA, Nov. 1993.

[951] J. L. Mannos and D. J. Sakrison, "The effects of visual fidelity criterion on the encoding of images," *IEEE Trans. Inform. Theory*, vol. IT-20, pp. 525–536, July 1974.

[952] J. O. Limb, "Distortion criteria of the human viewer," *IEEE Trans. on Systems, Man., and Cybernetics*, vol. SMC-9, pp. 778–793, Dec. 1979.

[953] J. E. Farell et al., "Perceptual metrics for monochrome image compression," *Society for Information Display Digest*, pp. 631–634, Dec. 1991.

[954] S. A. Rajala, "Impact of human visual perception of color on very low bit-rate image coding," *SPIE/VCIP*, vol. 2308, pp. 39–46, Chicago, IL, Sept. 1994.

[955] W. Xu et al., "Subjective rating of picture coding algorithms," *SPIE/VCIP*, vol. 2501, pp. 595–606, Taipei, Taiwan, May 1995.

[956] S. Westen, R. L. Lagendijk, and J. Biemond, "Perceptual image quality based on a multiple channel HVS model," *ICASSP '95*, vol. 4, pp. 2351–2354, Detroit, MI, May 1995.

[957] J. Villasenor, "Video coding: quality evaluation and system design," *DCC '95, Data Compression Conf.*, Snowbird, UT, March 1995.

[958] G. Qiu, M. R. Varley, and T. J. Terrell, "Image coding based on visual vector quantization," pp. 301–305, *5th ICIP '95*, Edinburgh, Scotland, July 1995.

[959] M. G. Albanesi and S. Bertoluzzu, "Human vision model and wavelets for high-quality image compression," pp. 311-315, *5th ICIP '95*, Edinburgh, Scotland, July 1995.

[960] R. Aldridge, et al., "Regency effect in the subjective assessment of digital-coded television pictures," *5th ICIP '95*, pp. 336–339, Edinburgh, Scotland, July 1995.

[961] T. P. O'Rourke and R. L. Stevenson, "Human visual system based subband image compression," *31st Annual Allerton Conf. on Commun. Control and Computing*, pp. 452–461, Monticello, IL, Sept. 1993.

[962] N. Jayant, J. D. Jonston, and R. Safranek, "Signal compression based on models of perception," *Proc. IEEE*, vol. 81, pp. 1385–1422, Oct. 1993.

[963] C. H. Chou and Y. C. Li, "A perceptually tuned subband image coder based on the measure of just noticeable-distortion profile," *IEEE Trans. CSVT*, vol. 5, pp. 467–476, Dec. 1995.

[964] K. Kotani et al., "Objective picture quality scale for color image coding," *ICIP '95*, vol. 3, pp. 133–136, Washington, DC, Oct. 1995.

[965] M. Yuen and H. R. Wu, "Reconstruction artifacts in digital video compression," *SPIE/IS&T*, vol. 2419, pp. 455–465, San Jose, CA, Feb. 1995.

[966] H. R. Wu, "Analysis of video reconstruction artifacts and quality metrics," *Australian Telecommun., Networks & Applications Conf.*, Sydney, Australia, Dec. 1995.

[967] A. Jacquin, "Perceptual quality evaluation of low-bit-rate model-assisted video," *Symp. on Multimedia Commun. and Video Coding*, Polytechnic Univ., Brooklyn, NY, Oct. 1995.

[968] W. Xu and G. Hauske, "Picture quality evaluation based on error segmentation," *SPIE/VCIP*, vol. 2308, pp. 1454–1465, Sept. 1994.

[969] S. Karunasekera and N. G. Kingsbury, "A distortion measure for image artifacts based on human visual sensitivity," *Proc. IEEE ICASSP '94*, vol. 5, pp. V.117–120, April 1994.

[970] CCIR, "CCIR Rec. 500-5, Method for subjective assessment of the quality of television pictures," pp. 166–189, Sept. 1992.

Subjective Assessment of Audio

[971] W. R. Daumer, "Subjective evaluation of several efficient speech coders," *IEEE Trans. Commun.*, vol. COM-30, pp. 655–662. Apr. 1982.

[972] Kitawaki and H. Nagabuchi, "Quality assessment of speech coding and speech synthesis systems," *IEEE Commun. Magazine*, vol. 26, pp. 36–44, Oct. 1988.

[973] S. Wang, A. Sekey, and A. Gersho, "An objective measure for predicting subjective quality of speech coders," *IEEE J. Selected Areas in Commun.*, vol. 10, pp. 819–829, June 1992.

[974] E. Zwicker and H. Fastl, *Psychoacoustics: Facts and Models*, Berlin, Germany: Springer, 1990.

[975] S. Bergman, C. Grewin, and T. Ryden, *The SR report on the MPEG/Audio subjective listening test*, Stockholm, Sweden, April/May 1991," ISO/IEC JTC1/SC2/WG11, June 1991.

[976] C. Grewin and T. Ryden, "Subjective assessments on low bit-rate audio codecs," *Proc. of the 10th International AES Conference*, pp. 91–102, London, UK, 1991.

[977] F. Feige and D. Kirby, "Report on the MPEG/audio multichannel formal subjective listening tests," ISO/IEC JTC1/SC29/WG11, MPEG94/063, March 1994.

[978] Annex H. Audio subjective test methods for low bitrate codec evaluations, *MPEG-4 testing and evaluation procedures documents*, final version, July 1995.

[979] IEEE standard 293, Recommended practice for speech quality measurements (info on subjective tests for speech quality and list of test sequences).

[980] ITU-T Rec. P. 80 (Guidelines for audio quality assessment).

[981] ITU-T Rec. BS.1116 (Standard audio test procedures).

BMA ME Hardware

[982] L. D. Vos, "VLSI-architectures for the hierarchical block-matching algorithm for HDTV applications," *SPIE/VCIP*, vol. 1360, pp. 398–409, Oct. 1990.

[983] L. D. Vos and M. Stegherr, "Parameterized VLSI architectures for the full-search block-matching algorithm," *IEEE Trans. Circuits, Systems*, vol. 36, pp. 1309–1316, Oct. 1989.

[984] L. D. Vos, M. Stegherr, and T. G. Noll, "VLSI architectures for the full-search block-matching algorithm," *ICASSP '89*, Glasgow, Scotland, pp. 1687–1690, May 1989.

[985] C. V. Reventlow, "System considerations and the system level design of a chip set for real-time TV and HDTV motion estimation," *J. of VLSI Signal Process.*, vol. 5, pp. 237–248, 1993.

[986] D. Bailey et al., "Programmable vision processor/controller," *IEEE Micro.*, pp. 33–39, Oct. 1992.

[987] T. Koga et al., "Motion compensated interframe coding for video conferencing," *NTC '81, National Telecommun. Conf.*, pp. G5.3.1–G5.3.5, New Orleans, LA, Nov.–Dec. 1981.

[988] W. Lee et al., "Real time MPEG video on a single chip multiprocessor," *SPIE/IS&T Symp. on Electronic Imaging Science & Technology*, vol. 2187, San Jose, CA, Feb. 1994.

[989] B. M. Wang, "An efficient VLSI architecture of hierarchical block-matching algorithm in HDTV applications," *Intl. Workshop on HDTV*, Ottawa, Canada, Oct. 1993.

[990] K. Müller et al., "A flexible real time HDTV motion vector estimation chipset based on phasecorrelation," *Intl. Workshop on HDTV*, Ottawa, Canada, Oct. 1993.

[991] T. Yoshino et al., "A 54 MHz motion estimation engine for real-time MPEG video encoding," *ICCE '94*, pp. 76–77, Chicago, IL, June 1994.

[992] J. Wiseman, "HDTV motion vector decoding with a TMS 320C30," *Intl. Conf. on SP Applications & Technology*, pp. 55–60, Dallas, TX, Oct. 1994.

[993] Z. He and M. L. Liou, "A new array architecture for motion estimation," *IEEE Workshop on Visual Signal Process. and Commun.*, Piscataway, NJ, Sept. 1994.

[994] L. A. de Barros and N. Demassieux, "A VLSI architecture for motion compensated temporal interpolation," *IEEE Workshop on Visual Signal Process. and Commun.*, Piscataway, NJ, Sept. 1994.

[995] B. M. Wang, J. C. Yen, and S. Chang, "Zero waiting cycle hierarchical block matching algorithm and its array architectures," *IEEE Trans. CSVT*, vol. 4, pp. 18–28, Feb. 1994.

[996] T. Komarek and P. Pirsch, "Array architectures for block matching algorithms," *IEEE Trans. Circuits, Systems*, vol. 36, pp. 1301–1308, Oct. 1989.

[997] S. K. Azim, "A low cost application specific video codec for consumer videophone," *IEEE CICC '94*, pp. 115–118, San Diego, CA, May 1994.

[998] M. Talmi and R. Horn, "Design of modular multi-DSP system for hierarchical motion estimation in combination with customized processors," *ICSPAT '95*, Boston, MA, Oct. 1995.

[999] E. Ogura et al., "A cost effective estimation processor LSI using a simple and efficient algorithm," *ICCE '95*, pp. 248–249, Chicago, IL, June 1995.

[1000] S. Desmet et al., "Classification based motion estimation in video coding," *J. of Visual Commun. and Image Representation*, vol. 5, pp. 370–378, Dec. 1994.

[1001] F. Dufaux and F. Moscheni, "Motion estimation techniques for digital TV: A review and a new contribution," *Proc. IEEE*, vol. 83. pp. 858–876, June 1995.

[1002] M. J. Chen, "A new block-matching criterion for motion estimation and its implementation," *IEEE Trans. CSVT*, vol. 5, pp. 231–236, June 1995.

[1003] T. Komarek and P. Pirsch, "VLSI architectures for hierarchical block matching algorithms," *IFIP Workshop*, pp. 168–181, Grenoble, France, Dec. 1989.

[1004] A. Pirson et al., "A Programmable motion estimation processor for full search block matching," *ICASSP '95*, pp. 3283–3286, Detroit, MI, May 1995.

[1005] Y. Huang and X. Zhuang, "An adaptively refined block matching algorithm for motion compensated video coding," *IEEE Trans. CSVT*, vol. 5, pp. 56–59, Feb. 1995.

[1006] S. Chang, J. H. Hwang, and C. W. Jen, "Scalable array architecture design for full search block matching," *IEEE Trans. CSVT*, vol. 5, pp. 332–343, Aug. 1995.

[1007] K. M. Nam et al., "A fast hierarchical motion vector estimation algorithm using mean pyramid," *IEEE Trans. CSVT*, vol. 5, pp. 344–351, Aug. 1995.

[1008] C. A. Papadopoulos and T. G. Clarkson, "Motion compensation using second order geometric transformation," *IEEE Trans. CSVT*, vol. 5, pp. 319–331, Aug. 1995.

[1009] H. Yeo and Y. H. Hu, "A novel modular systolic array architecture for full-search block matching motion estimation," *IEEE Trans. CSVT*, vol. 5, pp. 407–416, Oct. 1995.

[1010] L. D. Vos and M. Schobinger, "VLSI architecture for a flexible block matching processor," *IEEE Trans. CSVT*, vol. 5, pp. 417–428, Oct. 1995.

VQ Hardware/Software

[1011] *Real time digital TV encoder, low cost digital TV decoder and digital TV storage and play back system. VQ based compression. Prototype is built* by City Polytechnic of Hong Kong, Hong Kong. Contact: Dr. Chok Ki Chan, Dept. of Electrical Engineering, 83 Tat Chee Ave., CPHK, Kowloon, Hong Kong, Tel: 852-788-7718, Fax: 852-788-7791.

[1012] Picturetel, *Video codecs developed by Picturetel for video conferencing and videophone*, The Towers at Northwoods, 222 Rosemond Dr., Danvers, MA 01923, Tel: 508-762-5000, Fax: 508-762-5245, (Proprietary algorithm is based on hierarchical VQ.)

[1013] A. Netravali et al., "A codec for HDTV," *IEEE Trans. Consumer Electronics*, vol. 38, pp. 325–340, Aug, 1992. Also: *Technical Details – Digital Spectrum Compatible*, Zenith/AT & T, Sept. 23, 1991. (Quantizer selection pattern is implemented by VQ.)

[1014] *Channel compatible digicipher HDTV system*, submitted by MIT on behalf of ATA to FCC (FCC/HDTV), Apr. 3, 1992. (DCT coefficient coding is implemented by VQ.)

[1015] A. Rao et al., "A subband/VQ based HDTV coding scheme for satellite distribution and broadcast," *Intl. Workshop on HDTV*, Ottawa, Canada, Oct. 26–28, 1993. (Prototype hardware built by COMSAT.)

[1016] Intel's Indeo algorithm based on VQ, *PLV: Production Level Video*, (Highest-quality DVI motion video compression algorithm. Compression is off-line.) *RTV: Real time video* (On-line, symmetrical, DVI video compression algorithm. Both PLV and RTV are based on VQ.)

[1017] O. T. C. Chen, B. J. Sheu, and Z. Zhang, "An adaptive vector quantizer based on the Gold-washing method for image compression," *IEEE Trans. CSVT*, vol. 4, pp. 143–157, Apr. 1994. (VLSI architecture and circuit simulation.)

[1018] S. Panchanathan, "Computational RAM implementation of vector quantization for image compression," *IEEE Workshop on Visual Signal Process. and Commun.*, Piscataway, NJ, Sept. 1994.

[1019] A. Madisetti et al., "Architectures for integrated circuits for real-time vector quantization of images," *ICASSP '92*, pp. 667–680, San Francisco, CA, Mar. 1992.

[1020] H. Park and V. K. Prasanna, "Modular VLSI architectures for real-time full-search based vector quantization," *IEEE Trans. CSVT*, vol. 3, pp. 309–317, Aug. 1993.

[1021] IBM Corp: PhotoMotion & UltiMotion algorithms; Media Vision Corp: Motive algorithm; Sun Microsystems: Cell algorithm; Radius: Cinepak algorithm; Microsoft Corp.: Drivers for VQ algorithms including Cinepak, Indeo, and Motive in its video for windows software extensions.

[1022] A. A. Brahmbatt, "A VLSI architecture for real-time code book generation and encoder for a vector quantizer," *ICSPAT '95*, Boston, MA, Oct. 1995.

[1023] A. Chandrakasan, A. Burstein, and R. W. Brodersen, "A low-power chipset for a portable multimedia I/O terminal," *IEEE J. of Solid-State Circuits*, vol. 29, pp. 1415–1428, Dec. 1994.

[1024] T. M. Le and A. Panchanathan, "Computational RAM implementation of an adaptive vector quantization algorithm for video compression," *ICCE '95*, pp. 294–295, Chicago, IL, June 1995.

[1025] In G. 723 (dual-rate speech coder for multimedia telecommunication transmitting at 5.3 and 6.3 Kbps) parameters such as pitch predictor gain, LPC filter and LSP (line spectral pair) use VQ (predictive split vector quantizer: PSVQ).

[1026] W. S. Chen, "VLSI architecture for tree in VQ encoding," *Intl. Workshop on HDTV*, Taipei, Taiwan, Nov. 1995.

[1027] QuickTime by Apple includes VQ-based compression.

[1028] Cinepak, a QuickTime compatible algorithm by SuperMac Technology.

[1029] Video for Windows by Microsoft supports Intel's Indeo and Media Vision's Motive algorithm. (Both have VQ.)

[1030] R. Dianysian and R. L. Baker, "A VLSI chip set for real time vector quantization of image sequences," *IEEE ISCAS*, pp. 221–224, 1987.

[1031] R. K. Kolagotla, S. S. Yu, and J. F. Jaja, "VLSI implementation of a tree searched vector quantizer," *IEEE Trans. Signal Process.*, vol. 41, pp. 901–905, Feb. 1993.

[1032] D. B. Lidsky and J. M. Rabaey, "Low power design of memory intensive functions, case study: vector quantization," *In* J. Rabaey, P. M. Chau, and J. Eldon eds., *VLSI Signal Processing VII*, pp. 378–387, New York, NY: IEEE, 1994.

[1033] A. Chandrakasan, A. Burstein, and R. W. Brodersen, "A low-power chipset for a portable multimedia I/O terminal," *IEEE Journal of Solid-State Circuits*, vol. 29, pp. 1415–1428, Dec. 1994.

[1034] L. G. Chen et al., "Algorithm and VLSI design of a feature based classified vector quantizer for image coding," *Intl. Symp. on Communications*, Taipei, Taiwan, Dec. 1995.

[1035] H. Holzlwimmer, A. V. Brandt, and W. Tengler, "A 64 kbit/s motion compensated transform coder using vector quantization with scene adaptive codebook," *ICC '87*, Seattle, WA, pp. 151–156, June 1987.

Index

D

E

F

G

H

I

R

S

T

U

V

W

Z